Readings
in Animal
Behavior

Readings in Animal Behavior

Third Edition

Edited by
THOMAS E. McGILL
Williams College
Williamstown, Massachusetts

HOLT, RINEHART AND WINSTON

New York   Chicago   San Francisco   Atlanta
Dallas   Montreal   Toronto   London   Sydney

Library of Congress Cataloging in Publication Data

McGill, Thomas E   comp.
  Readings in animal behavior.

  Includes bibliographies and index.
  1. Animals, Habits and behavior of—Addresses,
essays, lectures.   2.   Psychology, Comparative—
Addresses, essays, lectures.   I.   Title.
QL751.6.M33   1977   591.5   76-49872
ISBN 0-03-089926-5

Acknowledgments

The articles comprising this collection are reprinted with the kind permission of the authors
and the following publishers and copyright holders:

ACADEMIC PRESS, INC., for permission to reprint P. P. G. Bateson, Specificity and the
origins of behavior. *Advances in the study of behavior,* 1976, *6,* 1–20; L. C. Drickamer,
Contact stimulation, androgenized females, and accelerated sexual maturation in female
mice. *Behavioral Biology,* 1974, *12,* 101–110. T. E. McGill, S. M. Albelda, H. H. Bible, and
C. L. Williams, Inhibition of the ejaculatory reflex in B6D2F$_1$ mice by testosterone propionate. *Behavioral Biology,* 1976, *16,* 373–378. Copyright 1974 and 1976 by Academic
Press, Inc. Reprinted by permission.

ALLEN PRESS, INC. for permission to reprint J. Alcock, The evolution of the use of tools
by feeding animals. *Evolution,* 1972, *26,* 464–473.

AMERICAN ASSOCIATION FOR THE ADVANCEMENT OF SCIENCE for permission
to reprint F. H. Bronson and C. Desjardins, Aggression in adult mice: Modification by
neonatal injections of gonadal hormones. *Science,* 1968, *161,* 705–706. G. M. Burghardt,
Chemical-cue preferences of inexperienced snakes: Comparative aspects. *Science,* 1967, *157,*
718–721. R. R. Capranica and L. S. Frishkopf, Encoding of geographic dialects in the
auditory system of the cricket frog. *Science,* 1973, *182,* 1272–1275. P. Chesler, Maternal
influence in learning by observation in kittens. *Science,* 1969, *166,* 901–903. J. F. Eisenberg,
N. A. Muckenhirn, and R. Rudran, The relation between ecology and social structure in
primates. *Science,* 1972, *176,* 863–874. S. T. Emlen, Celestial rotation: Its importance in the
development of migratory orientation. *Science,* 1970, *170,* 1198–1201. J. L. Gould, Honeybee recruitment: The dance-language controversy. *Science,* 1975, *189,* 685–693. C. R.
Gustavson, J. Garcia, W. G. Hankins, and K. W. Rusiniak, Coyote predation control by
aversive conditioning. *Science,* 1974, *184,* 581–583. R. A. Hinde and Y. Spencer-Booth,
Effects of brief separation from mother on rhesus monkeys. *Science,* 1971, *173,* 111–118.
R. R. Hoy and R. C. Paul, Genetic control of song specificity in crickets. *Science,* 1973,
180, 82–83. J. E. Lloyd, Aggressive mimicry in *Photuris* fireflies: Signal repertoires by
femmes fatales. Science, 1975, *187,* 452–453. D. L. Margules, M. J. Lewis, J. A. Dragovich,
and A. S. Margules, Hypothalamic norepinephrine: Circadian rhythms and the control of
feeding behavior. *Science,* 1972, *178,* 640–643. P. Marler and D. R. Griffin, The 1973
Nobel Prize for physiology or medicine. *Science,* 1973, *182,* 464–466. P. Marler and M.

Tamura, Culturally transmitted patterns of vocal behavior in sparrows. *Science*, 1964, *164*, 1483–1486. R. D. Nadler, Sexual cyclicity in captive lowland gorillas. *Science*, 1975, *189*, 813–814. D. R. Robertson, Social control of sex reversal in a coral-reef fish. *Science*, 1972, *177*, 1007–1009. W. R. Thompson, Influence of prenatal maternal anxiety on emotionality in young rats. *Science*, 1957, *125*, 698–699. N. Tinbergen, On war and peace in animals and man. *Science*, 1968, *160*, 1411–1418. C. L. Pratt and G. P. Sackett, Selection of social partners as a function of peer contact during rearing. *Science*, 1967, *155*, 1133–1135. G. P. Sackett, Monkeys reared in isolation with pictures as visual input: Evidence for an innate releasing mechanism. *Science*, 1966, *154*, 1468–1473. C. Walcott and R. P. Green, Orientation of homing pigeons altered by a change in the direction of an applied magnetic field. *Science*, 1974, *184*, 180–182. Copyright 1964, 1966, 1967, 1968, 1969, 1970, 1971, 1972, 1973, 1974, and 1975 by the American Association for the Advancement of Science.

AMERICAN MUSEUM OF NATURAL HISTORY for permission to reprint L. P. Brower and J. Van Zandt Brower, Investigations into mimicry. *Natural History Magazine*, 1962, *71*, 8–19. V. Geist, A consequence of togetherness. *Natural History* Magazine, October 1967, *76*, 24–30. N. Tinbergen, The shell menace. *Natural History* Magazine, August–September 1963, *72*, 28–35. Reprinted with permission from *Natural History* Magazine, April 1962, August–September 1963, and October 1967. Copyright © The American Museum of Natural History, 1962, 1963, and 1967.

AMERICAN PSYCHOLOGICAL ASSOCIATION for permission to reprint F. A. Beach, The snark was a boojum. *American Psychologist*, 1950, *5*, 115–124. B. A. Campbell and J. Jaynes, Theoretical note: Reinstatement. *Psychological Review*, 1966, *73*, 478–480. C. J. Erickson, Induction of ovarian activity in female ring doves by androgen treatment of castrated males. *Journal of Comparative and Physiological Psychology*, 1970, *71*, 210–215. W. Hodos and C. B. G. Campbell, *Scala Naturae*: Why there is no theory in comparative psychology. *Psychological Review*, 1969, *76*, 337–350. M. E. P. Seligman, On the generality of the laws of learning. *Psychological Review*, 1970, *77*, 406–418. R. E. Wimer, L. Symington, H. Farmer, and P. Schwartzkroin. Differences in memory processes between inbred mouse strains C57B1/6J and DBA/2J. *Journal of Comparative and Physiological Psychology*, 1968, *65*, 126–131. Copyright 1950, 1966, 1968, 1969, and 1970 by the American Psychological Association. Reprinted by permission.

AMERICAN SCIENTIST for permission to reprint T. H. Bullock, Seeing the world through a new sense: Electroreception in fish. *American Scientist*, 1973, *61*, 316–325. P. H. Klopfer, Mother love: What turns it on? *American Scientist*, 1971, *59*, 404–407. A. H. Meier, Daily hormone rhythms in the white-throated sparrow. *American Scientist*, 1973, *61*, 184–187. K. D. Roeder and A. E. Treat, The detection and evasion of bats by moths. *American Scientist*, 1961, *49*, 135–148.

AMERICAN SOCIETY OF ICHTHYOLOGISTS AND HERPETOLOGISTS for permission to reprint S. T. Emlen, Territoriality in the bull-frog, *Rana catesbeiana*. *Copiea*, 1968, *2*, 240–243.

BAILLIÈRE, TINDALL & CASSELL LTD. for permission to reprint J. L. Brown and R. W. Hunsperger, Neuroethology and the motivation of agonistic behavior. *Animal Behaviour*, 1963, *11*, 439–448. H. Kruuk, The biological function of gull's attraction towards predators. *Animal Behaviour*, 1976, *24*, 146–153.

THE COMPANY OF BIOLOGISTS LTD. for permission to reprint R. A. Hinde, *Energy models of motivation*, from *Symposium of the Society for Experimental Biology*, No. XIV, in *Models and Analogues in Biology*, 1960, 119–213, with the permission of the author and the Company of Biologists Ltd.

HOLT, RINEHART AND WINSTON for permission to reprint an excerpt from E. H. Hess, Ethology: An approach toward the complete analysis of behavior, in R. Brown, E.

Galanter, E. H. Hess, and G. Mandler, *New directions in psychology*. Copyright © 1962 by Holt, Rinehart and Winston, Inc. Reprinted by permission of Holt, Rinehart and Winston.

J. B. LIPPINCOTT COMPANY for permission to reprint J. M. Davidson, Activation of the male rat's sexual behavior by intracerebral implantation of androgen. *Endocrinology*, 1966, *79*, 783–794.

MACMILLAN (JOURNALS) LTD. for permission to reprint G. G. Eaton, R. W. Goy, and C. H. Phoenix, Effects of testosterone treatment in adulthood on sexual behaviour of female pseudo-hermaphrodite rhesus monkeys. *Nature, New Biology*, 1973, *242*, 119–120. E. A. Salzen and C. C. Meyer, Imprinting: Reversal of a preference established during the critical period. *Nature*, 1967, *215*, 785–786. B. Svare and R. Gandelman, Suckling stimulation induces aggression in virgin female mice. *Nature*, 1976, *260*, 606–608.

THE NEW YORK ACADEMY OF SCIENCES for permission to reprint D. A. Dewsbury, Comparative psychologists and their quest for uniformity, *Annals of the New York Academy of Sciences*, 1973, *223*, 147–167.

PERGAMON PRESS LTD. for permission to reprint A. Manning, Evolutionary changes and behaviour genetics. *Genetics Today*, 1964, Proceedings of the XI International Congress of Genetics, 807–813. The Hague, The Netherlands, September 1963, Pergamon Press, 1964.

PLENUM PUBLISHING CORPORATION for permission to reprint B. B. Gorzalka and R. E. Whalen, Effects of genotype on differential behavioral responsiveness to progesterone and 5 α1-dihydroprogesterone in mice. *Behavior Genetics*, 1976, *6*, 7–15. D. D. Thiessen, A move toward species-specific analyses in behavior-genetics. *Behavior Genetics*, 1972, *2*, 115–126.

JOHN WILEY & SONS, INC. for permission to reprint G. Lindzey, D. D. Thiessen, and A. Tucker, Development and hormonal control of territorial marking in the male Mongolian gerbil. *Developmental Psychology*, 1968, *1*, 97–99. J. P. Scott, J. M. Stewart, and V. J. DeGhett, Critical periods in the organization of systems. *Developmental Psychobiology*, 1974, *7*, 489–513. Copyright © 1968 and 1974 John Wiley & Sons, Inc. Reprinted by permission of John Wiley & Sons, Inc.

Preface

The science of animal behavior is rapidly becoming a distinct academic discipline. This is evidenced by the following: the growing number of colleges and universities that offer undergraduate courses in animal behavior; the formation of several graduate departments of animal behavior; the existence of research laboratories and institutes devoted exclusively to the study of the behavior of animals; recent "summer institutes" designed to train college teachers in the subject; and several scientific journals and "associations" that specialize in animal behavior.

Historically, new sciences developed by disengagement from larger disciplines. The science of animal behavior, however, is being formed in a slightly different way: by the recombination of parts of several sciences into one discipline.

The rather eclectic nature of the new science creates special problems for those engaged in teaching the subject. It is axiomatic that advanced undergraduates and graduate students should have experience in reading journal articles and other primary sources. However, it is very difficult to arrange for the availability of such source material in an area where major contributions have come from a great diversity of academic specialties. Endocrinologists, ethologists, geneticists, physiologists, psychologists, and zoologists, among others, have contributed to the field. Consequently, the reports have appeared in a bewildering variety of journals and books, and the instructor finds it difficult to arrange an adequate and representative "reserve shelf." *Readings in Animal Behavior* was designed to meet this need—to serve as an aid, either as core text or as collateral reading, in the teaching of advanced-undergraduate and graduate courses in animal behavior. It is hoped that the book may in this way contribute to the development of the science.

Three criteria were used in selecting the readings that are included in this volume. First, an effort was made both to secure a representation from a wide variety of disciplines and to obtain a broad sample of animal subjects. Second, whenever possible, contributions were selected that illustrate a *program* of research—the long-range work of one scientist or one laboratory. Finally, an attempt was made to include representative samples of both theoretical reviews and reports of specific experiments.

The stylistic idiosyncrasies of the various sources have been maintained in the reproductions that make up this volume. For example, if the summary or abstract preceded the article proper in the original journal, it does so in the reprinted version; if titles were omitted in the original reference list, they are omitted in the reproduction. This procedure should serve to familiarize the student with a variety of journal styles.

A difficult problem encountered in the preparation of a book of readings is organization, determining the categories for major sections and the ordering of studies within each section. It is obvious that many papers could logically be placed in any of several sections. Thus any particular arrangement must be arbitrary. Further, although a given arrangement may suit one instructor, others may find the sequence completely unworkable. For this reason, a simple alphabetical or chronological order was considered. But because this seemed a bit faint-hearted, if not downright cowardly, the decision was made to provide a scheme of classification for the material. It is the editor's sincere hope that at least some of his colleagues in the teaching of animal behavior will find the present organization useful.

Teachers who have previously used *Readings in Animal Behavior* will recall that the second edition contained a completely different, and more recent collection of readings than were found in the first edition. The present edition takes a more historical approach (perhaps as a result of the aging of the editor). About half the articles appeared in either the first or second editions, while the remainder are new selections.

As in the second edition, the brief introductions to the individual readings are included with the general introduction to each major section.

In the preparation of this book I have been impressed once more by the kindness, generosity, and willingness to help that characterize the scientists who wrote the original articles. I am very grateful to them.

Williamstown, Mass.   *T. E. McG.*
November 1976

Contents

Readings
in Animal
Behavior

PART 1

Introductory Readings

Why does an organism behave as it does? What determines its behavior? How can we account for the fact that within particular species certain behavior patterns appear identical from individual to individual, whereas others show large individual differences? The question of behavioral determination, or causality, is one of the most basic and recurrent problems in animal behavior. Historically, the answers to questions concerning the causes of behavior have taken two main forms. On the one hand were the hereditarians, who claimed that behavior is largely the result of genotype: "Like begets like." "He's a chip off the old block." On the other hand were the environmentalists, who maintained that most behavior is the product of experience: "As the twig is bent, so is the tree inclined."

Scientific and philosophical opinion as to which of these two schools of thought is "correct" has varied greatly over the years. Fortunately, with the passage of time, experimentation tended to replace mere speculation, and most (but not all) scientists came to the realization that this "nature–nurture controversy," as it came to be called, was largely a pseudoproblem. The wrong questions had been asked. To ask whether a given bit of behavior is learned or innate is both misleading and restrictive. It states an either / or dichotomy that does not allow for *interaction* between the variables or for the operation of variables that could be described as *neither* innate nor learned.

Two major branches of the science of animal behavior, which for the most part developed independently of one another, can trace their origins to different theoretical positions on the nature–nurture dichotomy. First, consider behavioral research in the United States during the last forty years. Under the combined influences of Watsonian behaviorism and the common interpretation of Freud's theory of personality, this research has been largely "environmental" in its approach. Most studies have been concerned with conditioning and learning (emphasized by the behaviorists) or with the effects of early experience on later behavior (emphasized by the Freudians).

In Europe, on the other hand, under the influence of Darwinism, a school of animal behavior known as Ethology has developed. The primary emphasis of the ethologist has been on "innate" patterns of behavior, their function and their evolution. The following table compares, in a rather crude and general fashion, these two major branches of the science of animal behavior.

	Comparative Psychology	Ethology
Geographical location	North America	Europe
Training	Psychology	Zoology
Typical subjects	Mammals, especially the laboratory rat	Birds, fish, insects
Emphasis	"Learning;" the development of theories of behavior	"Instinct;" the study of the evolution of behavior
Method	Laboratory work, control of variables, statistical analysis	Careful observation, field experimentation

It is obvious that the two approaches are complementary and that both are necessary (but probably not sufficient) for a complete understanding of behavior. After a period of sometimes vitriolic disagreement, these two "schools" have largely merged into one. But, as will be apparent from some of the readings included in the present volume, the merger is not yet completed.

The author of the first reading, Frank A. Beach, is a leading American researcher in the area of animal behavior. The importance of his contributions, however, extends beyond this discipline to all of modern psychology. Called the "comparative conscience" of American psychology, Beach has repeatedly sounded the warning that psychological theorizing is based on phylogenetic and empirical foundations that are narrow in the extreme. For those who agree with

Beach, the definition of psychology as "the study of learning in the white rat and the college sophomore" is without humor. "The Snark Was a Boojum" is recognized as a classic in the field of comparative psychology. Here Beach documents the sins of psychology and pleads for a diversion of psychological effort to the study of a broader range of problems in a greater variety of species. Most students will want to review animal taxonomy by consulting an introductory biology or zoology text in conjunction with their study of this paper.

Reading 2 is a critical article directed at the conceptual basis of comparative psychology. William Hodos and C. B. G. Campbell consider the concept of a continuous "phylogenetic scale" within the animal kingdom. This scale, originally proposed by Aristotle, conceives of an orderly evolutionary progression through the animal phyla. Thus man becomes the most "highly evolved" of all the animals. The authors attack this notion and make specific suggestions for the selection of animal subjects in comparative research.

In Reading 3, Donald A. Dewsbury responds to Hodos and Campbell and to other critics of comparative psychology. The article contains interesting discussions of the four problem areas in animal behavior that were originally proposed by Niko Tinbergen (see Reading 5). Then several other questions relevant to comparative psychology are considered. Dewsbury ends on a note of optimism, viewing "comparative psychology as a healthy and rapidly expanding discipline. . . . "

The next selection, although the work of a psychologist, presents a well-written and accurate introduction to the concepts of ethology. E. H. Hess has made important contributions to ethologically oriented research through his studies of imprinting and early perception in birds. He was one of the first to demonstrate that ethological findings could be further verified and quantified in a laboratory setting. The student who is unfamiliar with ethology will find his vocabulary greatly increased through a careful study of this paper. It is important that the terms and concepts of this reading be mastered, as they occur without introduction or definition in many of the later readings in this volume.

The fifth reading is different from other selections in the book. It is a news report describing the 1973 Nobel Prize in Physiology or Medicine that was awarded jointly to three animal behaviorists: Karl von Frisch, Konrad Lorenz, and Niko Tinbergen.

FRANK A. BEACH

1   The Snark Was a Boojum*

Those of you who are familiar with the writings of Lewis Carroll will have recognized the title of this address as a quotation from his poem "The Hunting of the Snark." Anyone who has never read that masterpiece of whimsy must now be informed that the hunting party includes a Bellman, a Banker, a Beaver, a Baker and several other equally improbable characters. While they are sailing toward the habitat of their prey the Bellman tells his companions how they can recognize the quarry. The outstanding characters of the genus *Snark* are said to be its taste which is described as "meager but hollow," its habit of getting up late, its very poor sense of humor and its overweening ambition. There are several species of Snarks. Some relatively harmless varieties have feathers and bite, and others have whiskers and scratch. But, the Bellman adds, there are a few Snarks that are Boojums.

When the Baker hears the word, Boo-

* Presidential address delivered before the Division of Experimental Psychology of the American Psychological Association, September 7, 1949.

jum, he faints dead away, and after his companions have revived him he explains his weakness by recalling for their benefit the parting words of his Uncle.

If your Snark be a Snark, that is right:
Fetch it home by all means—you may serve it
   with greens
And it's handy for striking a light.

But oh, beamish nephew, beware of the day,
If your Snark be a Boojum! For then,
You will softly and suddenly vanish away,
And never be met with again!

Much later in the story they finally discover a Snark, and it is the Baker who first sights the beast. But by great misfortune that particular Snark turns out to be a Boojum and so of course the Baker softly and suddenly vanishes away.

Thirty years ago in this country a small group of scientists went Snark hunting. It is convenient to personify them collectively in one imaginary individual who shall be called the Comparative Psychologist. The Comparative Psychologist was hunting a Snark known as Animal Behavior. His techniques were different from those used by the Baker, but he came to the same unhappy end, for his Snark also proved to be

5

a Boojum. Instead of animals in the generic sense he found one animal, the albino rat, and thereupon the Comparative Psychologist suddenly and softly vanished away. I must admit that this description is somewhat overgeneralized. A few American psychologists have done or are doing behavioral research that is broadly comparative. All honor to that tiny band of hardy souls who are herewith excepted from the general indictment that follows.

It is my aim, first, to trace the initial development and subsequent decline of Comparative Psychology in the United States. Secondly, I intend to propose certain explanations for the attitude of American psychologists toward this branch of the discipline. And finally I will outline some of the potential benefits that may be expected to follow a more vigorous and widespread study of animal behavior.

Instead of beginning with the uncritical assumption of a mutual understanding, let me define the basic terms that will be used. Comparative psychology is based upon comparisons of behavior shown by different species of animals including human beings. Comparisons between *Homo sapiens* and other animals are legitimate contributions to comparative psychology, but comparisons between two or more non-human species are equally admissible. Like any other responsible scientist the Comparative Psychologist is concerned with the understanding of his own species and with its welfare; but his primary aim is the exposition of general laws of behavior regardless of their immediate applicability to the problems of human existence. Now this means that he will not be content with discovering the similarities and differences between two or three species. Comparisons between rats and men, for example, do not in and of themselves constitute a comparative psychology although they may well represent an important contribution toward the establishment of such a field. A much broader sort of approach is necessary and it is the failure to recognize this fact that has prevented development of a genuine comparative psychology in this country.

Past and Current Trends

The history of comparative behavior studies in America is reflected in the contents of our journals that are expressly devoted to articles in this field. They have been the *Journal of Animal Behavior* and its successor, the *Journal of Comparative and Physiological Psychology*. Animal studies have, of course, been reported in other publications but the ones mentioned here adequately and accurately represent the general interests and attitudes of Americans toward the behavior of non-human animals. I have analyzed a large sample of the volumes of these journals, starting with Volume I and including all odd-numbered volumes through 1948. I have classified the contents of these volumes in two ways—first in terms of the species of animal used, and second in terms of the type of behavior studied. Only research reports have been classified; summaries of the literature and theoretical articles have been excluded from this analysis.

TYPES OF ANIMALS STUDIED

Figure 1 shows the number of articles published and the total number of species dealt with in these articles. The number of articles has tended to increase, particularly in the last decade; but the variety of animals studied began to decrease about 30 years ago and has remained low ever since. In other words, contributors to these journals have been inclined to do more and more experiments on fewer and fewer species.

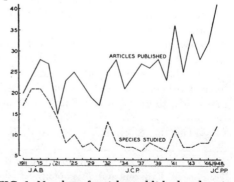

FIG. 1. Number of articles published and variety of species used as subjects.

6

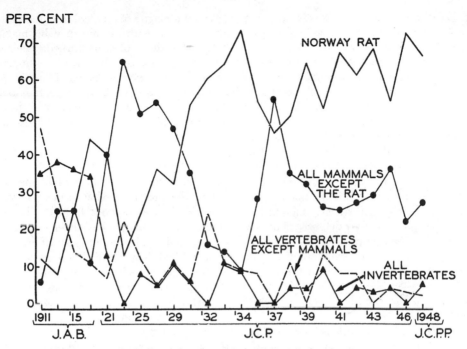

FIG. 2. Percent of all articles devoted to various phyla, classes or species.

Data represented in Figure 2 further emphasize the progressive reduction in the number of species studied. Here we see that the *Journal of Animal Behavior* contained nearly as many articles dealing with invertebrates as with vertebrates; but interest in invertebrate behavior fell off sharply after World War I and, as far as this type of analysis is capable of indicating, it never rose appreciably thereafter. The attention paid to behavior of invertebrates during the second decade of this century is also reflected in the policy of publishing annual surveys of recent research. Each volume of the *Journal of Animal Behavior* contains one systematic review devoted to lower invertebrates, another dealing with spiders and insects with the exception of ants, a third summarizing work on ants and a single section covering all studies of vertebrates.

Figure 2 shows that in the early years of animal experimentation sub-mammalian vertebrates, which include all fishes, amphibians, reptiles, and birds, were used as experimental subjects more often than mammals. But a few mammalian species

rapidly gained popularity and by approximately 1920, more work was being done on mammals than on all other classes combined. Now there are approximately 3,500 extant species of mammals, but taken together they make up less than one-half of one percent of all animal species now living. A psychology based primarily upon studies of mammals can, therefore, be regarded as comparative only in a very restricted sense. Moreover the focus of interest has actually been even more narrow than this description implies because only a few kinds of mammals have been used in psychological investigations. The Norway rat has been the prime favorite of psychologists working with animals, and from 1930 until the present more than half of the articles in nearly every volume of the journal are devoted to this one species.

During the entire period covered by this survey the odd-numbered volumes of the journals examined includes 613 experimental articles. Nine percent of the total deal with invertebrates; 10 percent with vertebrates other than mammals; 31 percent with mammals other than the rat; and

7

50 percent are based exclusively upon the Norway rat. There is no reason why psychologists should not use rats as subjects in some of their experiments, but this excessive concentration upon a single species has precluded the development of a comparative psychology worthy of the name. Of the known species of animals more than 96 percent are invertebrates. Vertebrates below the mammals make up 3.2 percent of the total; and the Norway rat represents .001 percent of the types of living creatures that might be studied. I do not propose that the number of species found in a particular phyletic class determines the importance of the group as far as psychology is concerned; but it is definitely disturbing to discover that 50 percent of the experiments analyzed here have been conducted on one one-thousandth of one percent of the known species.

Some studies of animal behavior are reported in journals other than the ones I have examined but the number of different animals used in experiments published elsewhere is even fewer. The six issues of the *Journal of Experimental Psychology* published in 1948 contain 67 reports of original research. Fifty of these articles deal with human subjects and this is in accord with the stated editorial policy of favoring studies of human behavior above investigations of other species. However, 15 of the 17 reports describing work on non-human organisms are devoted to the Norway rat.

During the current meetings of the APA, 47 experimental reports are being given under the auspices of the Division of Experimental Psychology. The published abstracts show that in half of these studies human subjects were employed while nearly one-third of the investigations were based on the rat.

Is the Experimental Psychologist going to softly and suddenly vanish away in the same fashion as his one-time brother, the Comparative Psychologist? If you permit me to change the literary allusion from the poetry of Lewis Carroll to that of Robert Browning, I will venture a prediction. You will recall that the Pied Piper rid Hamelin Town of a plague of rats by luring the pests into the river with the music of his magic flute. Now the tables are turned. The rat plays the tune and a large group of human beings follow. My prediction is indicated in Figure 3. Unless they escape the spell that *Rattus norvegicus* is casting over them, Experimentalists are in danger of extinction.

TYPES OF BEHAVIOR STUDIED

I trust that you will forgive me for having demonstrated what to many of you must have been obvious from the beginning—namely, that we have been extremely narrow in our selection of types of animals to be studied. Now let us turn our attention to the types of behavior with which psychologists have concerned themselves.

Articles appearing in our sample of volumes of the journals can be classified under seven general headings: (1) conditioning and learning; (2) sensory capacities, including psychophysical measurements, effects of drugs on thresholds, etc.; (3) general habits and life histories; (4) reproductive behavior, including courtship, mating, migration, and parental responses; (5) feeding behavior, including diet selection and reactions to living prey; (6) emotional behavior, as reflected in savageness and wildness, timidity and aggressive reactions; and (7) social behavior, which involves studies of dominance and submission, social hierarchies, and interspecies symbiotic relations.

In classifying articles according to type of behavior studied I have disregarded the techniques employed by the investigator. It is often necessary for an animal to learn to respond differentially to two stimuli before its sensory capacities can be measured; but in such a case the article was listed as dealing with sensory capacity rather than learning. The aim has been to indicate as accurately as possible the kind of behavior in which the experimenter was interested rather than his methods of studying it.

It proved possible to categorize 587 of the 613 articles. Of this total, 8.6 percent dealt with reproductive behavior, 3.7 per-

FIG. 3. Current position of many experimental psychologists.

cent with emotional reactions, 3.2 percent with social behavior, 3.0 percent with feeding, and 2.8 percent with general habits. The three most commonly-treated types of behavior were (1) reflexes and simple reaction patterns, (2) sensory capacities, and (3) learning and conditioning. Figure 4 shows the proportion of all articles devoted to each of these three major categories.

The figure makes it clear that conditioning and learning have always been of considerable interest to authors whose work appears in the journals I have examined. As a matter of fact slightly more than 50 percent of all articles categorized in this analysis deal with this type of behavior. The popularity of the subject has increased appreciably during the last 15 years, and only once since 1927 has any other kind of behavior been accorded as many articles per volume. This occurred in 1942 when the number of studies dealing with reflexes and

simple reaction patterns was unusually large. The temporary shift in relative emphasis was due almost entirely to a burst of interest in so-called "neurotic behavior" or "audiogenic seizures."

Combining the findings incorporated in Figures 2 and 4, one cannot escape the conclusion that psychologists publishing in these journals have tended to concentrate upon one animal species and one type of behavior in that species. Perhaps it would be appropriate to change the title of our journal to read "The Journal of Rat Learning," but there are many who would object to this procedure because they appear to believe that in studying the rat they are studying all or nearly all that is important in behavior. At least I suspect that this is the case. How else can one explain the fact that Professor Tolman's book *Purposive Behavior in Animals and Men* deals primarily with learning and is dedicated to the white

9

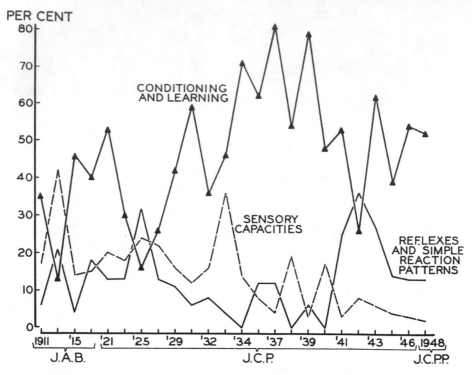

PER CENT

FIG. 4. Percent of all articles concerned with various psychological functions.

rat, "where, perhaps, most of all, the final credit or discredit belongs." And how else are we to interpret Professor Skinner's 457-page opus which is based exclusively upon the performance of rats in bar-pressing situations but is entitled simply *The Behavior of Organisms?*

Interpretation of Trends

In seeking an interpretation of the demonstrated tendency on the part of so many experimentalists to restrict their attention to a small number of species and a small number of behavior patterns, one comes to the conclusion that the current state of affairs is in large measure a product of tradition. From its inception, American psychology has been strongly anthropocentric. Human behavior has been accepted as the primary object of study and the reactions of other animals have been of interest only insofar as they seemed to throw light upon the psychology of our own species.

There has been no concerted effort to establish a genuine comparative psychology in this country for the simple reason that with few exceptions American psychologists have no interest in animal behavior *per se.*

Someone, I believe it was W. S. Small at Clark University in 1899, happened to use white rats in a semi-experimental study. The species "caught on," so to speak, as a laboratory subject, and gradually displaced other organisms that were then being examined. Psychologists soon discovered that rats are hardy, cheap, easy to rear, and well adapted to a laboratory existence. Because of certain resemblances between the associative learning of rats and human beings, *Rattus norvegicus* soon came to be accepted as a substitute for *Homo sapiens* in many psychological investigations. Lack of acquaintance with the behavioral potentialities of other animal species and rapid increase in the body of data derived from rat studies combined to progressively reduce the amount of attention paid to other mammals, to sub-mammalian vertebrates and to invertebrate organisms. Today the trend has reached a point where the average graduate student who intends to do a thesis problem with animals turns automatically to the white rat as his experimental subject; and all too often his professor is unable to suggest any alternative.

To sum up, I suggest that the current popularity of rats as experimental subjects is in large measure the consequence of historical accident. Certainly it is not the result of systematic examination of the available species with subsequent selection of this particular animal as the one best suited to the problems under study.

Concentration of experimental work upon learning seems to stem almost exclusively from the anthropocentric orientation of American psychology. Learning was very early accepted as embodying the most important problems of human behavior; and accordingly the majority of animal investigations have been concerned with this type of activity.

Advantages and Disadvantages of Concentration

I have no wish to discount the desirable aspects of the course which experimental psychology has been pursuing. There are many important advantages to be gained when many independent research workers attack similar problems using the same kinds of organisms. We see this to be true in connection with various biological sciences. Hundreds of geneticists have worked with the fruitfly, Drosophila. And by comparing, combining, and correlating the results of their investigations, it has been possible to check the accuracy of the findings, to accelerate the acquisition of new data, and to formulate more valid and general conclusions than could have been derived if each worker dealt with a different species. Something of the same kind is happening in psychology as a result of the fact that many investigators are studying learn-

ing in the rat, and I repeat that this is a highly desirable objective.

Another valuable result achieved by the methods currently employed in experimental psychology is the massing of information and techniques pertaining to rat behavior to a point which permits use of this animal as a pedagogical tool. A recent article in the American Psychologist reveals that each student in the first course in psychology at Columbia University is given one or two white rats which he will study throughout the semester. This, it seems to me, is an excellent procedure. The beginning student in physiology carries out his first laboratory exercises with the common frog. The first course in anatomy often uses the dogfish or the cat as a sample organism. And college undergraduates learn about genetics by breeding fruitflies. But the usefulness of the rat as a standardized animal for undergraduate instruction, and the preoccupation of mature research workers with the same, single species are two quite different things.

Advanced research in physiology is not restricted to studies of the frog and although many geneticists may confine their personal investigations to Drosophila, an even larger number deals with other animal species or with plants. As a matter of fact, the benefits that students can derive from studying one kind of animal as a sample species must always stand in direct proportion to the amount of information research workers have gathered in connection with other species. The rat's value as a teaching aid in psychology depends in part upon the certainty with which the student can generalize from the behavior he observes in this one animal; and this in turn is a function of available knowledge concerning other species.

There is another obvious argument in favor of concentrating our efforts on the study of a single animal species. It is well expressed in Professor Skinner's book, The Behavior of Organisms.

In the broadest sense a science of behavior should be concerned with all kinds of organisms, but it is reasonable to limit oneself, at

11

least in the beginning, to a single representative species.

I cannot imagine that anyone would quarrel with Skinner on this point and I am convinced that many of the psychologists currently using rats in their investigational programs would agree with him in his implicit assumption that the Norway rat *is* a "representative species." But in what ways is it "representative," and how has this "representativeness" been demonstrated? These questions lead at once to a consideration of the disadvantages of overspecialization in terms of animals used and types of behavior studied.

To put the question bluntly: Are we building a general science of behavior or merely a science of rat learning? The answer is not obvious to me. Admittedly there are many similarities between the associative learning of lower animals and what is often referred to as rôte learning in man. But the variety of organisms which have been studied, and the number of techniques which have been employed are so limited, it is difficult to believe that we can be approaching a comprehensive understanding of the basic phenomena of learning. It may be that much remains to be discovered by watching rats in mazes and problem boxes, but it is time to ask an important question. How close are we getting to that well-known point of diminishing returns? Would we not be wise to turn our attention to other organisms and to devise new methods of testing behavior before we proceed to formulate elaborate theories of learning which may or may not apply to other species and other situations.

Another very important disadvantage of the present method in animal studies is that because of their preoccupation with a few species and a few types of behavior, psychologists are led to neglect many complex patterns of response that stand in urgent need of systematic analysis. The best example of this tendency is seen in the current attitude toward so-called "instinctive" behavior.

The growing emphasis upon learning has produced a complementary reduction in the amount of study devoted to what is generally referred to as "unlearned behavior." Any pattern of response that does not fit into the category of learned behavior as currently defined is usually classified as "unlearned" even though it has not been analyzed directly. Please note that the classification is made in strictly negative terms *in spite of the fact that the positive side of the implied dichotomy is very poorly defined.* Specialists in learning are not in accord as to the nature of the processes involved, nor can they agree concerning the number and kinds of learning that may occur. But in spite of this uncertainty most "learning psychologists" confidently identify a number of complex behavior patterns as "unlearned." Now the obvious question arises: Unless we know what learning is—unless we can recognize it in all of its manifestations—how in the name of common sense can we identify any reaction as "unlearned"?

The fact of the matter is that none of the responses generally classified as "instinctive" have been studied as extensively or intensively as maze learning or problem-solving behavior. Data relevant to all but a few "unlearned" reactions are too scanty to permit any definite conclusion concerning the role of experience in the shaping of the response. And those few cases in which an exhaustive analysis has been attempted show that the development of the behavior under scrutiny is usually more complicated than a superficial examination could possibly indicate.

For example, there is a moth which always lays its eggs on hackberry leaves. Females of each new generation select hackberry as an oviposition site and ignore other potential host plants. However, the eggs can be transferred to apple leaves, and when this is done the larvae develop normally. Then when adult females that have spent their larval stages on apple leaves are given a choice of materials upon which to deposit their eggs, a high proportion of them select apple leaves in preference to hackberry. This control of adult behavior by the larval environment does not fit into the conventional pigeon-hole labeled "instinct," and neither can it be placed in the category of "learning." Perhaps we need more categories. Certainly we need more data on more species and more kinds of behavior.

Primiparous female rats that have been reared in isolation usually display biologically effective maternal behavior when their first litter is born. The young ones are cleaned of fetal membranes, retrieved to the nest, and suckled regularly. However, females that have been reared under conditions in which it was impossible for them to groom their own bodies often fail to clean and care for their newborn offspring. Observations of this nature cannot be disposed of by saying that the maternal reactions are "learned" rather than "instinctive." The situation is not so simple as that. In some way the early experience of the animal prepares her for effective maternal performance even though none of the specifically maternal responses are practiced before parturition.

It seems highly probable that when sufficient attention is paid to the so-called "instinctive" patterns, we will find that their development involves processes of which current theories take no account. What these processes may be we shall not discover by continuing to concentrate on learning as we are now studying it. And yet it is difficult to see how a valid theory of learning can be formulated without a better understanding of the behavior that learning theorists are presently categorizing as "unlearned."

Potential Returns from the Comparative Approach

If more experimental psychologists would adopt a broadly comparative approach, several important goals might be achieved. Some of the returns are fairly specific and can be described in concrete terms. Others are more general though no less important.

SPECIFIC ADVANTAGES

I have time to list only a few of the specific advantages which can legitimately be expected to result from the application of comparative methods in experimental psychology. In general, it can safely be predicted that some of the most pressing questions that we are now attempting to answer by studying a few species and by employing only a few experimental methods would be answered more rapidly and adequately if the approach were broadened.

Let us consider learning as one example. Comparative psychology offers many opportunities for examination of the question as to whether there are one or many kinds of learning and for understanding the rôle of learning in the natural lives of different species. Tinbergen (1942) has reported evidence indicating the occurrence of one-trial learning in the behavior of hunting wasps. He surrounded the opening of the insect's burrow with small objects arranged in a particular pattern. When she emerged, the wasp circled above the nest opening for a few seconds in the usual fashion and then departed on a hunting foray. Returning after more than an hour, the insect oriented directly to the pattern stimulus to which she had been exposed only once. If the pattern was moved during the female's absence she was able to recognize it immediately in its new location.

Lorenz's concept of "imprinting" offers the learning psychologist material for new and rewarding study. Lorenz (1935) has observed that young birds of species that are both precocial and social quickly become attached to adults of their own kind and tend to follow them constantly. Newly-hatched birds that are reared by parents of a foreign species often form associations with others of the foster species and never seek the company of their own kind. A series of experiments with incubator-reared birds convinced Lorenz that the processes underlying this sort of behavior must occur very early in life, perhaps during the first day or two after hatching, and that they are irreversible, or, to phase it in other terms, that they are not extinguished by

13

removal of reinforcement.

J. P. Scott's studies (1945) of domestic sheep reveal the importance of early learning in the formation of gregarious habits. Conventional learning theories appear adequate to account for the phenomena, but it is instructive to observe the manner in which the typical species pattern of social behavior is built up as a result of reinforcement afforded by maternal attentions during the nursing period.

The general importance of drives in any sort of learning is widely emphasized. Therefore it would seem worth while to study the kinds of drives that appear to motivate different kinds of animals. In unpublished observations upon the ferret, Walter Miles found that hunger was not sufficient to produce maze learning. Despite prolonged periods of food deprivation, animals of this species continue to explore every blind alley on the way to the goal box.

Additional evidence in the same direction is found in the studies of Gordon (1943) who reports that non-hungry chipmunks will solve mazes and problem boxes when rewarded with peanuts which the animals store in their burrows but do not eat immediately. Does this represent a "primary" drive to hoard food or an "acquired" one based upon learning?

Many experimentalists are concerned with problems of sensation and perception; and here too there is much to be gained from the comparative approach. Fring's studies (1948) of chemical sensitivity in caterpillars, rabbits and men promise to increase our understanding of the physiological basis for gustatory sensations. In all three species there appears to be a constant relationship between the ionic characteristics of the stimulus material and its effectiveness in evoking a sensory discharge. The investigations of Miles and Beck (1949) on reception of chemical stimuli by honey bees and cockroaches provides a test for the theory of these workers concerning the human sense of smell.

The physical basis for vision and the role of experience in visual perception have been studied in a few species but eventually it must be investigated on a broader comparative basis if we are to arrive at any general understanding of the basic principles involved. Lashley and Russell (1934) found that rats reared in darkness give evidence of distance perception without practice; and Hebb (1937) added the fact that figure-ground relationships are perceived by visually-naive animals of this species. Riesen's (1947) report of functional blindness in apes reared in darkness with gradual acquisition of visually-directed habits argues for a marked difference between rodents and anthropoids; and Senden's (1932) descriptions of the limited visual capacities of human patients after removal of congenital cataract appear to support the findings on apes. But the difference, if it proves to be a real one, is not purely a function of evolutionary status of the species involved. Breder and Rasquin (1947) noted that fish with normal eyes but without any visual experience are unable to respond to food particles on the basis of vision.

I have already mentioned the necessity for more extensive examination of those patterns of behavior that are currently classified as "instinctive." There is only one way to approach this particular problem and that is through comparative psychology. The work that has been done thus far on sexual and parental behavior testifies, I believe, to the potential returns that can be expected if a more vigorous attack is launched on a broader front.

We are just beginning to appreciate the usefulness of a comparative study of social behavior. The findings of Scott which I mentioned earlier point to the potential advantages of using a variety of animal species in our investigation of interaction between members of a social group. Carpenter's (1942) admirable descriptions of group behavior in free-living monkeys point the way to a better understanding of dominance, submission, and leadership.

One more fairly specific advantage of exploring the comparative method in psychology lies in the possibility that by this

14

means the experimentalist can often discover a particular animal species that is specially suited to the problem with which he is concerned. For example, in recent years a considerable amount of work has been done on hoarding behavior in the laboratory rat. The results are interesting, but they indicate that some rats must learn to hoard and some never do so. Now this is not surprising since Norway rats rarely hoard food under natural conditions. Would it not seem reasonable to begin the work with an animal that is a natural hoarder? Chipmunks, squirrels, mice of the genus *Peromyscus*, or any one of several other experimental subjects would seem to be much more appropriate.

And now, as a final word, I want to mention briefly a few of the more general facts that indicate the importance of developing comparative psychology.

GENERAL ADVANTAGES

For some time it has been obvious that psychology in this country is a rapidly expanding discipline. Examination of the membership roles of the several Divisions of this Association shows two things. First, that the number of psychologists is increasing at a prodigious rate; and second that the growth is asymmetrical in the sense that the vast majority of new workers are turning to various applied areas such as industrial and clinical psychology.

It is generally recognized that the applied workers in any science are bound to rely heavily upon "pure" or "fundamental" research for basic theories, for general methodology and for new points of view. I do not suggest that we, as experimentalists, should concern ourselves with a comparative approach to practical problems of applied psychology. But I do mean to imply that if we intend to maintain our status as indispensable contributors to the science of behavior, we will have to broaden our attack upon the basic problems of the discipline. This will sometimes mean sacrificing some of the niceties of laboratory research in order to deal with human beings under less artificial conditions. It may also mean

expanding the number of non-human species studied and the variety of behavior patterns investigated.

Only by encouraging and supporting a larger number of comparative investigations can psychology justify its claim to being a true science of behavior. European students in this field have justly condemned Americans for the failure to study behavior in a sufficiently large number of representative species. And non-psychologists in this country are so well aware of our failure to develop the field that they think of animal behavior as a province of general zoology rather than psychology. Top-rank professional positions that might have been filled by psychologically trained investigators are today occupied by biologists. Several large research foundations are presently supporting extensive programs of investigation into the behavior of sub-human animals, and only in one instance is the program directed by a psychologist.

Conclusion

If we as experimental psychologists are missing an opportunity to make significant contributions to natural science—if we are failing to assume leadership in an area of behavior investigation where we might be useful and effective—if these things are true, and I believe that they are, then we have no one but ourselves to blame. We insist that our students become well versed in experimental design. We drill them in objective and quantitative methods. We do everything we can to make them into first rate experimentalists. And then we give them so narrow a view of the field of behavior that they are satisfied to work on the same kinds of problems and to employ the same methods that have been used for the past quarter of a century. It would be much better if some of our well-trained experimentalists were encouraged to do a little pioneering. We have a great deal to offer in the way of professional preparation that the average biologist lacks. And the field of animal behavior offers rich returns to the psychologist who will devote himself to its exploration.

I do not anticipate that the advanced research worker whose main experimental program is already mapped out will be tempted by any argument to shift to an entirely new field. But those of us who have regular contact with graduate students can do them a service by pointing out the possibilities of making a real contribution to the science of psychology through the medium of comparative studies. And even in the absence of professorial guidance the alert beginner who is looking for unexplored areas in which he can find new problems and develop new methods of attacking unsettled issues would be wise to give serious consideration to comparative psychology as a field of professional specialization.

REFERENCES

1. BREDER, C. M. and P. RASQUIN: Comparative studies in the light sensitivity of blind characins from a series of Mexican caves. *Bulletin Amer. Mus. Natl. Hist.*, 1947, 89, Article 5, 325–351.
2. CARPENTER, C. R.: Characteristics of social behavior in non-human primates. *Trans. N. Y. Acad. Sci.*, 1942, Ser. 2, 4, No. 8, 248.
3. FRINGS, H.: A contribution to the comparative physiology of contact chemoreception. *J. comp. physiol. Psychol.*, 1948, 41, No. 1, 25–35.
4. GORDON, K.: The natural history and behavior of the western chipmunk and the mantled ground squirrel. *Oregon St. Monogr. Studies in Zool.*, 1943, No. 5, 7–104.
5. HEBB, D. O.: The innate organization of visual activity. I. Perception of figures by rats reared in total darkness. *J. gen. Psychol.*, 1937, 51, 101–126.
6. LASHLEY, K. S. and J. T. RUSSELL: The mechanism of vision. XI. A preliminary test of innate organization. *J. genet. Psychol.*, 1934, 45, No. 1, 136–144.
7. LORENZ, K.: Der Kumpan in der Umwelt des Vogels. *J. f. Ornith.*, 1935, 83, 137–213.
8. MILES, W. R. and L. H. BECK: Infrared absorption in field studies of olfaction in honeybees. *Proceed. Natl. Acad. Sci.*, 1949, 35, No. 6, 292–310.
9. RIESEN, A. H.: The development of visual perception in man and chimpanzee. *Science*, 1947, 106, 107–108.
10. SCOTT, J. P.: Social behavior, organization and leadership in a small flock of domestic sheep. *Comp. Psychol. Monogr.*, 1945, 18, No. 4, 1–29.
11. SENDEN, M. V.: *Raum- und Gestaltauffassung bei operierten Blindgeborenen vor und nach der Operation.* Leipzig: Barth, 1932.
12. TINBERGEN, N.: An objectivistic study of the innate behaviour of animals. *Biblio. Biotheoret.*, 1942, 1, Pt. 2, 39–98.

Psychological Review
1969, Vol. 76, No. 4, 337–350

2

SCALA NATURAE:
WHY THERE IS NO THEORY IN COMPARATIVE PSYCHOLOGY [1]

WILLIAM HODOS [2]   AND   C. B. G. CAMPBELL

Walter Reed Army Institute of Research   *Center for Neural Sciences, Indiana University*
Washington, D. C.

The concept that all living animals can be arranged along a continuous "phylogenetic scale" with man at the top is inconsistent with contemporary views of animal evolution. Nevertheless, this arbitrary hierarchy continues to influence researchers in the field of animal behavior who seek to make inferences about the evolutionary development of a particular type of behavior. Comparative psychologists have failed to distinguish between data obtained from living representatives of a common evolutionary lineage and data from animals which represent divergent lineages. Only the former can provide a foundation for inferences about the phylogenetic development of behavior patterns. The latter can provide information only about general mechanisms of adaptation and survival, which are not necessarily relevant to any specific evolutionary lineage. The widespread failure of comparative psychologists to take into account the zoological model of animal evolution when selecting animals for study and when interpreting behavioral similarities and differences has greatly hampered the development of generalizations with any predictive value.

Nearly two decades have passed since Beach (1950) presented his classic paper "The Snark was a Boojum" in which he deplored the decline of comparative psychology as a result of "excessive concentration upon a single species," namely, the albino rat. His paper appears to have stimulated a renewed interest in an animal psychology which is more broadly comparative than the rat learning studies which were prevalent in the 1940s and 1950s. Rhesus monkeys and White Carneaux pigeons have now been added to the animal psychologist's standard menagerie. Occasional studies of behavior in teleost fish, reptiles, and carnivores have also appeared in psychological journals and some attempts at comparison across species have been made. In addition, several text-

books and collections of readings in comparative psychology recently have been published. However, much of the current research in comparative psychology seems to be based on comparisons between animals that have been selected for study according to rather arbitrary considerations and appears to be without any goal other than the comparison of animals for the sake of comparison. This rather tenuous approach to research has apparently been brought about by the absence of any broad theoretical foundation for the field. Such a theoretical foundation, though partly or even totally incorrect, would at least have the virtue of encouraging a systematic study of animal behavior rather than the current haphazard manner of operation.

The purpose of this paper is to point out some of the factors which have hindered theoretical development in comparative psychology and to suggest some ways of remedying the situation. Many of the concepts that will be discussed are not novel; indeed, they would be regarded as rather elementary by students of such fields as systematic biology, paleontology, physical anthropology,

[1] The authors wish to express their gratitude to their colleagues and students for their helpful comments on this paper and to J. Z. Young and the Oxford University Press for their permission to reproduce the phylogenetic trees shown in Figures 1–4.

[2] Requests for reprints should be sent to William Hodos, Department of Experimental Psychology, Walter Reed Army Institute of Research, Walter Reed Army Medical Center, Washington, D. C. 20012.

etc. However, even a casual examination of the recent literature in the behavioral and neural sciences leads one to the conclusion that many experimenters are greatly misinformed about these fundamental concepts. As a result, a number of unjustified conclusions about behavioral and neural evolution have been drawn from the data of comparative studies.

The Scala Naturae

In attempting to find order in an apparently chaotic universe, Aristotle (1910, 1912a, 1912b; Ross, 1949) attempted various organizational schemes for classifying animals. These classifications were based on such characteristics as number of legs, whether or not the organism appeared to possess blood, whether they were oviparous or viviparous in their reproductive mechanisms, etc. Aristotle also proposed that the various categories of animals might be arranged on a graded scale of complexity or perfection, with man at the top. Although Aristotle did not advocate any such ranking of the animals within each category, later scholars expanded his suggestions so that there came to be general acceptance of the concept that all animals could be ranked on a unitary, graded, continuous dimension known as the *scala naturae* or Great Chain of Being (Lovejoy, 1936; Wightman, 1950). The lowest position on the *scala naturae* was occupied by sponges and other creatures considered to be essentially formless. At the intermediate levels were insects, fish, amphibians, reptiles, birds, and various mammals. At the top of the scale was man. Furthermore, as Lovejoy (1936) points out in his extensive treatment of the history of this subject, the *scala naturae* eventually became involved in theological formulations which considered that God was perfect and all other creatures were merely progressively less perfect copies. Thus, angels were somewhat imperfect copies, man more imperfect, apes still more imperfect, and so on, down the scale to the "formless" sponges.

The attractiveness of the notion of a *scala naturae* is attested to by its persistence throughout the centuries and its influence on contemporary scientific thought. The recent literature in the life sciences dealing with comparisons between different groups

of animals abounds with references to a hierarchy called the "phylogenetic scale," which appears to be the modern counterpart of the *scala naturae*. According to their relative positions on this scale, animals are designated as "subprimate" or "submammalian" or "higher animals" or "lower animals." However, there seems to be no compilation of the phylogenetic scale to which one could refer to answer such questions as "Is a porpoise a higher animal than a cat?" Nevertheless, in a recent textbook, Waters (1960) characterizes comparative psychology as "the study of behavior wherever exhibited along the phylogenetic scale . . . [p. 14]." Likewise, Ratner and Denny (1964), in a discussion of the evolution of behavior, state that "As one climbs the scale from fish to primates the principle seems best stated as follows: The higher the phyletic level, the greater the multiple determination of behavior [p. 680]." The meaning of such terms as "phyletic level" is usually not specified but the implication seems to be that an organism's phyletic level is determined by its proximity to man on the *scala naturae*. Apparently, these writers and numerous others have confused the *scala naturae* with another organizational arrangement of animals, one based on probable lines of evolutionary descent: the phylogenetic tree. However, as Simpson (1958a) has pointed out, the phylogenetic tree is a geneology. It is based on the data of paleontology and comparative morphology and represents the current state of knowledge about the course of evolution of the various animal species.[3] Like any other historical survey, it is subject to change with the acquisition of new data and by itself gives no indication of the relative status of the individuals listed with respect to any gradational arrangement. On the other hand, the *scala naturae* or phylogenetic scale is a hierarchical classification. While such a hierarchy can provide interesting information about relative performance and relative degrees of structural differentiation, it tells us nothing about evolutionary development since it is unrelated to specific evolutionary lineages. Thus, to say

[3] The term "phylogenetic tree" is used here in a generic sense since such trees, constructed by various evolutionary theorists, would differ in some details.

18

that amphibians represent a higher degree of evolutionary development than teleost fish is practically without meaning since they have each followed independent courses of evolution. Moreover, one can find characteristics in which teleosts exceed amphibians as well as vice versa. For example, the central nervous system of teleost fishes in many ways exhibits a greater degree of differentiation and specialization than does that of amphibians (Ariëns Kappers, Huber, & Crosby, 1960). Indeed the general absence of amphibians from recent comparative studies of learning or intelligence such as those of Bitterman (1965a, 1965b) suggests that their behavior patterns may be relatively inflexible. The difficulties encountered in the training of these organisms are discussed by McGill (1960) and van Bergeijk (1967).

An important feature of the *scale naturae* is the concept of a smooth continuity between living animal forms rather than the discontinuities implicit in the theory of evolution as a result of the divergence of evolutionary lines and the extinction of many intermediate forms. This continuity, which Lovejoy (1936) calls the "principle of unilinear gradation," has also had a profound influence on contemporary research design and theorization in the comparative life sciences. For example, in a paper on the evolution of learning, Harlow (1958) speculates that

simple as well as complex learning problems might be arranged into an orderly classification in terms of difficulty, and that the capabilities of animals on these tasks would correspond roughly to their positions on the phylogenetic scale [p. 283].

A survey of the literature reviews and research reports of the past 10 years might lead one to the conclusion that meaningful statements could be made about the evolution of some morphological, physiological, or behavioral characteristic by comparing goldfish, frog, pigeon, cat, and man. As Simpson (1958a) puts it,

some such sequence as dogfish-frog-cat-man is frequently taught as "evolutionary," *i.e.,* historical. In fact the anatomical differences among these organisms are in large part ecologically and behaviorally determined, are divergent and not sequential, and do not in any useful sense form a historical series [p. 11].

Another characteristic of the *scala naturae,*

which seems to be implicit in discussions of a phylogenetic scale, is the notion that man is the inevitable goal of the evolutionary process and that once he has evolved, the phylogenetic process ends. These ideas have been succinctly expressed in the following lines by Emerson:

Striving to be man, the worm
Mounts through all the spires of form.

While this view of the animal kingdom may have a certain amount of face validity, it too runs contrary to the currently accepted data on the course of evolutionary history which indicate that primates represent only one of the many lines of vertebrates which have evolved and survive today.

THE PHYLOGENETIC TREE

Figure 1 presents a phylogenetic tree showing the approximate times of origin and probable lineages of the various classes of living vertebrates and some related groups of animals (Young, 1962). The animals represented across the top form an approximation' of the *scala naturae* or phylogenetic scale. Note that the sequence of animals from left to right is completely arbitrary. A very different sequence would result from merely having some evolutionary lines branch to the left instead of the right. This would in no way alter the schematic representation of evolutionary lineages.

Several additional features of this phylogenetic tree should be noted. First, the evolutionary line of vertebrates leading to mammals, which begins in the Cambrian period, passes only through lobe-finned fishes (crossopterygians), amphibians, and reptiles. Second, the line of fishes which gave rise to amphibians evolved fairly early in fish evolution and followed a course of development quite separate from that of other fishes. The teleost fishes, so often used as a basis for evolutionary comparison with "higher vertebrates," are descendents of a line of development which is collateral to that giving rise to tetrapods. Therefore, no teleost fish ever was an ancestor of any amphibian, reptile, bird, or mammal. Likewise, birds represent another line of specialization from the reptiles and cannot be regarded as ancestral to mammals.

A similar situation exists in the phylo-

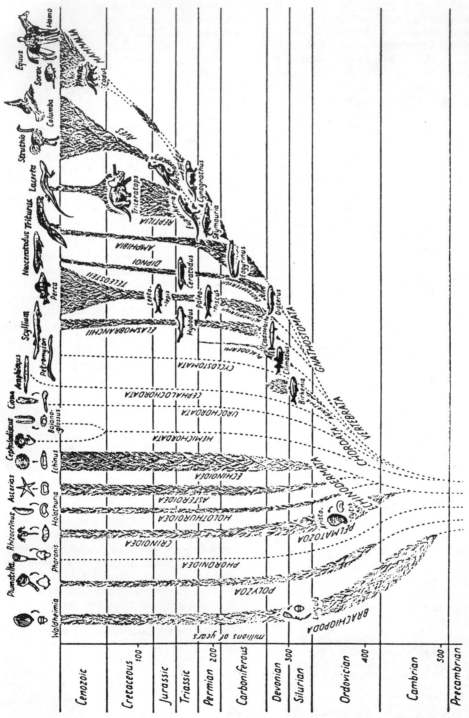

Fig. 1. A phylogenetic tree showing the probable times of origin and affinities of the vertebrates and some related groups of animals (Young, 1962).

20

genesis of mammals. As may be seen in Figure 2, which represents a phylogenetic tree of mammals (Young, 1962), primates evolved as a specialized branch of the insectivore line (i.e., shrews, moles, hedgehogs, etc.). Carnivores and rodents, which are frequently compared with primates, have followed independent courses of development from the primate line and from each other since the late Cretaceous or early Paleocene periods. Rats were never ancestral to cats nor were cats ancestral to primates; rather, each represents a different evolutionary lineage. Therefore, from the point of view of the phylogenesis of primate characteristics, the rat-cat-monkey comparison is meaningless.

Figure 3 represents a phylogenetic tree of primates (Young, 1962). The earliest primates, the prosimians, appear to have developed as a specialization of the line of the insectivores. The living prosimians (tarsiers, lorises, and lemurs) retain some insectivore characteristics (LeGros Clark, 1959). A comparison could therefore be made between living insectivores, prosimians, cercopithecid (Old World) monkeys, pongids (great apes), and hominids (men) which would give some clue to patterns of

evolution in the human lineage. Such a comparison was recently made by Diamond (1967) as an attempt to infer the course of evolution of neocortex in man.

Although primates are more closely related to each other than to other mammalian orders, there is still considerable variation and specialization within the primate order. Scott (1967) has recently warned that

Subhuman primates are not small human beings with fur coats and (sometimes) long tails. Rather, they are a group which has diversified in many ways, so that they are as different from each other as are bears, dogs and racoons in the order Carnivora. The fact that an animal is a primate therefore does not automatically mean that its behavior has special relevance to human behavior [p. 72].[4]

Evolutionary Inferences from the Study of Living Animals

Even though one may select animals for study that are descendants of a common evo-

[4] The term "subhuman" connotes relative position on the *scala naturae* or phylogenetic scale. A term more in keeping with the conceptual framework of the phylogenetic tree would be "nonhuman." Similar implications are carried in the term "subprimate" which should be replaced by "nonprimate" and "submammalian" which should be replaced by "nonmammalian."

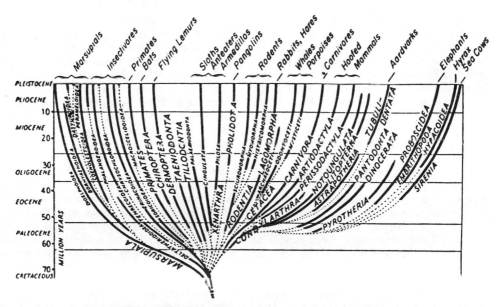

Fig. 2. A phylogenetic tree showing the probable times of origin and affinities of the orders of mammals (Young, 1962). (Common names have been added at the top.)

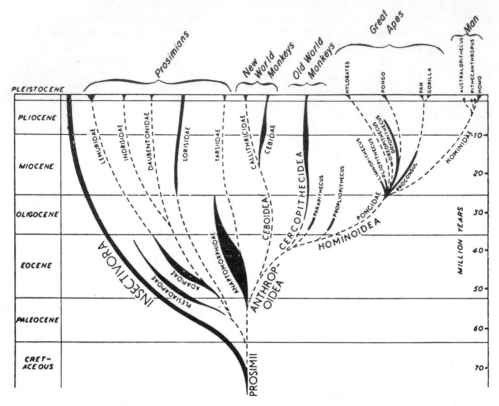

FIG. 3. A phylogenètic tree showing the probable times of origin and affinities of the primates (Young, 1962). (Common names have been added at the top.)

lutionary lineage, the question remains as to what can be learned about evolution, behavioral or otherwise, from the study of living animals. The rationale for such a study is based on the assumption that the living animals that have been selected are sufficiently similar to the ancestral forms that some inferences about the ancestral forms can be drawn. Our knowledge of ancestral vertebrates is based on fossilized skeletons. Soft tissues, including the central nervous system, do not usually fossilize; nor does behavior. Therefore, a living animal species which appears to be skeletally very similar to fossil ancestors, may in fact have undergone some change in its soft tissues or behavior. Thus, conclusions about evolution, even though they are based on the study of living animals selected because they are descendents of a continuous evolutionary lineage, do retain some degree of uncertainty. However, for the study of behavioral evolution, there are no alternatives. The behavior of ancestral organisms can only be

inferred from the behavior of living organisms which appear to be structurally similar and which inhabit similar environments. A method based on these assumptions has worked with reasonably good success in the study of morphological evolution and there seems to be good reason to believe that it will serve psychologists as well as it has served morphologists (Mayr, 1958; Simpson, 1958a).

Two general approaches can be used to collect data which are meaningful for the study of behavioral evolution. The first is the historical or "phylogenetic" approach similar to that used in comparative anatomy. In this instance, an attempt is made to infer changes in behavioral characteristics or patterns through time by comparing living animals which form a quasi-evolutionary series within a common lineage. An important consideration in such comparisons is the fact that the animals which have been chosen to represent ancestral groups are not the actual ancestors, but merely descendents of them. They are suitable for phylogenetic comparisons because they possess many characteristics which are primitive; that is, unchanged from ancestral forms (Simpson, 1961). The primitive behavior pattern of a group can often be inferred by comparing the behaviors of the living members of the group and looking for elements of the behavior pattern common to all. However, the greater the degree of diversity and specialization within a group, the greater the need for studying more variants. Furthermore, these elements should be sought in related groups and in living representatives of ancestral groups.

In using the "phylogenetic" method, the student of behavior has the same problems which face morphologists in the study of evolution. Some attempt must be made to determine which characteristics of the behavior are primitive, that is, derived from the behavior of the ancestors of the species, and which are specializations of the species studied. However, Simpson (1961) has cautioned that the concepts of primitive and specialized are meaningful only when related to a particular taxon, lineage, or phylogeny. Therefore, a behavioral characteristic which is found as a specialization in one animal group may be a primitive trait in another.[5]

In the case of living animals, the assumption is usually made that the more primitive characteristics a given species has, the more likely it is to resemble the ancestral form. However, the behaviorist has no behavioral fossil record to corroborate his decisions. He must decide on the basis of the properties of the behavior itself. Moreover, Klopfer and Hailman (1967), in their discussion of behavioral phylogenesis, conclude that

there is no good reason to assume that the most complex form has been evolved from the simpler one, since there are many cases of secondary simplification of morphological structures among animals. Also, one cannot infer direction of evolution from the relative abundance of variants; one cannot know a priori, whether the variation represents the end of an evolutionary process of adaptation (the commonest variant being the most advanced one) or merely the beginning (the commonest variant being the most primitive) [p. 187].

[5] See LeGros Clark (1959) and Simpson (1961) for a fuller discussion of the concepts of specialized and primitive.

A case in point is the study by Doty, Jones, and Doty (1967). These investigators compared learning set performance in four species of carnivores. They found that domestic cats (which rank with domestic dogs as the "standard" carnivores) acquired the learning set more slowly than did weasels, skunks, or ferrets. While these data are of considerable interest, there is, unfortunately, no way of concluding from the behavior itself whether the inferior learning set performance of the domestic cats represents a primitive state, comparable to that which presumably existed in ancestral carnivores or whether it represents a more recent secondary simplification of an older, more complex form of behavior.

One possible means of resolving this problem would seem to be to fall back on the assertion that behavior is subject to the same evolutionary principles as any other characteristic of the organism. As Nissen (1958) has argued

it seems just as logical and possible that an adaptive behavior should give selective value to a related structural character as that an adaptive anatomic feature should lend selective advantage to behavior which incorporates or exploits that structure [p. 186].

We might further assume that an animal which is morphologically primitive will be behaviorally primitive and one which has developed morphological specializations will have also developed appropriate behavioral specializations. This assumption seems justified in view of the strong emphasis which paleontologists have placed on the relations between form and function in survival (Colbert, 1958; Simpson, 1958b).

Attempts to set up quasi-evolutionary series of living animals for the purpose of determining trends in behavioral or nervous system evolution have usually been unsatisfactory because the animals chosen for the series were highly specialized species of widely divergent lineages. For example, von Békésy (1960) traced the course of evolution of the organ of Corti from bird to alligator to duckbill to man. Unfortunately, specific conclusions about the evolution of the organ of Corti are unjustified from this group of animals since birds are not ancestral to alligators or to mammals. Moreover, there are virtually no paleontological data on the monotremes, and their relationship to the other mammalian orders is obscure (Romer, 1966). Similarly, Bishop (1958), in a discussion of the evolution of the cerebral cortex, chose the brain of a snake to represent the reptile group from which mammals evolved. Snakes are among the most specialized of the living reptiles (Goin & Goin, 1962) and first appear in the fossil record in the Cretaceous period, 100 million years after the origin of the reptilian group which gave rise to the mammals (Romer, 1966).

Quasi-evolutionary series also often suffer from being too restrictive; that is, they attempt to represent all of vertebrate evolution with four or five species and sometimes even fewer. Bishop (1958), for example, based his evolutionary conclusions on three brains, one of which was the unlikely composite of a cat and a monkey.

Aside from the uncertainties inherent in using living animals to approximate actual evolutionary sequences and the necessity of choosing species which retain the most primitive traits, another problem arises. In some instances no living representatives of truly ancestral groups exist. Students of evolutionary patterns within the placental mammals, including the primates, however, are fortunate in that relatively unchanged direct descendants of the ancestral insectivores that gave rise to other placental mammals are living today. All of the living insectivores have diverged from ancestral insectivores to some degree, but the Erinaceidae (hedgehogs) are the most primitive and retain a relatively primitive eutherian status (Anderson & Jones, 1967). Study of this group of animals in some detail will probably be very profitable. Marsupials and placentals presumably had a common ancestry (Romer, 1966) and comparative studies of the primitive members of these two groups might allow some inferences to be made about the earliest mammals. Although monotremes were probably never ancestral to marsupials or placentals (Romer, 1966), they do retain many more reptilian characteristics than do the latter groups of mammals. The study of monotremes may therefore provide some insight into the behavioral characteristics of the therapsid reptiles from which all mammals evolved.

Comparative studies of behavioral or neural evolution of fishes have a particular advantage in that there are over 25,000 species of living fishes (Herald, 1961), a number of which closely resemble ancestral forms. Furthermore, fishes have undergone a considerable degree of evolution and occupy a great variety of ecological niches. In contrast, the living amphibians are relatively poor in number of species and diversity of terrain occupied and are generally regarded as a decadent class containing many relict groups (Darlington, 1957; Goin & Goin, 1962).

The crocodilians are living descendents of the archosaurian reptiles which also gave rise to birds (Romer, 1966) and appear to have remained relatively unchanged. Although the avian and crocodilian lineages are divergent, comparisons of the behavior and central nervous systems of these groups might be useful in view of their common archosaurian heritage.

There are unfortunately no surviving representatives of the reptile groups which gave rise to the mammals. These therapsid reptiles disappeared from the fossil record in the Jurassic period and since their lineage branched off from the stem reptiles very early, none of the living reptiles can be considered to be very closely related to mammals. Although turtles seem to be relatively unchanged since the Triassic period, they are highly specialized. *Sphenodon,* a surviving rhynchocephalian, also seems to be little changed since the Mesozoic, and the crocodilians mentioned previously are only slightly modified following their origin from thecodonts in the Triassic period (Romer, 1966). Comparative studies on the more primitive reptiles might allow some inferences about the behavioral repertoire of the ancestral reptiles from which the mammalian lineage is also derived. However, this would be a "weak inference" in the terminology of Klopfer and Hailman (1967) since it would rest on the assumption that the most frequently observed characteristics are primitive.

In his discussion of the evolution of intelligence, Bitterman (1965a, 1965b) compares the performance of teleost fish, turtles, pigeons, rats, and monkeys on probability and discrimination reversal learning tasks and characterizes their performance as "fishlike" or "ratlike." However, these comparisons do not permit generalizations to be made about the evolution of intelligence or any other characteristic of these organisms since they are not representative of a common evolutionary lineage. Figure 4 is a diagrammatic representation of two vertebrate phylogenetic trees. The left tree shows the approximate evolutionary relationships among the various animals studied by Bitterman in the visual reversal problem and the right tree shows the evolutionary relationships among the animals in the visual probability problem. The squares indicate the animals whose behavior was characterized by Bitterman as "ratlike" and the triangles indicate the animals whose behavior was characterized as "fishlike." Assuming for the moment that these animals are sufficiently similar to ancestral forms that some specific evolutionary generalizations could be made from them, one might conclude that the "ratlike" pattern in visual probability learning is limited to mammals. However, since no rat was ever an ancestor of any monkey, it is not clear whether the rat and monkey independently evolved the "ratlike" pattern or whether it was inherited from a common reptilian ancestor. The absence of this pattern in turtles would not support either possibility since turtles are so far from the line (or lines) of reptiles that were ancestral to mammals. In the case of the visual reversal learning data, the "ratlike" performance pattern is present in pigeons, rats, and monkeys, but absent in turtles. Again, no firm conclusions can be drawn. On the one hand, the assumption might be made that the "ratlike" pattern evolved independently in birds and mammals, since turtles, which share a remote common ancestry with birds and mammals in the stem reptiles (Romer, 1966), possess the presumably primitive "fishlike" pattern. On the other hand, the absence of the "ratlike" pattern in turtles may represent a secondary simplification of behavior. Still a third possibility must be considered in the analysis of Bitterman's data; that is, whether the "fishlike" pattern is a primitive characteristic at all. In the absence of data on the distribution of the "fishlike" pattern in am-

phibians and nonteleost fish, it is not possible to determine whether this pattern was independently acquired by teleosts and turtles or inherited from a common ancestor.

Similar problems attend Harlow's (1959) interpretation of evolutionary trends in learning set performance in primates. Harlow compares data from New World monkeys, Old World monkeys, chimpanzees, and humans. Old World monkeys and chimpanzees, although specialized in their own ways, are reasonable enough representatives of human ancestors (Romer, 1966) that meaningful conclusions about the phylogenesis of man's behavior may be drawn. However, the New World monkeys appear to have evolved from New World prosimians which are not closely related to the Old World prosimians from which are derived Old World monkeys, apes, and men. Consequently, New World monkeys should not be used as representatives of man's ancestors (Campbell, 1966; Romer, 1966). A more appropriate representative would have been one or more species of the Old World prosimians.

The comparative studies of Bitterman, Harlow, and others discussed in this paper have been selected to illustrate some of the complexities involved in the interpretation of evolutionary relationships, not as a critique of their research. Unless comparative psychologists are prepared to deal with the details and intricacies of the evolutionary history of vertebrates, meaningful descriptive statements or theoretical formulations of specific behavioral phylogenies will not be forthcoming.

ANALYSIS OF ADAPTATION

A second and equally important approach to the study of behavioral evolution is through the analysis of adaptation. This method is based on the study of living animals, selected because they possess differing degrees of specialization (adaptation) with respect to some particular characteristic such as development of sense organs or central nervous system, the amount of postnatal care given to offspring, complexity of courtship patterns, etc. Such animals need not be descendants of a common evolutionary lineage and consequently any conclusions drawn

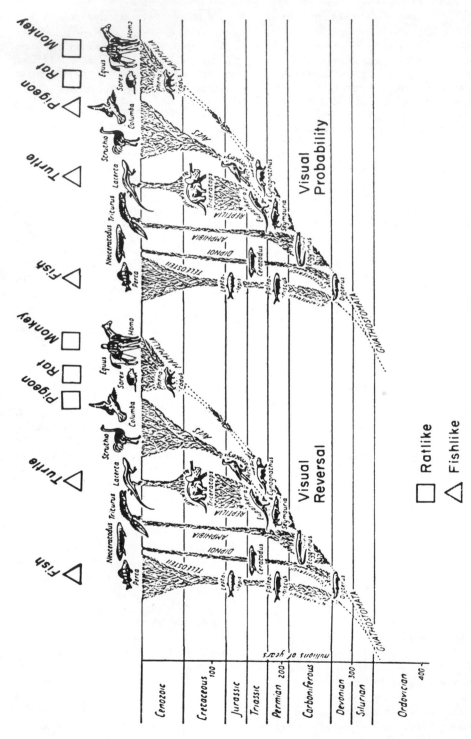

Fig. 4. Data from the table presented by Bitterman (1965a) plotted on segments of the phylogenetic tree shown in Figure 1 to indicate their relationship to approximate lineages. (The characterization of the behavior as "ratlike" or "fishlike" is Bitterman's notation.)

from such comparisons will be applicable only to general principles of adaptation and survival. They will give no direct clues to specific sequential patterns of evolutionary development in specific lineages. For example, Weiskrantz's (1961) description of the differential effect of lesions in the visual system in frogs, birds, rats, cats, dogs, monkeys, and man does not provide us with a picture of the sequential evolutionary history of the visual system and its function in the line of vertebrates leading to man; however, it is very useful in understanding the general relationship between the development of certain structural adaptations and their function in behavior. The same may be said of Woolsey's (1958) comparative studies of the sensory-motor cortex in rat, rabbit, cat, and monkey. They are extremely useful in relating structure and function, but are of no value in describing the sequence of evolutionary development in this system since each of these animals reached its respective degree of cortical development quite independently of any of the others.

Commenting on problems of the analysis of morphological evolution, Davis (1954) observed that most studies of comparative anatomy were concerned with determining the particular course of evolution rather than the reasons that one course was followed and not another. In his view, comparative studies should also be aimed at the mechanisms of adaptation. Similarly in comparative psychology, the study of analogous behavior in animals of divergent groups may be very useful in formulating generalizations about behavioral adaptation to specific problems of survival. Such generalizations might have broad applicability to a number of lineages of the phylogenetic tree and would greatly aid in the interpretation of data obtained through the phylogenetic approach.

Thus, the phylogenetic and adaptation approaches are not mutually exclusive. Data obtained through one approach can be used to augment conclusions based on the other. However, the experimenter must be clear as to which method of comparison he is using if he is to avoid the interpretive pitfalls described above.

CRITIQUE OF THE PHYLOGENETIC MODEL

King and Nichols (1960) object to the use of a phylogenetic model as a basis for the study of comparative psychology for three reasons: First, one apparently cannot predict behavior from one phylogenetic level to another. Second, the experimenter is required to make inferences about the behavior of extinct forms. Third, some taxonomic groups are differentiated only by characteristics which appear to be relatively unrelated to behavior. They suggest instead that comparative psychology develop its own system of classification based on behavior. However, such a classification would simply be a behavioral hierarchy and since it would not necessarily be related to evolutionary lineages, it would be nearly useless in understanding the evolution of behavior. Unless comparisons are made between organisms of a common evolutionary lineage, the relationship between the evolution of structure and the evolution of behavior will never become discernible. The apparent failure of the zoological model of evolution to lead to meaningful predictions of relative behavioral performance has been due to the fact that many comparative psychologists, neuroanatomists, physiologists, etc., have applied the inappropriate phylogenetic scale model rather than the appropriate phylogenetic tree model. Furthermore, they have generally failed to recognize the important taxonomic principle that there are diverse sources of similarity (both morphological and behavioral) among various organisms. Only one of these sources is inheritance from common ancestors and is usually referred to as "homology." On the other hand, similarities may also be due to the independent evolution of similar characteristics by more or less closely related forms. This similarity, not due to inheritance from a common ancestor, is referred to as "homoplasy." Homoplasy is a generic term which includes such forms of similarity as parallelism, convergence, analogy, mimicry, etc.[6] When nonhomologous characters serve similar *functions,* whether or not they are similar in appearance, they are referred to as "analogous." Thus, the hands of a racoon and a man are homologous as anterior pentadactyl appendages and homoplastic in their particular appearance as hands. They are not homologous as hands since they were evolved independently of each other. Finally, they are analogous in their functions as prehensile organs.

[6] For a more extensive discussion of homoplasy and homology, see Simpson (1961).

There would then seem to be little point in looking for parallels between a hierarchy of behavioral complexity and the phylogenetic tree unless the behaviors were homologous. Behaviors that are convergent or analogous may provide insights into general mechanisms of survival and adaptation, but will not shed any light on the specific behavioral history of any particular group of organisms.

Another deterrent to theoretical development in comparative psychology has been the typological approach to behavior. As Mayr (1968) has described it, "When the learning psychologist speaks of The Rat or The Monkey, or the racist speaks of The Negro, this is typological thinking [p. 597]." The typological approach carries with it the implication that the particular species being investigated is a generalized representative of the entire order or class when in fact that species may be highly specialized and not at all representative. Simpson (1958a, 1958b) has also discussed this point. Bitterman (1965a, 1965b) has regrettably used typological designations of "ratlike" and "fishlike" as a shorthand characterization of the behavior of the teleosts, reptiles, birds, and mammals which he has studied in probability learning and reversal learning problems. Unfortunately, this typology implies that the "fishlike" behavior of the particular fishes which Bitterman studied is representative of most fishes or possibly all fishes. Considering the diversity of specialization in teleost fishes, one would not be surprised to find several or even many species which had evolved the capacity for behavior analogous to that which Bitterman has called "ratlike."

CONCLUSIONS

Beach (1950) has suggested that any experiments in which a nonhuman species is compared with another nonhuman species or with humans should be considered as being within the realm of comparative psychology. His main point is that the term comparative psychology should be reserved for experiments in which organisms of different species are *compared*. We would like to expand this definition by further suggesting that any experiments carried out with nonhuman subjects, in which no attempt is made to compare these subjects to other species of animals or humans, be regarded as simply animal psychology. In other words, the term comparative psychology should be reserved for experiments in which interspecies comparisons are made.

Schneirla (1949) has stated that

The general problem of the animal psychologist is the nature of behavioral capacities on all levels of accomplishment. He must contrive to understand how each animal type functions as a whole in meeting its surrounding conditions: what its capacities are like and how they are organized . . . [p. 245].

This would seem to be a reasonable goal for animal and comparative psychologists alike. The difference between the two would be the emphasis of the latter on the similarities and differences between various taxonomic groups of organisms. However, Ratner and Denny (1964) warn that "capricious comparisons" will add little to our understanding of systematic differences and similarities among species. This is not to say that questions such as "Do pigeons acquire learning sets faster than rats?" are without meaning if there is a specific reason for comparing those particular organisms and the suggestion is not made that the outcomes of the comparisons necessarily imply anything about the evolution of these behaviors. However, our opinion is that if comparative psychology is ever to develop a theoretical model capable of predicting behavior, experimenters will have to specify their independent variable more precisely than "species difference." Similarly, an experimenter who reports differences in visual acuity in several species that have varying degrees of differentiation of the visual cortex might be criticized on the grounds that these animals also vary in the near point of accommodation which could be a confounding variable. This example illustrates the point that comparative psychologists have the same problems as other experimenters in determining that their chosen independent variable is indeed the only relevant variable operating.

What then can we hope for as attainable goals for comparative psychology? First, as Schneirla (1949) has suggested, would be a description of the behavioral capacities of organisms throughout the animal kingdom. A second would be the search for systematic trends in behavior which hopefully would vary reliably with other taxonomic indexes, since form and function in nature are inextricably interrelated. A third would be an attempt to reconstruct the historical development of behavior as best as can be done from the data of paleontology and the study of living organisms which resemble, as closely as possible, ancestral forms of particular evolutionary lines. A fourth 'goal would be the analysis of the general mechanisms of adaptation and survival. These goals can best be attained by ridding ourselves of concepts like the phylogenetic scale, higher and lower animals, nonrepresentative behavioral typologies, and other notions which have had the effect of oversimplifying an extremely complex field of research.

REFERENCES

ANDERSON, S., & JONES, J. K. *Recent mammals of the world.* New York: Ronald, 1967.

ARIËNS KAPPERS, C. U., HUBER, G. C., & CROSBY, E. C. *The comparative anatomy of the nervous system of vertebrates, including man.* New York: Hafner, 1936. (Republished: 1960.)

ARISTOTLE. *Historia animalium.* (Trans. by D. W. Thompson) Oxford: Clarendon, 1910.

ARISTOTLE. *De partibus animalium.* (Trans. by W. Ogle) Oxford: Clarendon, 1912. (a)

ARISTOTLE. *De generatione animalium.* (Trans. by A. Platt) Oxford: Clarendon, 1912. (b)

BEACH, F. A. The snark was a boojum. *American Psychologist,* 1950, 5, 115–124.

BISHOP, G. H. The place of cortex in a reticular system. In H. H. Jasper et al. (Eds.), *Reticular formation of the brain.* Boston: Little, Brown, 1958.

BITTERMAN, M. E. The evolution of intelligence. *Scientific American,* 1965, 212, 92–100. (a)

BITTERMAN, M. E. Phyletic differences in learning. *American Psychologist,* 1965, 20, 396–410. (b)

CAMPBELL, B. G. *Human evolution.* Chicago: Aldine, 1966.

COLBERT, E. H. Morphology and behavior. In A. Roe & G. G. Simpson (Eds.), *Behavior and evolution.* New Haven: Yale University Press, 1958.

DARLINGTON, P. J. *Zoogeography: The geographical distribution of animals.* New York: Wiley, 1957.

DAVIS, D. D. Primate evolution from the viewpoint of comparative anatomy. *Human Biology,* 1954, 26, 211–219.

DIAMOND, I. T. The sensory neocortex. In W. D. Neff (Ed.), *Contributions to sensory physiology.* Vol. 2. New York: Academic Press, 1967.

DOTY, B. A., JONES, C. N., & DOTY, L. A. Learning set formation by mink, ferrets, skunks and cats. *Science,* 1967, 155, 1579–1580.

GOIN, C. J., & GOIN, O. B. *Introduction to herpetology.* San Francisco: Freeman, 1962.

HARLOW, H. F. The evolution of learning. In A. Roe & G. G. Simpson (Eds.), *Behavior and evolution.* New Haven: Yale University Press, 1958.

HARLOW, H. F. Learning set and error factor theory. In S. Koch (Ed.), *Psychology: A study of a science.* Vol. 2. New Haven: Yale University Press, 1959.

HERALD, E. S. *Living fishes of the world.* New York: Doubleday, 1961.

KING, J. H., & NICHOLS, J. W. Problems of classification. In R. H. Waters, D. A. Rethlingshafter, & W. E. Caldwell (Eds.), *Principles of comparative psychology.* New York: McGraw-Hill, 1960.

KLOPFER, P. H., & HAILMAN, J. P. *An introduction to animal behavior: Ethology's first century.* Englewood Cliffs, N. J.: Prentice-Hall, 1967

LEGROS CLARK, W. E. *The antecedents of man.* Edinburgh: Edinburgh University Press, 1959.

LOVEJOY, A. O. *The great chain of being.* Cambridge: Harvard University Press, 1936.

MAYR, E. Behavior and systematics. In A. Roe & G. G. Simpson (Eds.), *Behavior and evolution.* New Haven: Yale University Press, 1958.

MAYR, E. The role of systematics in biology. *Science,* 1968, 159, 595–599.

McGILL, T. E. Response of the leopard frog to electric shock in an escape-learning situation. *Journal of Comparative and Physiological Psychology,* 1960, 53, 443–445.

NISSEN, H. W. Axes of behavioral comparison. In A. Roe & G. G. Simpson (Eds.), *Behavior and evolution.* New Haven: Yale University Press, 1958.

RATNER, S. C., & DENNY, M. R. *Comparative psychology.* Homewood, Ill.: Dorsey, 1964.

ROE, A., & SIMPSON, G. G. (Eds.). *Behavior and evolution.* New Haven: Yale University Press, 1958.

ROMER, A. S. *Vertebrate paleontology.* Chicago: University of Chicago Press, 1966.

ROSS, W. D. *Aristotle.* London: Methuen, 1949.

SCHNEIRLA, T. C. Levels in the psychological capacities of animals. In R. W. Sellars, V. H. McGill, & M. Farber (Eds.), *Philosophy for the future.* New York: Macmillan, 1949.

SCOTT, J. P. Comparative psychology and ethology. *Annual Review of Psychology,* 1967, 18, 65–86.

SIMPSON, G. G. The study of evolution: Methods and present status of theory. In A. Roe & G. G. Simpson (Eds.), *Behavior and evolution.* New Haven: Yale University Press, 1958. (a)

SIMPSON, G. G. Behavior and evolution. In A. Roe & G. G. Simpson (Eds.), *Behavior and evolution.* New Haven: Yale University Press, 1958. (b)

SIMPSON, G. G. *Principles of animal taxonomy.* New York: Columbia University Press, 1961.

VAN BERGEIJK, W. Anticipatory feeding behavior in the bullfrog (*Rana catesbeiana*). *Animal Behaviour,* 1967, 15, 231–238.

VON BÉKÉSY, G. Experimental models of the cochlea with and without nerve supply. In

G. L. Rasmussen & W. F. Windle (Eds.), *Neural mechanisms of the auditory and vestibular systems.* Springfield, Ill.: Charles C Thomas, 1960.

WATERS, R. H. The nature of comparative psychology. In R. H. Waters, D. A. Rethlingshafter, & W. E. Caldwell (Eds.), *Principles of comparative psychology.* New York: McGraw-Hill, 1960.

WEISKRANTZ, L. Encephalization and the scotoma. In W. H. Thorpe & O. L. Zangwill (Eds.), *Current problems in animal behaviour.* Cambridge: Cambridge University Press, 1961.

WIGHTMAN, W. P. D. *The growth of scientific ideas.* Edinburgh: Oliver & Boyd, 1950.

WOOLSEY, C. N. Organization of somatic sensory and motor areas of the cerebral cortex. In H. F. Harlow & C. N. Woolsey (Eds.), *Biological and biochemical bases of behavior.* Madison: University of Wisconsin Press, 1958.

YOUNG, J. Z. *The life of vertebrates.* Oxford: Oxford University Press, 1962.

(Received July 12, 1968)

3 COMPARATIVE PSYCHOLOGISTS AND THEIR QUEST FOR UNIFORMITY

Donald A. Dewsbury

*Department of Psychology
University of Florida
Gainesville, Florida 32601*

Four years ago, I wrote that comparative psychologists seemed troubled by the state of their discipline, although the discipline itself appeared to be in rather sound condition.[1] Since that time, I can detect no worsening in the state of the discipline. Nevertheless, the literature of self-criticism has reached epidemic proportions. Perhaps a peak was reached when one comparative psychologist's musings on the "fall" of comparative psychology[2] were met with the reply of another[3] that "comparative psychology has never had the elevation from which to fall." (p. 858.) While both may be exaggerating to dramatize a point, these two assessments are indicative of much writing in comparative psychology.

Ewer,[4] an ethologist, did little to help the intellectual climate when she referred to psychologists who do not wish to consider evolutionary questions as "a set of pig-heads who have merely progressed from studying what rats do in mazes to recording how many intromissions they make before ejaculation." (p. 805.)

The major point that I wish to make in this paper is simple. There are many ways in which to conduct legitimate and important research in animal psychology. The quest for uniformity of approach is fruitless. I shall attempt to integrate some of the criticism into a positive and consistent approach. I shall stress the importance of evolutionary questions and consider comparative and noncomparative research in relation to the kind of question being asked and the methods being used.

The Quest for Uniformity

A central theme in the literature of self-criticism is an apparent frustration with the diverse methods used by different comparative psychologists. The critics are in quest of uniformity. Lockard,[2] for example, detects a "scientific revolution," in the wake of which lies "a confused scatter of views of nature, problems and methods." (p. 168.)

Rather than bemoan this "scatter," I would look for the advantages of the diversity of method and approach in comparative psychology and would attempt to systematize them. As I wrote earlier,[1] "It may be disquieting to the mental health of the comparative psychologists to be unable to isolate clear, progressive trends in his discipline. However, this may be a sign of a healthy, vigorous, and more mature comparative psychology." (p. 37.)

It is necessary to maintain some standards if we are to have a respectable and productive discipline. I believe that for each study we may ask that the investigator (1) state clearly the type of question being asked in his research, (2) indicate why that question is worth asking, and (3) show that the method used is appropriate to the question being asked. Beyond that, each scientist should have the freedom to go his own way. Comparative psychology is a delightfully complex assortment of different individuals who are using a variety of methods to answer a set of questions about animal behavior.

The various critics wish to see comparative psychology changed through a revision of some current methodological and conceptual practices. While I disagree with Professors Lockard and Ewer on a number of tactical issues, we all seek similar changes. I give Lockard credit for giving these ideas more exposure than any other psychologist of the last few years. I prefer to work to add new methods to the existing ones, however, and to recognize that each is appropriate for asking certain types of questions.

If new species are to be studied and new questions are to be asked in comparative psychology, it will more likely be because the working animal psychologist recognizes that these approaches aid him in answering questions in which he is interested and raise important new questions. I believe that Schreier[5] is correct in writing, "Until more behavioral facts, and especially principles based on those facts, can be brought to bear on the question, I suspect that things will continue very much the way they have during the past 40 years." (p. 682.) What we need are more data, not more polemics.

The Problem Areas of Comparative Psychology

If we are to detect any order in the complex picture that I have portrayed, we must consider the questions that are under consideration in the study of animal behavior. As is so often the case, Niko Tinbergen perceives the issues with particular clarity. Tinbergen[6] proposed that, once behavior is adequately described, there are four basic problem areas in the study of animal behavior. These are the problems of development, immediate causation or mechanism, evolutionary history, and adaptive significance. These four problem areas are as characteristic of the study of animal behavior carried out by psychologists as they are of such study carried out by zoologists. Many polemical arguments can be traced to failures to distinguish differences in the questions being asked by different scientists. The manner in which such failures have affected semantic and conceptual issues in relation to the nature-nurture problem has been illustrated by Lehrman.[7] Later I shall try to show how a similar confusion over the problem areas has caused some difficulties regarding the selection of species for study.

Development and Causation

The problems of development and immediate causation have been much studied in psychology. Few would quarrel with the importance of studying the interaction of genetic and environmental factors in producing the total

31

behavioral repertoire of the individual animal. The area is popular and is alive with much exciting research.[8,9] Similarly, the problem of immediate causal mechanisms has engaged many researchers. Causation is studied through the manipulation and recording of environmental and behavioral events, with or without manipulations inside the skin of the organism. Studies of causal mechanisms are so widespread in psychology that they hardly need elaboration or defense.

Evolutionary History

The first of the two evolutionary problem areas is that of evolution as history. Hodos and Campbell[10] term this the historical, or phylogenetic, approach. Lockard[2] sees this method as stemming from the "principle of phylogenetic relatedness," but terms it "the comparative method." While I do not wish to quibble over terminology, there are many comparative methods, of which this approach is but one. To call this one approach the only comparative approach is to open the door to many sorts of confusion. Therefore, I hope that Lockard's usage will not become widespread.

The historical approach has been widely used in classical ethology. Hinde and Tinbergen[11] summarized various ways in which displays can evolve from other behavioral patterns and the processes through which such transformations might occur. The work of Dilger[12,13] on lovebirds of the genus *Agapornis* is an example of the kind of insight that can be gained when a whole series of closely-related species is studied with this approach.

Central to this approach has been the concept of homology, the notion that behavioral patterns can be traced to recent common ancestors. The importance of distinguishing homology from homoplasy has been underscored by Campbell and Hodos.[14] The difficulties involved in applying the homology concept to behavior have been reviewed by Atz.[15] Nevertheless, the data of the ethologists make such usage appear quite reasonable.

There are relatively few examples of successful use of the historical approach in comparative psychology. This may be in part related to the tendency of comparative psychologists to make broad phyletic comparisons (see below). The attempt of Allison and VanTwyver[16] to trace the evolutionary history of sleep patterns provides one of the best examples of the application of the historical method. Another is the attempt of Heffner, Ravizza, and Masterton[17] to reconstruct the evolution of auditory sensitivity by studying a series of species from marsupials to man. Unfortunately, the emphasis of comparative psychologists on phyletic comparisons of learning ability (by Bitterman,[18] for example) has not proved successful.[19]

Study of the evolution of behavior from an historical perspective might provide important information for comparative psychology. The path is a difficult one, however, and there are few examples of success. Concentration on groups of species within restricted taxa and on species selected for taxonomic reasons may be critical.

Teleonomy

I am more sanguine about the potential of the study of adaptive significance, or teleonomy,[20,21] in becoming an important part of comparative psychology.

The study of teleonomy is essentially identical to the "analysis of adaptation" of Hodos and Campbell[10] and the "ecological method" of Lockard.[2] While serious attempts to study adaptive significance in comparative psychology are just beginning, there are enough examples of the successful use of this approach to reveal the potential gain to be derived from its widespread use.

Within the area of teleonomy, some studies are directed at the long-term survival value of a pattern, while others are directed at its more immediate function.[22]

Methods. There are at least three methods of studying teleonomy: the experimental method, the method of adaptive correlation, and the behavior-genetic method.[22,23] In the experimental method, one manipulates an independent variable and measures the effect of that variable on the animals in question. This method has been used by ethologists in teleonomic studies. Thus Tinbergen[24] was able to demonstrate the function of eggshell removal by gulls in preserving camouflage. In studies by deRuiter[25] on countershading in caterpillars and by Blest[26] on eye-like markings in moths, similar methods were used.

With the method of adaptive correlation, several species are compared along a number of dimensions such as ecological characteristics, morphological characteristics, and patterns of behavior. An attempt is made to infer the function of a character from the nature of other characters with which it appears to be consistently associated. For example, a number of adaptations are clearly associated with flying, with an aquatic habitat, with life in a cold environment, and so on.

The method of adaptive correlation produces tentative hypotheses on relationships, which can never be as satisfying to the methodological purist as can the results of experiments. The problem of teleonomy is so important, however, that some tolerance of ambiguity may be warranted. Once a correlation is proposed on the basis of observation of several species, the behavior of additional, as yet unstudied, species can be predicted. Confirmation of the prediction tends partially to validate the proposed correlation. Thus the study of adaptation need not be entirely postdictive (as was implied by Adler and Tobach[27]).

The classical use of the method of adaptive correlation is that of Cullen,[28] who found that the breeding behavior of cliff-dwelling kittiwakes could be differentiated from the breeding behavior of other gulls with respect to 32 characters, each apparently adapted to nesting on cliffs.

The behavior-genetic method of studying adaptation is based on the application of theory from biometrical genetics to behavior. Bruell[29,30] considered the possible interpretation of results from a diallel cross study. According to this theory, Henderson[31] has said:

> . . . When a substantial fraction of genetic variance can be attributed to dominance, the characteristic being studied has probably been subjected to selection pressure, and therefore is relevant to the survival of the organism. Furthermore, if the dominance obtained is largely unidirectional, the performance of hybrids relative to inbreds indicates the direction of optimum value for the characteristic being studied. (p. 777–778.)

Bruell[29,30] presents data relevant to this theory.

Other behavior-genetic methods, for example the selection method used by Crossley,[33] can provide information about adaptive function.

Examples. I shall consider some examples of the utility of teleonomically-oriented research, because I believe that it is only through the demonstration of its utility that the approach can come to be accepted.

33

These examples fall roughly into two classes: those in which the object of the research deals with teleonomic considerations, and those in which teleonomic considerations are helpful in dealing with other questions. First I shall consider seven examples of the former.

1. Allison and VanTwyver[16] differentiated mammalian species into "good sleepers" and "poor sleepers." Good sleepers tended either to be predatory or to have secure sleeping places, while poor sleepers tended to be subject to predation at all hours. This correlation can be tested by studying sleep in additional species, in order to see whether sleep patterns continue to appear related to prey-predator relationships.

2. Glickman and Sroges[33] studied curiosity in over 300 zoo animals, which represented over 100 species. Patterns of exploration of novel objects were related not only to taxonomic position, but to such variables as the type of food eaten and the method of obtaining food. A knowledge of the life-style of the species in their natural habitats was essential to an understanding of the rather substantial differences observed in their exploration of novel objects.

3. A number of behavior-genetic investigations have produced evidence consistent with the theory that relates adaptive significance to directional dominance. Henderson[31] found that both brain weight *per se* and increases in brain weight with environmental enrichment showed heterotic inheritance. Collins[34] found directional dominance for fast learning in an avoidance situation. We have recently completed a diallel cross study of the copulatory pattern of inbred strains of rats. There was a directional dominance for fast copulation, in the sense that smaller numbers of mounts and intromissions preceded ejaculation and there was a shortening of some intervals between copulations. Further examples of teleonomically-oriented behavior-genetic research can be found in Wilcock.[23]

4. Study of different species of the rodent genus *Peromyscus* has produced a number of insights into the adaptive significance of behavioral patterns.[35-37,90] For example, the behavioral patterns of the cotton mouse, *Peromyscus gossypinus*, and the Florida mouse, *Peromyscus floridanus,* are differentiated in the laboratory with respect to digging, nest-building, and climbing. Whereas cotton mice are widely distributed geographically, Florida mice are not. Their differences of behavior are consistent with the ability of cotton mice to move into and exploit new habitats. Thus there is some indication of the role that behavioral differences which are measureable in the laboratory can play in adaptation in the natural habitat.

5. One of the characteristics of a species most critical to our understanding of its behavior is the kind of social organization it adopts. Various authors have attempted to study the adaptive significance of different modes of social organization.[38-42] A wide array of factors, including the nature and degree of predation, the type and method of collecting food, the need for camouflage, and so on, all appear to be correlated with social organization. For example, Eisenberg and colleagues[42] say that diurnal, leaf-eating primates tend to have a social structure with a one-male system, a small home range, and with chorusing behavior to effect spacing.

6. The work of Adler[43] provides an example of the experimental study of the adaptive function of behavior as conducted in a psychology laboratory. Adler found that the probability of a female rat's becoming pregnant was a function of the number of vaginal intromissions (without sperm transfer) that she received prior to the male's ejaculation. Thus, the male's complex copulatory pattern is apparently adapted to trigger critical neuroendocrine and other responses in the

34

female that function to increase the probability of pregnancy. This experiment provides important insight into the role of the rat's copulations without sperm transfer, and has opened up a number of exciting areas of research.

7. The program of research that my students and I are conducting is aimed primarily toward an understanding of the function of the various patterns of mammalian copulatory behavior. The first thing needed was a behavioral taxonomy that would permit us to describe and classify these patterns. The schema I proposed[44] features four questions, each of which can be answered with a yes or no: (1) Is there a "lock," or mechanical tie, between penis and vagina? (2) Is there intravaginal thrusting? (3) Are multiple intromissions prerequisite for ejaculation? and (4) Do multiple ejaculations occur in a single episode? This system produces a total of 2^4, or 16 possible patterns. In reviewing the literature on mammalian copulation, examples of many of these patterns were found.

Our laboratory approach to copulatory behavior stresses the study of species within a particular restricted taxon, the muroid rodents, particularly the family Cricetidae. We have had 24 species in our laboratory, and have observed copulation in 17 of them thus far. The correlates of copulatory patterns that have emerged thus far have been morphological. For example, simple baculum muroid rodents[45] that lock tend to have a relatively thicker glans penis than those that do not lock. Locking species also tend to have a reduced complement of certain accessory reproductive glands, as described by Arata.[46] These correlations are preliminary and are based on study of a small number of species. However, they permit us to predict the pattern that should be displayed by new species which we are about to study. Thus these are correlations that can be confirmed or refuted by further research. We plan to work toward making similar correlations between copulatory patterns and ecological variables. We hope that this approach may represent a prototype for some empirical testing of proposed correlations between behavioral patterns and other characteristics of species.

The research reported in each of these examples is directed toward teleonomic considerations and can serve to broaden our understanding of the relevant behavioral patterns: sleep, exploratory behavior, digging, nest-building, climbing, social behavior, and copulatory behavior. Functional considerations can be of considerable help to our understanding of behavior, in the laboratory as well as in the field.

Information about the adaptive function of behavior can be important, even when the primary question of interest deals with immediate causation or development. This has received increased recognition in psychological journals in recent years. I shall consider some examples of this approach.

Our understanding of the physiological mechanisms that are correlated with overt behavior may be facilitated by consideration of ecological factors. For example, in attempting to understand the termination of reinforcing intracranial stimulation, Schiff and colleagues[47] introduced an approach "which is characterized by an appreciation of the ecological determinants of animal behavior, and explanation in terms of mechanisms of adaptation rather than specific theories of motivation."

Warren[48] advocates increased application of ethological methods and increased comparative study in attempts to understand the prefrontal cortex. He writes, "Only by studying many species in a variety of situations can we hope to differentiate specific and general lesion effects and to establish unequivocal similarities across species." Regarding the function of the prefrontal cortex in primates, Warren speculates that "an important selection pressure leading to the elaboration of the prefrontal cortex in rhesus monkeys was the advantage resulting from more complex and highly coordinated social behavior."

In studying the responses of single units to visual stimuli in frogs, Lettvin and colleagues[49] related the stimuli that were effective in firing one group of units to natural stimuli in the life of the species. "Bug perceivers" were found to respond maximally to small convex objects that were moving toward the center of the visual field.

Perhaps most persuasive for the majority of animal psychologists is the surge of papers that relate learning to adaptive function. Bolles[50] has shown how diverse data on avoidance learning can be integrated when they are considered in relation to species-typical defense reactions. Seligman[51] has questioned general process learning theory by stressing the phylogenetically adapted preparedness to form certain associations that appears in many species. Garcia and colleagues[52] have demonstrated the occurrence of learning with exceptionally long conditioned stimulus–unconditioned stimulus intervals, but only in situations of particular significance to survival in the rat's habitat. Glickman and Schiff[53] proposed a theory of reinforcement based on the importance of appropriately elicited species-typical responses in adaptation. Rozin and Kalat[54] have suggested the possibility that specific learning abilities are phylogenetically adapted to survival in different habitats for different species. Warren,[19] in reviewing literature on comparative studies of learning in vertebrates, calls for greater concern with adaptive function, and writes: "Each surviving species has become adapted to survive as a species in a particular niche by whatever means that responded most adequately to selection pressures. Specific learning capacities must have been subjected to the same kind and degree of selection as any other trait of the organism." This surge of contemporary thought in which learning is related to specific adaptations is having an appreciable impact on much of animal psychology.

Theory in Comparative Psychology

One of the criticisms of comparative psychology has been that it is without theory. The subtitle of a 1969 paper by Hodos and Campbell[10] is *Why There is No Theory in Comparative Psychology*. Beach[55] points out that comparative psychology "has never generated any theories or even broad points of view, nor has it taken serious account of evolutionary theory." (p. 801.) Eaton[56] similarly notes the lack of attention to evolutionary theory in comparative psychology. The theories cited by Caldwell,[57] in a chapter on "Theoretical Foundations of Comparative Psychology," were drawn primarily from areas of psychology outside the comparative area.

Some interesting theoretical analyses are available with regard to questions of development and mechanisms, particularly physiological mechanisms, within the animal behavior/comparative psychology area. It is to be hoped that these will continue to develop on a firm empirical base. Many important advances can be made in the absence of broad, integrating theories. It seems to me, however, that a broad synthesis of the diverse questions that concern comparative psychologists can only come within the context of evolutionary theory. It is the modern synthetic theory of evolution that unifies much of biology. Evolutionary theory may have to be elaborated and developed along specialized lines in comparative psychology. I do not, however, foresee the development of integrative nonevolutionary theories in comparative psychology. Evolutionary theory appears to hold the greatest promise for broad synthesis in this field. As Warren[58] has replied to the query of Hodos and Campbell, comparative psychology has a theory: it is the same modern, synthetic theory of evolution that unifies all of the other areas of the biological sciences.

If we are ever to have a true "biopsychology," we shall have to include those areas at the interface between psychology and whole animal biology (for example, ethology, ecology, evolution, and mammalogy), in addition to the interfaces between psychology and the physiological and anatomical areas of biology.

Much has been written about evolution in psychology. It is time for comparative psychologists to utilize evolutionary considerations as a basis for their laboratory behavior as well as for their writing behavior.

How Should We Select Species for Study?

The major questions on the selection of species for study can be broken down into two areas: (1) Should we be comparative, and study more than one species or population? and (2), Which animals should we use? The answer to the second question is partially a function of the answer to the first.

Should We Be Comparative?

Ultimately, and in the very long run, I believe that all animal research must be comparative. Our findings will have to be examined across different species. While few would deny this, it is of little direct consequence to the individual researcher. Within the time span of an individual and his research program, the need for comparative study varies as a function of the problem area and the method used. I have attempted to portray various approaches to the study of behavior in TABLE 1. In TABLE 1, I am defining noncomparative research as any research in which the genotype is not varied. Thus, studies that remain within a species but utilize different strains, breeds, subspecies, and so on, are considered comparative.

TABLE 1

APPROPRIATE PROBLEM AREAS, METHODS, AND LEVELS
IN COMPARATIVE PSYCHOLOGY

Problem Area	Noncomparative Study	Comparative Study		
		Genetic Level	Species Level	Phyletic Level
Development	yes	yes	yes	yes
Causal Mechanisms	yes	yes	yes	yes
Evolutionary History	yes?	yes?	yes	yes
Teleonomy— Method: behavior genetic	no	yes	yes	no?
adaptive correlation	no	yes	yes	yes
experimental	yes	yes	yes	yes

Evolutionary history is usually studied through comparisons of different species. As some inferences about phylogeny can be made from ontogeny, however, and as some estimates of the past operation of selection pressures can be made from behavior-genetic data, the study of evolutionary history does not appear to demand comparisons between species in all cases. The need for comparative study with regard to the problem of adaptation varies with the method used. Many important experiments on teleonomy can be conducted with a single genotype. However, the method of adaptive correlation and the behavior-genetic method require comparative data as defined here. Problems of development and causal mechanisms can be studied with either a comparative or a noncomparative approach. Thus, while comparative study may be very important in each of the four problem areas, many valid and important studies can be conducted in a noncomparative framework.

The advantages of concentration on one species include standardization of methods, replicability, and the utilization of a body of accumulated information. These advantages were fully understood and discussed by Beach[59] and elaborated upon by Skinner.[60] I am impressed by the great generality of the principles of genetics that were developed from studies of peas and fruit flies, and of the principles of operant psychology that were developed from studies of a small number of species; the latter are applicable to the study of language in chimpanzees,[61] to creativity in porpoises,[62] and to a wide variety of human applied situations. I am impressed by the generality of the research on the effects of early hormone manipulations, which was done primarily on rats. While the comparative method appears to be essential to the study of evolutionary history, and may be very useful in other problem areas, we need not all work within an immediate framework.

Which Species Should We Study If We Select Just One?

One variety of comparative psychology produces data that may be of comparative interest, although its primary purpose is to study processes rather than to make any explicit comparisons.[48] A single, particular species is selected, solely because it has some particular advantage as a research animal. There are many advantages in using animals with relatively simple nervous systems and animals near that part of evolutionary history that led to man.[63] Other species have particularly accessible cochleas or electric organs, which make them particularly good subjects for research. "With such a comparative approach, one would choose the squid in preference to the rat for the study of electrical properties of the axon; the love bird in preference to the rat for the study of the evolution of behavior; but the rat in preference to many other organisms for the study of certain olfactory functions." (Dewsbury,[64] p. 54.)

Availability, ease of maintenance, and the existence of a substantial literature on behavior, disease, and methodology are factors that can make one species preferable to another. The latter factors have been important in the widespread popularity of domesticated forms of *Rattus norvegicus*, which I shall call laboratory rats.

Lockard[63] has argued that laboratory rats are an "indefensible choice" for behavioral research. He refers to albino rats variously as freaks, unnatural animals, and degenerates. Lockard documents the changes that have occurred in the domestication of Norway rats. He points out that the selection pressures of the laboratory differ from those of the natural habitat. Thus, laboratory rats and feral Norway rats differ along a number of dimensions. Lockard goes on to infer that laboratory rats are not only different, they are inferior.

I would agree that if one wishes to generalize to the natural habitat or to perform certain experiments on adaptive function, wild *Rattus norvegicus* are the animals of choice. I fail, however, to see any support for the generalization that because laboratory rats differ from wild Norway rats, they are inferior and are therefore indefensible choices for research. They are different. To anyone who works with wild rodents, they seem sluggish and not very alert. But so do turtles, and they are a perfectly viable species. Laboratory rats have evolved in a manner appropriate to their present habitat, the laboratory. They are different, not degenerate.

Lockard[63] divides animal research into that concerned with "natural functions" and that concerned with learning and motivation, and he considers each separately. I study one of the natural functions: sexual behavior. I have observed copulatory behavior in 17 species of muroid rodents. Many of these animals were wild-trapped. I see no way in which the basic copulatory pattern of rats is "degenerate" in relation to that of these other species. Indeed, the basic pattern is virtually identical to that found in some nondomesticated species.[65] Not all species respond to experimental manipulations in exactly the same way. However, I know of no data to suggest that laboratory rats perform in an atypical manner, outside the range of normal species differences. I have been studying 17 species of muroid rodents because I wish to use the method of adaptive correlation to study adaptive significance. In my laboratory we also are studying development[66] and causal mechanisms.[67] We have used the diallel cross method to study adaptive significance. For all of this research we have used laboratory rats and shall continue to use laboratory rats. I see no necessity at present to change to using inconvenient species when dealing with these questions.

The claim that laboratory rats are unsuitable for learning studies appears equally unsupported. Boice[68] considers "the unfounded, yet popular assumption that the laboratory rat is degenerate and dull compared to his wily progenitor in the wild." (p. 177.) Price,[69] who has studied learning in both varieties, concludes that "the results did not support the hypothesis that the learning ability of the domestic rat has declined during domestication." (p. 54.)

That selective breeding alters laboratory strains has been documented by Lockard[63] and Price.[70] That captivity itself can affect behavior, even in wild-born rodents, has been documented by Kavanau[71] and Wagner.[72] For these and other reasons, the researcher who confines himself to laboratory rats, or to any of a variety of other laboratory species, must be very careful about the kinds of generalization that he makes.

Lockard's concern regarding substantial strain, supplier, and laboratory variation in rat populations is well justified. While his suggestion that the American Psychological Association establish a standard strain is innovative and would be an excellent solution to the problem, I am not optimistic. The problems of selecting a strain, in a climate in which each investigator has his pet strain, would be monumental. The problems of standardization and distribution would be many.

Needless controversy can arise when different researchers fail to recognize that they are asking different questions, and that therefore the appropriate species of choice will differ. This is illustrated in the controversy between Hawkins[73] and Kavanau.[74]

I believe that there is much to be gained by comparative research in all four problem areas of animal behavior. Much very important research remains to be done on laboratory rats, however. It should not be disparaged. If we had it all to do over again, we might not select rats. But we need not throw away years of important research. If rats are neither a bad habit nor a defensible choice, they are a defensible habit.

How Should We Select Species for Comparative Research?

The problem of deciding upon the species to be used for comparative analysis involves first selecting a level of analysis, and then selecting a group of animals at the appropriate level.

At What Level Should We Work? The problems of selecting a level* for comparative analysis have been treated systematically by King.[77],[78] King distinguishes three levels of comparative analysis: the genetic level, the species level, and the phyletic level. At the genetic level, subjects are drawn from a single species but are of varying genotypes. At the species level, small groups of closely related species are compared. At the phyletic level, comparisons are made among genera, families, orders, classes, and phyla. Almost any of the problems can be studied at any level (see TABLE 1). Different methods, however, produce different results and hence different generalizations. The broader the range of comparisons, the more generality there is, but the more difficult it is to isolate variables responsible for species differences.

Development and causal mechanisms can be studied at any level. Evolutionary history is best studied at the species level[12] or the phyletic level.[16],[17]

Teleonomy can be studied at any level, but the level of choice varies with the method used. The behavior-genetic method is appropriate at the genetic level as well as the species level. In the latter, the behavior of hybrids of closely related species is an important source of behavioral data. The method of adaptive correlation can be used at the phyletic level,[16] the species level,[35] or when working with different subspecies, for example, at the genetic level.[78],[79] The experimental method appears to be appropriate at any level of analysis.

The ethologists and many American zoologists generally prefer to work at the species level (for example, Hinde and Tinbergen,[11] and King[78].) It provides a powerful method of uncovering the principles by which adaptation and evolutionary history may have worked.

Wilcock[23] extols the virtues of the genetic level of analysis, and suggests that "comparative psychology lives on under an assumed name — psychogenetics."

Comparative psychologists have traditionally emphasized the phyletic level.

I return to my main point. There are many useful methods in comparative psychology. The quest for uniformity is fruitless. What is important is that the method be appropriate to the question being asked. The conceptions I have discussed and summarized in TABLE 1 are almost certainly incomplete and will require modifications. However, they may help to underscore the diversity of the questions and of the approaches available for use. If this is the case, they are a useful step.

How Should We Select a Group for Study? Selection of a group of animals

---

* Note that the word "level" is used by King to refer to the level of analysis. It should not be confused with the concept of "levels of organization," as used by Schneirla and his students (for example, see Tobach[75]). It also differs from the "levels of comparative research" of Ratner and Denny.[76]

for study is more complex than, but fundamentally the same as, the selection of a single species. One must select a group of species that will serve as an adequate research preparation and on which good research can be conducted. The group of species can be selected because work with them has promise of leading to broad psychological and biological principles.

As with the selection of level, the selection of species varies with the question under consideration. Hodos and Campbell[10] and Hodos[80] have pointed out that when using the phyletic level of comparison to study problems of evolutionary history, one must be careful to select as historically meaningful a group of contemporary species as can be found. Historical continuity is also critical when one is studying evolutionary history at the species level.

Historical continuity is not a critical factor in the selection of species for work on questions of mechanism, development, or teleonomy. When studying teleonomy by the method of adaptive correlation, one would select species because they are either similar or divergent on some criterion variable with which one hopes to correlate behavior. This may be an ecological variable, such as habitat; a morphological variable, such as the location of the eyes; and so on. One may wish to work within a restricted taxon. Any of the above considerations might provide a rationale for the selection of groups of species for comparative study of questions of mechanism or development.

Selection of Species in Relation to Man

Implicit in my statement that the value of a species for research is a function of the broad principles such research might generate, is a belief that such value is not necessarily a function of the proximity of that species to man. Often laboratory animals have been treated as surrogates for humans, and attempts have been made to study problems of human interest in other species. In some areas this has produced very valuable information. It has its place. There is another strategy, however; this has been elaborated by Beach:[81]

> ... What I am proposing differs from current practice in animal experimentation in that animals are not to be used as substitutes for people, and the kinds of problems investigated are not to be exclusively or even principally derived from human behavior or psychology. (p. 2.)
> ... If we remove man from the central point in a comparative science of behavior, this may, in the long run, prove to be the very best way of reaching a better understanding of his place in nature and of the behavioral characteristics which he shares with other animals as well as those which he possesses alone or which are in him developed to a unique degree. (p. 17.)

Lehrman[82] makes a similar point with his "natural history orientation" and cautions:

> The value of comparison comes not from the merging of different levels into a misleading unified conception of behavior but from the development of an evolutionary perspective which enables us to appreciate the emergence of new qualities without neglecting the underlying continuities and their transformations. (p. 468.)

Ratner and Denny,[77] Hodos and Campbell,[10] and Lockard[2] all discuss the dangers of "capricious comparisons," particularly those between rats and men. Lockard discusses the issue of direct generalizability on several occasions. He suggests that if we wish to generalize from nonhuman subjects to humans, we should use appropriate species. He states that "apes are relevant by relatedness, wolves for ecological reasons." (p. 177). I would agree that if one must make direct generalizations about complex behavior from nonhumans to humans, it is better to do so from apes and wolves than from rats. I agree with Beach, however, that in the long run we shall be better off if we strive for broad principles rather than for direct generalizations to man. It would be a mistake to follow the anthropologists, who are interested in animal behavior yet confine their studies virtually exclusively to primate behavior. I believe that the real value of comparative psychology will come from establishing an evolutionary perspective within which we can view the behavior of many species, including humans.

What's In a Name?

The literature of self-criticism in comparative psychology even reaches so far as to question the name of comparative psychology itself. Hodos and Campbell[10] believe that "the term comparative psychology should be reserved for experiments in which interspecies comparisons are made." They suggest that "any experiments carried out with nonhuman subjects, in which no attempt is made to compare species of animals or humans, be regarded as simply animal psychology." (p. 349.) Lockard[2] again is more outspoken: "The modern phrase is animal behavior, and it has the advantage of disconnecting current work from the past of comparative psychology." (p. 176.) Adler and Tobach[27] respond by pointing out that "introduction of such labels as animal behavior, neurobiology, or biopsychology does not change the substance of a field of investigation. Scientific disciplines may change their names, but their focal problems will remain the same." (p. 857.)

Perhaps we ought to masquerade under two names in the interest of confusion. Thus Rose[83] classified comparative psychology and animal behavior as two of the twelve areas of psychology. What a great boon to our egos to realize that we represent one-sixth of all psychology!

I agree with Adler and Tobach that changing a name will not necessarily change an area. Debate over a name is not worthy of polemics.

Nevertheless, I would like to enter a brief statement in favor of the term "comparative psychology." As I was growing up around New York City, I spent many enjoyable hours in Madison Square Garden. The name puzzled me, as Madison Square Garden was on Eighth Avenue (not on Madison), was not square, and bore little resemblance to a garden. I then learned that several indoor stadia had been built in succession, and that with each move the name had been carried to the new facility. The tradition of playing in "the Garden" was more important than the precise fit of the name to the arena. The same decision was made in the latest move.

Unlike Lockard, I am not ashamed of comparative psychology's past. I am rather proud of the tradition established by E. L. Thorndike, Robert Yerkes, C. Lloyd Morgan, Leonard Carmichael, Karl Lashley, Calvin Stone, C. R. Carpenter, T. C. Schneirla, Frank Beach, and others. There are some things that I want to do differently from the methods chosen in the past. However, I see no more need to change the name of comparative psychology than a geneticist sees to change the name of his discipline, although the study of DNA differs rather substantially from the methodology used by Mendel.

Comparative psychology may be difficult to define in precise terms. This represents a problem for the mental health of the comparative psychologist, but may not be indicative of any problem in comparative psychology itself.[1] In reality, animal behavior is not that much easier to define. Although precise definitions are difficult, examination of the content material covered in courses in comparative psychology reveals considerable agreement as to what content is included.[84] Most psychologists appear to have similar views of the subject matter of the comparative psychology/animal behavior area, even though they are unable to verbalize precise definitions.

If my choice of terms were to remain, we must recognize that comparative psychology would include many studies that are not explicitly comparative in nature. I believe this is as manageable a situation now as it has been for the rest of our century. Like the owners of Madison Square Garden, I am reluctant to discard a name that is associated with a great tradition, even though we may be in a period of great change. Tradition is one thing that is in a name.

A Diagnosis of Comparative Psychology

It is difficult to diagnose a discipline. One popular method of diagnosing the health of comparative psychology has been to count the number of species used in the papers appearing in the *Journal of Comparative and Physiological Psychology*.[5,18,59,85] Invariably it is found that relatively few species are used.

It is easy to jump from the finding that few species are used in the papers appearing in the *Journal of Comparative and Physiological Psychology* to the conclusion that comparative psychology is in a poor state of health. This conclusion contains at least two questionable assumptions. The first is that absolute numbers are the critical factor. I believe that impact and scientific value outweigh sheer numbers of papers. Comparative study always will be a small part of psychology, and appropriately so. However, it plays a very important part in maintaining evolutionary perspective in psychology. This function cannot be assessed through a simple frequency count.

The second assumption is that the *Journal of Comparative Physiological Psychology* is the logical place to look for papers in comparative psychology. This journal has been given some rough treatment by students of animal behavior. Driver and Corning[86] refer to it as a place "where the main concern is with what *experimenters* do and not with what *animals* do." (p. 47.) Boice[3] writes, "The *Journal of Comparative and Physiological Psychology* is clearly oriented to physiological papers, and its editorial board, where it is not obviously physiological, is not very comparative either." (p. 858.) Lorenz[87] wrote:

> I must confess that I strongly resent it not only from a terminological viewpoint, but also in the interests of the very hard-working and honest craft of really comparative investigators when an American journal masquerades under the title of 'comparative psychology', although to the best of my knowledge, no really comparative paper has been published in it. (p. 239.)

Some of this criticism appears to be exaggerated for the sake of dramatizing a point. The *Journal of Comparative and Physiological Psychology* is an excellent journal, with methodological standards second to none in the animal behavior field. Traditionally, however, it has been oriented toward studies of development and causation and has never appeared overly sympathetic toward comparative studies or evolutionary questions. This may be a very reasonable and defensible editorial policy. All journals must limit their coverage. However, it would seem

43

to make the *Journal of Comparative and Physiological Psychology* an illogical place to search for papers of the sort in question, at least in 1973. There are now over a dozen journals in which comparative research by psychologists might be published. The Animal Behavior Society meetings also contain many such papers.[88] A practical count of the number of species employed in comparative psychology would be very difficult, precisely because the literature is, of necessity, so spread out through a myriad of journals.

I have no objective manner in which to assess the vitality of comparative psychology. Like all basic sciences, it is under some attack from without. It is also facing attack from within. Nevertheless, I cannot but be optimistic about the state of the art. There are many very exciting developments in comparative psychology.[89] I view the flux and debate, not as a sign of decline, but as symptomatic of a convergence of new and exciting ideas which are meeting head on in the battleground of comparative psychology. Influences are coming from many directions and viewpoints, and techniques that were not anticipated just a few years ago are coming to the fore. I see no reason to alter my diagnosis of four years ago[1] in regarding "comparative psychology as a healthy and rapidly-expanding discipline . . . in the midst of . . . vigorous growth." (p. 37.) Indeed, it is an interesting and exciting time to be a comparative psychologist.

REFERENCES

1. DEWSBURY, D. A. 1968. Comparative psychology and comparative psychologists: an assessment. J. Biol. Psychol. 10(1): 35–38.
2. LOCKARD, R. B. 1971. Reflections on the fall of comparative psychology: is there a message for us all? Amer. Psychologist 26: 168–179.
3. BOICE, R. 1971. On the fall of comparative psychology. Amer. Psychologist 26: 858–859.
4. EWER, R. F. 1971. Review. Animal Behaviour 19: 802–807.
5. SCHREIER, A. M. 1969. *Rattus* revisited. Amer. Psychologist 24: 681–682.
6. TINBERGEN, N. 1963. On aims and methods of ethology. Z. Tierpsychologie 20: 410–429.
7. LEHRMAN, D. S. 1970. Semantic and conceptual issues in the nature-nurture problem. *In* Development and Evolution of Behavior. L. R. Aronson, E. Tobach, D. S. Lehrman & J. S. Rosenblatt, Eds.: 17–52. W. H. Freeman & Co., Publishers. San Francisco, Cal.
8. MOLTZ, H. 1971. The Ontogeny of Vertebrate Behavior. Academic Press Inc. New York, N. Y.
9. TOBACH, E., L. R. ARONSON & E. SHAW. 1971. The Biopsychology of Development. Academic Press Inc. New York, N. Y.
10. HODOS, W. & C. B. G. CAMPBELL. 1969. *Scala naturae*: why there is no theory in comparative psychology. Psychological Rev. 76: 337–350.
11. HINDE, R. A. & N. TINBERGEN. 1958. The comparative study of species-specific behavior. *In* Behavior and Evolution. A. Roe & G. G. Simpson, Eds.: 251–268. Yale University Press. New Haven, Conn.
12. DILGER, W. C. 1960. The comparative ethology of the African parrot genus *Agapornis*. Z. Tierpsychologie 17: 649–685.
13. DILGER, W. C. 1962. The behavior of lovebirds. Sci. Amer. 206: 88–98.
14. CAMPBELL, C. B. G. & W. HODOS. 1970. The concept of homology and the evolution of the nervous system. Brain, Behavior, and Evolution 3: 353–367.

15. ATZ, J. W. 1970. The application of the idea of homology to behavior. *In* Development and Evolution of Behavior. L. R. Aronson, E. Tobach, D. S. Lehrman & J. S. Rosenblatt, Eds.: 53–74. W. H. Freeman & Co., Publishers. San Francisco, Cal.

16. ALLISON, T. & H. B. VANTWYVER. 1970. The evolution of sleep. Natural Hist. **79**(2): 56–65.

17. HEFFNER, H. E., R. J. RAVIZZA & B. MASTERTON. 1969. Hearing in primitive mammals, IV: bushbaby (*Galago senegalensis*). J. Auditory Res. **9**: 19–23.

18. BITTERMAN, M. E. 1965. Phyletic differences in learning. Amer. Psychologist **20**: 396–410.

19. WARREN, J. M. 1973. Learning in vertebrates. *In* Comparative Psychology: A Modern Survey. D. A. Dewsbury & D. A. Rethlingshafer, Eds. McGraw-Hill Book Company. New York, N. Y.

20. PITTENDRIGH, C. S. 1958. Adaptation, natural selection, and behavior. *In* Behavior and Evolution. A Roe & G. G. Simpson, Eds.: 390–416. Yale University Press. New Haven, Conn.

21. WILLIAMS, G. C. 1966. Adaptation and Natural Selection. Princeton University Press. Princeton, N. J.

22. BEER, C. G. 1973. Species-typical behavior and ethology. *In* Comparative Psychology: A Modern Survey. D. A. Dewsbury & D. A. Rethlingshafer, Eds. McGraw-Hill Book Company. New York, N. Y.

23. WILCOCK, J. 1972. Comparative psychology lives on under an assumed name — psychogenetics! Amer. Psychologist **27**: 531–538.

24. TINBERGEN, N. 1963. The shell menace. Natural Hist. **72**: 28–35.

25. DeRUITER, L. 1956. Countershading in caterpillars. Arch. Neerl. Zool. **11**: 285–341.

26. BLEST, A. D. 1957. The function of eyespot patterns in the Lepidoptera. Behaviour **11**: 209 256.

27. ADLER, H. E. & E. TOBACH. 1971. Comparative psychology is not dead. Amer. Psychologist **26**: 857–858.

28. CULLEN, E. 1957. Adaptations in the kittiwake to cliff-nesting. Ibis **99**: 275–302.

29. BRUELL, J. H. 1964. Inheritance of behavioral and physiological characters of mice and the problem of heterosis. Amer. Zoologist **4**: 125–138.

30. BRUELL, J. H. 1967. Behavioral heterosis. *In* Behavior-Genetic Analysis. J. Hirsch, Ed.: 270–286. McGraw-Hill Book Company. New York, N. Y.

31. HENDERSON, N. D. 1970. Brain weight increases resulting from environmental enrichment: a directional dominance in mice. Science **169**: 776–778.

32. TINBERGEN, N. 1965. Some recent studies of the evolution of sexual behavior. *In* Sex and Behavior. F. A. Beach, Ed.: 1–33. John Wiley & Sons, Inc. New York, N. Y.

33. GLICKMAN, S. E. & R. W. SROGES. 1966. Curiosity in zoo animals. Behaviour **26**: 151–188.

34. COLLINS, R. L. 1964. Inheritance of avoidance conditioning in mice: a diallel study. Science **143**: 1188–1190.

35. LAYNE, J. N. 1969. Nest-building in three species of deer mice (*Peromyscus*). Behaviour **35**: 288–303.

36. LAYNE, J. N. 1970. Climbing behavior of *Peromyscus floridanus* and *Peromyscus gossypinus*. J. Mammalogy **51**: 580–591.

37. LAYNE, J. N. & L. M. EHRHART. 1970. Digging behavior of four species of deer mice (*Peromyscus*). Amer. Museum Novitates 2429 (16 pp).

38. ETKIN, W. 1964. Types of social organization in birds and mammals. *In* Social Behavior and Organization Among Vertebrates. W. Etkin, Ed.: 256–297. University of Chicago Press. Chicago, Ill.

39. CROOK, J. H. 1970. Social organization and the environment: aspects of contemporary social ethology. Animal Behaviour 18: 197—209.

40. ITO, Y. 1970. Group and family bonds in animals in relation to their habitats. *In* Development and Evolution of Behavior. L. R. Aronson, E. Tobach, D. S. Lehrman & J. S. Rosenblatt, Eds.: 389—415. W. H. Freeman & Co., Publishers. San Francisco, Cal.

41. FISLER, G. F. 1969. Mammalian organizational systems. Los Angeles County Museum Contrib. Sci. 167: 1—32.

42. EISENBERG, J. F., N. A. MUCKENHIRN & R. RUDRAN. 1972. The relation between ecology and social structure in primates. Science 176: 863—874.

43. ADLER, N. T. 1969. Effects of the male's copulatory behavior on successful pregnancy of the female rat. J. Comp. Physiol. Psychol. 69: 613—622.

44. DEWSBURY, D. A. 1972. Patterns of copulatory behavior in male mammals. Quart. Rev. Biol. 47: 1—33.

45. HOOPER, E. T. & G. G. MUSSER. 1964. The glans penis of neotropical cricetines (family Muridae) with comments on classification of muroid rodents. Misc. Publ. Museum Zool. Univ. Mich. 123 (57 pp).

46. ARATA, A. A. 1964. The anatomy and taxonomic significance of the male accessory reproductive glands of muroid rodents. Bull. Florida State Museum 9: 1—42.

47. SCHIFF, B. B., B. RUSAK & R. BLOCK. 1971. The termination of reinforcing intracranial stimulation: an ecological approach. Physiol. Behavior 7: 215—220.

48. WARREN, J. M. 1973. Evolution, behavior and the prefrontal cortex. Acta Neurobiologiae Experimentalis. In press.

49. LETTVIN, J. Y., H. R. MATURANA, W. S. McCULLOCH & W. H. PITTS. 1959. What the frog's eye tells the frog's brain. Proc. Inst. Radio Engrs. 47: 1940.

50. BOLLES, R. C. 1970. Species-specific defense reactions and avoidance learning. Psychological Rev. 77: 32—48.

51. SELIGMAN, M. E. P. 1970. On the generality of the laws of learning. Psychological Rev. 77: 406—418.

52. GARCIA, J., F. R. ERVIN & R. A. KOELLING. 1966. Learning with prolonged delay of reinforcement. Psychonomic Sci. 5: 121—122.

53. GLICKMAN, S. E. & B. B. SCHIFF. 1967. A biological theory of reinforcement. Psychological Rev. 74: 81—109.

54. ROZIN, P. & J. W. KALAT. 1971. Specific hungers and poison avoidance as adaptive specializations of learning. Psychological Rev. 78: 459—486.

55. BEACH, F. A. 1971. Review. Animal Behaviour 19: 799—801.

56. EATON, R. L. 1970. An historical look at ethology: a shot in the arm for comparative psychology. J. Hist. Behavioral Sciences 6: 176—187.

57. CALDWELL, W. E. 1960. Theoretical foundations of comparative psychology. *In* Principles of Comparative Psychology. R. H. Waters, D. A. Rethlingshafer & W. E. Caldwell, Eds.: 378—404. McGraw-Hill Book Company. New York, N. Y.

58. WARREN, J. M. Unpublished observations.

59. BEACH, F. A. 1950. The snark was a boojum. Amer. Psychologist 5: 115—124.

60. SKINNER, B. F. 1966. The phylogeny and ontogeny of behavior. Science 153: 1205—1213.

61. PREMACK, D. 1971. Language in chimpanzee? Science 172: 808—822.

62. PRYOR, K. 1969. The porpoise caper. Psychology Today 3(7): 47—49; 64.

63. LOCKARD, R. B. 1968. The albino rat: a defensible choice or a bad habit. Amer. Psychologist 23: 734—742.

64. DEWSBURY, D. A. 1965. Electric fishes and their potential for psychological research. Worm Runner's Digest 7: 54–66.
65. DEWSBURY, D. A. 1971. Copulatory behavior of old-field mice (*Peromyscus polionotus subgriseus*). Animal Behaviour 19: 192–209.
66. DEWSBURY, D. A. 1969. Copulatory behavior of rats as a function of prior copulatory experience. Animal Behaviour 17: 207–213.
67. DEWSBURY, D. A., H. N. DAVIS, JR. & P. E. JANSEN. 1972. Effects of monoamine oxidase inhibitors on the copulatory behavior of male rats. Psychopharmacologia 24: 209–217.
68. BOICE, R. 1971. Laboratizing the wild rat (*Rattus norvegicus*). Behavior Res. Methods and Instrumentation 3: 177–182.
69. PRICE, E. O. 1972. Domestication and early experience effects on escape conditioning in the Norway rat. J. Comp. Physiol. Psychol. 79: 51–55.
70. PRICE, E. O. 1972. Novelty-induced self-food deprivation in wild and semi-domestic deermice (*Peromyscus maniculatus bairdii*). Behaviour 41: 91–104.
71. KAVANAU, J. L. 1967. Behavior of captive white-footed mice. Science 155: 1623–1639.
72. WAGNER, M. 1971. Laboratory living produces a different animal. Psychologische Beitr. 13: 79–88.
73. HAWKINS, J. D. 1964. Wild and domestic animals as subjects in behavior experiments. Science 145: 1460–1461.
74. KAVANAU, J. L. 1964. Wild and domestic animals as subjects in behavior experiments. Science 145: 1461–1462.
75. TOBACH, E. 1970. Some guidelines to the study of the evolution and development of emotion. *In* Development and Evolution of Behavior. L. R. Aronson, E. Tobach, D. S. Lehrman & J. S. Rosenblatt, Eds.: 238–253. W. H. Freeman & Co., Publishers. San Francisco, Cal.
76. RATNER, S. C. & M. R. DENNY. 1964. Comparative Psychology, Research in Animal Behavior. The Dorsey Press. Homewood, Ill.
77. KING, J. A. 1963. Maternal behavior in *Peromyscus*. *In* Maternal Behavior in Mammals. H. L. Rheingold, Ed.: 58–93. John Wiley & Sons, Inc. New York, N. Y.
78. KING, J. A. 1970. Ecological psychology: an approach to motivation. Nebraska Symp. Motivation 18: 1–33.
79. EHRHART, L. M. 1972. Comparative studies of temperament and activity in three subspecies of *Peromyscus polionotus*. Paper presented at meetings of the American Society of Mammalogists. Tampa, Fla.
80. HODOS, W. 1970. Evolutionary interpretation of neural and behavioral studies of living vertebrates. *In* The Neurosciences Second Study Program. F. O. Schmitt, Ed.: 26–39. Rockefeller University Press. New York, N. Y.
81. BEACH, F. A. 1960. Experimental investigations of species-specific behavior. Amer. Psychologist 15: 1–18.
82. LEHRMAN, D. S. 1971. Behavioral science, engineering, and poetry. *In* The Biopsychology of Development. E. Tobach, L. R. Aronson & E. Shaw, Eds.: 459–471. Academic Press Inc. New York, N. Y.
83. ROSE, R. M. 1972. Supply and demand for psychology Ph.Ds in graduate departments of psychology: 1970 and 1971 compared. Amer. Psychologist 27: 415–421.
84. BERMANT, G. 1965. Modern courses in comparative psychology. Paper presented at the American Association for the Advancement of Science. Berkeley, Cal.
85. CASSEL, C. A. 1971. The snark: twenty years later. Paper presented at the American Psychological Association. Washington, D. C.
86. DRIVER, P. M. & W. C. CORNING. 1968. The nature of psychology. J. Biol. Psychol. 10(1): 47–73.

87. LORENZ, K. 1950. The comparative method in studying innate behavior patterns. Symp. Soc. Exp. Biol. 4: 221–268.

88. HARLESS, M. 1971. On the demise of comparative psychology. Amer. Psychologist 26: 1034.

89. DEWSBURY, D. A. & D. A. RETHLINGSHAFER. 1973. Comparative Psychology: A Modern Survey. McGraw-Hill Book Company. New York, N. Y.

90. KING, J. A. 1968. Psychology. *In* Biology of *Peromyscus* (Rodentia). J. A. King, Ed. American Society of Mammalogists. Lawrence, Kans.

E. H. HESS

4 Excerpt from: Ethology: An Approach toward the Complete Analysis of Behavior

DEVELOPMENT OF MODERN ETHOLOGY

In 1898 C. O. Whitman, an American zoologist at the University of Chicago, wrote the sentence that initiated the birth of modern ethology: "Instincts and organs are to be studied from the common viewpoint of phyletic descent" (p. 328). Thus it was he, through the study of orthogenic evolution as opposed to random mutation while cataloguing morphological characteristics of many species of pigeons, who recognized and discovered endogenous movements—movements whose origin came from within—as being a very distinct phenomenon of behavior and therefore able to be systematically studied and evaluated just like any morphological taxonomic characters. It was his contention that "as the genesis of organs takes its departure from the elementary structure of protoplasm, so does the genesis of instinct proceed from the fundamental properties of protoplasm. Primordial organs and instincts are alike few in number and generally persistent" (p. 329). He also criticized the notion that instinctive actions have clock-like regularity and inflexibility, calling it greatly exaggerated; actually, he contended, instinctive movements have only a low degree of variability under normal conditions, the most machinelike instincts always revealing some degree of adaptability to new conditions; thus he echoed Darwin's views.

At very nearly the same time, 1910, Oskar Heinroth, a German zoologist studying ducks and geese, independently came to very similar conclusions. While Whitman's observations on the taxonomic value of innate behavior had been more or less incidental, Heinroth was the first to demonstrate with empirical evidence that the concept of homology applied just as much to movement patterns as to morphological characters, so that these could be used to reconstruct phylogenetic relationships.

Heinroth did not set up any hypotheses about the newly discovered homologies of movement patterns. He described them, studied them, and pointed out the similarities and dissimilarities between various species, genera, and families.

Wallace Craig (1918), Whitman's student, took the first step from the purely descriptive to the analytic stage when he formulated a certain lawfulness in these behavior patterns. He observed that an animal does not only react to a stimulus, but searches, by appetitive behavior, for a certain stimulus situation which allows a consummatory act to run off. Craig defined appetitive behavior as follows: "appetite (or appetence) is a state of agitation which continues as long as a certain stimulus which may be called the appeted stimulus is absent" (p. 91). Craig went on to explain that when a consummatory reaction is released "the appetitive behavior ceases and is succeeded by a state of relative rest" (p. 91). Aversive behavior was similarly defined, but the period of rest comes when the aversive stimulus has been removed.

These consummatory acts were the same as the stereotyped species-specific movement coordinations described by Whitman and Heinroth. In most cases this consummatory act is at the end of a long chain of behavior patterns involved in the appetitive behavior. In some cases, however, the consummatory act occurs so quickly that the appetitive stage is not apparent, but by withholding from the animal the opportunity to perform the consummatory act, appetitive behavior is clearly seen. The fixity of these consummatory acts was shown by the fact that sometimes a species-specific consummatory act that is of selective value in nature is made by the animal in an inappropriate situation, as pointed out by Darwin in the case of dogs trying to cover their excrement in a bare concrete floor. Thus the confusion between desired ends (goals) and the species-preserving function of behavior was eliminated through this analysis, since Craig's standpoint was that the discharge of the consummatory act, not survival value, is the goal of appetitive behavior.

Craig pointed out that the appetitive and aversive factors in instinctive behaviors showed that mere chain reflexes were not the only constituents of such instinctive behaviors. An appetite, furthermore, is accompanied by a certain *readiness to act*, and many of the behavior patterns performed during the appetitive behavior are not innate or completely innate, but must be learned by trial. Thus a cat out looking for food may go to a certain place because he once caught a mouse there, or he may find that he must surmount some unexpected obstacles. But the end action—the consummatory act—of the series of behaviors in the appetitive chain is always innate, as shown by the occurrence of incipient consummatory acts during the appetitive behavior, when the adequate stimulus for the consummatory action has not yet been received.

Craig's scheme was only a beginning that Konrad Lorenz, Heinroth's student, took up and further developed, thus initiating the present period of ethology as a science. In 1935 he described the behavior of birds, drawing from his observations on tame or partly tame birds and paying special attention to the formation of social relationships. In 1937 and 1950 Lorenz elaborated on Craig's behavior analysis by proposing his scheme of *action-specific energy*. He found that the ease with which a given stimulus elicits or releases the corresponding behavior through the innate releasing mechanism (IRM) is dependent on how long it has been since the animal has last given that response. Thus action-specific energy, that is, energy for a particular action, is being produced continuously in an animal's central nervous system, but is held in check by some inhibitory mechanism until the appropriate stimulus releases this energy to certain muscular systems and the reaction takes place. The longer the animal has gone without performing the action in question, the more easily this behavior can be triggered off or released. In fact, in the case of a very prolonged absence of relevant stimulation, this behavior can go off without there being any observable stimulus present. This special case he called the *Leerlaufreaktion* or *vacuum activity*. Similarly, if the relevant stimulation is repeatedly given, the animal's response can decrease to the point where he gives none at all. Lorenz's conceptualization of the way in which a readiness to react is built up and then dissipated through reaction was an extension of Craig's scheme of appetitive behavior seeking the discharge of the consummatory act. One and the same physiological event are probably responsible for the occurrence of appetitive behavior as well as for the raising and lowering of the threshold for response to the stimuli.

Lorenz was careful to point out that this action-specific energy concept did not have anything to do with the exhaustion of or recuperation of muscular systems.

Niko Tinbergen, through his studies on the releasers for innate behavior patterns, further enlarged on the Craig-Lorenz scheme of behavior. Like Lorenz, he postulated (1950) the removal of a central inhibition when an innate behavior pattern is released. He demonstrated, furthermore, that behavior patterns in themselves can function as releasers, as Lorenz had suggested (1935), this finding being based on his studies on the behavior of the three-spined stickleback (1942). Tinbergen (1950) proposed that there was a hierarchical order of appetitive behaviors and consummatory acts. Both Tinbergen and G. P. Baerends (1956) have demonstrated that the lawfulness of behavior in intact higher organisms and the neurophysiological events in lower organisms are parallel.

Thus inspired by the work of Lorenz and Tinbergen, ethology began to expand into the neurophysiological bases of behavior. That these neurophysiological bases existed in fact and had a clear correlation with behavorial events was shown by the work of Erich von Holst (1933 *et seq.*), P. Weiss (1941a, b) W. R. Hess (1927, 1949, 1954, 1956), and K. D. Roeder (1955). Von Holst and Hess demonstrated the existence of neurological organization underlying various behavior patterns by stimulating a particular brain region of birds and cats with electrodes. The elicited patterns of behavior in response to electrical brain stimulation were the species-specific behavior patterns which ethologists had already observed in those particular species. Hess's elicitation in integrated threat, fighting, sleeping, and eating behaviors upon electrical stimulation of the cat's brain are examples of the behaviors observed. The *integration* of these behaviors was a particularly important finding; for example, cats not only went to sleep but also searched about for a place to sleep. Other areas of neurophysiological research have included von Holst's studies on the relationship between the rhythms of nervous impulses and of muscular movements, Weiss's proof of central nervous determination of primitive movement coordinations, and Roeder's demonstration that behavior is created spontaneously within the central nervous system. Recently, von Holst and von St. Paul have published a monumental paper on the interaction of different behavior patterns through the method of simultaneous electrical stimulation of adult chicken brains

(1960). Their findings represent a landmark in ethological theory and evidence, and will accordingly be discussed in detail later.

There are many other modern ethologists who have made significant contributions to the growth of their discipline. Indeed, it would be a staggering task to attempt to treat all of them adequately.

Ethological Concepts

FIXED ACTION PATTERNS

The fixed action pattern, or more simply, fixed pattern, is one of the most fundamental concepts of modern ethology. In fact, ethology was founded on the discovery of this phenomenon by Heinroth and Whitman, who were the first ones to describe it.

The fixed action pattern is defined as a sequence of coordinated motor actions that appears without the animal having to learn it by the usual learning processes. The animal can perform it without previous exercise and without having seen another species member do it. The fixed action pattern is *constant in form*, which means that the sequence of motor elements never varies.

These patterns are not equivalent to the well-known reflexes, nor are they chain reflexes; this statement will be discussed in connection with neurophysiological evidence for ethological tenets. There are many differences between the fixed action patterns and reflexes; however, an adequate discussion of these differences is beyond the scope of this paper. It may be mentioned, nevertheless, that the frequency with which the fixed action pattern is performed depends in part on how long it is since it was last performed; this is in accordance with Lorenz's postulation of specific action potential. This is not the case with reflexes. The animal engages in appetitive behavior that ends in the performance of the fixed action pattern, or consummatory act. No such appetitive behavior ending in the discharge of a reflex has been found. What is more, as we shall see later on, rhythmic movements such as those of locomotion are not necessarily equivalent to chain reflexes.

The appearance of such fixed action patterns in animals isolated from their own species is clear evidence of their genetic fixedness. Even though experimentally isolated animals may never have had any experience with the particular objects or situations involved, they will still perform the fixed pattern when the appropriate situation arises, as demonstrated by squirrels that were reared in isolation and were never given any objects to handle. Nevertheless, the squirrels attempted to bury nuts or nut-like objects in a bare floor upon their first encounter with them, making scratching movement as if to dig out earth, tamping the object in the floor with the nose, and, finally, making complete covering movements in the air (Eibl-Eibesfeldt, 1956a).

The fixed action pattern, appearing as it does in animals that have been isolated from their own species and deprived of experience relevant to this behavior, is therefore a constant characteristic of the species in question, being based, presumably, on specific central nervous system mechanisms that are inherited, just as are other morphological and physiological characters. In line with this, they are, furthermore, characteristics of genera and orders right up to the highest taxonomic categories. The fact that they are always to be found in more than one species proves their taxonomic value. They can be used, in fact, to differentiate between very closely related species.

The fixed action patterns are quite resistant to phylogenetic change through evolution, more so than morphological characters. For example, a morphological structure (such as a colored skin patch or an enlarged feather) connected with a fixed action pattern may often appear after the fixed action pattern has come into existence in the species, or the fixed action pattern may remain but the morphological structure may disappear. The first case is illustrated by the mandarin duck's pointing to an enlarged and colored wing feather during courtship. This pointing has derived, through the process of ritualization, from wing-preening, which presumably occurred before the wing feathers had become enlarged and specially colored. The last case is illustrated by the behavioral rudiments that have been observed in animals.

Krumbiegel (1940) compared the behavior of long-tailed and short-tailed monkeys and found that when a long-tailed monkey runs along a branch, its tail moves from the right to the left and back again, thus achieving balance. These same compensatory movements are to be observed in the short stumps of the short-tailed monkeys, even though there is obviously no value at all in these movements for balancing. Similarly, hornless domestic cattle and goats attempt to fight with their heads in the same way as their horned relatives.

Finally, mutant drosophila flies with no wings still perform the wing-preening movements typical of the species (Heinz, 1949).

Not only may the fixed pattern persist even though the relevant morphological structures have disappeared, but it also may persist even though the biologically appropriate situation no longer exists in the normal environment of the species. The injury-feigning ruse of a number of birds when their nest containing young is approached by an intruder or a predator is well known. On the Galapagos Islands, there are no mammalian predators, and most of the birds no longer show this distraction behavior when their nest is approached by man. However, the Galapagos dove still does this, running and fluttering from the nest as if it had a broken wing, in order to distract the attention of the intruder from the nest (Eibl-Eibesfeldt, personal communication).

In the same fashion, a kind of ontogenetic development, similar to Haeckel's well-known "phylogenetic recapitulation"

of organs, has been observed in some cases where a primitive behavior pattern precedes the more recent one during ontogeny of the individual. There are some species of birds (the lark, the raven, and the starling) belonging to the *Passeres* whose original form of ground locomotion was hopping, but that now live on the ground and run. It is most interesting to find that the juveniles of these species hop in biped fashion before they run (Lorenz, 1937).

It is therefore clear that the fixed action patterns are extremely conservative in the evolution of a species. In only one case has rapid phylogentic change in fixed action patterns been observed. This happens in the sexual behavior of closely related species that live in the same territory. In such cases the courtship behavior of these two species must of necessity become differentiated from each other if crossbreeding is not to occur. Thus changes take place in the motor patterns and vocalizations by means of which the sexual partner of the same species is selected during courtship. For example, closely related songbirds inhabiting the same ecological niche, as P. Marler (1957) has shown, have very different songs and courtship movements, whereas other fixed patterns that do not play a role in the ethological barrier between the species have remained the same.

Evolutionary changes that occur in fixed action patterns are similar to those that occur in morphological or physiological characteristics. Thus selection pressure may favor the development of a clear-cut execution of a fixed pattern that has much survival value, or cause it to disappear when it has no survival value. In some cases, of course, behavioral rudiments or vestigial behavior may remain even though there is little survival value.

A phylogenetic change during evolution is the altering of the function of a fixed action pattern in a species that has changed its manner of living. This was demonstrated by one of Tinbergen's group at Oxford, Esther Cullen (1957), who analyzed the behavior of a gull species, the Kittiwake, which now lives on cliffs rather than at the shore. Many fixed patterns changed their function completely, but were still the same in form, so that their homologies with the patterns of the gull group to which this particular species belonged were quite apparent. For example, the young, unlike those of other gulls, cannot run away from strange adults, and consequently have developed an attack-inhibiting appeasement gesture. In related gulls this gesture appears only in sexually mature animals during mating.

Ethologists have postulated that there are two stages in the evolution of motor patterns. In the first, there is a quantitative increase or decrease of motor elements—perhaps actual disappearance in some cases—and in the second, there is a coordinative coupling and disengagement of almost unalterable basic motor units. These basic motor units were shown by Lorenz's film analyses of courtship in ducks. This coupling of previously independent motor elements into a fixed sequence is part of the phylogenetic process that J. S. Huxley, an English naturalist, called ritualization in 1914.

Another phylogenetic process in ritualization is one in which a recently formed fixed pattern may become motivationally autonomous of the situation that originally aroused it, or dependent on another motivation. An example of this process is to be seen in the different forms of the female's "inciting" movement during courtship in different species of swimming ducks (Anatinae). In the original form of this behavior, the female attempts to separate the male from the group by inciting fights between her partner and other males. To do this she runs toward the strange male, but at a certain point fear overtakes her, and she runs back toward her mate. When she is close enough to him, however, aggressiveness again takes hold of her, and she stops and turns toward the strange male. This results in her standing at the point of equilibrium, near her mate, but stretching her neck toward the other male, making

inciting movements. Here the angle between her body and her stretched-out neck is a function of her position and those of the two males. This behavior constitutes the unritualized form of incitement. Now, in some other species, the behavior has become so fixed that the female simply stands near her mate and moves her head back over the shoulder, regardless of where the strange male is. This shows that the movement is now performed solely as a courtship movement by the female. The interpretation that this behavior constitutes ritualized incitement is supported by the fact that still other species perform actions which are intermediate between these two extremes when in a similar situation (Lorenz, 1941).

Still another phylogenetic process in the ritualization of behavior is one in which the fixed pattern is no longer performed in different degrees of intensity, but in one intensity only. Thus the degree of motivation is not expressed in the intensity of the behavior, but in how often the behavior is repeated, much as the urgency of a telephone call may be recognized by the frequency of ringing rather than in loudness. All fixed patterns with a single or typical intensity function as means of communication between species members; selection pressure would in these cases operate to enforce a quite simple and unmistakable form of the movement (Morris, 1957).

Another very interesting evolutionary process in the development of fixed patterns is to be found in fighting behavior between members of the same species. Here actions that injure the opponent have been removed to the end of the sequence of fighting behavior by raising the threshold for its release to a very high level. This results in the development of ceremonial "tournaments" with a very small likelihood of actual bodily harm, a development that has clear survival value to the species, since fighting behavior will maintain its function of spacing out members of a species without causing injury to species members.

The result of ritualization, in summary,

is that actions come to be performed in a mimetically exaggerated way in special situations. The same species may be observed to perform both forms of the fixed action pattern, ritualized and unritualized. As different and related species perform ritualized and unritualized forms of these movement patterns, as well as forms intermediate to these extremes, the homologies between these forms can be used to reconstruct phylogenetic relationships.

During the past few years ethologists have expanded from the phylogenesis of these fixed action patterns to their physiological bases. At the present time very little is known regarding their neuroanatomical bases, but it seems probable that they share a common physiological base, and the investigations so far conducted strongly indicate that they are based on inherited and structured neurophysiological mechanisms. For the time being, nevertheless, the fixed action pattern is defined in purely functional terms.

RELEASERS AND THE INNATE RELEASING MECHANISM

An animal, like man, does not give a particular response to all of the stimuli it perceives; most of them are only *potential* stimuli, to some of which it may learn to respond. Instinctive behavior, in particular, is evoked in response to only a few of the stimuli in an animal's environment; these stimuli are called *sign stimuli,* or *releasers* of the behaviors which they elicit.

An example of these facts is to be found in the behavior of the common tick, which was described in detail by von Uexküll (1909). The tick does not respond to the sight of a host, but when an odor of butyric acid from a mammal strikes the tick's sensory receptors, the tick drops from the twig to the host, finds a spot on the skin which is about 37° centigrade, and begins to drink blood. Only a few stimuli elicit the tick's behavior, this behavior being without doubt innate.

But the simplicity of releasers can sometimes lead animals into grave situations. For instance, a patient tick climbs up a slippery twig to waylay its prey, a nice, juicy mammal. When it has reached the end of the twig, it is above a rock on which a fat, perspiring man has been sitting. The rock therefore emanates the typical odor of butyric acid, and is just the right temperature. So the tick jumps, and lands on the rock—whereupon, in trying to suck from the rock, it breaks its proboscis.

This description shows clearly that only a few stimuli elicit the tick's behavior, this behavior being without doubt innate.

Another example is the carnivorous water beetle *Dytiscus marginalis.* It does not react to the sight of prey—even though it has perfectly well developed compound eyes, as is easily demonstrated when a tadpole in a glass tube is presented to it —but to the chemical stimuli emanating from the prey through the water. If a meat extract solution is put into the water, the beetle engages in frantic searching behavior, clasping inanimate objects (Tinbergen, 1951). These two examples of the few stimuli that release innate behavior patterns make it easy to understand why a male robin will attack a bundle of red feathers but not a dummy of a male, perfect in all respects except that it lacks the characteristic red breast (Lack, 1939).

Animals react automatically to sign stimuli, with little insight; the behavior is just run off. The lack of insight is well demonstrated by the hen's failure to rescue a chick that it can see struggling under a glass bell, but that it cannot hear. It will come to its rescue immediately, however, if it does not see him but can hear his distress cries (Brückner, 1933).

Sign stimuli, or releasers, furthermore, almost always release only one *reaction,* an example being the fact that the smell of butyric acid can release only jumping in a tick. This property differentiates sign stimuli from an acquired picture, where the response seems to depend on a configurational stimulus or Gestalt complex, which may be so altered by small changes or additions so that the original is not recognized. A tame bird, for example, may become quite frightened when it sees its keeper wear glasses or a hat for the first time. In addition, the response given to an acquired stimulus may be changed; a dog may learn to avoid a particular place instead of approaching it.

Sign stimuli also differ from *Gestalten* in that when several of them (usually attached to the same object) that produce the same response are present, they do not interact in determining the animal's response; instead, their effects are completely additive. This is in contrast to the Gestalt psychologists' view that an object is *more* than the sum of its component parts. The additiveness of releaser stimuli was demonstrated by Seitz (1940) when he studied the fighting response of the male cichlid fish *Astatotilapia Strigigena* (Pfeffer). He found that the following stimuli released fighting behavior: (1) silvery blueness, (2) dark margin, (3) highness and broadness by means of fin erection, (4) parallel orientation to the opponent, and (5) tail beating. All of these are characteristics normal to an intruding and attacking male. Seitz found that each of these, presented singly, would elicit the fighting response about equally. This remarkable finding provides a fundamental distinction between releasers and acquired *Gestalten*, since only the whole is responded to in a *Gestalt*. Furthermore, if two of these stimuli were presented simultaneously, the elicited fighting response was twice as great. The intensity of a reaction, therefore, can depend not only on what sign stimuli are present, but how many are present. Seitz called this phenomenon the "law of heterogeneous summation" (*Reizsummen regel*).

Weidmann and Weidmann (1958) recently tested the law of heterogeneous summation quantitatively and found it to hold, in most cases, in a strict arithmetical sense. They counted the number of pecks made by black-headed gull chicks, while they were begging, at cardboard models. If they presented a round and a rectangular model to the chick, one would receive more pecks than the other. But if they made both models red, each received the same increase in the total number of pecks.

Although sign stimuli often consist of a quality such as butyric acid for the tick, or red feathers for the male robin, and differ from *Gestalten* in many ways, most releasers, like *Gestalten*, consist of relational characteristics between stimuli. Indeed, the registration of relationships between stimuli is a fundamental attribute of perception, since the strength or distinctiveness of a perception is to a large part dependent on the contrast or relationship between, or distinctiveness of, stimuli.

For example, a male stickleback reacts with fighting behavior to the red on a rival male. But this red must be on the rival's belly, for if the red is on the back, the stickleback will not attack (Tinbergen, 1951).

Similarly, Tinbergen and Kuenen (1939) showed that the young thrush's gaping response was to a particular stimulus relationship. They constructed a round cardboard model with two different-sized

FIG. 1. The two double-headed models used by Tinbergen and Kuenen in their study of the stimuli eliciting the gaping response in young thrushes. The body of the model on the left is 4 cm.; the body of the model on the right is 8 cm. The small heads are 1 cm.; the large heads are 3 cm. When the left model is presented, the small head is found to elicit gaping, but when the right model is presented, the large head now elicits gaping. This shows that the ratio between the size of the head and of the body is the stimulus quality releasing gaping in young thrushes. (From TINBERGEN, N., and D. J. KUENEN, *Zeitschrift für Tierpsychologie*, 1939, 3, Verlag Paul Parey, Berlin.)

heads attached to it. Tinbergen and Kuenen found that the thrushes directed their gaping toward the larger of the two heads. They then took off the two heads and placed them on a smaller round cardboard. The young thrushes now directed their gaping toward the smaller head, thus showing that it was the relationship between the sizes of the head and body that elicited gaping.

Since releasers can consist of relational qualities, these relations can be made even stronger in some cases through transposition or exaggeration. For instance, Koehler and Zagarus (1937) studied egg recognition on the ringed plover. If a normal egg having dark brown spots on a light brown background was presented together with another egg having black spots on a white background, the birds preferred the black and white eggs, because the spots contrasted more strongly with the background. More remarkable was their preference for abnormally large eggs, even ones they could not sit on. A herring gull behaves similarly (Baerends, 1957, 1959, and Kruijt, 1958). Such stimuli that release a response stronger than that released by the natural stimulus are *superoptimal,* or supernormal.

Magnus (1958) also convincingly demonstrated superoptimality of releasers, this time in the silver-washed fritillary butterfly, *Argynnis paphia L.* Magnus first found that the color yellow-orange, the same as in the female's wings, released courtship responses in the male. The female also flutters its wings, and the resulting alternation between color and dark releases the male's courtship behavior, as Magnus showed by placing yellow-orange and dark strips on a revolving cylinder. Surprisingly, Magnus further found that increasing the speed of the rotating cylinder so that the color and dark alternation was more rapid than the rate used by the female resulted in greater effectiveness in eliciting courtship. The greater the speed of rotation of the cylinder, the greater the courtship responses, right up to the physiologically demonstrated flicker fusion frequency for the species when the color and dark alternation was so rapid that it could not be seen. This was indeed a very dramatic demonstration of supernormal releasers.

The susceptibility of animals to superoptimal releasers provides us with an insight as to the reason for the development of bizarre morphological structures in some animals such as the peacock. These strange structures are used in courtship of the female. It is clear, also, that parasitic birds capitalize on this phenomenon, since the young parasite is usually larger and more babyish than are the host's own young, so that it is actually preferred, with resultant neglect of the host's own young (Cott, 1940). Heinroth described the situation accurately when he called the young cuckoo a "vice" of its foster parents (1938).

When the different stimuli that act as releasers are examined, it is apparent that they are all very clear and simple in character—the color red on the belly or breast, a head-body size ratio, an odor, a movement, and so on. At the same time, they are unmistakable distinguishing characteristics of the appropriate biological situation—characteristics that are highly improbable in any other biological situation. This fact, when coupled with the selection pressures that must operate so as to bring about the ability to recognize without fail certain biological situations essential for the survival of the individual and of the species, accounts for their reliability in eliciting the required and adaptive response in natural conditions.

Their great simplicity, in fact, means that they can be imitated for certain purposes. The fisherman utilizes a releaser when he places a silvery lure into the water in order to attract and catch a pike (Baerends, 1950). There are some flowers which give off a putrid odor that attracts flies, and fertilization of these flowers is thus carried out by the flies. Still other flowers look like a female bumblebee's body and are fertilized in the male bumblebee's attempt to copulate with it.

Releasers, or sign stimuli, are also used for interspecies communication. An example of this is to be found in the fish symbioses studied by Eibl-Eibesfeldt (1955b). There are certain large species of fish in the Caribbean, the groupers (*Epinephelus striatus*), that allow their teeth and gills to be cleaned of particles and ectoparasites by the neon goby, *Elacatinus oceanops*. The fish wishing to be cleaned makes a movement inviting the cleaner fish to enter its mouth and pick its teeth, and makes another movement signaling it to come out from the mouth. The principle of imitation is also to be found in this case (Eibl-Eibesfeldt, 1959). These fish do not normally allow any other fish to come near it, and in the Indopacific oceanic regions they permit only fish looking and behaving like cleaners of that area, the *Labroides dimidiatus*, to approach it. However, there is another species, *Aspidontus taeniatus*, which has taken advantage of this by having a certain coloration and by imitating the movements of the cleaner fish. When they come close to the large fish, they fall on it, biting out chunks of flesh, and the fish must flee. However, these parasites do occasionally pick ectoparasites from larger fish.

Now that we have considered the nature of the releaser, or sign stimulus, let us complete the picture by taking a look at the *innate releasing mechanism,* known for short as the I.R.M. As has been earlier mentioned, an inhibitory block has been postulated by Lorenz (1937) and Tinbergen (1950) to prevent the continuous discharge of internally produced energy. The I.R.M. functions to remove this inhibition when it receives sensory impulses arising from a releaser stimulus. This brings us to an important rule in evaluating the relative effectiveness of a given stimulus as a releaser for a fixed action pattern, in the light of the fading of reactivity due to repeated stimulation, and reaction recovery during a period in which no performance of the reaction takes place (see above). If the ani-

mal is under low motivation, a strong releaser is required in order to elicit a reaction; whereas if the animal is highly motivated, a weak stimulus is sufficient to elicit a response of the same strength. Therefore, it is necessary to test the potential effectiveness of different releasers in relation to each other. This can be done by making certain that the animal is under the same motivation when exposed to each of them, or by exposing the animal to a standard stimulus just prior to the presentation of the stimulus in question. Thus the difference in responsiveness between the different stimuli and the antecedent standard stimulus is the basis for evaluation. Only in these ways can it be determined whether a stimulus is a normal, subnormal, or supernormal releaser.

The I.R.M. operates as a receptor of key stimuli and must be adapted to the world as it exists; this is necessary to make sure it will respond only to stimuli that unfailingly characterize a particular biological situation, and no other. If a particular response is elicited by several sign stimuli belonging to a certain biological situation, then the presence of most of them in that situation guarantees the elicitation of the response in question. A pike cannot attach any releaser or sign stimuli to the fish on which it preys in order to differentiate it from a fisherman's lure; therefore the species must adapt the I.R.M. accordingly. However, within a species or where interspecies communication takes place, natural selection can quickly result in the evolution of special sign stimuli or releasers that will be understood easily by the reacting animal.

Not only morphological structures but also behavioral patterns performed by another animal may function as releasers, as Tinbergen has shown in his study of the courtship behavior of the three-spined stickleback (1942). In such cases there is a reaction chain, each animal's action serving as a releaser for the other animal's subsequent response. Thus, the appearance of the female initiates the male's zigzag dance toward the nest, which in turn releases a following response in the female. The male, in responce to her having followed him, shows her the nest entrance, and she reacts by swimming into it. On seeing her in the nest, the male touches her tail with a quivering motion, with the result that she lays the eggs and then swims out. The male then swims into the nest and fertilizes the eggs. Reaction chains, where they exist, enable the male to select a female of the same species. As Baerends (1950) points out, they also serve to select fully mature and healthy females, since any others would fail to respond correctly after the first couple of links in the chain.

Another feature of the I.R.M. is that in some cases its selectivity for sign stimuli may increase during the life of the individual. Such a process occurs in all members of the species, not in just one individual.

The increased selectivity may be of either of two types:

1. *Narrowing of the range of stimuli evoking a particular response through the dropping out of individual stimuli.* This occurs by means of habituation or one-trial negative conditioning. For example, a bird becomes habituated to the motion of leaves, other members of its species, etc., with the result that it no longer responds with fright to them. Its fright behavior in the face of predators, however, is absolutely unaffected. Similarly, a toad will at first snap at all small objects, but after a single unhappy experience will avoid bees and wasps.

2. *Selection and strengthening of one releaser out of a large range of potential releasers, with the result that only this releaser is responded to and not any of the others.* This occurs in the socialization process of many animals, and this instance is known as *imprinting*. Soon after hatching, for example, a gosling has the disposition to follow almost any moving object and to behave as if that object were its mother. After this experience only this object, and no other, is treated as if it were the mother. Thus a gosling imprinted to a green box

will not have anything to do with its real mother, but will stay close to the green box. This shows, of course, that the socialization process in the natural situation occurs in the same fashion. The subject of imprinting and its difference from ordinary types of association learning will be discussed in a later section.

In both cases of the increased selectivity of the I.R.M. during the life of the individual, it is clear that it results in behavior which has greater survival value.

SIMULTANEOUS AROUSAL OF
DIFFERENT DRIVES

Most often people think of behavior as being influenced by one drive at a time. Actually, however, behavior can most frequently be seen to be influenced by or be a result of more than one activated drive, thus making the analysis of behavior quite complex. There are several types of such behaviors: successive ambivalent behavior, simultaneous ambivalent behavior, redirected behavior, and displacement behavior.

In the case of successive ambivalent behavior, it is seen that the animal alternates between incompletely performed movements belonging to the conflicting drives. When two males of a territory-owning species (such as the stickleback or herring gull) meet each other at the common boundary of their two territories, each is influenced by the other in two ways. First, because the other is so near the territory, the owner is roused to attack. But the stranger also causes hesitance and avoidance, for it is not in the owner's territory, but just outside it, and furthermore in its own territory. Each male is therefore in the same conflict, one which usually finds expression in an ambivalent reaction: attack and retreat in quick and repeated alternation.

Simultaneous ambivalent behavior, however, is also to be found. In such a case both tendencies are simultaneously, rather than alternately, aroused. An example of

58

this is the threat posture of the cat, which according to Leyhausen (1956), results from the animal retreating more rapidly with the front paws than with the rear ones, or advancing with the rear paws while retreating with the front ones. Lorenz (1953) has shown that fear and aggression occur simultaneously in the dog, with at least nine different expressions possible as a result of mixing three different degrees—high, medium, and low—of either drive.

Redirected behavior is another type in which two conflicting behaviors manifest themselves. This occurs when one behavior is inhibited or suppressed by another motivation. For example, an animal feeling aggressive towards a member of its own species (its mate, for instance) may at the same time be inhibited in its attack. Hence, the aggression may be redirected, and the animal will attack another animal or even an object if no third animal is available. Thus, prairie falcons defending their nest from a human intruder may swoop down toward him, but at the last moment fear overtakes the falcon: it makes a sudden swerve to the side and flies to attack bypassing birds.

Since cases in which behavior is determined by only one drive are relatively rare, the different behaviors of animals result from independently varying sets of motivations. In accordance with this knowledge, Oehlert (1958) demonstrated the existence of a mechanism of sex recognition that is widely distributed throughout the animal kingdom. In *Cichlasoma biocellatum* and *Geophagus brasiliensis*, two species of cichlids studied by Oehlert, and perhaps in many other species in which the sexes do not have a distinctive sexual dimorphism, the only sexual difference in behavior consists of the fact that three drives—sex, aggression, and fear—which are always activated simultaneously when two strange fish meet, can be mixed in different ways in males and females. In the males just about every possible mixture and superposition of sexual and aggressive behavior elements can be made, but the flight drive,

even when minimal, will immediately inhibit the sexual drive. In the female, however, flight behavior can mix very easily with sexual behavior, whereas aggressiveness immediately suppresses sexuality. This difference between males and females is enough to guarantee the formation of male-female pairs. In the same way, fear suppresses male sex behavior and aggression inhibits female sex behavior in grouse and crows (A. A. Allen, 1954, and Lorenz, 1931).

Still another type of behavior may occur when two motivations are in conflict with each other. This is the *displacement activity* discovered independently in 1940 by both Tinbergen and Kortlandt. Very often in conflict situations the animal may show behavior patterns that do not belong to either of the two conflicting drives, but that are completely different. At first such "irrelevant" behavior is surprising, but after further study it is found that a particular irrelevant act is often typical of a particular set of conditions. Since the intensity of the displacement act is correlated with strength of the conflicting motivations, Tinbergen proposed that the irrelevant or displacement act is activated by the energy from the conflicting drives. Thus such displacement acts were *allochthonous*, as distinct from the case in which they are motivated by their own drive, when they are *autochthonous*.

For example, fighting domestic cocks may suddenly peck at the ground as if they were feeding (Lorenz, 1935, Tinbergen, 1939). In this case the irrelevant feeding results from a conflict between aggressiveness and fear. Quite often such pecking on the ground has been observed in other birds. The prairie horned lark does this during territorial fights (Pickwell, 1931), and so does the lapwing between attacks of an intruder disturbing it at its nest. Male snow buntings will peck at the ground during boundary disputes (Howard, 1920).

Many other types of displacement activities have been observed. The three-spined stickleback makes nest-building movements (displacement sand digging) during boundary fights with another male (Tinbergen and van Iersel, 1947), and herring gulls pick up nesting material during boundary disputes. Sexual movements such as wing-flapping and gurgling, which normally occur in courtship behavior, are shown by cormorants when fighting, and sky larks burst into violent song just after having escaped from a predator (Tinbergen, 1952). Cormorants also perform sessions of "false brooding" during nest fights (Kortlandt, 1940a), and the avocet may suddenly sit down during exciting hostile encounters (Makkink, 1936). Even sleep may occur as a displacement activity; this was first discovered by Makkink (1936) in the avocet and has been observed since then in various other waterbirds such as the oyster catcher, turnstone, and common sandpiper, and in male snow buntings, always in an aggressive situation.

It is apparent even from this short review of displacement activities that they often occur when the fighting and escape drives are simultaneously aroused. They are also very common in sexual situations. They can, of course, be aroused by combinations of other drives.

From what has been found concerning displacement activities, it is clear that such activities are usually dependent, at least in part, on internal impulses. The so-called innate "comfort movements" such as preening, shaking the feathers, wiping the bill, bathing, and so forth, most often depend on internal impulses and also appear as displacement activities in both hostile and sexual situations. Displacement preening is quite common during courtship, and displacement bathing has been observed regularly in the sheld duck when copulation was interrupted for some reason (Makkink, 1931). Scratching of the fur is a very common displacement activity in mammals up to the primates (Portielje, 1939, and Tinbergen, 1940).

Another fact in considering the nature of displacement activities is that they are more likely to occur when conflicting drives are relatively intense. For example, in the boundary clash between males of territory-owning species, ambivalent behavior consisting of attack and retreat in alternation occurs as long as the conflicting drives of aggressiveness and fear are not too strong. But when these drives are very intense, displacement activities appear. Tinbergen (1952) therefore hypothesized that displacement activities were outlets through which strong but thwarted drives could express themselves in motion.

In sexual situations, displacement activities may occur as a result of sudden cessation of external stimulation emanating from the partner, or from conflict. Also, it has been suggested (Verwey, 1930, Kortlandt, 1940b) that copulation in birds brings about a situation which they normally avoid, that is, bodily contact with another individual. Therefore, the precoition situation may evoke in each individual a tendency to keep away from or to attack the partner as well as the expected one of approach.

However, no experimental evidence has been found that supports Tinbergen's hypothesis. Recently, Andrew (1956), Sevenster (1958) and van Iersel and Bol (1958), by studying grooming in birds, displacement fanning in the three-spined stickleback, and displacement preening in terns, respectively, suggested that inhibitions exerted by one or both of the conflicting systems on an activity are temporarily removed as a result of the conflict between these drives. These authors hypothesize that if the two conflicting motivations are balanced, then not only their own motoric manifestations, but also the inhibiting effect on the third motor pattern is removed.

This new conceptualization agrees with a number of well-known facts regarding displacement activities. For example, displacement activities occur particularly when specific motivations that strongly inhibit other motivations and dominate the entire organism, such as flight, escape, and copulation, come into conflict. Furthermore, the fixed patterns that occur as displacement activities are almost always those that are repeated many times daily and that do not usually require a high motivation level. Also, the form of the displacement movement has often been observed to be correlated with the body position into which an animal is forced by his primary conflict: a movement for which this body position is the basis is made in a large number of cases. This last is a point made by Tinbergen (1952) in connection with his theory.

Sevenster (1958) supported his hypothesis by the finding that the presence of adequate external stimuli can facilitate the appearance of certain types of displacement activities. For example, if food is thrown to fighting cocks, then displacement feeding increases. Sevenster could increase the fanning of the nest by the stickleback as a displacement activity by bringing about the appropriate conflict situation and at the same time maintaining a high CO_2 concentration in the water. Van Iersel and Bol (1958) also found that displacement bill-shaking, which occurs in sandwich terns when in conflict between escape and incubation drives, was increased in rainy weather. These phenomena would be hard to explain if displacement were based only on irradiation of energy, channeled strictly in a certain direction by the structure of the central nervous system.

On the other hand, there are several facts that could be better explained by the earlier hypothesis than by the new theory. The new theory, for example, fails to explain why in the majority of cases where there is a particular conflict, only a particular displacement activity is performed. This is an important point, because in closely related species the same small motor patterns are performed, but in the same conflict situation different specific displacement activities occur. Furthermore, according to the inhibition theory, the *balance* between the two conflicting motivations is responsible for the occurrence of displacement activity, whereas it has been shown that the intensity of the displacement act is proportional to the absolute drive level of the conflicting motivations. Lastly, if the inhibition hypothesis were correct, then displacement activity would have to be performed with the same *irregularity* as its autochthonous forms, as, for instance, in the case of a bird; it does not preen itself ceaselessly when the preening drive is not under inhibition by other drives.

Finally, these different types of mixed-drive behavior—successive and simultaneous ambivalence, redirection and displacement activities—are not mutually exclusive and may often be found to exist in the same behavior sequence. For example, as Baerends (1958) has shown, the courtship of the male cichlid fish, *Cichlasoma meeki*, has successive ambivalence in the zigzag dance (the *zigs* being incipient attacks and the *zags* being incipient leading-to-the-nest), redirection in its incipient attacks toward plants, and displacement activity in a peculiar quivering movement interpreted as displacement digging.

HIERARCHICAL ORGANIZATION OF BEHAVIOR

Lorenz and Craig distinguished between two types of behavior: the appetitive and aversive behaviors, and the consummatory act. Appetitive behavior consists of initially variable searching behavior that becomes more and more specific until the simpler and more stereotyped consummatory act, in response to a releasing stimulus situation, is performed. Aversive behavior has some similarity with appetitive behavior, but consists of behavior that continues until a disturbing situation is removed, and the animal reaches a state of equilibrium. Here the goal is not the discharge of specific behavior patterns, but the cutting off of appetitive behavior that in this case is undirected locomotion.

The distinction between appetitive behavior and the consummatory act is not an absolute one; there are many forms intermediate to them. But these concepts serve to mark the extremes, since appetitive behavior is variable and plastic, whereas the consummatory act is relatively fixed and stereotyped.

Appetitive behavior is characterized in three ways. The first is by its motor pattern, usually one of locomotion. The others are its orientation component and the stimuli to which the animal is particularly responsive. Since the first and second components may remain the same while the third changes, as seen when a hungry squirrel (1) climbs (2) up trees (3) looking for cones; and when a squirrel motivated to build nests (1) climbs (2) up trees (3) looking for twigs and bark, appetitive behavior is usually classified according to the third component. Therefore, in the first case, the squirrel is showing appetitive behavior for food, while in the second case it is showing appetitive behavior for nesting material.

The consummatory act itself is made in response to one or more releasers and is composed of an orientation component, the *taxis,* and a motor component, the *fixed action pattern.* These two aspects of the consummatory act were pointed out by Lorenz and Tinbergen (1938) during a study of the egg-rolling response of the greylag goose. But before we discuss this, let us briefly examine the nature of the taxes, since the fixed action pattern has already been discussed in detail.

Taxes are oriented locomotory reactions of motile organisms. When exposed to a source of stimulation, the body as a whole, or a particular part, is oriented in line with the source of stimulation. Movement toward the source is said to be positive, while movement away from the source is negative. There are many types of taxes recognized by ethologists, and some of the most important are klinotaxis, tropotaxis, telotaxis, and transverse orientation (Fraenkel and Gunn, 1940).

Klinotaxis consists of the animal traversing a short distance, stopping, making a turning movement from one side to the other, and then finally moving toward one side according to whether it is nearer or farther away from the source of stimulation. The animal then travels a short distance further, and then repeats the whole procedure. In this way an animal gradually gets nearer or farther away from the source of stimulation. Tropotaxis is like klinotaxis except that there are no trial movements, but a gradual turning toward or away from the stimulation. Telotaxis, however, is direct locomotion toward the source of stimulation, without trial movements, and looks as if it were goal-directed. Transverse orientation may or may not involve locomotion, and is illustrated by the light compass reaction shown by bees navigating toward or away from the hive, and by the dorsal light reaction of fishes swimming always belly down, that is, belly away from the light.

Instinctive behavior is complex, being composed of several elements: reflexes, taxes, and fixed action patterns or instinctive movements. In the egg-rolling reaction of the greylag goose we find two separate components in this behavior. If an incubating goose is presented with an egg outside of her nest, she will reach out with her neck toward it and roll it back into her nest with her bill, balancing it in its course with little sideways movements. If, after it has begun this action, the egg is taken away, the goose will continue to perform the action to the very end. However, the bill is moved directly, without the balancing movements, toward the nest. Hence the movement of the bill toward the nest is a pure fixed action pattern, being released but not guided, while the balancing movements are guided by continuous external stimulation from the egg. Tinbergen (1951) has compared the taxic component and the fixed action pattern with the steering and propulsory mechanisms of a ship, respectively.

Another example of the simultaneous presence of fixed action patterns and of taxic movements is to be found in the wing-cleaning behavior of the common fly. The fly moves its legs over and under the wings, making little scrubbing movements. If, however, the wings are removed, the fly still moves its legs around in the region where the wings had been, but will not make the tiny scrubbing movements. This shows that the action of passing the legs over and under the wings is a fixed action pattern, while the little scrubbing movements were guided by the stimulation offered by contact with the wings (Heinz, 1949).

Often the same stimuli both steer and release the reaction, and in some cases the taxic steering component is absent in certain consummatory acts such as swallowing or blinking. Although the distinction between the fixed action pattern and the taxis is not absolute, the isolation of the fixed action pattern has contributed a great deal to the analysis of behavior.

However useful the distinction between appetitive behavior and the consummatory act, the Lorenz-Craig scheme of appetitive behavior leading to the consummatory act has been found to be a rather simplified case. Normally, particular appetitive behavior does not usually end in a consummatory act, but, rather, leads to a stimulus situation that initiates another and more specific appetitive behavior (Baerends, 1941 and 1956, and Tinbergen, 1950). A chain of appetitive behaviors gives rise to a temporal sequence of "moods" or readiness to act, anchored in the central nervous system. We shall illustrate this by considering the reproductive behavior of the male stickleback.

In spring the gradual increase in length of day is responsible for bringing the fish into a reproductive motivation. It does not immediately acquire a red belly, but first begins migrating into shallow fresh water. The rise in water temperature, as it goes further inland, together with the visual stimulation arising from suitable territory consisting of heavily vegetated sites, then releases the entire reproductive behavior. Only then, after it has established its territory, does the male stickleback acquire its characteristic red belly. At this point it begins to react to particular stimuli that previously had no effect on it. If males caught during migration are put together in a bathtub, they all will remain in a school except one who establishes a territory by the drain plug chain. This chain is the only structured element in the bathtub; it is for this reason that a territory is established there. Furthermore, it provides enough territory for only one male. This male will then fight at the appearance of a stranger, begin to build a nest with suitable material, and court passing females. Since the male's behavior depends principally on the stimulus situation, this makes difficult the prediction as to precisely what it will do. Fighting, for example, is released by the stimulus *red belly* on a male intruding into its territory, but it cannot be predicted which of the five known fighting movements will occur, each being, again, dependent on still further and highly specific stimuli. If the stranger bites, the territory owner will bite in turn, or if the stranger beats with its tail, the owner makes the same response. Fleeing will elicit chasing, and threatening will elicit threatening in turn. In summary, the stimuli emanating from suitable territory will activate the fighting, building, and mating drives. The more specific stimulus, *red-bellied male*, activates only a general readiness to fight; the specific movements made are dependent on even more specific stimuli.

Thus one can organize different levels of integrations into a hierarchical system. In this way it is clear that there are chains of behavior tendencies that are connected in higher and lower levels of integration. The adaptive advantages of the mood hierarchy rather than a stereotyped series of single fixed action patterns lie in its adaptability to several situations. For example, a hunting peregrine falcon initially seeks prey of any kind, this behavior being of a rather general appetitive type. If it sees a group of prey birds, it dives down toward them and isolates a single bird from the swarm. If it has met with success, it performs the very specific behavior patterns of killing the prey, pulling out its feathers, and eating it (Tinbergen, 1950). The adaptability of the mood hierarchy lies in the fact that if a falcon should happen to meet a single bird by accident, the previous part of the chain will drop out, and it immediately proceeds to kill, pluck, and eat it.

It seems evident that a structural organization within the central nervous system must exist which parallels the lawfulness of behavior, particularly the mood hierarchy. Tinbergen (1950) has illustrated this in the diagram shown in Figure 2, which, of course, represents only functional units and does not attempt to represent anatomy or localization of functions. The number and kind of appetitive links, as well as the innate releasing mechanisms

(IRM) that respond to certain stimulus situations and the chain of built-in fixed patterns that are also their final goals or consummatory acts, are rigorously derived from empirical observations of behavior.

Close relationships exist between Tinbergen's results and those of P. Weiss (1941b), who showed a similar hierarchical organization of the central nervous system mechanism in motor processes that are almost exclusively below the level of integration of the goal-forming fixed patterns. Since Weiss had specified that the highest level of his hierarchy was the behavior of the animal as a whole, Tinbergen assumed that the two hierarchies were physiologically similar, and therefore he added Weiss's diagram below a horizontal line

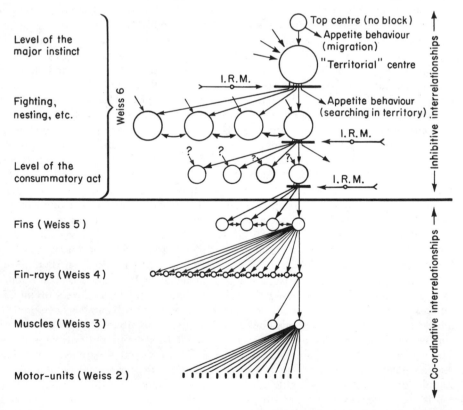

FIG. 2. Tinbergen's diagram of the hierarchical organization of instincts, superimposed on Weiss's diagram of the hierarchical organization of central nervous mechanisms, Weiss's highest level incorporating the same phenomena as those placed in Tinbergen's hierarchy. (From TINBERGEN, N., 1950.)

under his own hierarchical organization, making Weiss's uppermost level equivalent to his own system.

In this scheme Tinbergen ignored a complication that he recognized: some of the same behavior patterns at the lowest level, for instance those of locomotion, can be among the final effecting links in several hierarchical organizations of a higher order as a common final pathway, so that under certain circumstances the chains diverge and then converge again. Tinbergen's results and hypothesis are highly valuable because they provide a link to a real physiology of the central nervous system. Further evidence regarding this will be discussed later when we return to neurophysiological evidence for ethological concepts.

REFERENCES

ALLEN, A. A. Sex rhythm in the ruffed grouse (Bonasa umbellus) and other birds. Auk, 1954, 51, 180–199.

ANDREW, R. J. Normal and irrelevant toilet behaviour in Embenza, Spp. Brit. J. Anim. Behaviour, 1956, 4, 85–96.

BAERENDS, G. P. Fortpflanzungsverhalten und Orientierung der Grabwespe, Ammophila campestris Jur. Tijdschr. Ent., 1941, 84, 68–275.

BAERENDS, G. P. Specialization in organs and movements with a releasing function. Symp. Soc. Exp. Biol., 1950, 4, 337–360.

BAERENDS, G. P. Aufbau des tierischen Verhaltens. Kukenthal's Hb. Zool., 1956, 10, 1–32.

BAERENDS, G. P. The ethological concept "releasing mechanism" illustrated by a study of the stimuli eliciting egg-retrieving in the herring gull. Anat. Record, 1957, 128, 518–519.

BAERENDS, G. P. Comparative methods and the concept of homology in the study of behaviour. Arch. Néer. Zool., 1958, 13, Suppl. 1, 401–417.

BAERENDS, G. P. The ethological analysis of incubation behaviour. Ibis, 1959, 101, 357–368.

BRUCKNER, G. H. Untersuchungen zur Tiersoziologie, insbesondere der Auflösung der Familie. Zs. Psychol., 1933, 128, 1–120.

COTT, H. B. Adaptive coloration in animals. London: Methuen, 1940.

CRAIG, W. Appetites and aversions as constituents of instincts. Biol. Bull., 1918, 34, 91–107.

CULLEN, ESTHER Adaptations in the Kittiwake to cliff-nesting. Ibis, 1957, 99, 275–302.

EIBL-EIBESFELDT, I. Uber Symbiosen, Parasitismus, und andere besondere zwischenartliche Beziehungen bei tropischen Meeresfischen. Zeit. f. Tierpsychol., 1955b, 12, 203–219.

EIBL-EIBESFELDT, I. Angeborenes und Erworbenes in der Technik des Beutetötens (Versuche am Iltis, Putorious putorius L.). Z. Säugetierkunde, 1956a, 21, 135–37.

EIBL-EIBESFELDT, I. Der Fisch Aspidontus taoniatus als Nachahmer des Putzers Lubroides dimidiatus. Zeit. f. Tierpsychol., 1959, 16, 19–25.

FRAENKEL, G. S., and D. I. GUNN The orientation of animals. Oxford: Clarendon Press, 1940.

HEINROTH, O. Beiträge zur Biologie, namentlich Ethologie und Psychologie der Anatiden. Verh. 5th Int. Ornith. Kongr., 1910, 589–702.

HEINROTH, O. Aus dem Leben der Vögel. Berlin: Springer, 1938.

HEINZ, H.-J. Vergleichende Beobachtungen über die Putzhandlungen bei Dipteren im allgemeinen und bei Sarcophaga carnaria L. im besonderen. Zeit. f. Tierpsychol., 1949, 6, 330–371.

HESS, W. R. Stammganglien-Reizversuche. Vortrag Dtsch. Physiol. ges. Bonn, 1927.

HESS, W. R. Das Zwischenhirn: Syndrome, Lokalisationen, Funktionen. Basel: Schwabe, 1949.

HESS, W. R. Das Zwischenhirn. Syndrome, Lokalisationen, Funktionen. Mongr. II. Erweiterte aufl. Basel: Schwabe, 1954.

HESS, W. R. Hypothalamus and Thalamus. Experimental-dokumente. Stuttgart: Thieme, 1956. (Parallel English and German text.)

HOLST, E. v. Weitere Versuche zum nervösen Mechanismus der Bewegung beim Regenwurm. Zool. Jb., 1933, 53, 67–100.

HOLST, E. v. Uber den Prozess der zentralnervösen Koordination. Pflüg. Arch. ges. Physiol., 1935, 236, 149–158.

HOLST, E. v. Versuche zur Theorie der Relativen Koordination. Pflüg. Arch. ges. Physiol., 1936a, 237, 93–121.

HOLST, E. v. Uber den "Magnet-Effekt" als koordinationender Prinzip im Rückenmark.

Pflüg. Arch. ges. Physiol., 1936b, **237**, 655–682.

HOLST, E. v.   Bausteine zu einer vergleichende Physiologie der lokomotorischen Reflexe bei Fischen. II Mitteilung. Z. vergl. Physiol., 1937a, **24**, 532–562.

HOLST, E. v.   Vom Wesen der Ordnung im Zentralnervensystem. Naturwiss., 1937b, **25**, 625–631, 641–647.

HOLST, E. v.   Die Auslösung von Stimmungen bei Wirbeltieren durch "punktförmige" elektrische Errengung des Stammhirns. Naturwiss., 1957, **44**, 549–551.

HOLST, E. v., and U. v. ST. PAUL   Das Mischen von Trieben (Instinkbewegungen) durch mehrfache Stammhirnreizung beim Huhn. Naturwiss., 1958, **45**, 579.

HOLST, E. v., and U. v. ST. PAUL   Vom Wirkungsgefüge der Triebe. Naturwiss., 1960, **18**, 409–422.

HOLST, E. v., and U. v. ST. PAUL   Electrically controlled behavior. Sci. Amer., 1962, **206**, No. 3, 50–59.

HOWARD, H. E.   An introduction to the study of bird behaviour. Cambridge: Cambridge University Press, 1929.

HUXLEY, J. S.   The courtship habits of the great crested Grebe (Podiceps cristatus); with an addition to the theory of sexual selection. Proc. Zool. Soc. London, 1914, 491–562.

IERSEL, J. v., and A. C. BOL   Preening of two tern species. A study on displacement activities. Behaviour, 1958, **13**, 4–89.

KOEHLER, O., and A. ZAGARUS   Beiträge zum Brutverhalten des Halsbandregenpfeifers (Charadrius h. hiaticula L.). Beitr. Fortpfl.-Biol. Vögel, 1937, **13**, 1–9.

KORTLANDT, A.   Eine Ubersicht der angeborenen Verhaltensweisen des mitteleuropäischen Kormorans (Phalacrocorax Carbosinensis), ihre Funktion, ontogenetische Entwicklung und phylogenetische Herkunft. Arch. Néer. Zool., 1940a, **4**, 401–442.

KORTLANDT, A.   Wechselwirkung zwischen Instinkten. Arch. Néer. Zool., 1940b, **4**, 442–520.

KRUIJT, J. P.   Speckling of herring gull eggs in relation to brood behaviour. Arch. Néer. Zool., 1958, **12**, 565–567.

KRUMBIEGEL, I.   Die Persistenz Physiologischer Eigenschaften in der Stammesgeschichte. Zeit. f. Tierpsychol., 1940, **4**, 249–258.

LACK, D.   The behavior of the robin. II. Proc. Zool. Soc. London, 1939, **109**, 200–219.

LEYHAUSEN, P.   Verhaltensstudien an Katzen. Beiheft 2 zur Zeit. f. Tierpsychologie, 1956.

LORENZ, K. Z.   Beiträge zur Ethologie sozialer Corviden J. f. Ornith., 1931, **79**, 67–120.

LORENZ, K. Z.   Der Kumpan in der Umwelt des Vogels. J. f. Ornith., 1935, **83**, 137–213, 289–413.

LORENZ, K. Z.   Uber die Bildung des Instinktbegriffes. Naturwiss., 1937, **25**, 289–300, 307–318, 324–331.

LORENZ, K. Z.   Vergleichende Bewegungsstudien an Anatiden. J. f. Ornith., 1941, **89**, 194–294.

LORENZ, K. Z.   Die angeborenen Formen möglicher Erfahrung. Zeit. f. Tierpsychol., 1943, **5**, 335–409.

LORENZ, K. Z.   The comparative method in studying innate behaviour patterns. Symp. Soc. Exp. Biol., 1950, **4**, 221–268.

LORENZ, K. Z.   Die Entwicklung der vergleichenden Verhaltensforschung in den letzten 12 Jahren. Verh. Dtsch. Zool. ges. Freiburg, 1952, 36–58. Leipzig: Akad. Verlag. 1953.

LORENZ, K. Z., and N. TINBERGEN   Taxis und Instinkthandlung in der Eirollbewegung der Graugans. Zeit. f. Tierpsychol., 1938, **2**, 1–29.

MAGNUS, D.   Experimentelle Untersuchungen zur Bionomie und Ethologie des Kaisermantels Argynnis paphia L. (Lep. Nymph). I. Uber optische Auslöser von Angliegereaktionen und ihre Bedeutung für das Sichfinden der Geschechter. Zeit. f. Tierpsychol., 1958, **15**, 397–426.

MAKKINK, G. F.   Die Kopulation der Brandente (Tadorna tadorna L.). Ardea, 1931, **20**, 18–22.

MAKKINK, G. F.   An attempt at an ethogram of the European Avocet (Recurvirostra avosetta L.) with ethological and psychological remarks. Ardea, 1936, **25**, 1–62.

MARLER, P.   Specific distinctiveness in the communication signals of birds. Behaviour, 1957, **9**, 13–39.

MORRIS, D.   "Typical intensity" and its relation to the problem of ritualisation. Behaviour, 1957, **11**, 1–13.

OEHLERT, B.   Kampf und Paarbildung bei einigen Cichliden. Zeit. f. Tierpsychol., 1958, **15**, 141–174.

PICKWELL, G. B. The prairie horned lark. *Trans. Acad. Sci. St. Louis*, 1931, **27**, 1–153.

PORTIELJE, A. F. J. Triebleben bzw. intelligente Aeusserungen beim Orang-Utan *(Pongo pigmaeus* Hoppius*). Bijdr. Dierk.*, 1939, **27**, 61–114.

ROEDER, K. D. Spontaneous activity and behavior. *Sci. Mon. Wash.*, 1955, **80**, 362–370.

SEITZ, A. Die Paarbildung bei einigen Cichliden. I. Die Paarbildung bei *Astatotilapia strigigena* (Pfeffer). *Zeit. f. Tierpsychol.*, 1940, **4**, 40–84.

SEVENSTER, P. A causal analysis of a displacement activity. Unpublished, 1958.

TINBERGEN, N. On the analysis of social organization among vertebrates, with special reference to birds. *Amer. Midl. Nat.*, 1939, **21**, 210–234.

TINBERGEN, N. Die Uebersprungbewegung. *Zeit. f. Tierpsychol.*, 1940, **4**, 1–40.

TINBERGEN, N. An objectivistic study of the innate behaviour of animals. *Biblioth. Biother.*, 1942, **1**, 39–98.

TINBERGEN, N. The hierarchical organisation of nervous mechanisms underlying instinctive behaviour. *Symp. Soc. Exp. Biol.*, 1950, **4**, 305–312.

TINBERGEN, N. *The study of instinct.* London: Oxford University Press, 1951.

TINBERGEN, N. "Derived" activities; their causation, biological significance, origin, and emancipation during evolution. *Quar. Rev. Biol.*, 1952, **27**, 1–32.

TINBERGEN, N., and J. VAN IERSEL "Displacement reactions" in the three-spined stickleback. *Behaviour*, 1947, **1**, 56–63.

TINBERGEN, N., and D. J. KUENEN Uber die auslösenden und richtungsgebenden Reizsituationen der Sperrbewegung von jungen Drosseln *(Turdus m. merula* und *T. e. ericetorum* Turton). *Zeit. f. Tierpsychol.*, 1939, **3**, 37–60.

UEXKÜLL, J. v. *Umwelt und Innenwelt der Tiere.* Jena: 1909. Second edition, Berlin, 1921.

VERWEY, J. Die Paarungbiologie des Fischreihers. *Zool. Jb. Physiol.*, 1930, **48**, 1–120.

WEIDMANN, R., and U. WEIDMANN An analysis of the stimulus situation releasing food-begging in the black-headed gull. *Brit. J. Anim. Behaviour,* 1958, **6**, 114.

WEISS, P. Autonomous versus reflexogenous activity of the central nervous system. *Proc. Amer. Phil. Soc.*, 1941a, **84**, 53–64.

WEISS, P. Self-differentiation of the basic patterns of co-ordination. *Comp. Psychol. Monogr.*, 1941b. **17**, 1–96.

WHITMAN, C. O. Animal behavior. *Biol. Lect. Marine Biol. Lab. Wood's Hole, Mass., 1898.* Boston: 1899. 285–338.

5 The 1973 Nobel Prize for Physiology or Medicine

The 1973 Nobel prize for Physiology or Medicine has been awarded jointly to three zoologists: Karl von Frisch, 86 years old, of the University of Munich; Konrad Lorenz, 69 years old, of the Max Planck Institute for Behavioral Physiology at Seewiesen, near Munich; and Nikolaas Tinbergen, 66 years old, of the Department of Zoology at Oxford University, for their discoveries concerning organization and elicitation of individual and social behavior patterns. The award is a new departure for the Nobel Committee of the Karolinska Institute, acknowledging for the first time major advances in our understanding of sociobiology, especially in the area of behavioral science known as ethology. At a time when studies of learning in animals were generally conducted in the laboratory, thereby posing problems largely irrelevant to their natural biology, these three men discovered in the natural behavior of animals both learned and innate patterns, exquisitely adapted to their particular phylogenetically determined ways of life. At one stroke they explained some of the most remarkable examples of the fine control of elaborate patterns of behavior by external stimuli known to science, sometimes learned, sometimes not, while leaving in no doubt the crucial importance of genetic differences in understanding the development of behavior.

Karl von Frisch, inspired pioneer of comparative physiology, has opened our eyes to several unsuspected "sensory windows" through which animals view the world, and to complex and versatile communication behavior controlled by insect nervous systems formerly thought capable only of rigid mechanical responses. Stimulated by a distinguished family background in Vienna, including the physiologist Sigmund Exner, his boyhood enthusiasm for biology matured through studies with Richard von Hertwig, whom he later succeeded as professor of zoology at Munich. Shortly before World War I von Frisch demonstrated that, contrary to prevailing scientific opinion, fish and honeybees could discriminate colors. After the war he turned to experiments on olfaction and showed that bees could distinguish among dozens of odors, including the scents of closely related flowers. His thorough experiments in the 1920's settled in the affirmative the long-standing question whether fish could hear. Unsophisticated in the best sense, these experiments have been amply confirmed in later years with appropriate monochromators and hydrophones. An ardent Darwinian who successfully defended his views at his oral examination in philosophy against a professor who did not believe in evolution, von Frisch was motivated by a naturalist's faith that phenomena such as the colors and scents of flowers, or the Weberian ossicles of catfish, must have an adaptive biological significance.

In 1923 he described as a simple language the round and waggle dances of honeybees. In that heyday of behaviorism he observed simply that round dances occurred when foraging bees brought sugar solutions into the hive from artificial feeders, whereas waggle dances accompanied the gathering of pollen. But in 1944 he found the real "Rosetta Stone" to decipher the language of bees: Round dances mean a food source nearby, waggle dances one at some distance. More important, the direction of the straight portion of the waggle dance points the way to the food, and its duration signals the distance. On a horizontal surface the dancing bee points directly toward the food, but ordinarily the dances take place inside a dark hive on a vertical surface. Here *straight up* corresponds to the direction of the sun, which serves as a directional reference point. But if the sun is obscured by broken clouds, the bees use instead the plane of polarization of light from patches of blue sky. Thus behavioral experiments that had stemmed from earlier studies of sensory physiology disclosed a new sensory channel.

Von Frisch also demonstrated that odors are very important to identify the exact food source, and we now know that sounds or vibrations are also involved in the communication process. Bees dance only when the colony is in severe need of something, but dances are used not only for food but also for water when it is needed in hot weather to cool the hive. The most remarkable use of the dances was discovered by Martin Lindauer, one of von Frisch's leading students. When a colony of bees is swarming, scouts fly out from the teeming cluster of bees that have left their former hive and search for a cavity where thousands of bees can fly to establish a new colony. When a scout has located a suitable cavity, she signals its location by the same dance pattern used for food. Individual bees exchange information about the suitability and location of various cavities, sometimes the same bee acting alternately as transmitter and receiver of information.

Questions have been raised about the accuracy with which information is actually transmitted, and about the relative importance of the dances, odors, and sounds or vibrations. Philosophers and linguists may debate whether the term language is appropriate. But, for behavioral scientists, the revolutionary discovery was that an insect sometimes communicates with fellow members of a closely integrated society by flexible, iconic, graded gestures about distant objects that are urgently needed by the social group as a whole. Behavioral continuity between animals and men extends even to fruitful comparisons between animal communication and human language.

Konrad Lorenz, acknowledged founder of the science of ethology, derived his insights into the causation and organization of behavior from studying fish and birds. At Altenburg in Austria, the house of his father, a Viennese orthopedist, was always full of animals and birds. A precocious naturalist, Lorenz developed early what became a lifelong passion for raising both wild and domestic animals by hand, and for living with them in the closest quarters, and so gaining insights into the relation between genome and experience in ontogeny. Medical training at the University of Vienna was followed by excursions, inspired by Ferdinand Hochstetter, Karl Bühler, and others, into comparative anatomy, psychology, and philosophy. One senses early tension between the attractions of a career in medicine and academia, and fascination with the beauty and diversity of animals. During a two-semester stint in the Columbia Medical School in New York in 1922, he is said to have spent more time studying the inhabitants of the New York Aquarium than at lectures. Comparative ethology was deemed an inappropriate pursuit in the department of anatomy, so to his M.D.

degree he added, in 1936, a Ph.D. in zoology at the University of Munich, and remained in that department until 1941. The major features of his theory of behavior were laid during that period. After World War II, under the aegis of the Institute for Marine Biology, a Max Planck Institute was established at Büldern in Westphalia for Lorenz' group, and, in 1958, it became the Max Planck Institute for Behavioral Physiology, at Seewiesen in Bavaria.

Ethological findings derive much of their force and generality from insightful use of comparative techniques and subjects selected appropriate to the problem. If Lorenz has a totem animal, it is surely the greylag goose in which, with his revered teacher Oskar Heinroth, he discovered imprinting, an especially rapid and relatively irreversible learning process with an optimal critical period early in the gosling's life. Imprinting has repercussions not only on what constitutes an acceptable parental object, or companion as Lorenz called it, but also on what becomes an appropriate sexual companion when the gosling grows up, one of many findings that have proved heuristically valuable in psychoanalysis and psychiatry.

This and other discoveries were incorporated in the panorama of ethological theory presented in 1935, and translated into English soon afterward by Margaret Nice, that was at once a treatise on the social behavior of animals and how the structure of a society relates to its component parts, and a manifesto for the objective analysis of the natural behavior of animals. A central conception complementary to that of imprinting, is the "innate release mechanism." These were visualized as genetically determined sensory mechanisms that predispose an organism to be especially responsive to stimuli, from the environment or from companions, that have assumed special valence in the course of evolution of behavioral adaptations for survival and reproduction. They match behaviors evolved for social communication that generate key "releasing schemata" or "sign stimuli," in turn evoking or guiding particular patterns of behavior in the respondent.

A series of germinal papers over the next 15 years defined more sharply inadequacies of purely reflexive and behavioristic theories of behavior, demonstrating that endogenous changes in motivation to perform certain activities and endogenous changes in responsiveness to different kinds of stimuli cannot be omitted from a behavioral theory if it is to have any general validity.

Some of his viewpoints as expressed in the popular book *On Aggression*, which suggests an endogenous motivation to seek out opportunities for fighting in fish, and perhaps in man as well, proved highly controversial. However one senses deeper roots to the outrage with which some react to analogies between animal and human behavior. In the introduction to the 1970 translations of his work, Lorenz reflects wryly

The fact that the behaviour not only of animals, but of human beings as well, is to a large extent determined by nervous mechanisms evolved in the phylogeny of the species, in other words, by "instinct", was certainly no surprise to any biologically-thinking scientist. It was treated as a matter of course, which, in fact, it is. On the other hand, by emphasizing it and by drawing the sociological and political inferences I seem to have incurred the fanatical hostility of all those doctrinaires whose ideology has tabooed the recognition of this fact. The idealistic and vitalistic philosophers to whom the belief in the absolute freedom of the human will makes the assumption of human instincts intolerable, as well as the behaviouristic psychologists who assert that all human behaviour is learned, all seem to be blaming me for holding opinions which in fact have been public property of biological science since *The Origin of Species* was written.

The young Niko Tinbergen, an avid naturalist from his boyhood in the sand dunes and pine forests of Hulshorst in Holland, saw the intricacies of insect behavior, specifically that of digger wasps hunting other insects and provisioning nest burrows with the corpses, as a testing ground for hypotheses about the sensory control of behavior. An opportunity while a graduate student in zoology at the University of Leiden to participate in 1931–32 in an Arctic expedition added snow buntings, phalaropes, and Eskimo sled dogs to a growing list of animals into whose behavior Tinbergen was to cast profound evolutionary insights. Returning to join the zoology faculty at Leiden, a seminal meeting with Lorenz in 1936, followed by a 6-month visit to Altenberg, gave rise to their only joint paper, in 1938, on the egg-rolling behavior of the greylag goose, and to more than 30 years of mutual cooperation, criticism, and stimulation that brought the new science of ethology into full flower. When he went to Oxford University in 1951, where he became a professor of animal behavior, he left the seeds of ethology firmly planted in Holland, and Groningen and Leiden continue as fertile research centers where aspiring students become versed in ethological discipline. One senses environmental imprinting in Tinbergen's choice of research sites while he was at Oxford. The sands of the Farne Islands, Scolt Head, and most recently Ravenglass are the Hulshorsts of England, permitting expansion of a research theme, already broached in Holland, on the social behavior of gulls, which led to fundamental insights into relationships between the behavior and ecology of animals.

From a theoretical framework established in his 1942 paper on "an objectivistic study of the innate behavior of animals," Tinbergen and his colleagues concentrated on the stimulus control of behavior. In both laboratory and field conditions, with butterflies, fish, and birds as subjects, he demonstrated that, by using inanimate models whose properties are systematically varied, experimental demonstration can replace intuitive judgment in deciding which elements of a stimulus complex control a response. New insights into how signaling behavior originates in the course of evolution were summarized, together with a general development of ethological theory in his 1951 book, *The Study of Instinct*, which introduced many English-speaking readers to the subject. Many patterns of social behavior, often with a signaling function, were understood as the outcome of social conflicts, a point of view that Tinbergen, with his wife Elizabeth, has since applied to the genesis of autistic behavior in children.

Perhaps most distinctive in the breadth of Tinbergen's research is his frontal attack in the 1950's and '60's on the problem of adaptiveness, which was for so long the subject of judgments from zoologists' armchairs. However, Tinbergen and his associates demonstrated that one can actually measure in animals preyed upon by others the cost or benefit of such traits as the color of a moth's wings or a bird's eggs, the spines of a three-spined stickleback, habits such as a gull's removal of egg shells from the nest after young have hatched, or living on the edge of a gull colony rather than in the center. The studies of gull behavior illustrate beautifully how an ecological decision made in phylogeny can reverberate through many aspects of the biology of a species. With von Frisch and Lorenz, Tinbergen has expressed the view that ethological demonstrations of the extraordinarily intricate interdependence of the structure and behavior of organisms are relevant to understanding the psychology of our own species. Indeed, this award might be taken not only as fitting recognition of the outstanding research accomplishments of these three zoologists, but also as an appreciation of the need to review the picture that we often seem to have of human behavior as something quite outside nature, hardly subject to the principles that mold the biology, adaptability, and survival of other organisms.

P. MARLER
D. R. GRIFFIN

Rockefeller University,
New York 10021

PART 2

Behavior Genetics

Behavior genetics is a relatively new branch of the science of animal behavior. As the name implies, it is concerned with the effects of genotype on behavior. But it should be noted at once that the behavior geneticist is not studying "instinct" as classically defined. Instead he or she is concerned with the role that genetic differences play in the determination of *behavioral differences* within a population. An example may clarify this distinction. Suppose that all females of a given bird species build identical nests, in identical ways, at a precise time of the year. This would suggest a challenging series of studies for those interested in the development of such an extremely stable behavioral sequence. But, since there are no behavioral differences to work with, the techniques of behavior genetics could not be used. On the other hand, if a variety of nests is built, in different ways, and at different times of the year, the raw material for behavior-genetic investigations exists. By appropriate breeding experiments, the behavior geneticist could study the effects of genotype on the behavioral differences in nest building.

The behavior geneticist, then, requires variation in the traits to be studied. Traits, whether morphological, physiological, or behavioral, vary within a population in two major ways. First, they may vary *qualitatively*, that is, in a discontinuous fashion. This means that the organisms can be placed in mutually exclusive categories on the basis of the trait. Examples are red and white flowers, straight and curled *Drosophila* wings, normal and fizzled chicken feathers. Such qualitatively varying traits were studied by Mendel as he established the principles of genetics.

Second, traits may be said to vary *quantitatively*, or continuously, over a broad range within a population. Height and weight offer good examples. Such traits usually can be shown to result from the interaction of many genes with each other and with the environment. A branch of genetics, known as quantitative genetics, studies such traits, primarily through the use of complex statistical procedures. Because most behavioral traits vary quantitatively rather than qualitatively, the techniques of quantitative genetics are used frequently in behavior genetics.

In most experiments it is possible to specify two classes of variables: independent and dependent. An independent variable is systematically changed by the experimenter; it is varied *independently* of the other variables in the experiment. (These "other variables," ideally, are controlled and held as constant as possible for all groups throughout the course of the experiment.) A dependent variable is that which is under investigation, that which is being measured. Most hypotheses state that changes in an independent variable will result in changes in a dependent variable. Thus most experiments can be entitled "The Effects of _____ (independent variable) on _____ (dependent variable)." In experiments in behavior genetics, genotype is usually the independent variable, and some measurable aspect of behavior is the dependent variable. Obviously, if genotype is to be used as an independent variable, the experimenter must be able to manipulate and vary it while holding other variables constant.

The two most frequently used methods of manipulating genotype are selective breeding and the use of inbred strains. In a typical selection experiment, males and females with high values of the trait in question are mated. At the same time, other males and females with low scores are mated. Genetic effects on behavior are demonstrated if discrete populations result from several generations of selective breeding.

A different experimental design is used if the population consists of inbred strains. Inbreeding is defined as the mating of close relatives, usually brothers and sisters. The objective of inbreeding is not to arrive at a particular value for a particular trait as in selective breeding, but rather to establish the highest possible degree of *homozygosity* in the population. Chromosomes usually occur in pairs, so that each gene has a "partner" at the corresponding locus of the homologous chromosome. Homozygosity simply means that the members of each pair of genes are identical; if Gene A occurs on one chromosome, its partner must also be Gene A. Inbreeding produces genetically similar animals, so that all same-sexed members of an inbred strain have the same, or nearly the same, genotype. The members of the strain may then be described as genetically "homogeneous."

The first reading in this section is concerned with memory processes in two different strains of inbred mice. Here Richard E. Wimer and his colleagues clearly demonstrate that conclusions from behavior-genetic experiments are specific to the populations studied and to the testing procedures used. These limitations to the behavior-genetic approach are not always recognized by those concerned with such items as racial differences in human potential.

Reading 7 involves the response of two outbred strains of housemice, and their reciprocal F_1 hybrids, to treatment with particular ovarian hormones. The paper demonstrates the first step in classical behavior-genetic analysis: comparison of parental strains with F_1's to determine mode of inheritance. But the work is more important for the mechanistic questions that it raises. What physiological differences underlie the genetically determined behavioral differences?

Reading 8 is an example of behavior-genetic research using naturally occurring wild populations, in this case crickets. Note that once again the results permit the formation of hypotheses regarding underlying physiological differences.

Behavior is a product of the evolutionary process. Furthermore, what an animal does, its behavior, can be critical in the determination of its survival to reproductive age. Therefore, behavior-genetic research is important in the study of evolution, and vice versa. This section closes with two readings concerned with this topic. In Reading 9, D. D. Thiessen, a psychologist, asks that "behavior genetics move toward a closer association with other biological disciplines and cast its experiments and interpretations within an evolutionary context." In Reading 10, Aubrey Manning, Secretary-General of the International Ethological Conference, considers how "behavior evolves in relation to the behavioral results of genetic changes."

6

Journal of Comparative and Physiological Psychology
1968, Vol. 65, No. 1, 126–131

DIFFERENCES IN MEMORY PROCESSES BETWEEN INBRED MOUSE STRAINS C57BL/6J AND DBA/2J[1]

RICHARD E. WIMER, LAWRENCE SYMINGTON, HERTHA FARMER, AND PHILIP SCHWARTZKROIN

The Jackson Laboratory, Bar Harbor, Maine

2 experiments on differences between strains of house mice in characteristics related to processes of memory trace formation are reported. In Experiment 1, performances on active shock-escape and passive shock-avoidance tasks were observed under massed and distributed practice conditions; on both tasks strain DBA/2J mice were superior when learning trials were massed, while strain C57BL/6J mice were superior when trials were distributed. In Experiment 2, immediate posttrial etherization facilitated learning of both tasks for strain DBA/2J mice, but had no effect on strain C57BL/6J.

Recent evidence indicates that substantial differences in processes of learning and memory may exist among diverse genetic stocks of animals within the same species. Specifically, what has been established clearly so far is that different animal stocks may vary in (*a*) optimal temporal distribution of practice trials for learning (McGaugh, Jennings, & Thomson, 1962) and in (*b*) how and when they are affected by posttrial treatment with agents (ECS or a drug) thought to affect memory (Breen & McGaugh, 1961; McGaugh, Thomson, Westbrook, & Hudspeth, 1962; McGaugh, Westbrook, & Thomson, 1962; Ross, 1964; Stratton & Petrinovich, 1963; Thomson, McGaugh, Smith, Hudspeth, & Westbrook, 1961). These studies raise many broad questions concerning (*a*) the scope of individual differences in optimal distribution of practice and in the processes of memory trace formation, (*b*) the possibility of relating these two phenomena, and (*c*) the nature of their biological bases.

The present experiments are devoted to the establishment of a genetically associated difference in optimal distribution of practice for two learning tasks for mice and the search for correlated differences in the consolidative process as indicated by differential effects of posttrial etherization upon memory.

METHOD

Subjects

The house mouse (*Mus musculus*) was selected over other species for study because of the greater number of genetically varying animal stocks this species affords. The Jackson Laboratory alone maintains 25 production and research mouse strains, 6 F_1 hybrids derived from them, and has over 100 mutant stocks available for special purposes (Green, 1964; Lane, 1966), and many more stocks are available elsewhere (Jay, 1963). On the basis of a preliminary survey of small numbers of animals from nine inbred mouse strains, DBA/2J and C57BL/6J were selected for thorough investigation. All mice used were males about 6 wk. old at time of testing.

Apparatus

The apparatus used was a modification of one devised by Maatsch (1959). A translucent box with a grid floor and a plastic shelf was constructed of ¼-in. sheet plastic, with outside dimensions of $7 \times 6\frac{3}{4} \times 6\frac{3}{4}$ in. The grid floor, mounted 1 in. above the base, consisted of $\frac{3}{32}$-in. stainless-steel rods separated by ¼ in. measured between centers. The box was covered with a removable plastic lid. Other pieces of equipment were a Grason-Stadler shock generator and a Meylan electric timer measuring in .01-min. intervals.

For the active task, a shelf 1½ in. wide running completely around the inside walls of the apparatus was mounted $2\frac{5}{8}$ in. above the grid floor.

[1] This investigation was supported in part by Public Health Service Research Grants MH 01775 and MH 11327, by Public Health Service Graduate Training Grant No. CRT 5013 from the National Cancer Institute, and by National Science Foundation Science Education Grants GE2888, GE040, and GE113. The invaluable assistance of William Mace is gratefully acknowledged. This investigation was reported in part at the 1966 meetings of the American Psychological Association, New York City.

For the passive task, a shelf 1 in. wide and 5⅝ in. long was inserted along one wall only, 1 in. above the floor; this shelf could be rotated about a pivot point at one end so that it lay virtually flat against one of the walls of the box when in a vertical position.

General Procedures

Active shock-escape learning. Learning trials were preceded by a familiarization period in the apparatus, during which S was placed for .50 min. on the shock-escape shelf followed by .50 min. on the grid floor. Immediately following familiarization, the first training trial was begun by activating the shock generator while S was on the grid and measuring latency until S escaped to the shelf. The shock generator was set initially to deliver .10 ma. at 340 v. ac. Whenever S froze or was relatively inactive, current level was momentarily increased by a factor of 10. If S failed to escape to the shelf by .50 min., the current was increased to .13 ma. This was followed by additional current level increases to .16 ma. and .20 ma. at successive .50-min. intervals if S was still on the grid. Animals remaining on the grid more than 2 min. on the first trial were discarded. The same shock level adjustment procedure was followed on the second and third trials, with the exception that maximum shock level was not raised above the maximum necessary to produce shock escape on the preceding trial. A final current level of .1 ma. on the fourth trial was reached with all but a few Ss. On all trials following the first, each S was placed directly on the grid floor and the shock was turned on with a delay period of less than .02 min. Training was continued for six trials.

The learning measure on each trial was time in .01-min. units to escape from grid floor to shelf. An S's learning score, which summarized performance on the critical learning trials and was amenable to simple statistical analysis and graphic presentation, was the sum of the learning measures for Trials 3, 4, 5, and 6. (Trial 2 was not included because of very high variability.)

Passive shock-avoidance learning. The S was placed on the narrow shock-avoidance platform and latency of descent to the grid floor, i.e., placing all four paws on the grid floor, was measured. When S descended, the latency timer was stopped and the shock-avoidance shelf was rotated into its vertical position. Shock was then delivered through the grid floor for .07 min. at an intensity of .10 ma. at 340 v. ac. Training was continued for six training trials, or until S remained on the shock-avoidance shelf for the full session length of 2.0 min.; Ss attaining criterion prior to the sixth trial were discontinued and given credit for 2.0 min. on remaining allotted trials. The learning score was the sum of the learning measures for Trials 2–6. (The Trial 2 score, not used for the active task, had to be used here because an occasional S reached criterion on this trial.)

EXPERIMENT 1

The purpose of this experiment was to establish a strain difference in optimal

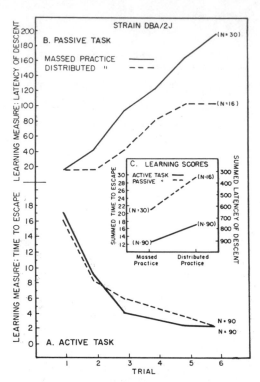

FIG. 1. Learning measures and learning scores for active and passive tasks under conditions of massed or distributed practice for strain DBA/2J.

temporal distribution of practice trials. Learning performances of the two strains were studied under highly massed and distributed conditions for both the active and the passive tasks.

Method

Under the active condition, 360 Ss of the strains DBA/2J and C57BL/6J were tested with an intertrial interval of 5–40 sec. (massed) or 24 hr. or more (distributed). For the passive task, 92 Ss were used. The intertrial interval was either approximately 2 min. or 24 hr. or more.

Results

The learning measures for each strain considered separately are presented in Figures 1 and 2, together with the corresponding learning scores (latencies combined across trials). Analyses of variance for the learning scores were carried out separately for the active and the passive task. For the active task, mice of strain DBA/2J performed better under the conditions of massed practice ($F = 16.0$,

$df = 1/178$, $p \leqslant .001$), while mice of strain C57BL/6J performed better under the distributed condition ($F = 8.8$, $df = 1/176$, $p \leqslant .01$). Results for the passive task were similar: strain DBA/2J performed better under the massed-practice condition ($F = 13.1$, $df = 1/44$, $p \leqslant .001$) while mice of strain C57BL/6J performed better under the distributed condition ($F = 16.7$, $df = 1/44$, $p \leqslant .001$). Note that scales for time to escape shock on the active task and time to descend from the platform on the passive task are plotted in opposite directions for graphs of learning scores, so that points for high levels of performance for both tasks are low on the ordinate and points for low levels of performance for both tasks are high. Thus graphs of learning scores of strains with similar optimal distribution of practice characteristics for active and passive tasks will show lines with similar slopes for the two tasks.

EXPERIMENT 2

Results of Experiment 1 indicate that there is a substantial strain difference in optimal distribution of practice for learning. Experiment 2 was devoted to a search for evidence that this difference in optimal distribution between strains DBA/2J and C57BL/6J is based upon a difference in some aspect of the processes of memory trace formation. One source of such evidence is the establishment of differences between them in direction, magnitude, and time course of effect on memory produced by agents which facilitate or impede the processes of memory trace formation. Post-trial etherization is one of several treatments which has been reported to produce forgetting in both rats (Pearlman, 1966; Pearlman, Sharpless, & Jarvik, 1961) and mice (Abt, Essman, & Jarvik, 1961; Essman & Jarvik, 1961). It was selected for use in the present study because of (a) established efficacy with mice, and because (b) duration of ether administration can be adjusted easily to assure approximately comparable levels of anesthesia for various strains.

Method

Mice of strains DBA/2J and C57BL/6J learned either the active or the passive task under one of three different ether treatments. The Ss of the immediate-ether groups were etherized immediately

following a trial; Ss in the delayed-ether groups were put in holding cartons for 2 or 4 hr. and then etherized; a third group of Ss received no ether. Etherization was achieved by placing S in a bell jar containing cotton saturated with ether and allowing it to remain there until righting response was lost and breathing became very deep and irregular (about .58 min.). Then S was removed and observed until the righting response returned.

For the active task, there was an interval of 24 hr. or more between trials, and the delayed-ether interval was always 2 hr. There were 20 Ss from each strain in each treatment group. For the passive task, there was some indication of either a longer time course of memory trace formation or a proactive effect of etherization; so replications of the experiment were carried out with an altered ether-delay interval of 4 hr. (R2), or with a changed intertrial interval of 72 hr. or more (R3), as well as with the 2-hr.-ether-delay and 24-hr.-intertrial interval (R1) used for the active task. About 40 Ss from each strain were used in each replication.

Results

Active task. Comparisons of the three ether treatments for strains DBA/2J and C57BL/6J on the active task are presented

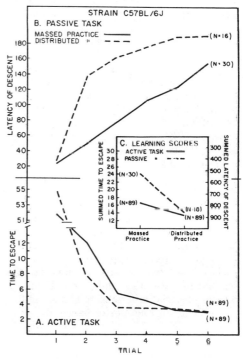

FIG. 2. Learning measures and learning scores for active and passive tasks under conditions of massed or distributed practice for strain C57BL/6J.

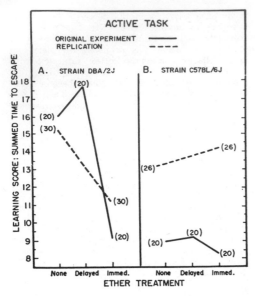

FIG. 3. Summed latency in hundredths of a minute for Trials 3–6 to escape from shock grid to platform. (Numbers in parentheses are sample sizes.)

in Figure 3. Analysis of variance for strain DBA/2J indicated a substantial treatment difference ($F = 4.8$, $df = 2/57$, $p \leqslant .05$) which subsequent t tests showed to be due to significantly lower learning scores of the group receiving immediate posttrial etherization compared to that of the no-ether and delayed-ether groups, which did not differ from each other. Analysis of variance of the learning scores for strain C57BL/6J indicated no difference among the treatment groups ($F = .25$, $df = 2/57$, $p > .05$).

Because several previous studies had shown decreased learning to result from posttrial etherization, it was decided that the study should be repeated. A different E, using only the immediate-etherization and no-ether procedures of the original experiment, tested approximately 60 mice each of the two strains. Results for this replication are also shown in Figure 3. Again, posttrial etherization had a facilitative effect on the performance of strain DBA/2J ($F = 7.2$, $df = 1/58$, $p \leqslant .01$), but no effect on the performance of strain C57BL/6J ($F = .26$, $df = 1/50$, $p > .05$).

Passive task. For the passive task, R1 in Figure 4 shows results for the same ether-delay time and the same intertrial

interval as that used for the active task. Analysis of variance of the R1 results for DBA/2J indicated a significant ether treatment effect ($F = 5.27$, $df = 2/33$, $p \leqslant .05$) which subsequent t tests showed to be due largely to superior performance of the immediate etherization group to that of the no-ether group ($p \leqslant .01$). However, the crucial delayed-etherization control group was intermediate in performance and not quite significantly different from either of the other two groups. Results for strain C57BL/6J indicated no significant ether treatment effects ($F = .38$, $df = 2/33$, $p > .05$). As in strain DBA/2J, however, there was an indication that delayed etherization might have a facilitative effect on retention. It was the possibility of longer time course of memory trace formation on a passive task which led to replication R2. The possibility that ether was having a general facilitative effect on the next day's performance was tested in replication R3. The results of the three experiments did not differ significantly for either strain, and were combined for analysis of variance. When this was done, it was clear that etherization treatments had substantial effects on the performance of Ss of strain DBA/2J ($F = 9.98$, $df = 2/105$, $p \leqslant .001$), but not on Ss of strain

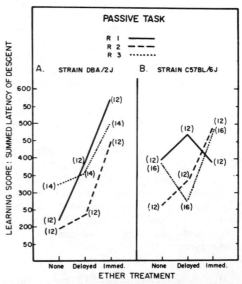

FIG. 4. Summed time to descend from shelf to grid floor in hundredths of a minute for Trials 2–6. (Numbers in parentheses are sample sizes.)

C57BL/6J ($F = 1.78$, $df = 2/111$, $p >$ .05). Subsequent t tests for strain DBA/2J show performance of the immediate-etherization group to be superior to that of both the delayed-etherization and no-ether groups, which did not differ from each other.

Contrary to expectation, then, the results indicate that immediate posttrial etherization has no effect on retention in strain C57BL/6J and actually facilitates retention in strain DBA/2J for both the active shock escape and the passive shock-avoidance tasks.

DISCUSSION

The objective of the present experiments was to establish genetically associated differences in learning performance for mice under different temporal distributions of practice trials, and to identify some possibly related difference in processes involved in memory trace formation.

The results of Experiment 1 clearly show that different inbred strains of house mice can vary substantially in optimal distribution of practice for learning. Indeed, such differences among mice may be common, for a survey of nine inbred strains resulted in identification of at least two which are quite different. Results for the active and passive tasks were consistent, indicating that differences in optimal distribution of practice may be broad trait-like characteristics involving many different tasks. However, there is already evidence that different strain characterizations for optimal distribution of practice can be obtained, possibly because of the relative contributions of such factors as fatigue, stress, and conflict. Thus, Bovet, Bovet-Nitti, and Oliverio (1966), studying shuttling with a temporally complex practice schedule consisting of varying intervals of time between successive long sets of massed trials, found DBA performance superior under their distributed condition. Unfortunately, detailed comparison of our study with that of Bovet et al. (1966) cannot be performed because of the complexity of their task and practice conditions.

In view of the literature on the effects of ether on memory showing increased forgetting or no effect (Bures & Buresova,

1963; Ransmeier & Gerard, 1962), the present finding of substantially improved retention in strain DBA/2J on both active and passive tasks came as a distinct surprise. Though it produces a reduction both in amplitude and in frequency of electrical activity of the nervous system, which some have thought likely to interfere with processes of memory trace formation (Pearlman et al., 1961), there are also possible mechanisms for a facilitative effect of etherization. Briefly, they are (a) increased ease of alteration in neuronal membrane structure, (b) rise in brain glucose, and (c) altered release of brain transmitter substances such as norepinephrine and acetylcholine (Wimer, 1968). It seems not unreasonable to us to assume that altered activity of transmitter substances is the initial process on which the enduring changes in form or function in nervous system which are learning are based, and that the third consequence of etherization is highly likely to be involved in its observed effects on memory.

Our observed strain difference in the ether facilitation effect, as well as the discrepancy between our findings and those of Es who have found that ether produces forgetting, are probably due at least in part to biological differences between the stocks of animals used. Thus, ether might have diverse effects upon different stocks depending upon, say, their varying characteristic levels of transmitter activity. There have been reports (e.g., Bennett, Crossland, Krech, & Rosenzweig, 1960; Kurokawa, Machiyama, & Kato, 1963; Roderick, 1960; Sudak & Maas, 1964) demonstrating differences between brains of different stocks in the availability of transmitter substances and related enzymes. Though this literature is still small, there is a strong impression that extensive biochemical diversity between various stocks of animals exists, and that many discrepancies between them, in result from various treatments, may be expected. Some have been noted or suggested for various stocks already (e.g., Prien, Wayner, & Kahan, 1963).

Since our findings involve only two inbred strains—which must be considered as two large sets of genetically identical individuals—some aspects of the results must be regarded with caution. Both the

81

consistency between results for the active and the passive task and the apparent relation between (a) strain characteristics of optimal distribution of practice trials and (b) effects of posttrial etherization on retention may be based on fortuitous association by fixation of genetically independently controlled systems (Falconer, 1960). Given that the two strains differ in two characteristics, a relation must be apparent, though the basic processes the phenomena represent may be unrelated save in the trivial sense that they exist within the same organisms.

REFERENCES

ABT, J. P., ESSMAN, W. B., & JARVIK, M. E. Ether-induced retrograde amnesia for one-trial conditioning in mice. *Science,* 1961, **133,** 1477–1478.

BENNETT, E. L., CROSSLAND, J., KRECH, D., & ROSENZWEIG, M. R. Strain differences in acetylcholine concentration in the brain of the rat. *Nature,* 1960, **187,** 787–788.

BOVET, D., BOVET-NITTI, F., & OLIVERIO, A. Short and long term memory in two inbred strains of mice. *Life Sci.,* 1966, **5,** 415–420.

BREEN, R. A., & McGAUGH, J. L. Facilitation of maze learning with posttrial injections of picrotoxin. *J. comp. physiol. Psychol.,* 1961, **54,** 498–501.

BURES, J., & BURESOVA, O. Cortical spreading depression as a memory disturbing factor. *J. comp. physiol. Psychol.,* 1963, **56,** 268–272.

ESSMAN, W. B., & JARVIK, M. E. Impairment of retention for a conditioned response by ether anesthesia in mice. *Psychopharmacologia,* 1961, **2,** 172–176.

FALCONER, D. S. *Introduction to quantitative genetics.* New York: Ronald Press, 1960.

GREEN, E. L. *Handbook of genetically standardized Jax mice.* The Jackson Laboratory, Bar Harbor, Maine, 1964.

JAY, G. E. Genetic strains and stocks. In W. J. Burdette (Ed.), *Methodology in mammalian genetics.* San Francisco: Holden-Day, 1963. Pp. 83–123.

KUROKAWA, M., MACHIYAMA, Y., & KATO, M. Distribution of acetylcholine in the brain during various states of activity. *J. Neurochem.,* 1963, **10,** 341–348.

LANE, P. *Lists of mutant genes and mutant-bearing stocks of the mouse.* Production Department, The Jackson Laboratory, Bar Harbor, Maine, June 1, 1966. (Mimeo)

MAATSCH, J. L. Learning and fixation after a single shock trial. *J. comp. physiol. Psychol.,* 1959, **52,** 408–410.

McGAUGH, J. L., JENNINGS, R. D., & THOMSON, C. W. Effect of distribution of practice on the maze learning of descendants of the Tryon maze bright and maze dull strains. *Psychol. Rep.,* 1962, **10,** 147–150.

McGAUGH, J. L., THOMSON, C. W., WESTBROOK, W. H., & HUDSPETH, W. J. A further study of learning facilitation with strychnine sulphate. *Psychopharmacologia,* 1962, **3,** 352–360.

McGAUGH, J. L., WESTBROOK, W. H., & THOMSON, C. W. Facilitation of maze learning with posttrial injections of 5-7-diphenyl-1-3-diazadaman-tan-6-ol (1757 I.S.). *J. comp. physiol. Psychol.,* 1962, **55,** 710–713.

PEARLMAN, C. A., JR. Similar retrograde amnesic effects of ether and spreading cortical depression. *J. comp. physiol. Psychol.,* 1966, **61,** 306–308.

PEARLMAN, C. A., SHARPLESS, S. K., & JARVIK, M. E. Retrograde amnesia produced by anesthetic and convulsant agents. *J. comp. physiol. Psychol.,* 1961, **54,** 109–112.

PRIEN, R. F., WAYNER, M. J., JR., & KAHAN, S. Lack of facilitation in maze learning by picrotoxin and strychnine sulphate. *Amer. J. Physiol.,* 1963, **204,** 488–492.

RANSMEIER, R. E., & GERARD, R. W. Effects of temperature and metabolic factors on rodent memory and EEG. *Amer. J. Physiol.,* 1962, **203,** 782–788.

RODERICK, T. H. Selection for cholinesterase activity in the cerebral cortex of the rat. *Genetics,* 1960, **45,** 1123–1140.

ROSS, R. B. Effects of strychnine sulphate on maze learning in rats. *Nature* (London), 1964, **201,** 109–110.

STRATTON, L. O., & PETRINOVICH, L. Post-trial injections of an anti-cholinesterase drug and maze learning in two strains of rats. *Psychopharmacologia,* 1963, **5,** 47–54.

SUDAK, H. S., & MAAS, J. W. Central nervous system serotonin and norepinephrine localization in emotional and non-emotional strains of mice. *Nature,* 1964, **203,** 1254–1256.

THOMSON, C. W., McGAUGH, J. L., SMITH, C. E., HUDSPETH, W. J., & WESTBROOK, W. H. Strain differences in the retroactive effects of electroconvulsive shock on maze learning. *Canad. J. Psychol.,* 1961, **15,** 69–74.

WIMER, R. Bases of a facilitative effect upon retention resulting from posttrial etherization. *J. comp. physiol. Psychol.,* 1968, **65,** in press.

(Received January 30, 1967)

Effects of Genotype on Differential Behavioral Responsiveness to Progesterone and 5α-Dihydroprogesterone in Mice

Boris B. Gorzalka[1,2] **and Richard E. Whalen**[1,3]

Received 31 July 1974—Final 9 Jan. 1975

Thirty female CD-1 mice, 30 female Swiss-Webster mice, 45 hybrid female mice of the strain $SWCD1F_1$, and 45 hybrid female mice of the strain $CD1SWF_1$ were ovariectomized and administered estradiol benzoate once weekly for 6 weeks. Estrogen injections were followed 2 days later by injections of progesterone, dihydroprogesterone (DHP), or oil and the animals were tested for receptivity 7 hr later. Over the six tests, there was a progressive increase in the frequency of lordosis responses in all strains following progesterone treatment. However, lordosis scores varied widely across animals within strains. Following DHP treatment, lordosis frequency was not increased in the Swiss-Webster strain. Females in the other strains did show a progressive increase in lordosis frequency over weeks. The data indicate that the hybrid strains develop the potential to respond to DHP and thus behave like the CD-1 strain, suggesting that sensitivity to DHP is a dominant trait.

KEY WORDS: lordosis; progesterone; dihydroprogesterone; strain differences in hormone sensitivity; progestins and mating.

INTRODUCTION

It has been well documented that genetic differences in male sexual behavior exist within species (Valenstein *et al.*, 1955; Whalen, 1961; McGill and Blight, 1963; Vale and Ray, 1972). However, the physiological differences which underlie the observed differences in sexual behavior remain unknown. Apparently differences in output of testicular hormones are not responsible for much of the variance in male sexual performance (Valenstein *et al.*, 1955; Champlin *et al.*, 1963). While little is known about the mechanisms underlying genetic differences in male sexual behavior, even

Supported by Grant HD-00893 to R. E. W. from the National Institute of Child Health and Human Development.

[1] Department of Psychobiology, University of California at Irvine, Irvine, California.
[2] Present address: Department of Psychology, University of Western Ontario, London, Ontario, Canada.
[3] Requests for reprints should be sent to Richard E. Whalen, Department of Psychobiology, University of California, Irvine, California 92664.

less is known about the regulation of female sexual behavior. The present study, therefore, sought information on hormone–genotype interactions in the control of female sexual behavior.

Female sexual receptivity in rodents can be abolished by ovariectomy and can be reinstated by exogenous estrogen and progesterone treatment (Young, 1961). Estrogen alone restores minimal levels of receptivity but interacts synergistically with progesterone to restore full receptivity. 5α-Dihydroprogesterone (DHP), the 5α-reduced metabolite of progesterone, is almost as effective as progesterone in facilitating sexual receptivity in estrogen-primed rats (Whalen and Gorzalka, 1972). In mice, however, the facilitatory effects of DHP vary with the genotype. DHP is almost as effective as progesterone in CD-1 mice but is ineffective in facilitating sexual receptivity in Swiss-Webster mice (Gorzalka and Whalen, 1974). The sensitivity of the CD-1 mouse to DHP and the insensitivity of the Swiss-Webster mouse to DHP hold for a variety of DHP doses and testing times (Gorzalka, 1974). As the first step in a genetic analysis, the present report examines the effects of reciprocal cross-breeding in mice on responsiveness to progesterone and DHP. The availability of hormone-sensitive and hormone-insensitive strains obtained through genetic analyses may further an understanding of the genetic mechanisms by which hormones regulate behavior.

METHODS

Subjects were 30 female CD-1 mice, 30 female Swiss-Webster mice, 45 hybrid female mice of the strain SWCD1F$_1$ (Swiss-Webster female × CD-1 male), and 45 hybrid female mice of the strain CD1SWF$_1$ (CD-1 female × Swiss-Webster male). Following weaning, animals were housed in individual cages and were maintained on *ad libitum* food and water. Females were ovariectomized at 60 days of age. Weekly hormone administration commenced 2 weeks following surgery. All hormones were given subcutaneously in 0.1 ml peanut oil.

For 6 consecutive weeks, females of each strain received weekly injections of 10 μg estradiol benzoate. All estrogen injections were followed 2 days later by 500 μg progesterone, 500 μg DHP, or the oil vehicle. Animals were tested weekly for sexual receptivity 7 hr after administration of progesterone, DHP, or the oil vehicle.

Behavioral testing began several hours after the onset of the dark period of a reversed lighting cycle (12 hr light, 12 hr dark). For mating, females were introduced to the home cages of sexually experienced, vigorous males of the same strain. Receptivity scores were calculated by determining the ratios of female lordotic responses to mounts with pelvic thrusting by the male and were expressed as a lordosis quotient (LQ). A test was continued until the male had mounted the experimental female ten times.

RESULTS

As shown in Fig. 1 and Table I, there was a progressive increase in lordosis responses in all strains following progesterone treatment. By the

Table I. Mean Lordosis Quotients (Lordosis Responses/Mounts with Pelvic Thrusting × 100) and Measures of Variability of Mice Treated and Tested Weekly

	Week	Estrogen plus progesterone				Estrogen plus DHP				Estrogen plus oil			
		Mean	Range	SD	SE	Mean	Range	SD	SE	Mean	Range	SD	SE
CD-1	1	15.0	0–40	17.2	5.7	11.0	0–40	15.2	5.1	9.0	0–50	16.0	5.3
	2	36.0	0–80	25.9	8.6	23.0	0–40	17.0	5.7	12.0	0–50	19.9	6.6
	3	45.0	10–80	23.2	7.7	30.0	0–70	25.8	8.6	12.0	0–60	23.0	7.7
	4	51.0	0–80	27.3	9.1	31.0	0–70	21.8	7.3	15.0	0–70	25.1	8.4
	5	55.0	10–80	22.7	7.6	44.0	10–80	25.5	8.5	13.0	0–60	22.6	7.5
	6	62.0	20–80	22.5	7.5	54.0	20–90	25.5	8.5	17.0	0–50	18.3	6.1
S^F	1	8.0	0–30	10.3	3.4	8.0	0–40	13.2	4.4	7.0	0–40	13.4	4.5
	2	13.0	0–30	11.6	3.9	15.0	0–60	19.0	6.3	7.0	0–30	12.5	4.2
	3	19.0	0–40	11.0	3.7	13.0	0–60	19.5	6.5	9.0	0–50	17.3	5.8
	4	37.0	0–60	18.9	6.3	16.0	0–60	20.1	6.7	14.0	0–60	19.0	6.3
	5	53.0	0–80	25.4	8.5	18.0	0–40	16.2	5.4	15.0	0–60	18.4	6.1
	6	55.0	0–80	24.6	8.2	18.0	0–60	19.3	6.4	12.0	0–70	21.5	7.2
S^{W}CD1F$_1$	1	11.3	0–30	8.4	2.2	8.7	0–20	9.1	2.4	6.7	0–20	7.2	1.9
	2	22.0	0–40	13.7	3.7	10.7	0–20	7.9	2.1	7.3	0–20	8.0	2.1
	3	29.3	10–50	11.7	3.1	21.3	0–40	15.1	4.0	9.3	0–40	12.3	3.3
	4	44.0	10–70	18.0	4.8	36.0	10–70	24.4	6.5	11.3	0–50	15.1	4.0
	5	60.0	20–80	19.3	5.2	48.0	10–80	26.5	7.1	15.3	0–50	13.3	3.6
	6	63.3	30–80	16.9	4.5	52.7	10–80	29.8	8.0	14.0	0–40	11.8	3.2
CD1SWF$_1$	1	8.7	0–30	9.1	2.4	7.3	0–20	6.0	1.6	7.1	0–20	7.1	1.9
	2	12.7	0–30	11.6	3.1	12.0	0–20	6.8	1.8	7.3	0–20	8.0	2.1
	3	20.7	0–40	12.7	3.4	16.0	0–40	13.5	3.6	8.7	0–40	11.8	3.2
	4	40.0	0–80	23.3	6.2	30.7	0–70	20.8	5.6	14.7	0–50	17.2	4.6
	5	52.0	10–90	22.4	6.0	39.3	10–80	25.5	6.8	14.0	0–50	17.2	4.6
	6	63.3	20–90	21.0	5.6	48.7	20–90	25.2	6.7	14.7	0–40	15.5	4.1

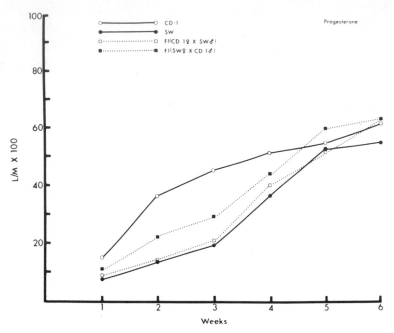

Fig. 1. Effects of progesterone on sexual receptivity (lordosis responses/mounts with pelvic thrusting ×100) in estrogen-primed, ovariectomized Swiss-Webster (SW), CD-1, SWCD1F₁, and CD1SWF₁ mice.

sixth week of testing, the mean LQ was 60.9 and all strains were exhibiting similar levels of sexual receptivity. An analysis of variance for the Swiss-Webster and CD-1 parent strains indicated no significant strain differences, a significant increase over weeks (F = 43.20, df = 5, 90, p < 0.001), and no significant interaction. Analysis of variance for the SWCD1F₁ and CD1SWF₁ hybrid strains indicated no significant strain differences, a significant increase over weeks (F = 90.86, df = 5, 140, p < 0.001), and no significant interaction. Furthermore, there were no significant differences in sexual receptivity between the four strains during the sixth week of testing.

The frequency distributions of lordosis scores following progesterone treatment are shown in Fig. 2. For all strains, there appeared to be a wide distribution of lordosis scores. Mean LQs for all weekly tests ranged from 0 to 60 in CD-1 mice, 0 to 50 in Swiss-Webster mice, and 10 to 60 in the hybrid mice. LQs during the sixth week of testing ranged from 20 to 90 in CD-1 mice, 0 to 80 in Swiss-Webster mice, 30 to 80 in SWCD1F₁ mice, and 20 to 90 in CD1SWF₁ mice.

As shown in Fig. 3, there was a progressive increase in lordosis responses in the CD-1, SWCD1F₁, and CD1SWF₁ strains following DHP treatment. By the sixth week of testing, these three strains were exhibiting similar levels of sexual receptivity (mean LQ 52.0). However, DHP was relatively ineffective in facilitating sexual receptivity in Swiss-Webster mice, and by the sixth week of testing the LQ of this group was 16.0. Levels of sexual receptivity in estrogen-primed Swiss-Webster mice receiving DHP

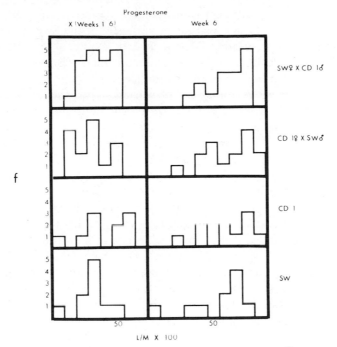

Fig. 2. Frequency distribution (*f*) of lordosis scores during weeks 1–6 ($\bar{X}$) and during week 6 following progesterone treatment in estrogen-primed, ovariectomized Swiss-Webster (SW), CD-1, SWCD1F₁, and CD1SWF₁ mice.

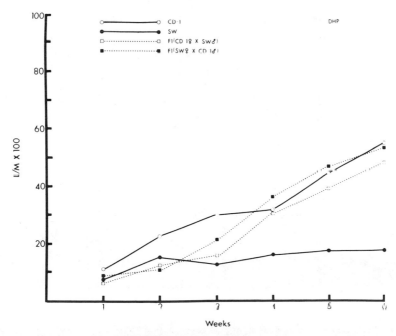

Fig. 3. Effects of 5α-dihydroprogesterone (DHP) on sexual receptivity (lordosis responses/mounts with pelvic thrusting × 100) in estrogen-primed, ovariectomized Swiss-Webster (SW), CD-1, SWCD1F₁, and CD1SWF₁ mice.

were not significantly different from levels of receptivity in mice receiving only estrogen (Table I). An analysis of variance for the Swiss-Webster and CD-1 parent strains receiving DHP indicated significant strain differences ($F = 4.75$, df = 1, 18, $p < 0.05$), significant increases over weeks ($F = 16.53$, df = 5, 90, $p < 0.001$), and a significant interaction ($F = 6.89$, df = 5, 90, $p < 0.001$). An analysis of variance for the SWCD1F$_1$ and CD1SWF$_1$ hybrid strains receiving DHP indicated no significant strain differences, significant increases over weeks ($F = 55.59$, df = 5, 140, $p < 0.001$), and no significant interaction. Furthermore, there were significant differences in sexual receptivity among the four strains during the sixth week of testing ($F = 4.59$, df = 3, 46, $p < 0.01$). A Newman–Keuls analysis revealed that the Swiss-Webster strain scored significantly lower than the three other strains ($p < 0.01$). However, the CD-1, SWCD1F$_1$, and CD1SWF$_1$ strains were not significantly different from each other.

The frequency distributions of lordosis scores following DHP treatment are shown in Fig. 4. For the CD-1, SWCD1F$_1$, and CD1SWF$_1$ strains, there appeared to be a wide distribution of lordosis scores representing at least two populations for each strain. However, for the Swiss-Webster strain, there was a relatively narrow distribution of lordosis scores skewed toward minimal levels of sexual receptivity. Mean LQs for all weekly tests ranged from 0 to 50 in CD-1 mice, 0 to 30 in Swiss-Webster mice, and 0 to 50 in the hybrid mice. LQs during the sixth week of

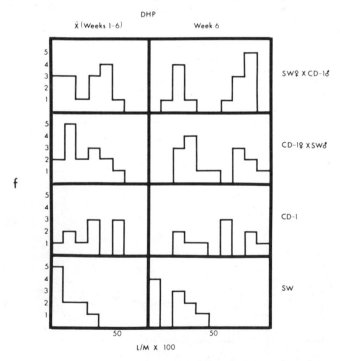

Fig. 4. Frequency distribution (f) of lordosis scores during weeks 1–6 ($\bar{X}$) and during week 6 following 5α-dihydroprogesterone (DHP) treatment in estrogen-primed, ovariectomized Swiss-Webster (SW), CD-1, SWCD1F$_1$, and CD1SWF$_1$ mice.

testing ranged from 20 to 90 in CD-1 mice, 0 to 40 in Swiss-Webster mice, 10 to 80 in SWCD1F₁ mice, and 20 to 90 in CD1SWF₁ mice.

Table I suggests that there was a slight, progressive increase in lordosis responses following treatment with only estrogen. The strains did not appear to vary in this regard. By the sixth week of testing, the mean LQ for all strains was 14.4. An analysis of variance for the Swiss-Webster and CD-1 parent strains receiving estrogen indicated no significant strain differences, significant increases over weeks ($F = 2.52$, df $= 5, 90$, $p < 0.05$), and no significant interaction. An analysis of variance for the SWCD1F₁ and CD1SWF₁ hybrid strains receiving estrogen indicated no significant strain differences, significant increases over weeks ($F = 4.92$, df $= 5, 140$, $p < 0.001$), and no significant interaction. Furthermore, there were no significant differences in sexual receptivity among the four strains during the sixth week of testing.

DISCUSSION

The present results confirm that the CD-1 mouse is behaviorally responsive to progesterone and DHP while the Swiss-Webster mouse is responsive to progesterone but not to DHP (Gorzalka and Whalen, 1974). It now appears that hybrids of these two strains are phenotypically similar to CD-1 mice. In their behavioral responses to estrogen, estrogen plus progesterone, and estrogen plus DHP, SWCD1F₁ and CD1SWF₁ mice were indistinguishable from CD-1 mice. The similarity of the SWCD1F₁ and CD1SWF₁ strains would suggest that maternal and paternal effects do not contribute appreciably to DHP sensitivity. The similarity of the CD-1 and hybrid strains would suggest that sensitivity to DHP is a dominant trait. Perhaps the ability to respond to more than one progestin represents an advantageous adaptation.

The distribution of lordosis scores for animals receiving DHP during week 6 (Fig. 4) suggests the possibility of more than one population of scores. That is, females of the CD-1, SWCD1F₁, and CD1SWF₁ strains can be subdivided into populations which are relatively insensitive to DHP and populations which are relatively sensitive to DHP. All Swiss-Webster females, however, fall within a population which is relatively insensitive to DHP.

These strain differences in sensitivity to DHP may provide a useful means for analyzing the physiological mechanisms through which progestins regulate behavior. For example, since a fundamental action of hormones in many tissues may be to produce specific proteins (Villee, 1971), the DHP-sensitive and DHP-insensitive strains may differ in their potential to induce DHP-specific proteins in the brain. The many weeks of progestin stimulation required to elicit maximal receptivity in mice are suggestive of an induction process in which the progestin stimulates the production of a specific protein in the brain much as progesterone stimulates the synthesis of the protein avidin in the chick oviduct (O'Malley, 1967). The insensitivity of the Swiss-Webster mouse to DHP may therefore reflect a failure of DHP to induce proteins in this strain.

If progestins indeed produce specific neural proteins in the mouse, these could be brain receptor proteins specific for the hormone. Evidence that the CD-1 mouse, but not the Swiss-Webster mouse, shows an increase in brain concentration of radioactivity following weekly DHP priming and acute DHP-H³ administration is consistent with this hypothesis (Gorzalka, 1974). However, it is conceivable that the correlation between behavioral responsiveness to DHP and brain retention of radioactivity is fortuitous in the parent strains. The SWCD1F₁ and CD1SWF₁ hybrid strains, which are behaviorally similar to the CD-1 strain only in responsiveness to DHP yet are similar to both parent strains in genetic constitution, provide a means for further testing the relationship between progestin-induced behavior and brain retention of progestins. Ultimately, an analysis of the nature of DHP-induced proteins may contribute to a general understanding of neural mechanisms of progestin action.

REFERENCES

Champlin, A. K., Blight, W. C., and McGill, T. E. (1963). The effects of varying levels of testosterone on the sexual behaviour of the male mouse. *Anim. Behav.* 11:244–245.

Gorzalka, B. B. (1974). Neural mechanisms of progesterone action. Unpublished doctoral dissertation, University of California, Irvine.

Gorzalka, B. B., and Whalen, R. E. (1974). Genetic regulation of hormone action: Selective effects of progesterone and dihydroprogesterone on sexual receptivity in mice. *Steroids* 23:499–505.

McGill, T. E., and Blight, W. C. (1963). The sexual behaviour of hybrid male mice compared with the sexual behaviour of males of the inbred parent strains. *Anim. Behav.* 11:480–483.

O'Malley, B. W. (1967). *In vitro* hormonal induction of a specific protein (avidin) in chick oviduct. *Biochemistry* 6:2546–2551.

Vale, J. R., and Ray, D. (1972). A diallel analysis of male mouse sex behavior. *Behav. Genet.* 2:199–209.

Valenstein, E. S., Riss, W., and Young, W. C. (1955). Experiential and genetic factors in the organization of sexual behaviour in male guinea pigs. *J. Comp. Physiol. Psychol.* 48:397–403.

Villee, C. A. (1971). Effects of sex hormones on the genetic mechanism. In McKearns, K. W. (ed.), *The Sex Steroids,* Appleton-Century-Crofts, New York, pp. 273–294.

Whalen, R. E. (1961). Strain differences in sexual behaviour of the male rat. *Behaviour* 18:199–204.

Whalen, R. E., and Gorzalka, B. B. (1972). The effects of progesterone and its metabolites on the induction of sexual receptivity in rats. *Horm. Behav.* 3:221–226.

Young, W. C. (1961). The hormones and mating behaviour. In Young, W. C. (ed.), *Sex and Internal Secretions,* Williams and Wilkins, Baltimore, pp. 1173–1240.

8

Genetic Control of Song Specificity in Crickets

Ronald R. Hoy and Robert C. Paul

Reprinted from
6 April 1973, Volume 180, pp. 82-83

Abstract. *The calling song of male field crickets is composed of stereo-typed rhythmic pulse intervals, which are predictable expressions of genotype. Females identify conspecific males by their song. Two species of crickets were found to exhibit species-specific song preference, and hybrids between them preferred hybrid calls over either parental call. These results imply genetic control of song reception as well as transmission.*

Acoustical communication in field crickets is extremely important to these nocturnally active species. The stereotyped calling song of adult male field crickets attracts females (which do not sing) for mating (*1, 2*). In previous experiments, measurements were made of the ability of a female cricket to orient toward the sound by means of tympana on her forelegs and to walk to it. We were able to quantify the phonotaxis by measuring the "phonomotor response," a cricket's turning tendency while it is walking in place on a Y-maze in a sound field. Our experimental paradigm is the optomotor studies (*3, 4*).

Teleogryllus commodus and *T. oceanicus* are northern and southern Australian field crickets, respectively, which overlap geographically in southern Queensland. These crickets differ markedly in rhythmic structure of their calling songs. Exhaustive measurements of rhythmic pulse intervals show that these two species produce viable hybrids whose calling song is distinctly different from either parental song. Most rhythmic elements of hybrid song are intermediate between parental ones; this suggests polygenic control (*5, 6*). We showed that genetic differences that

cause song changes in males also apparently alter responsiveness to song in females.

We measured locomotor response to sound by placing tethered females upon a Y-maze globe, which they suspended in midair (see cover). This styrofoam maze consisted of three straight run paths of 10.5 mm interconnected by two Y choice points. Recordings of cricket song were played from symmetrically placed high-fidelity speakers (Daltronix) on a female's left and right; each speaker was situated 94 cm from the maze and deviated from her longitudinal body axis by 40°. Although her position in space with respect to each sound source was fixed, she could walk freely on the maze (which rotated in space beneath), and when she inevitably came to a choice point she had to choose the right or left arm of the maze. Her behavior at choice points with respect to the sound source was measured by her turning tendency, which we termed r and defined as the number of turns toward the sound source divided by the total number of turns.

Sound stimuli were provided by playing tape loops of recorded calls. In a typical experiment sound was played through either the right or left speaker for 20 choices, whereupon the origin of the sound was immediately switched to the opposite speaker for another 20 choices. A Uher 4400 tape recorder was used both to record and play back calls.

Audiospectrograms of the tape-recorded calls were made on a Kay 7030 sonograph, and the frequency spectra of the calls agreed with data of LeRoy (6) for *T. commodus* and *T. oceanicus*. The intensity of the calls measured at

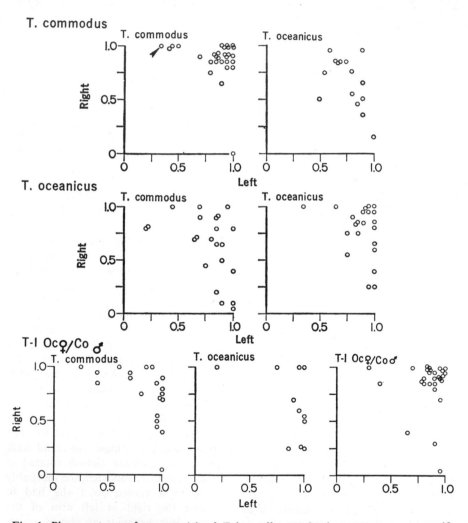

Fig. 1. Phonomotor performance (r) of *Teleogryllus* species in response to conspecific and heterospecific calling songs. Top scattergrams refer to the performance of *T. commodus* when played recorded songs of *T. commodus* and *T. oceanicus*. Middle scattergrams show performance of *T. oceanicus*. The bottom set refers to performance of hybrid females (formed from a cross of *T. oceanicus* females and *T. commodus* males) when played recorded calling songs of both parental species as well as that of sibling male hybrids. Each scattergram summarizes phonomotor performance as a function of sound from the cricket's right and left. Each point represents a different animal. The r value for left-directed sound can be read from the horizontal axis, and that for right-directed sound can be read from the vertical axis. This graphic representation does not convey a functional relationship between r values for right and left, which vary independently of each other.

the maze was 69 db relative to 2×10^{-4} μbar. The 147 adult, virgin females used in these experiments varied in adult age from 1 to 6 weeks; older females were less likely to run on a maze. Hybrid crickets (T-1) resulted from crossing virgin *T. oceanicus* females with *T. commodus* males; several hundred offspring grew to adulthood, and 60, isolated at the final instar, were studied. All experiments were performed in an anechoic chamber [7] at $24° \pm 2°C$. We recorded calling songs at several different temperatures so that we could match recording and playback temperatures within 2°C. All animals supported the maze when it was presented, but they usually would not walk unless stimulated acoustically.

Each female was required to make 40 choices, 20 with sound played from the right speaker and 20 from the left speaker. An *r* value (turns toward the sound divided by 20) was obtained for each speaker. Values of *r* could be between 0 and 1.0; $r = 0$ if no turns were toward the sound, $r = 1.0$ if all were toward the sound, and $r = 0.5$ if there was no preference. The *r* values for right and left sound directions are displayed conjointly in each scattergram of Fig. 1. Each point represents the performance of a different, individual female for 20 choices in each direction. The relative clustering or scattering of data points can be seen.

We arbitrarily defined a strong phonomotor response as one having values of 0.75 or greater (15 of 20 turns directed toward sound) for each of the sound directions. By this standard, maze behavior can be explained in terms of species-specific song preference (Table 1). This is also seen when the clustering of data points for homospecific song is compared to the relative scatter of points for heterospecific song. Statistical significance was demonstrated by applying the *G*-test of independence. Preference for homospecific song is significant at the 1 percent confidence level

in all three groups. Thus, in the phonomotor experiment, as in previous phonotaxis experiments, female crickets are attracted to homospecific calling song. Also, hybrid females show an equally strong preference for the calling song of hybrid males over either parental call. To our knowledge this is the first such demonstration in field crickets [8].

The arrow in Fig. 1 shows that some animals turned consistently in one direction regardless of the direction of sound. Thus one *T. commodus* animal turned to the right in all choices with sound from the right, and when sound was switched to the left the animal continued to turn right in 70 percent of choices. We have evidence [9] that such animals are "right-biased" (or "left-biased") and turn right (or left) when presented nondirectional sound; such locomotor bias was previously observed in an optomotor experiment [4]. To exclude biased animals from our sample, we chose the symmetrical criterion of *r* values of 0.75 or greater for both right and left sound directions (Table 1).

Female hybrids can not only identify the song of the sibling hybrid males, but find it more attractive than parental

Table 1. Comparison of phonomotor response of 147 female crickets to conspecific and heterospecific calling songs. The criterion performance was $r \geq 0.75$ both for sound played from the right and for sound from the left (20 choices for direction).

Calling song	Females at criterion/ total tested	Females at criterion (%)
T. oceanicus on maze		
T. oceanicus	14/22	63.6
T. commodus	4/22	18
T. commodus on maze		
T. oceanicus	3/15	20
T. commodus	21/28	75
Hybrid (T-1) on maze		
T. oceanicus	3/11	27.3
T. commodus	8/21	38
Hybrid (T-1)	21/28	75

song. This has interesting physiological implications. Perhaps female crickets carry genetically conferred sensory templates (10) of the species calling song. Genetic hybridizations would result in hybrid templates that are revealed only in song discriminations, since females do not announce their genetic identity as do calling males. What are some possible features of this template? It must contain information about the rhythmic elements of the song, since frequency modulation does not occur in cricket songs (11), although sensory elements may be tuned to the carrier frequency of their species song (12). This raises the question of the nature of the "species filter," particularly at the level of neurophysiological mechanisms. Are such filters single cells or networks of several cells? The capacity of the nervous system to abstract species-specific features of acoustical calling signals has been demonstrated in the bullfrog, and the existence of mating call detectors has been proposed (13). Although it is premature to speculate on the neuronal loci under genetic control in crickets, coupling of the male's song generator and a female's species-specific sensory template through a common set of genes presents an attractive mechanism for speciation in crickets and is relevant to the evolution of communication in these species (11).

We believe that the phonomotor technique is a valid measure of a female's normal acoustical response, that is, of phonotaxis. The two techniques are not exactly comparable because a tethered cricket on the maze maintains a fixed angle with respect to a constant-intensity sound source, whereas the free-walking animal in a phonotaxis experiment, once she localizes the sound source, usually turns to directly face the source and runs straight toward the increasingly louder signal (2, 9). The phonomotor method allows us to measure the relative attraction of one song compared to another. If the species-specific song is not presented, however, the cricket often will be attracted to heterospecific songs; they do not simply ignore or avoid songs of other species, and the same is true of free-walking animals (9). This technique is not limited to behavioral measurements. It should be feasible to record from neurons during the maze behavior. Neural units that code information about rhythmic elements of song have been described (14), and they have been monitored during apparently normal behavior of crickets (15). Coupled with further genetic experiments and neurophysiological analysis, such efforts should facilitate our understanding of the genetic control of behavior.

RONALD R. HOY
ROBERT C. PAUL

*Department of Biology,
State University of New York,
Stony Brook 11790*

References and Notes

1. R. D. Alexander, in *Animal Sounds and Communication*, W. E. Lanyon and W. N. Tavolga, Eds. (American Institute of Biological Sciences, Washington, D.C., 1960), pp. 38–92; J. Regen, *Pflüger's Archiv. Gesamte Physiol. Menchen Tiere* **155**, 193 (1913).
2. R. K. Murphey and M. D. Zaretsky, *J. Exp. Biol.* **56**, 335 (1972).
3. W. Reichardt, in *Sensory Communication*, W. A. Rosenblith, Ed. (M.I.T. Press, Cambridge, Mass., 1961), pp. 303–319.
4. D. M. Wilson and R. R. Hoy, *Z. Vergl. Physiol.* **58**, 136 (1968).
5. R. R. Hoy and D. R. Bentley, *Amer. Zool.* **10**, 482 (1970); D. R. Bentley and R. R. Hoy, *Anim. Behav.*, in press; D. R. Bentley, *Science* **174**, 1139 (1971).
6. Y. LeRoy, *C. R. Acad. Sci.* **259**, 892 (1964).
7. Model 402, Industrial Acoustics Co., New York, N.Y.
8. A. C. Perdeck [*Behavior* **12**, 1 (1958)] presents that hybrid grasshoppers are attracted to hybrid songs.
9. R. R. Hoy and R. C. Paul, unpublished results.
10. M. Konishi, *Z. Tierpsychol.* **20**, 349 (1963).
11. R. D. Alexander, *Evolution* **16**, 443 (1962).
12. J. J. Loftus-Hills, M. J. Littlejohn, K. G. Hill, *Nature* **233**, 184 (1971).

13. R. R. Capranica, *The Evoked Vocal Response of the Bullfrog* (M.I.T. Press, Cambridge, Mass., 1965); L. S. Frishkopf, R. R. Capranica, M. H. Goldstein, *Proc. IEEE* **56**, 969 (1968).
14. M. D. Zaretsky, *Nature* **229**, 195 (1971); J. F. Stout, *Amer. Zool.* **10**, 502 (1970).
15. J. F. Stout, *Z. Vergl. Physiol.* **74**, 26 (1971).
16. We thank C. Walcott for anechoic facilities; D. R. Bentley for crickets; M. D. Zaretsky for his manuscript (2); R. Radermann for preliminary maze studies; G. Loustalet for illustrations; and C. Walcott, A. Carlson, D. Smith, and M. Mendelson for reading the manuscript. Supported by an NSF center of excellence award and NIH grant 1 RO1 NS10176-01.

12 May 1972; revised 31 October 1972

9

Behavior Genetics, Vol. 2, No. 2/3, 1972

A Move Toward Species–Specific Analyses in Behavior Genetics[1]

D. D. Thiessen[2]

Received 3 May 1971

It is essential that behavior genetics move toward a closer association with other biological disciplines and cast its experiments and interpretations within an evolutionary context. In my opinion, behavior genetics has been too preoccupied with the extent of genetic variability and may, in many cases of high heritability, be dealing with genetic junk. The species as a unit of behavioral response, and as the outcome of genetic polishing, deserves more consideration. Adaptation is always the crux of natural selection and offers the best hope of understanding the evolution of behavior and the restriction of genetic variability. Moreover, it is essential to understand the overwhelming significance of regulatory mechanisms of gene action in natural selection and to relate these to behavioral speciation. Examples for these arguments are discussed here.

INTRODUCTION

The study of social behavior in the Mongolian gerbil (*Meriones unguiculatus*) and of single gene effects on behavior in mice (*Mus musculus*) has convinced me that evolutionary theory holds the principal key to the understanding of most behavior. While indeed we may be seeking answers to the right questions—questions about variability, heritability, and gene–environ-

---

Research from this laboratory described in this paper was made possible by Grant MH 14076–04 and Research Scientist Development Award MH 11, 174–04 to D. D. Thiessen from the National Institutes of Mental Health.

[1] This paper was originally presented at the First Annual Meeting of the Behavior Genetics Association, Storrs, Connecticut, April 8, 1971.

[2] Department of Psychology, University of Texas, Austin, Texas.

ment interactions—I suspect that our strategies and tactics are often restrictive, if not self-defeating.

One cannot, it seems to me, fully understand genetic variations, mutant forms, and recombinant types without first identifying species-specific behaviors, without first exploring gene flow between natural populations, and without first defining gene relations to environmental adaptation. Nor can one learn the intricacies of artificial selection, balanced polymorphisms, and heritability without first knowing the diversity of environmental demands, the characteristics of isolating mechanisms, and the structure of social systems. Finally, it is nearly impossible to investigate critical physiological processes intervening between DNA action and behavior without reference to the evolution of regulatory genes, gene canalization, and homeostasis, and without an understanding of convergent and divergent evolution. In short, in order to understand the genetics of behavior we must attend to the demands of natural selection and the function of behavior in gene transmission and species survival. Ernst Mayr cast the challenge for behavior genetics when he said,

> There are vast areas of modern biology, for instance, biochemistry and the study of behavior, in which the application of evolutionary principles still is in the most elementary state. (Mayr, 1970, 7)

We must agree, unfortunately, that Mayr is correct and that at best we have given lip service to the importance of evolution in the structuring of behavior. Behavioral evolution has been left primarily to the ethologists, who, in spite of their impressive accomplishments, lack genetic sophistication and an appreciation of laboratory techniques, and species-specific analyses have been the province of comparative psychologists, who for the most part are unconcerned with individual differences and ecological adaptation. Each discipline has of course made experimental and theoretical progress, but rarely have their views been sufficiently broad to include the significant principles of all the separate disciplines. It is the field of behavior genetics, in my opinion, that can forge links between the various disciplines and provide unification within the broad framework of evolutionary principles.

Paradoxically, the very advances that gave behavior genetics its separate character are the same that compel it to search for new directions of accomplishment and propel it in the direction of evolutionary principles. The last two decades were devoted to three major goals: (1) demonstrating unequivocally the influence of gene action on behavior, (2) establishing the validity of single gene and polygene models of behavioral analysis, and (3) convincing other life science areas of the reality and pervasiveness of gene–behavior interactions. The accomplishments are a matter of record and can be ignored only by the foolhardy. Almost every behavior of interest has been demonstrated to have some genetic base (Manosevitz et al., 1969). Mendelian and related analyses do apply to behavior as well as to morphological and biochemical traits. And indices of gene–environment interactions are as plentiful as the number of studies attempted. A bigger task remains, however, and that is the task of relating behavioral observations to ecological and phylogenetic considerations and completing bridges between behavioral and other biological disciplines.

For now, let me consider the areas that appear to deserve and require the most attention and those that should lead us to a better understanding of evolutionary principles of behavior. My first point concerns what I consider to be the present general preoccupation with individual differences and the need to look more at the unity of the species. The second point considers the utility of organizing research around traits that are diagnostic of species and environmental adaptation. Finally, the third point considers available opportunities for the investigation of regulatory processes of gene action. In total, I hope to demonstrate the value of accepting the species as the unit of behavioral analysis and the utility of viewing species-specific behavior in evolutionary terms.

THE PREOCCUPATION WITH VARIABILITY AND THE CONCEPT OF GENETIC JUNK

Clearly, genetic variation in animals is so pervasive that, with the possible exception of monozygotic twins, it is proper to consider that every individual of nearly every species is unique. Electrophoretic studies of proteins bear this out (Mayr, 1970). In *Drosophila* (Lewontin and Hubby, 1966) and mice (Selander *et al.*, 1969), for example, 30–50% of the loci tested were polymorphic for an average of three or four alleles. Even this estimate is conservative, as only about one-third of the amino acid substitutions are electrophoretically detectable. The extreme estimate is that even closely related species differ in the majority of their genes (Shaw, 1970). Yet, because of gene neutrality, gene canalization, stabilizing selection, and convergent evolution, little of this distinctiveness is ever expressed.

Gene canalization is so restrictive that even sibling species that are reproductively isolated and that presumably differ by a significant number of genes are often similar or identical morphologically. Four species of North American thrushes of the genus *Catharus* (*C. fuscescens*, *C. guttatus*, *C. ustulatus*, and *C. minimus*) are similar enough visually to confuse each other as well as man (see Table I). Nevertheless, a careful analysis of habitat preference and song characteristics substantiates the species designation

Table 1. Behavioral Differences in Sibling Species of North American Thrushes (Genus *Catharus*)[a]

Behavioral characteristic	C. fuscescens	C. guttatus	C. ustulatus	C. minimus
Breeding range	Southernmost	More northerly	Boreal	Arctic
Breeding habitat	Bottomland woods	Coniferous woods	Mixed tall and coniferous woods	Stunted fir and spruce
Flight song	Absent	Absent	Absent	Present
Hostile call	*beer-pheu*	*chuck-seecep*	*peep-chuck-burr*	*beer*

[a] From Dilger (1956).

(Dilger, 1956). A small number of behaviors, in fact, often appear as the major determinants of species specificity and reproductive isolation.

A study of morphologically similar fireflies of the genus *Photuris* further illustrates the diagnostic value of behavior as an index of specificity. At first glance, it appears that only one or a few species of *Photuris* exist, yet Barber (1951) was able to define 18 species on the basis of flash signals that differed in color (yellow, green, or reddish), intensity, frequency, and pattern. Close observation revealed that these species not only varied in communication signals but in habitat preference and breeding season as well. Thus several behavioral traits act to differentiate the species, even though the number of these is no doubt few relative to the amount of imbedded genetic variation.

King and Jukes (1969) have recently argued, in contradistinction to traditional views, that many mutations that differentiate individuals and species are neutral in effect on biochemical activities and that their presence should not be considered as *prima facie* evidence for natural selection. I would further like to add the point that extreme genetic variability is most likely associated with traits with little adaptive value (also see Falconer, 1960). With those exceptions where survival depends on heterozygosity, traits of a critical nature have generally been selected free of extensive variability that might be harmful to their adaptive expression. The variability that we see in the laboratory, therefore, and that which we can manipulate in artificial selection experiments is probably of little immediate relevance and can in some sense be considered genetic junk. This is not to say that variability cannot be the forerunner of adaptability or that the selected phenotype is not of interest, but only that critical aspects of adaptation and species specificity are often devoid of significant variability.

What needs to be stressed instead of the extent of genetic and phenotypic variability is the critical nature of specialized genes—genes that characterize species, genes that lack variable expression, and genes that insure reproductive fitness. In some cases, major adaptive changes are related to simple genetic systems, and often a behavioral change is the most obvious and significant. It is these "switch" genes—genes that specify major adaptive transitions—that demand our attention and promise to provide significant insights into evolutionary processes. For it may be the case, as Ernst Mayr has said, that

> The larger the number of genes that contribute to the shaping of phenotypic trait, a "character," the less likely it is that such a character will be modified through natural selection. (Mayr, 1970, p. 367)

Or, to put it another way, polygenic traits respond slowly to selection pressures, and their change may lag behind adaptive needs. Major gene effects, on the other hand, can be moved to an adaptive level of functioning quickly and fixed at that level in relatively few generations.

Several illustrations point to the crucial nature of switch genes in the modification of adaptive behaviors. The classic analysis is that of industrial melanism in moths (Kettlewell, 1955), where one or two major genes convey cryptic coloring to moths so that they match the darkened and polluted environment. The selection for a dark morph took less than 100 years, and now that industrial pollution is being reduced the light morph is reestablishing itself (Cook *et al.*, 1970). Similarly, the ecdysone-dependent metamorphosis of *Diptera* species (e.g., *Drosophila*) seems to depend upon the activation of a few key genes on loci of chromosomes I and IV (Beerman, 1965). And in the laboratory it has been found that a host of characters leading to adaptation can be grounded mainly on a simple major gene difference. A particularly elegant experiment by de Souza *et al.* (1970) illustrates this point:

> In this instance larvae of *Drosophila willistoni* originally had one place to live in the population cages, the food cups. The environment outside the cups was inhospitable, larvae died of starvation and dehydration. However, the situation was changed in time. Genetic variants appeared that conferred higher resistance to dehydration outside the cups. These larvae had a faster rate of development, needed less food, and preferred a solid dry environment to pupate. The environment out of cups of food, which was a lethal environment, became available for the populations. Mayr (1963) has said that "a shift into a new niche or adaptive zone is almost without exception, initiated by a change in behavior. The other adaptations to the new niche, particularly the structural ones, are acquired secondarily." The larvae able to survive and pupate outside the cups are intolerant of high moist food. The behavior of the larvae, together with their capacity to survive away from food, permitted the population to colonize a new ecological niche. (de Souza *et al.*, p. 185)

A single major gene was found responsible for this transition to a new adaptive zone and led to almost complete reproductive isolation.

Finally, the simple but beautifully designed experiments of Julius Adler (1969) should be mentioned. For the study of chemotaxis in *Escherichia coli* bacteria, a capillary tube containing a solution of a chemical attractant is pushed into a suspension of bacteria on a slide and the number of bacteria attracted is counted. Using this straightforward technique combined with single gene mutants, Adler was able to conclude that the peripheral membrane of *E. coli* contains at least five chemoreceptors which direct movements of flagella toward different chemical substances. There receptors respond to galactose, glucose, ribose, aspartate, and serine—chemicals of vital interest to bacteria. Clearly, single switch genes are of great significance for the sensory–motor components of adaptation in *E. coli*.

Thus major changes in adaptation can be regulated by one or a few genes operating through species-specific behaviors. Some adaptations are bound to be more significant than others. Whenever we find a large transition in life style, we can be certain that we are dealing with critical genes and genetic junk, and in many cases the number of relevant genes will be small. Those facets of behavioral transition that are likely to be the most informative include metamorphosis from larvae to adult, development of sexual dimorphism, transition from nonflying to flying stages, behavioral selection of different habitats, establishment of reproductive isolation, seasonal variations in behavior, and the evolvement of mimicry. The study of isolating mechanisms in sympatric species, in particular, should be extremely rewarding, as they involve the exaggeration of all aspects of life that preserve the integrity of the species and prevent gametic wastage or maladapted hybrids (see Table II).

So, while I am in accord with the notion that genetic variation is extensive, a prerequisite for adequate adaptation and evolution, and at times

Table II. Behavioral Isolating Mechanisms in Sympatric Populations

Mechanisms that prevent interspecific crosses (premating barriers to gametic wastage)
  A. Differential habitat selection and niche specialization
  B. Assortative mating (homogamy)
  C. Pair formation and internal fertilization
  D. Incongruous social signals
  E. Uncoordinated mating patterns (including seasonal)
  F. Social exclusion
    1. Competition for resources
    2. Social class differences
    3. Territorial barriers

Mechanisms that reduce viability of interspecific crosses (postmating barriers with gametic wastage)
  A. Lack of maternal care
  B. Failure of young to imprint
  C. Agonistic reactions between individuals
  D. Lack of ecological adaptation (including incongruous social signals)

---

essential for survival, I would nevertheless emphasize the importance of investigating simple polymorphic mechanisms that canalize and restrict genetic expression and insure that species uniformity is preserved and that ecological demands are met. The restriction of genetic variation for traits of fitness appears to be the rule rather than the exception.

CONVERGENT EVOLUTION AND THE SUPREMACY OF FUNCTION OVER GENOTYPE

There is little investigative hope of constructing a phylogenetic tree to express the evolutionary trends of behavior. Evolution has not been progressive or linear and has not occurred at uniform rates. In any case, ancestral species are nearly all fossilized or show specializations beyond expectations from phylogenetic relations (Hodos and Campbell, 1969; Thiessen, 1970). There is more hope, it seems to me, in dealing directly with species specializations and treating them as evolutionary reflections of ecological demands. It is in this area of investigation that many mechanisms of behavior are likely to unfold:

> In the case of specialized adjustments, generalizations do not always rest on the invariance of structure–function relations, but rather on the adaptiveness of the response, regardless of genotype or mechanism. Classic Mendelian analyses are hardly relevant to the clarification of control mechanisms, as a near infinite sample of genes and gene products can manage the same solution. Evolution is very opportunistic in the sense that it will take advantage of any genetic variance which will satisfy the same environmental requirement. (Thiessen, 1970, p. 101)

Organisms, without exception, must adapt to variable yet prepotent selection factors such as gravity, climate, food and oxygen supplies, shelter requirements, predators, photoperiodicities, and the like. Clinal variations (variations that are gradated), such as those that show systematic changes with latitude, are especially good evidence for specialized adaptations along ecological gradients. Clines are evident for the majority of continental species and are more apparent in sendentary species that have no alternative but to adapt to environmental demands. For example, as latitude increases in northerly or southerly directions, the following general trends occur in morphology, biochemistry, and behavior:

1. Body size increases (Bergmann's rule).
2. The tail, ears, bills, and limbs become relatively short (Allen's rule).
3. The relative length of hair increases.
4. Wings become more pointed.
5. The relative size of the heart, pancreas, liver, kidney, stomach, and intestines increase.
6. There is a reduction in the pigments phaeomelanins and eumelanins (Gloger's rule).
7. Relative oxygen consumption and metabolic needs decrease, and general activity diminishes.
8. Migratory instincts become stronger.
9. Larger and warmer nests are constructed (King's rule).
10. Home ranges become larger, and territorial behavior is more pronounced.
11. Photoperiodic rhythms become more evident.

None of these "clinal laws" could have been predicted from phylogenetic relationships, but all become obvious when climatic demands are considered. The primary demands are related to needs to conserve body heat, to compete more successfully for limited or seasonal food supplies, and to find protection from predators and changeable conditions of weather. The fact that so many species show these trends suggests that there has been convergent evolution toward those biological features most apt to guarantee species survival around the world.

Convergent evolution, of course, need not be tied to clinal variations. It is often local in character, reflecting the peculiar needs and niche specifications of the species or population. Territorial behavior and scent marking, for example, are evident in at least 13 of the 19 mammalian orders living the world over. In those few species studied to any degree, such as Maxwell's duiker, European rabbit, sugar glider, golden marmoset, golden hamster, and Mongolian gerbil (see Ralls, 1971), it appears that scent marking is related to similar systems of social organization and depends upon an identical hormone base (see Table III). Evidently, territoriality acts to conserve basic commodities and is intricately linked to hormones of sex and aggression. In any case, convergent evolution has moved many species toward a territorial system of behavior and has oftentimes capitalized on olfactory signals and reproductive hormones in its regulation. It is evident that in order to understand the function of territoriality we must understand those environmental factors that demand its expression.

Likewise, cryptic shading and coloration, morphological disguises, and warning signals show convergent evolution and local adaptation in innumerable species (Cott, 1940; Portmann, 1959; Wickler, 1968). In all cases investigated, it is apparent that morphology, color, behavior, and background correspond to produce the best possible adaptation for that species regardless of genetic descent. The study of such relations has only begun.

Table III. Examples of Convergent Evolution for Scent Marking

Species	Distribution	Gland characteristics	Behavioral characteristics
Cephalophus maxwelli (Maxwell's duiker)	Central West Africa	Preorbital gland	Objects and conspecifics marked, especially by dominant male
Oryctologus cuniculus (European rabbit)	Europe and North Africa	Apocrine chin gland	Gland and marking more prominent in male and are androgen dependent
Petaurus breviceps (sugar glider)	Australia and New Guinea	Frontal and sternal glands	Gland and marking more prominent in male; used to demark territories and for recognition
Leontideus rosalia (golden lion marmoset)	South America	Sebaceous glands on sternal and gular areas	Gland and marking more prominent in dominant male and become functional at puberty
Mesocricetus aurotus (golden hamster)	East Europe and West Asia	Sebaceous gland on flanks	Gland and marking more prominent in dominant male and become functional at puberty; used to demark territories
Meriones unguiculatus (Mongolian gerbil)	Northeast Asia	Sebaceous gland on ventral area	Gland and marking are androgen dependent and more prominent in dominant male; become functional at puberty; used to demark territories

Hence *function* and not mechanism is the key to the understanding of a great deal of convergent evolution, which implies that our attention must be directed toward the outcome of evolution rather than simply toward the genetic structure or physiology underlying a particular phenotype. Obviously, function cannot be completely understood in laboratory investigations where the individual and species stand stripped of their most salient environmental influences. However inconvenient, the behaviorist must move his observational acuity to the geographical site of natural selection and speciation. It is there that the natural adaptations of the species are displayed, and there where relevant laboratory experiments can be formulated.

REGULATORY GENES AND SPECIES-SPECIFIC BEHAVIOR

Higher mammals perhaps have enough DNA for more than 5 million functional genes, yet protein studies would indicate that not more than 10–50 thousand of these are structural genes, genes directly concerned with enzyme formation. Moreover, most of these structural genes are common to a wide array of species and function in approximately the same way. Sturtevant has emphasized the functional equivalence of gene action in this way:

> The more recent comparative biochemical data . . . favor the idea of the great stability of genetic systems, since they show essential identity of some of the gene-controlled basic biochemical pathways in bacteria, fungi, and vertebrates. (Sturtevant, 1965, p. 115)

The near universality of the DNA code itself and the species identity of energy-storing compounds such as adenosine triphosphate emphasize the fundamental identity of life systems.

Although the basic machine of life is similar, of course, its expression shows prodigious diversity, within and between species. Much of this diversity must be due to the modification of basic biochemical processes and not due to differences in chemical forms. In other words, the greatest proportion of phenotypic variance, at least in mammalian species, is probably due to regulatory rather than structural genes—genes that activate, deactivate, or otherwise alter the expression of a finite number of structural genes. Support is added to this view by the observation that the total DNA content of diploid species, which all possess the same fundamental biochemicals, increases substantially and regularly from fish through amphibians, reptiles, and mammals (Britten and Davidson, 1969).

The best-known model for gene regulation is of course that proposed by Jacob and Monod (1961), who determined with the bacterium *E. coli* that the synthesis of galactosidase by a structural gene is under the control of a single regulator gene responsive to the amount of lactose, the substrate of galactosidase, in the cytoplasm. This singularly important observation led to the generalization that gene action is open to modification by environmental factors and hence added the dynamic character to gene action necessary to account for variable expression and homeostatic reactions.

Much behavior that we see may be controlled by regulatory genes open to processes of canalization, early and later experiences, and natural selection. As Britten and Davidson see it,

> Any evolutionary changes in the phenotype of an organism require, in addition to changes in the producer genes [structural genes],[3] consistent changes in the regulatory system. Not only must the changes be compatible with the interplay of regulatory processes in the adult, but also during the events of development and differentiation. At higher grades of organization, evolution might indeed be considered principally in terms of changes in the regulatory systems. (Britten and Davidson, 1969, pp. 355–356.)

Not many clear examples of regulatory actions exist for behavior. We have considered one, the molting pattern of *Diptera* species. Here, under the influence of the inductor hormone, ecdysone, the entire life style of the organism changes abruptly during metamorphosis from that of a wormlike animal to that of a fully developed fly. In our own laboratory, we have found that the territorial scent-marking response of the male Mongolian gerbil is androgen dependent and can be elicited by small amounts of testosterone implanted directly into the preoptic area of the hypothalamus (Thiessen *et al.*, 1968; Thiessen and Yahr, 1970). When genes are prevented from templating RNA by adding actinomycin D, an antibiotic which binds DNA and prevents its action, the hormone implanted in the brain is no longer effective in producing the behavior. Similarly, we have evidence that ribonuclease, which destroys RNA, and puromycin, which disrupts protein synthesis, are also effective in attenuating or blocking the hormone response. Magnesium pemoline, on the other hand, stimulates higher RNA synthesis and to some degree activates

[3] My interpretation.

103

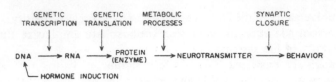

Fig. 1. Hormone–gene flow.

territorial marking and competes with testosterone for receptor sites. Our working model schematized in Fig. 1 supposes that testosterone normally evokes territorial behavior by activating specific segments of DNA, thus initiating a chain of biochemical activities that culminates in territorial marking. Further examination of this model may show that most hormone-behavior relations depend upon gene-regulatory processes, and the concept may extend to many chemical–response relations. Hopefully, such a model is relevant to any concept of environmental determination of gene action and will eventually add a substantial increment to our knowledge of genetic variability and behavior.

In summary, let me stress that behavior genetics is a part of evolutionary biology and that its viability depends upon that relation. While it is theoretically possible to describe gene–behavior associations without reference to their evolutionary origins (and admitting that physiological mechanisms can be studied in the same way), coherence and meaning can only follow when the full implications of natural selection are appreciated. Ultimately, our attention must be on factors that restrain phenotypic variability as well as those that exaggerate it, and those factors that give species their identity. It is necessary, therefore, to emphasize the primary importance of ecological adaptation and concentrate more on regulatory processes that permit differential genetic expression. Above all, it seems to me, the species requires a higher status in our conceptual values as a unit of behavioral response.

REFERENCES

Adler, J. (1969). Chemoreceptors in bacteria. *Science* **166**: 1588–1597.
Barber, H. S. (1951). North American fireflies of the genus *Photuris*. *Smithsonian Misc. Coll.* **117**: 1–58.
Beerman, W. (1965). Cytological aspects of information transfer in cellular differentiation. In Bell, E. (ed.), *Molecular and Cellular Aspects of Development*, Harper, New York, pp. 204–212.
Britten, R. J., and Davidson, E. H. (1969). Gene regulation for higher cells. *Science* **165**: 349–357.
Cook, L. M., Askew, R. R., and Bishop, J. A. (1970). Increasing frequency of the typical form of the peppered moth in Manchester. *Nature* **227**: 1155.
Cott, H. B. (1940). *Adaptive Coloration in Animals*, Methuen & Co., London.
de Souza, H. M. L., da Cunha, A. B., and dos Santos, E. P. (1970). Adaptive polymorphism of behavior evolved in laboratory populations of *Drosophila willistoni*. *Am. Naturalist* **104**: 175–189.
Dilger, W. C. (1956). Hostile behavior and reproductive isolating mechanisms in the avian genera *Catharus* and *Hylocichla*. *Auk* **73**: 313–353.
Dobzhansky, T. (1970). *Genetics of the Evolutionary Process*, Columbia University Press, New York.
Falconer, D. S. (1960). *Introduction to Quantitative Genetics*, Ronald Press, New York.

Hodos, W., and Campbell, C. B. G. (1969). Scala naturae: Why there is no theory in comparative psychology. *Psychol. Rev.* **76**: 337–350.

Jacob, F., and Monod, J. (1961). Genetic regulatory mechanisms in the synthesis of proteins. *J. Mol. Biol.* **3**: 318–356.

Kettlewell, H. B. D. (1955). Selection experiments on industrial melanism in Lepidoptera. *Heredity* **9**: 323–342.

King, J. L., and Jukes, T. H. (1969). Non-Darwinian evolution. *Science* **164**: 788–798.

Lewontin, R. C., and Hubby, J. L. (1966). Amount of variation and degree of heterozygosity in natural populations of *Drosophila pseudoobscura. Genetics* **54**: 595–609.

Manosevitz, M., Lindzey, G., and Thiessen, D. D. (1969). *Behavioral Genetics: Method and Research*, Appleton-Century-Crofts, New York.

Mayr, E. (1970). *Populations, Species and Evolution*, Harvard University Press, Cambridge.

Portmann, A. (1959). *Animal Camouflage*, University of Michigan Press, Ann Arbor.

Ralls, K. (1971). Mammalian scent marking. *Science* **171**: 443–449.

Selander, R. K., Hunt, W. G., and Yang, S. Y. (1969). Protein polymorphism and genic heterozygosity in two European sub-species of the house mouse. *Evolution* **23**: 379–390.

Shaw, C. R. (1970). How many genes evolve. *Biochem. Genet.* **4**: 275–283.

Sturtevant, A. H. (1965). *A History of Genetics*, Harper, New York.

Thiessen, D. D. (1970). Philosophy and method in behavior genetics: Its relation to biology. In Gilgen, A. R. (ed.), *Contemporary Scientific Psychology*, Academic Press, New York, pp. 71–114.

Thiessen, D. D., and Yahr, P. (1970). Central control of territorial marking in the Mongolian gerbil. *Physiol. Behav.* **5**: 275–278.

Thiessen, D. D., Friend, H. C., and Lindzey, G. (1968). Androgen control of territorial marking in the Mongolian gerbil (*Meriones unguiculatus*). *Science* **160**: 432–443.

Wickler, W. (1968). *Mimicry in Plants and Animals*, McGraw-Hill, New York.

10

Reprinted from
GENETICS TODAY
Proceedings of the XI International Congress of Genetics
The Hague, The Netherlands, September, 1963
PERGAMON PRESS
OXFORD · LONDON · EDINBURGH · NEW YORK
PARIS · FRANKFURT
1964

EVOLUTIONARY CHANGES AND BEHAVIOUR GENETICS

AUBREY MANNING

Department of Zoology, University of Edinburgh, Scotland

IN this paper I shall try to consider the evidence we have on the manner in which behaviour evolves in relation to the behavioural results of genetic changes. I shall confine my attention to instinctive behaviour. Only in this context can we speak meaningfully of the inheritance and evolution of behaviour patterns themselves. The evolution of the cerebral cortex through the vertebrate series has resulted in greatly increased behavioural potentialities and flexibility, but it has been roughly paralleled by a reduction in the repertoire of inherited behaviour. At present we can only talk about the evolution of learning abilities in the most general terms, but we can do more for instinctive patterns. These are usually rigid and easily recognized and altogether bear comparison with morphological features.

We are faced with formidable problems of analysis at present. So many different things can affect the performance of a behaviour pattern. To put it naïvely, we must consider all

105

the chain of mechanisms between sense organs and muscles, with genes operating at every link in the chain. Even if we can classify behaviour into sensible units, we are usually unable to do more than express the effects of genetic changes in purely behavioural terms and mechanism eludes us. Nevertheless, there is a consistency in these effects which does give us a meaningful insight into the way behaviour evolves.

Considering first the more directly genetical evidence, work has proceeded in three main ways:

(1) The comparison of behaviour between animals whose genotypes differ, as near as possible, only at a single locus.

(2) The comparison of inbred lines.

(3) The production of behaviourally divergent lines by deliberate selection.

All the evidence leads to the same conclusion. The basic form of instinctive behaviour patterns is very stable, but almost any genetic change produces effects of a quantitative type and alters the frequency with which the patterns are performed.

For example, in her study on the effects of the mutant *yellow* on the sexual behaviour of *Drosophila melanogaster*, Bastock (1956) showed that it reduced the frequency with which males performed certain patterns, but their form was identical with that of normal flies. By selection for speed of mating—also in *D. melanogaster*—Manning (1961, 1963) was able to alter the behaviour of males in a similar manner. Ewing (1961) describes quantitative differences in behaviour between lines of *D. melanogaster* selected for body-size criteria. The changes are subtle and certainly not due merely to the mechanical effects of changed size. Inbred lines of guinea pigs (Goy and Jakway, 1960; Jakway, 1960) and mice (McGill, 1962) differ quantitatively in a number of measures of latency and performance frequency of elements of sexual behaviour. By selection, Wood-Gush (1960) was able to alter the levels of sexual responses in domestic cockerels.

In all these examples changes to the genotype have changed thresholds of performance somewhere within the system. In few cases can we say much more than this, but sometimes analysis has proceeded a bit further and gives some clues on the action of genes. The Maudsley reactive and non-reactive strains of rats which show large differences in their emotional responses to a new and rather frightening environment, have proved to differ in their thyroid activity (Feuer and Broadhurst, 1962). Here the genes may operate, at least partially, to affect behaviour through the endocrine balance. Wood-Gush, Goy and Jakway, on the other hand, did not find that hormone levels are responsible for the differences between their lines, although these lines showed differing responsiveness to exogenous hormones.

I must now turn from this brief sample of the more directly genetical work and examine the nature of micro-evolutionary changes. The evidence comes from the comparative behaviour studies of ethologists. These mostly concern birds, fish and arthropods and it is sexual behaviour which has attracted most attention because of its conspicuous nature and relatively rapid evolution.

Some of the commonest features of micro-evolution are easily related to what we know of the effects of genetic changes. Thus, closely related species generally show the same instinctive repertoire, but differ in the frequency with which the elements are performed. *Drosophila* offers many good examples. Brown (1962, 1963) describes the repertoire of courtship patterns common to the *obscura* group. The males' "rowing" movement with the legs is common in *miranda*, moderately so in *obscura*, but is extremely rare, though still occasionally seen in *pseudoobscura* and *persimilis*. In the *melanogaster* group, the sibling pair *melanogaster* and *simulans* differ markedly in frequency of the "scissoring" wing movement in the males' display (Manning, 1959). Scissoring is not normally seen in *melanogaster* but can be evoked in certain circumstances.

Differences between performance frequencies suggest effects on behavioural thresholds and some micro-evolutionary changes demonstrate this most vividly. The herring-gull (*Larus argentatus*) and the lesser black-backed gull (*L. fuscus*) have virtually identical alarm calls, but the latter species requires stronger stimuli to produce them; walking into a mixed colony will often cause the herring-gulls to call, but not their relatives (Goethe, 1954). Blest (1957) describes a similar case from the defence displays of Saturnioid moths. The aposematic species display at a light touch, the more cryptic species only after vigorous prodding.

Sometimes a species may apparently lack a pattern which is found among its close relatives, the *Drosophila* examples given above are a case in point. Nevertheless, it is very unlikely that behaviour patterns can disappear so easily. Their threshold may be so raised that they never normally appear, but the requisite neural mechanisms will still be there. In general, hybrids show the same behaviour patterns as the parent species but at intermediate frequencies. Hinde's (1956) work with Cardueline finch hybrids and that of Clark, Aronson and Gordon (1954) with Xiphophorin fish show this, and indicate that multiple loci are involved in determining pattern frequency. Sometimes, as in Hörmann-Heck's (1957) work with cricket (*Gryllus*) hybrids, species differences in frequency appear to be due to a single locus. Most of the observations on hybrids tell us nothing about the actual neural organization of the patterns or its inheritance. Certainly the development of an instinctive pattern must depend upon numerous genes. Hybridization is normally possible only between close relatives who possess very similar behaviour repertoires. A fine analysis of a pattern into its smallest separately "viable" units, such as one might expect to pick up in F_2 hybrids, is difficult. Bits of a behaviour pattern or malformed behaviour patterns are not so easily recognized as their morphological equivalents.

Changes to the sequence with which behaviour patterns are performed may well be another result of threshold changes, but we know too little about the organization of sequences to be confident. However it is interesting that whilst patterns themselves are so stable, sequence control does appear to be sensitive to genetic changes. The courtship displays of ducks, for example, consist of a number of easily recognized patterns (Lorenz, 1941). These patterns are similar from species to species but as well as showing different performance frequencies, species may differ in the sequence of patterns they perform. Species X may have A–B–C as its commonest pattern sequence; species Y, A–C–B. The work of Ramsay (1961) and van der Wall (unpublished, quoted by Hess, 1962) shows that species hybrids usually perform the patterns quite normally but often in odd sequences. Sometimes F_1 and F_2 hybrids perform sequences never seen in either parent species. Clearly the breakdown of the naturally evolved genotype upsets sequence control before the control of the individual patterns. Apart from the duck family, sequence changes have played some part in the behavioural radiation of *Drosophila*. Within the *melanogaster* group, some species normally perform a courtship sequence we may denote as A–B–C–D, others, using homologous patterns, have the order A–C–D–B.

So far I have been considering species differences in relation to threshold changes. Some of the other common types of evolutionary divergence are less obviously related to them.

(i) Species may differ in the speed with which homologous movements are performed. This is seen in the defence displays of moths (Blest, 1957), the threat displays of gulls (Tinbergen, 1959), and probably the wing vibration displays of *Drosophila* species.

(ii) Species may differ in the relative "emphasis" given to the various parts of a homologous pattern. Most of the gulls have a "long-call" pattern in which the head is moved upwards and backwards on an extended neck whilst calling. The degree to which the head is thrown back during the call varies between species; it is least marked in the western gull (*Larus occidentalis*) and most emphasised in the common gull (*L. canus*), (Tinbergen, 1959).

The *obscura* group of *Drosophila* vary in the degree to which the trailing-edge of the males' wing is drooped during the wing-vibration display. In *D. obscura* this drooping is most marked and there are detectable differences in this respect between *D. pseudoobscura* and *D. persimilis*, whose displays have often been regarded as identical (Brown, 1962 and *in press*). In fiddler-crabs of the genus *Uca*, the males show a rhythmic waving of an enlarged claw during courtship. Species vary in the relative emphasis given to "up-down" and "side-to-side" components of the wave, and also in the degree to which they raise and lower the whole body on the walking legs in time with the wave (Crane, 1957).

The form of a pattern in hybrids tends to be intermediate between the parental types. Hinde's (1956) finch hybrids sometimes had distinct but quite serviceable display patterns, half-way between those of the parents. However the work of Dilger (1959) on the F$_1$ hybrids between two species of parrot (*Agapornis*) shows how the intermediate between the two very distinct parental patterns for gathering nest material produces a very inadaptive result.

(iii) Species may differ in the relative importance of the various sensory modalities involved in the perception of displays. Within the *melanogaster* group, *Drosophila auraria* will not mate in the absence of light, *D. simulans* and *D. rufa* show reduced mating, but *D. melanogaster* is unaffected (Spieth and Hsu, 1950). Part of this variation probably relates to the degree by which the different females are stimulated by visual aspects of their males' courtship. Klopfer (1959) found some variation in the responsiveness of newly hatched ducklings to sound or visual stimuli from an object to which they were being imprinted. Ducklings of the surface-nesting species were most responsive to visual stimuli but those of the hole-nesting woodduck (*Aix sponsa*) would respond to sound alone. Subsequently Klopfer and Gottlieb (1962) have discovered that responsiveness to sound and visual stimuli varies even within a single brood of the mallard (*Anas platyrhynchos*). They suggest that this behavioural polymorphism reflects a genetic polymorphism and have begun a selective breeding programme.

The above are but three of many possible types of behavioural change; Blest (1961) considers them all more fully. At first sight they appear very heterogeneous, but even if we know next to nothing about the mechanisms involved, there seems no reason to suppose that gene action is fundamentally different in each category. It is reasonable to suggest that all these changes have been produced by the accumulation of small threshold changes. For example, the change of emphasis within a pattern could result from threshold changes on the motor side of a mechanism such that particular muscle groups came into action earlier, or later, for a shorter time or a longer one, and so on. In this way small quantitative changes could summate to produce the great diversity of variants on a common "ancestral" pattern which we see in most groups. Again, changes to the importance of one stimulus modality with respect to another could be produced by threshold changes on the sensory side of the organization, perhaps accompanied by changes to the sense organs themselves.

There is no doubt that all natural populations show considerable variability for genes which affect behaviour and that this variability has been the raw material upon which selection has operated. In general I think the correspondence between gene effects and micro-evolutionary changes in behaviour is a good one, which is as it should be.

The relationship between behaviour and evolution is not all in one direction. Changes to behaviour, perhaps themselves the result of genetic changes, may in turn affect the course of evolution. For example, genetic differences between populations may alter their preferences for particular conditions of light, temperature or humidity and thus influence their choice of habitat. This may lead to changes in the degree to which they are isolated ecologically from one another.

Genetic changes also affect the sexual isolation between populations. This is an important and well-worked field but I have space to discuss only one aspect of it. This is the repercussions which the gradual "quantitative" evolution of behaviour have had on the development of sexual isolation. The shifts of performance frequency which will so commonly be the initial behavioural result of genetic divergence, are probably inadequate to prevent hybridization unless backed up by some more positive identification mark. This is often provided by colour patterns in birds, reptiles and fish, whilst among insects changes in scent are often important. Within the subgenus *Sophophora* of *Drosophila* where hybrids are very rare, sexual isolation is usually based on the female's ability to identify the scent of her own males and much less on her "sampling" their courtship displays. This is not to deny, of course, that selection will operate to make females most responsive to courtship of their own male's type, but they can identify males before courtship has proceeded far. It is not surprising that in the two cases where we have behavioural data on the effects of selection for increased sexual isolation between strains or species (Koopman, 1950; Pearce, 1962) it is discrimination which has been altered and not the courtship patterns.

A comparable example of behavioural differences being supported by chemical ones is provided by Hunsaker's (1962) work with lizards of the genus *Sceloporus*. He has shown that a species-specific pattern of head-bobbing is one of the factors which cause females to approach conspecific males but this selection is reinforced by chemical discrimination.

If most types of behavioural change accompanying the divergence of populations are too imprecise for reliable identification, those occurring in species with acoustical signals may be better. Quite small quantitative changes to the motor patterns which make up the mating calls of birds, frogs and crickets, can result in detectable sound differences. In crickets, for example, changes to pulse length, to the interval between pulses and to the amplitude of pulses within a series, all these and more have been developed to avoid interspecific confusion (Alexander, 1962). Sympatric species tend to be more different in their calling songs than allopatric ones. Blair (1955) has demonstrated a similar divergence of sexual call notes in the two sibling frog species, *Microhyla olivacea* and *M. carolinensis*. Samples taken from those parts of their range where only one species lives sound very similar, but where the two occur together in the same breeding ponds their calls are distinct. The three species of *Phylloscopus* warblers which often inhabit the same areas in Europe have similar alarm notes and courtship displays, but markedly different songs.

Sometimes it appears that isolation is based solely upon the response of females to the calls of their own males. Blair (1955) says that other isolating mechanisms between the *Microhyla* species are weak, and Perdeck (1958) found a similar situation in two sibling grasshopper species. However sound differences may also be backed up by chemical differences between species in some crickets (Alexander, 1962), just as plumage differences often reinforce song differences among birds.

My argument has been that the early stages of behavioural divergence, as opposed to any associated morphological or chemical changes, are often only quantitative and indistinct. I feel therefore that it behoves us to be cautious when suggesting that the evolution of elaborate courtship displays within a group has been dictated by the need for sexual isolation between overlapping species. The genus *Drosophila*, where there is the likelihood that several species will gather to mate on a common food source, is a case in point. Here we have an elaborate courtship evolved, but their quite close relatives the Tachinids and Anthomyids include a large number of species which similarly associate on food yet the males have no courtship at all. They mount and attempt to copulate with any fly of roughly the right size and females apparently distinguish their own males on contact and repel others. Sexual isolation, based presumably on scent differences, is as effective here as in *Drosophila*, and as we have seen it is doubtful whether female *Drosophila* discriminate against foreign males on the basis of their courtship displays.

It is, of course, unrealistic to separate completely the purely behavioural aspects of courtship from scent, colour or any of the other specially evolved releasers involved in reproduction. One function of the behaviour patterns may be to display conspicuously some patch of colour or to expose some scent gland and thereby enhance discrimination based on these structures. Clearly the advantages of sexual isolation have often been a factor in the evolution of courtship, but so have the need for behavioural synchrony, the appeasement of conflicting tendencies and perhaps sexual selection.

So far I have been considering the evolution of instinctive behaviour solely in the conventional terms of selection operating on small, undirected mutations. Certainly much of the adaptiveness of behaviour has been attained in this way but I want, in conclusion, to consider another source of adaptiveness, that of genetic assimilation (Waddington, 1961). The starting point is the organism's ability to make an adaptive response to an environmental stimulus. If this stimulus is consistent between generations and the response has survival value, selection will favour a more rapid and complete response. Eventually the threshold may be so lowered that the response occurs even in the absence of any exogenous stimulus. In behavioural terms, this means that patterns which were originally learnt become inherited. If selection is consistent the speed of learning will increase and the response itself will become more precisely adaptive. Eventually the response becomes encoded within the nervous system of the developing individual who requires only the correct environment to produce it fully perfected at the first exposure. These ideas are discussed by Ewer (1956) who suggests that imprinting responses, where, for example, a parent figure or a food plant are learnt extremely rapidly and perhaps at a single exposure, may represent an intermediate stage in the evolution of completely inherited behaviour.

Waddington (1961) has most elegantly demonstrated experimentally the genetic assimilation of acquired morphological and physiological characters. To do the same for behaviour is difficult, but a start has been made using as the acquired response the type of "larval conditioning" which Thorpe (1939) first showed that *Drosophila* make to contaminants in their culture medium (Moray and Connolly, 1963). Clearly genetic assimilation and the selection of random mutations are not completely distinct processes. Both must have played a part in the evolution of instinctive behaviour. I believe that once the arthropod level of organization and learning ability has been reached, assimilation may well be important for the incorporation of novel motor patterns into the instinctive repertoire which are subsequently modified in a more quantitative way.

The study of behaviour genetics is only just beginning, but already we can report some progress. If we must remain vague on mechanism, yet we can certainly consider the details of behaviour's evolution in genetical terms. Further work will be of value, not only for its elucidation of evolutionary mechanisms, but for information on the organization of behaviour within the nervous system which we get by investigating how genes affect it.

REFERENCES

ALEXANDER, R. D. (1962) Evolutionary change in cricket acoustical communication. *Evolution* **16**, 443-467.

BASTOCK, M. (1956) A gene mutation which changes a behavior pattern. *Evolution* **10**, 421-439.

BLAIR, W. F. (1955) Mating call and stage of speciation in the *Microhyla olivacea—M. carolinensis* complex. *Evolution* **9**, 469-480.

BLEST, A. D. (1957) The evolution of protective displays in the Saturnioidea and Sphingidae (Lepidoptera). *Behaviour* **11**, 257-309.

BLEST, A. D. (1961) The concept of ritualization. In Thorpe, W. H. and Zangwill, O. L. *(Eds.) Current Problems in Animal Behaviour*. Cambridge, pp. 102-124.

BROWN, R. G. B. (1962) A comparative study of mating behaviour in the *Drosophila obscura* group. D. Phil. Thesis, Univ. of Oxford.

BROWN, R. G. B. (1963). *In press. Behaviour.*

CLARK, E., ARONSON, L. R. and GORDON, M. (1954) Mating behaviour patterns in two sympatric species of Xiphophorin fishes: their inheritance and significance in sexual isolation. *Bull. Amer. Mus. Nat. Hist., N.Y.* **103**, 135-226.

CRANE, J. (1957) Basic patterns of display in fiddler crabs (Ocypodidae, genus *Uca*). *Zoologica* **42**, 69-82.

DILGER, W. C. (1959) Nest material carrying behaviour of F_1 hybrids between *Agapornis fischeri* and *A. roseicollis*. *Anat. Rec.* **134**, 554.

EWER, R. F. (1956) Imprinting in animal behaviour. *Nature* **177**, 227-228.

EWING, A. W. (1961) Body size and courtship behaviour in *Drosophila melanogaster. Anim. Behav.* **9**, 93-99.

FEUER, G. and BROADHURST, P. L. (1962) Thyroid function in rats selectively bred for emotional elimination. II. Differences in thyroid activity. *J. Endocrinol.* **24**, 253-262.

GOETHE, F. (1954) Vergleichende Beobachtungen über das Verhalten der Silvermowe *(Larus a. argentatus)* und der Heringsmowe *(Larus f. fuscus). Proc. XI Int. Orn. Congr.* 557-582.

GOY, R. W. and JAKWAY, J. S. (1960) The inheritance of patterns in sexual behaviour in female guinea pigs. *Anim. Behav.* **7**, 142-149.

HINDE, R. A. (1956) The behaviour of certain Cardueline F_1 inter-species hybrids. *Behaviour* **9**, 202-213.

HESS, E. H. (1962) Ethology. In Brown, R., Salanter, E., Hess, E. H. and Mandler, G. *New Directions in Psychology*. Holt, Rinehart, Winston, New York, pp. 159-266.

HÖRMANN-HECK, S. VON (1957) Untersuchungen über den Erbgang einiger Verhaltensweisen bei Grillen-bastarden (*Gryllus campestris* L. x *Gryllus bimaculatus* De Geer). *Z. Tierpsychol.* **14**, 137-183.

HUNSAKER, D. (1962) Ethological isolating mechanisms in the *Sceloporus torquatus* group of lizards. *Evolution* **16**, 62-74.

JAKWAY, J. S. (1960) The inheritance of patterns of mating in the male guinea pig. *Anim. Behav.* **7**, 150-162.

KLOPFER, P. H. (1959) An analysis of learning in young Anatidae. *Ecology* **40**, 90-102.

KLOPFER, P. H. and GOTTLIEB, G. (1962) Learning ability and behavioural polymorphism within individual clutches of wild ducklings. *Z. Tierpsychol.* **19**, 183-190.

KOOPMAN, K. F. (1950) Natural selection for reproductive isolation between *Drosophila pseudoobscura* and *Drosophila persimilis. Evolution* **4**, 135-148.

LORENZ, K. (1941) Vergleichende Bewegungstudien an Anatinen. *J. Orn.* **89**, 194-294.

MANNING, A. (1959) The sexual behaviour of two sibling *Drosophila* species. *Behaviour* **15**, 123-145.

MANNING, A. (1961) The effects of artificial selection for mating speed in *Drosophila melanogaster. Anim. Behav.* **9**, 82-92.

MANNING, A. (1963) Selection for mating speed in *Drosophila melanogaster* based on the behaviour of one sex. *Anim. Behav.* **11**, 19, 341-350.

McGILL, T. E. (1962) Sexual behaviour in three inbred strains of mice. *Behaviour* **19**, 341-350.

MORAY, N. and CONNOLLY, K. (1963) A possible case of genetic assimilation of behaviour. *Nature* **199**, 358-360.

PEARCE, S. (1962) Evolution of mating behaviour in *Drosophila* under artificial selection. M/S. of paper read to *Brit. Ass. Adv. Sci.* 1962.

PERDECK, A. C. (1958) The isolating value of specific song patterns in two sibling species of grasshoppers (*Chorthippus brunneus* Thunb. and *C. biggutulus* L.). *Behaviour* **12**, 1-75.

RAMSAY, A. O. (1961) Behaviour of some hybrids in the mallard group. *Anim. Behav.* **9**, 104-105.

SPIETH, H. T. and HSU, T. C. (1950) The influence of light on the mating behavior of seven species of the *Drosophila melanogaster* group. *Evolution* **4**, 316-325.

THORPE, W. H. (1939) Further experiments on pre-imaginal conditioning in insects. *Proc. Roy. Soc. B.* **127**, 424-433.

TINBERGEN, N. (1959) Comparative studies of the behaviour of gulls. *(Laridae)*: a progress report. *Behaviour* **15**, 1-70.

WADDINGTON, C. H. (1961) Genetic assimilation. *Adv. Genet.* **10**, 257-293.

WOOD-GUSH, D. G. M. (1960) A study of sex drive of two strains of cockerels through three generations. *Anim. Behav.* **8**, 43-53.

111

PART **3**

Neural and Hormonal Control of Behavior

The readings in Part 2 presented many examples of the effects of genotype on behavior. Related to the material in that section is the important question, "How does genotype affect behavior?" While the answers to this question are largely unknown and probably immensely complex, it is possible, nevertheless, to make some general statements about the intermediate steps whereby genes affect behavior. First, it is generally agreed that genes control the production of enzymes and enzyme systems. Enzymes in turn mediate the complex metabolic activities necessary for the growth and differentiation of cells into tissues and organs. The resulting morphology and physiology are the substrates on which behavior ultimately depends. These substrates are, of course, not immutable, and we shall cover many examples of environmental and behavioral effects on physiology and anatomy.

Most important for the study of the physiology of behavior are the nervous system and the endocrine system. Therefore, the readings that follow are con-

113

cerned primarily with brain function and the influence of hormones on the brain and on behavior.

Behavioral physiology is concerned with relations between the internal processes of the body and behavior and thus serves as a bridge between purely physiological, and purely behavioral, research. Even those behaviorists who are not directly concerned with physiological processes in their own experiments, recognize the importance of this discipline. For behavioral theories must not violate physiological principles (just as physiological theories must take into account the facts of biochemistry). When such violations occur, the behavioral theory is probably in need of revision. Reading 11 poses such a problem.

The study of biological clocks in plants and animals has been an active area of research for many years. Many of the clocks exhibit circadian (roughly 24-hour) rhythms. In Reading 12, a circadian rhythm in the control of feeding behavior by hypothalamic norepinephrine is described. The results are surprising and emphasize the need for further study of biological rhythms in relation to the physiological control of behavior. Reading 13 considers the changes in daily hormone rhythms over the annual reproductive cycle of the white-throated sparrow. It is these changes that "control the orderly sequence of migratory and reproductive conditions" in this species.

Reading 14 also reports on behavioral cyclicity, only this time the phenomenon was found where, theoretically at least, it should not have existed.

The next few pages deal with hormone–behavior interactions, one example being hormones' direct action on particular parts of the nervous system. Reading 15 by Julian Davidson provides a carefully controlled study of such an effect. Davidson's work shows that in the adult rat certain neurons in the brain are sensitive to the presence of testosterone. Implants of testosterone in other parts of the brain are ineffective in activating castrated males, as are implants of cholesterol in the testosterone-sensitive region.

Reading 16 shows that behavioral stimulation (suckling) is necessary and sufficient to induce aggressive behavior in virgin female mice. A similar example is also provided in the next reading. The late D. S. Lehrman was involved for several years in research designed to elucidate the intricate relations among hormones, behavior, and external stimulation in ring doves. In Reading 17, one of Lehrman's students, Carl J. Erickson, describes the effectiveness of castrated male ring doves (receiving various amounts of androgen) in inducing changes in oviduct weights and ovulation frequencies in female ring doves.

As we have seen, hormones can affect both morphology and behavior. Reading 18 presents a relatively unusual example of both such effects. Here Gardner Lindzey, Delbert Thiessen, and Anne Tucker show that both territorial marking and the size of the scent-marking gland in male gerbils are under hormonal control.

What determines the organization of the nervous system? An hypothesis directed at a partial answer to this question was formulated in the 1950s by the late W. C. Young. On the basis of his work on the effects of hormonal manipulation in the developing organism, Young hypothesized that early hormones function to organize particular parts of the nervous system. In the adult, the same hormones activate the previously organized mechanisms. Readings 19 and 20 stem directly from Young's research. Reading 19 describes the sexual behavior of female pseudohermaphrodite Rhesus monkeys receiving androgen treatment in adulthood, while Reading 20 shows that neonatal hormone treatment can modify the aggressiveness of adult mice.

Reading 21 from the Editor's laboratory reports an unusual, perhaps genotype-specific, behavioral response to androgen treatment. Wouldn't it be interesting if humans responded as B6D2F1 mice do?

JERRAM L. BROWN and
ROBERT W. HUNSPERGER

11   Neuroethology and the Motivation of Agonistic Behaviour*

I. Introduction

The aim of this paper is to bring to the attention of ethologists some findings on the neural bases of agonistic behaviour and to interpret them in relation to some commonly held ethological concepts of motivation. It is hoped that in doing so a better understanding between conventional ethologists and those who experiment directly with the central nervous system will result. Both of these groups depend on each other's findings for the understanding of their own problems; and both share common goals in the study of the mechanisms of behaviour.

The central problems confronting both ethology and the neurological sciences concern the mechanisms by which information

* Based on a paper presented at the 1961 International Ethological Conference. This investigation was supported in part by a U.S. Public Health Service fellowship (No. MF-11, 884) from the National Institute of Mental Health, P.H.S.

coming from outside the nervous system is received and evaluated, and by which responses are selected and programmed to result in co-ordinated and typically adaptive effector performance throughout ontogeny and phylogeny.

The methods of investigating these problems characteristic of ethology and the neurological sciences are fundamentally different. Ethological techniques are generally confined to manipulations and observations external to the organism. In contrast, the neurological sciences investigate directly neural structure and function.

Although ethologists may study such internal phenomena as releasing mechanisms and motivation, their method is essentially that of *drawing correlations between externally observable events* and using these correlations to characterize phenomena which are internally mediated.

For example, some ethological studies have concentrated on effector performance and the "internal motivation" of it, such as the study of Baerends, Brouwer & Water-

117

bolk (1955) on the sexual behaviour of the male guppy *(Lebistes reticulatus)*. In this study test females of standardized sizes were presented to the males and the resulting behaviour was correlated with the colour pattern of the male (which reflected its "internal motivation"). Both the colour pattern of the male and his behaviour to the test female may be considered effector responses which were correlated with each other.

A more complex example of concentration on effector performance is the factor analysis of the behaviour of the bitterling *(Rhodeus amarus)* performed by Wiepkema (1961) in which the occurrences of many types of behaviour were correlated with each other and the correlations mathematically attributed to a relatively small number of common factors.

In contrast to the methods used in conventional ethology a common experimental procedure for neuroethology is to make an alteration of some part of the nervous system, for instance, by activating or inactivating a specific part of it, and then to correlate changes in the behaviour with the alteration. The most common means of activation for neuroethology at the present time is the electrical stimulation of circumscribed small areas of the brain. By this means specific neural areas may be implicated in the mechanisms of the behaviour resulting from their stimulation. The neurobehavioural work on the cat will be considered as a specific example of this general approach.

The methods used by the authors are based on the original technique of Hess (1932, 1957), further developed in its electrical part by Wyss (1945, 1950, 1957) and Hunsperger & Wyss (1953).

The term, *agonistic behaviour,* includes all types of behaviour thought to contain elements of *aggressiveness, threat,* or *fear.*

II. Neurobehavioural Investigations of Agonistic Behaviour in the Cat

The programme of research in this laboratory on neural mechanisms in agonistic behaviour of the cat attempts to correlate various aspects of agonistic behaviour with both anatomical and physiological properties of the brain. This paper is based primarily on the findings of Hess & Brügger (1943), Hunsperger (1956), Fernandez de Molina & Hunsperger (1959, 1961), and Brown, Hunsperger & Rosvold (in preparation). The relevant literature including contributions by other workers has been reviewed by Hunsperger (1959).

A. AGONISTIC BEHAVIOUR ELICITABLE BY LOCALIZED STIMULATION

A considerable range of agonistic behaviour patterns may be elicited in the cat by electrical (or chemical; MacLean & Delgado, 1953) stimulation of small areas in the brain. The elicited behaviour depends on the site of stimulation in the brain, the intensity of activation, and the environment. It generally follows one of four patterns. These are (1) threat alone, (2) threat followed by attack, (3) threat followed by escape, (4) escape alone.

Some typical components of threat alone as elicited through brain stimulation are listed below (not in order of appearance, intensity, or other classification).

Opening of eyes
Pupillodilatation
Piloerection
Folding down of ears
Folding back of ears
Crouching
Forward rotation of whiskers
Lowering of the head
Protrusion of claws
Arching of back
Straightening and stiffening of legs
Faster and deeper respiration
Urination, defaecation
Growling and yowling
Shrieking
Hissing
Standing up
Tail quivering
Tail whipping
Erection of tail base

It is not uncommon to obtain every component in this list from a single brain locus; however, partial coverage of the list is more often obtained.

These components may be fully integrated into a behaviour which cannot be distinguished from the normal behaviour which cats show towards other cats, and sometimes toward other, larger species, such as dogs or humans. A second normal cat placed together with a cat stimulated to show such threat behaviour, reacts to the stimulated cat as it would under normal circumstances.

The behaviour described above is referred to as threat behaviour because it has that general function in social communication.

In the pattern, threat followed by attack, the cat performs threat as described above followed by an attack if a dummy is present or by a short forward rush and an explosive forward extension of the forepaws with claws protruding when no suitable object is present. Attacks on dummies commonly consist of striking the dummy's face with one or both forepaws with sufficient force to knock it over. But in some instances the cat has leaped upon the dummy, bitten it in the nape and ears, and used the hind feet in knocking the dummy over. Thus, the attacks vary in their execution, some appearing more aggressive than others.

In threat followed by escape the cat performs threat as described above followed by jumping off the experimental table, often accompanied by hissing.

In escape alone the cat first looks in all directions and then jumps off the table without any previous growling, hissing, or other actions especially characteristic of threat. The pupils may be dilated and there may be piloerection. When prevented from escaping by enclosure on the table, the cat runs rapidly back and forth looking for an exit.

B. NEUROANATOMICAL LOCALIZATION OF THREAT, ATTACK, AND ESCAPE BEHAVIOUR

No two items in the list of components of threat behaviour above have identical patterns of anatomical localization in the brain, but they all overlap in the general region from which threat behaviour is elicitable. Growling and hissing are generally characteristic of the threat pattern and the areas from which they have been elicited have been plotted in Fig. 1.

Fig. 1 (from Fernandez de Molina & Hunsperger, 1959) represents a parasagittal section of the cat brainstem with, superimposed upon it, some structures of the forebrain lying more laterally, such as the amygdala and hippocampus. It is a diagrammatic summary of many localization experiments and provides an overall view of the agonistic behaviour system in relation to the brain as a whole. In black and cross hatching are shown the areas from which threat behaviour was obtained, using as criteria growling and/or hissing integrated in a reaction involving other threat components. The areas from which escapes were obtained are indicated by vertical lines. Stimulation in other brain areas has not established that any other areas are so intimately concerned with the integration of threat behaviour. These results are in general agreement with those of other authors, such as Nakao (1958; see review by Hunsperger, 1959).

The figure illustrates that anatomically the system for threat behaviour is *not unitary* but *multirepresentational*. It has principal representation at three brain levels: the midbrain, the hypothalamus, and the amygdala.

There is no justification for the argument that the effects are primarily dependent on activation of areas lying outside the threat system. Indeed, the evidence is that activation of the neurons located in these areas is directly responsible for the threat behaviour elicited. Each of these three areas where threat behaviour is elicitable is an area of high cell density, and with the exception of the amygdala, long myelinated fibres are generally absent. The co-ordinated nature of the elicited behaviour, the absence of agonistic responses from stimulations in other brain areas, and the anatomical evidence (including lesion studies; see below) have led to acceptance among

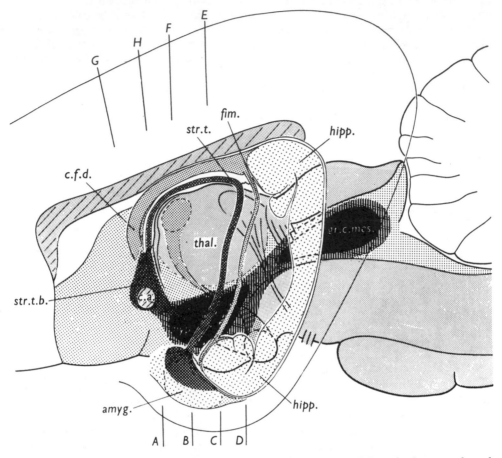

FIG. 1. Sagittal section through the brain stem of the cat with amygdala and other more lateral forebrain structures superimposed illustrating the threat and escape areas of brainstem and forebrain. This schematic representation is based on histological examination of over 800 sites of stimulation (from FERNANDEZ DE MOLINA and HUNSPERGER, 1959.)

 Active field of amygdala, stria terminalis, and stria terminalis bed, continuing into:

 Active field of the hypothalamus and the central gray of the midbrain. Black: inner zone yielding hissing; hatched: outer zone yielding escape.

amyg.	= amygdala.
c.a.	= anterior commissure.
c.f.d.	= descendent column of fornix.
fim.	= fimbria.
gr. c. mes.	= midbrain central gray substance.
hipp.	= hippocampus.
hyp.	= hypothalamus.
str. t. b.	= stria terminalis bed.
thal.	= thalamus.

120

neurophysiologists and neuroanatomists of the concept that these areas are integrative in function, concerned with motivation, and depend on the population of neurons located in these areas for their properties.

The threat behaviour patterns elicited from the amygdala, hypothalamus, and midbrain have many elements in common. such as lowering of the head, laying back of the ears, piloerection, pupillodilatation, deepened, more rapid respiration, hunching of the back, hissing, and growling. Differences in the behaviour elicited from these three areas are present, however. The most striking of these is in the vocalizations: from the midbrain hissing is the predominant vocalization; from the amygdala growling predominates although hissing has also been elicited in addition to growling from about half of the points which gave growling. From the hypothalamus the reactions are usually characterized by mixed growling and hissing, thus intermediate behaviourally and anatomically between midbrain and amygdala.

Since growling is correlated more with aggression, and hissing with defence (Leyhausen, 1956), the possibility arises that this difference in behaviour elicitable from anterior and posterior ends of the threat system might reflect an anterior-posterior gradient in aggressiveness of the threat. It would be interesting to know if a similar anterior-posterior gradient in aggressiveness of the elicited threat behaviour occurs in other species.

Another anatomical finding of importance to ethology is that the areas from which escapes may be elicited are located adjacent and overlapping with those for threat. This is consistent with the principle often encountered in neuroanatomy that similar functions are represented close together in the brain. Escape and threat in the cat may both function in self-protection, whether from other cats or other species. In the brainstem escape reactions are elicitable from regions just anterior and posterior to the threat areas in the hypothalamus and midbrain and also to a lesser extent from a thin layer surrounding and inter-

connecting the threat areas (Hunsperger, 1956). Combinations of threat and escape are often obtained from the border areas. In the amygdala the escape responses are obtained from areas which are also closely related anatomically to those for threat responses (Fernandez de Molina & Hunsperger, 1959; Ursin, 1960).

C. THREAT BEHAVIOUR AS A FUNCTION OF INTENSITY OF ACTIVATION

In addition to being dependent on the *locus* of stimulation the nature of the elicited threat behaviour is also greatly influenced by the *intensity* of stimulation. Raising the intensity of stimulation either through frequency or voltage has the effects of (1) activating the area under direct stimulation more intensely and/or (2) activating a larger area. The more conspicuous effects of increasing the voltage are seen not only in latencies but also in the intensities or rates of expression of the component elements of threat and in the behaviour as a whole. For example, with increasing voltage pupillodilatation and piloerection may become extreme, the ears fully lowered, the rate of growling or hissing may increase, and the loudness and pitch of growling may increase; at near maximum activation the legs are rigidly extended, the tail rigid or whipping, and explosive attacks with the forepaws may be made. The elicited behaviour may also change in character, for example, from growling alone to growling and hissing.

D. MODIFICATION OF CENTRALLY ELICITED BEHAVIOUR BY EXTERNAL ENVIRONMENT

Of the various types of physiological and behavioural changes which can be elicited by central stimulation in the vertebrates ranging from the movement of a finger in man to whole series of natural integrated behaviour patterns in many species it is the latter type which is of special interest in regard to motivation. The neural areas from which natural integrated behaviour may be elicited do not seem to be just motor co-ordination centres, for stimulation of these areas appears to cause real changes in the pre-disposition of the animal. When humans are stimulated in the amygdala, for example, "the strong emotional response is conscious and integrated in the thinking of the patient. . . ." (Heath, Monroe & Mickle, 1955). Similar findings have been reported for the human hypothalamus and midbrain (Sem-Jacobsen & Torkildsen, 1960; Heath & Mickle, 1960).

The basic mood or *motivation, which may be considered as neural activity,* is rather specific to the locus stimulated and appears to be relatively independent of the presence or absence of particular objects in the environment. However, such objects may play a role in the orientation of the behaviour and may bring out some external expression of the elicited motivation which might not otherwise have appeared. Just as the neurosurgeon can characterize the emotions of a patient during stimulation better by talking with him, the neuroethologist can characterize the motivation underlying the elicited behaviour better by varying the external environment through the presence of stuffed or live animals during stimulation and other means.

The independence of the basic motivation elicited during stimulation of areas which elicit threat and escape from various aspects of the external situation is striking. The threat and escape reactions appear in the "neutral" experimental room and require no special factors in the environment. They have appeared in a blind cat and in cats which could not see the experimenters as readily, or more so, as in normal cats in full view of the experimenters.

Sex, age, the individual temperament of the cat (tameness, wildness, submissiveness) and the particular behaviour in which the cat was engaged just before stimulation influence the occurrence of centrally elicited threat and escape relatively little compared to the location of the electrode, in the experience of the authors. The elicited threat and escape behaviour take precedence over apparently all other forms of activity when the appropriate areas of the brain are stimulated. The adaptive value

of this attribute of agonistic behaviour is obvious. It also makes agonistic behaviour one of the easier to study of the categories of complex behaviour patterns which may be elicited by brain stimulation.

In many experiments the cat has been stimulated while it was in the presence of a stuffed cat, dog, fox, or pigeon. In order for the dummy to have an effect upon threat behaviour some part of the neural substrate for threat behaviour had apparently to be activated electrically. The dummy had no effect alone. In almost all cases where the presence of the dummy strengthened threat behaviour or brought in an attack it was with electrode placements where stimulation was capable of eliciting elements of threat behaviour even in the absence of any dummies. In a few cases elements of threat and sometimes attacks were brought in by stimulation in the presence of the dummy from loci which elicited escape but did not elicit threat behaviour. In these it seemed likely because of the anatomical location of the electrodes next to the threat zone and the nature of the behaviour which was elicited that part of the threat substrate was being activiated but at subthreshold levels.

The behaviour towards the dummies during stimulation at loci where escape but not elements of threat were elicited consisted primarily of investigative sniffing, especially in the nasal and facial regions (Nasenkontrolle). Unlike the agnostic responses, this sniffing waned rapidly with a few repetitions using the same dummy. The loci lay mostly lateral and posterior to the hypothalamic threat substrate.

Where there was an effect of the dummy on threat behaviour, it consisted of more extreme development of the whole pattern, sometimes with shortened latencies and quickened rates, and in some cases attacks on the dummy with the forepaws.

In order for the elicited threat behaviour to be directed toward the dummy, the species of dummy seems not to be critical. Strong threat reactions with much hissing and growling and apparent fear of the opponent have been given toward the stuffed fox, dog, cat, and pigeon, and also toward a live guinea pig and a live kitten. The motivation elicited by the electrical stimulus, therefore, does not depend in our experience on the species of dummy but primarily on the location of the electrode.

It was at first thought that the strengthening effect was dependent on the cat's perception of the dummy as an enemy. However, in several cases cats directed startling threats and attacks at dummies which they seemed to completely ignore before stimulation. Recent evidence of a preliminary nature indicates that with weak stimulation in the threat zone the perception of the dummy as an enemy (as judged by the cat's behaviour when not stimulated) and the strengthening effect of the dummy on the elicited threat behaviour may wane together. But with strong stimulation in the same area the dummy may acquire new meaning to the cat as a dangerous enemy even when it was ignored beforehand. Thus, the strengthening effect of the dummy may require that the cat regard it as an enemy; but when such regard is absent, it may be generated by the electrical stimulus itself in conjunction with the dummy. Long after-reactions of growling toward the dummy which sometimes occurred are evidence for this view.

In some ways the presence of the dummy had an effect similar to the simultaneous stimulation of a second area eliciting hissing and associated threat elements (see below). This was apparent in the shortened latencies and fuller, more intense development of the threat sequence.

E. INTERRELATIONS

As the nervous system functions on the principle of interdependence of cells and of populations of cells, dynamic aspects of the neural activities mediating the motivation of agonistic behaviour are of primary importance. Unfortunately, however, they are poorly understood. For example, it is largely unknown whether the threat areas, or parts of them, in amygdala, hypothalamus, and midbrain are normally differentially activated according to different

motivations and what spatio-temporal patterns of activation occur during different motivations. Also the exact neural paths of input and output to these areas, whether inhibitory or excitatory, are poorly known. Since so little is actually known about these questions only a few aspects can be mentioned.

Some aspects of interdependence have, however, been studied. The work in which a lesion in one area was combined with stimulation of another area (Hunsperger, 1956; Fernandez de Molina & Hunsperger, 1961) illustrates some aspects of the interdependence of the various parts of the threat system. After acute lesions at some levels of the system the threat or escape behaviour elicited from remaining levels was eliminated. The elimination of threat behaviour elicited in one area after lesioning in another area was strikingly dependent on the location of the lesion. Damage to the anterior part of the threat system (amygdala, bilateral) had no effect on the threshold of the threat and escape behaviour elicited more posteriorly; while damage to the posterior parts (midbrain and hypothalamus) eliminated threat behaviour elicited anterior to the lesion immediately after coagulation. Two weeks later it could again be elicited without apparent change in character but at a higher threshold. These studies also indicate that threat behaviour may be elicited from the midbrain area independently from the anterior parts of the system.

Work using the technique of simultaneous stimulation of two brain areas concerned with agonistic behaviour (domestic fowl, von Holst & von Saint Paul, 1958, 1960; cat, Hunsperger & Brown, 1961) has not reached a stage where generalizations concerning the whole threat system can be made. However, studies in this laboratory indicate that within the hypothalamus of the cat are areas which when stimulated strengthen threat and escape reactions elicited and areas which weaken them. Generally when one response may be elicited by each of two electrodes, that response, such as growling, hissing, piloerection, ear-flattening, or escaping, is strengthened by simultaneous stimulation with the two electrodes. Threat behaviour elicited from the hypothalamus is generally strengthened and changed in character by simultaneously elicited escape behaviour, and in these combinations escape behaviour is often delayed.

Since areas for "pure" attack have not been found in the cat brain, it has not been possible to perform the interaction experiment so long desired by ethologists, namely the simultaneous activation of "pure" attack and escape mechanisms. The failure to find these areas is of significance in itself, however, and will be discussed below.

III. Attack-Escape Theory

The neural bases of agonistic behaviour have not been investigated in detail for many species; and until the results of studies in other vertebrate classes are published, the mammals, especially the cat, remain the principle source of information. Knowledge of the neural bases of agonistic behaviour in the cat is only suggestive of the situation in other species; but because of the existence of basic homologies in the anatomy of the brainstem among vertebrate classes it is reasonable to suppose that *some* of the *general* properties of the neural bases of agonistic behaviour will be found to be similar in other species, despite difference in detail. Evidence for basic similarities between bird and mammal in the brainstem organization of certain physiological and behavioural regulatory mechanisms has been given by Åkerman et al., (1961). The more general implications of the works discussed above will be considered as an aid in improving the working hypothesis underlying ethological motivation studies.

Attack-escape theory has apparently never received formal definition in its entirety, nor has it been customarily referred to in print by any name. Rather, it has developed in stages and is best known through the investigations in which it has been applied. The theory became strongly influential on ethological analyses of the motivation of agonistic behaviour through the works of Tinbergen (1952), Hinde (1952), Moynihan (1955) and others. No attempt to assign credit for each development of the theory to individual authors will be made in this paper.

In each of the following four sections we shall attempt briefly to first state the principal conclusions of attack-escape theory, second, mention the types of behavioural observations which have given rise to each conclusion, third, examine the neurological correlations which may be made with these observations, and fourth, compare the neurological experimental results with the conclusions of attack-escape theory.

The principal inferences characterizing attack-escape theory follow:

(*a*) The motivation of behaviour may be understood through the use of *unitary drive concepts.*

(*b*) Threat behaviour has *dual* rather than unitary motivation; the two drives being those for *attack* and *escape.*

(*c*) Threat behaviour is caused to appear by the simultaneous activation and *mutual inhibition* of attack and escape drives.

(*d*) The motivation of different types of threat and other behaviour patterns of a species may be *characterized* by the intensities and relative strengths of attack and escape drives.

(*a*) *Unitary drives.* Ethological motivation analyses of agonistic behaviour have been in recent years frequently founded on unitary drive concepts. These allow behaviour to be interpreted in terms of attack, escape, sex, parental, hunger drives, etc. The dangers of unitary drive concepts in ethology have been ably reviewed by Hinde (1959), and the reader is referred to that paper for a detailed discussion. In ethology such concepts have been inferred primarily from the common observation that different behaviour patterns with similar functions (e.g. obtaining of food) frequently occur in close spatio-temporal proximity with each other. This observation has led to the postulating of a common causal mechanism for the behaviour patterns concerned, and this mechanism has in some cases been hypothesized to be a neural "centre". For the central nervous system concepts of unitary centers have a history of inadequacy in the explanation of various physiological and behavioural functions. Fig. 1 reveals that anatomically the concept of central nervous centres for attack and escape has little meaning in the cat. In any case the assumption that unitary mechanisms exist is unnecessary for the investigation of neural mechanisms, and it seems equally unnecessary in purely behavioural investigations.

(b) Dual motivation of threat. That threat behaviour has dual rather than unitary motivation has been concluded from inferences that both attack and escape drives are active during threat. Such inferences concerning the existence of sub-threshold activation of attack and escape drives during threat have been made from three types of observation (Tinbergen, 1959). These are:

(1) occurrence of threat in close temporal proximity with the acts of attack and escape.

(2) occurrence of threat in spatial and temporal contexts where attack and escape drives would both be expected to be sub-threshold, and

(3) the involvement of acts and postures in threatening which are related to or identical with those used in actual attack and escape.

Similar observations can be made in the cat. For example, growling occurs frequently in temporal proximity with attack and hissing with defence and escape (Leyhausen, 1956). Gradations in posture and facial expression between attack behaviour and various kinds of threat behaviour have been shown by Leyhausen (ibid.). Threat and escape behaviour in the cat have in common piloerection and pupillodilatation.

The postulating of unitary attack and escape drives to explain such observations is unnecessary. For such close relationships between threat, attack and escape may be correlated simply with the extensive coincidence of the neural areas where threat and attack may be elicited and the considerable overlap of the neural areas where threat and escape may be elicited.

It has been established that there are in the cat brain definite areas from which threat behaviour may be consistently elicited. This fact appears to contradict the theory that threat behaviour results from simultaneous activation and conflict of neural mechanisms for attack and escape, since *only one area needs to be directly stimulated, not two*. In birds threat can be elicited with one electrode also (von Holst & von Saint Paul, 1958). Furthermore, no anatomical or physiological provisions have been shown in any species for the motivation of threat behaviour by means of antagonistic interaction between neural mechanisms for pure attack and escape. Thus, although the occurrence of threat behaviour may be correlated with simultaneous activation of hypothetical attack and escape drives in some species, it can in the cat and chicken also be correlated with activation of an experimentally demonstrable neural substrate for threat behaviour.

(c) Mutual Inhibition. The inference that both attack and escape drives are active during threat was drawn from the three types of observations discussed above (temporal proximity, spatio-temporal contexts, postures). The failure of attack and escape to appear during threat even though active has been attributed to their mutual inhibition and conflict. The normally observed mutual exclusion of the acts of attack and escape has probably been most responsible for the concept that they are mutually inhibitory or conflicting. For the present we cannot investigate this particular conflict in the cat, if it exists, with neurological methods because it has not been possible to elicit an attack which is not preceded by threat. Stimulation in the threat-attack zone of the hypothalamus often retards escape elicited from the hypothalamus. But stimulation in the escape zone generally strengthens and prolongs simultaneously elicited threat behaviour. In the two cases investigated to date stimulation in the escape zone hastened simultaneously elicited attacks preceded by threats. Present results, therefore, indicate that *mutual* inhibition between threat-attack and escape zones does not occur.

(d) Characterization of Threat Displays. Characterization of the specific threat displays of a species according to the intensities and relative strengths of attack and escape drives has become widespread in recent ethological literature. Such characterizations have been inferred through the three types of observations mentioned above.

For example, from the observations of Leyhausen (1956), on the cat it can be deduced that growling occurs typically and more often than hissing in temporal proximity with attack and that hissing occurs more often than growling in temporal proximity with escape and defence. From these observations the inference could be made according to attack-escape theory that the attack drive was higher during growling than the escape drive and the escape drive higher during hissing than the attack drive.

These observations may also be correlated with experimental findings for the brain: Attacks elicited by brain stimulation are characteristically preceded by growls and hisses. In contrast, escapes elicited from regions of overlap of escape and threat are characteristically preceded by hisses without growls. In addition, growls are characteristically elicited from the amygdala and hypothalamus while hisses are characteristically elicted from the midbrain and hypothalamus. This evidence suggests that the spatial pattern of neural activation within the threat-attack system is different during these two types of threat vocalization. Furthermore, the extensive overlap of the areas where growling and attack may be elicited correlates with the close temporal relationships of growling and attack in normal behaviour. And the overlap of the areas where hissing

and escape may be elicited correlates with the close temporal relationships of hissing and escape in normal behaviour. Thus, for the cat different types of threat and their temporal relationships with attack and escape can be correlated with different spatial patterns of central nervous activity.

In summary, when attack-escape theory as discussed above is compared with present knowledge of the structure and function of the nervous system, little direct agreement is found. However, many of the behavioural observations for which attack-escape theory has been invoked may now be correlated directly with the structure and function of the brain.

IV. Terminology

"Pure Attack". Attacks are easily elicited by brain stimulation in the cat, but "pure attacks" are not. According to attack-escape theory "pure attacks" should be those attacks which have no elements of fear or threat motivation in them. In the cat this might be equivalent to catching a mouse or to attacking in situations where the cat has learned not to be afraid. Such confident attacks do not result from stimulation of the threat-attack zone under the conditions of our experiments. On the contrary, stimulation there has caused cats to threaten in apparent self-defence such harmless objects as a live guinea pig and a small kitten and to attack the guinea pig defensively.

In discussing "pure attack" in the cat Leyhausen (1956) wrote, "Das reine, durch teilweise Ueberlagerung mit anderen Verhaltensweisen nicht gestörte Angriffsverhalten des Katers sieht man nur selten". [Only rarely does one see in the tomcat purely aggressive behaviour that is not partially modified by or combined with other kinds of social behaviour.] However, he also referred to the same behaviour as in "extreme threat position" ("in extremer Drohstellung"). From his own statements, photographs, and descriptions it is clear that this is a threat posture rather than a

126

"pure attack". Moreover, it is not the extreme threat posture obtained by stimulation of the threat zone in the cat brain but a partial development of it. It is similar to some attacks made on dummies during stimulation of the border area between threat-attack and escape zones.

Some clarification of the ontogenesis of "pure attack" may be gained by consideration of established social hierarchies among groups of vertebrates. Under these conditions supplanting attacks are often seen devoid of external signs of fear or threat. The absence of fear evinced in these attacks is specific to the opponent or the location and depends on previous establishment and stabilization of a dominant-subordinant relationship. If the opponent is slow in leaving, a mild threat may occur; and if it stays, strong threat followed by fighting often results. It would appear probable from this relationship that the more hesitant the attack and the more resistance of the opponent, the more activation of a threat system in the attacker.

Conditioning to specific opponents or situations is probably necessary for such attacks, and they could probably also be conditioned in the cat. It is known that mice can be trained to attack or not (Scott, 1958: 18) and rats to be killers of mice or not (Heimstra & Newton, 1961). In 14 of 16 such killer rats bilateral amygdalectomy eliminated the killing response (Karli, 1956) thus implicating a structure known to be part of the threat-attack system in the cat in a learned attack behaviour. In man anger and attack have also been eliminated by ablation of temporal lobe structures including the amygdala (cases reviewed by Ursin, 1960). Work on the cat amygdala in relation to attack is in progress in this laboratory.

Although much has been said about a "threat system" or "threat-attack system", *one should not infer that it is a unitary system and that all varieties of threat behaviour in the cat are merely the result of* quantitatively different levels of activation of a unitary threat system. The term "threat system" has been used in reference to the basic similarities in the behaviour patterns elicited from all parts of it. It should be stressed, however, that important differences also exist in the behaviour elicited from various places in the threat system. Virtually no two component elements of threat behaviour have exactly the same pattern of localization in the brain; and just which combination of elements is elicited appears to vary from place to place within the threat-attack zone.

All factors which affect motivation in the vertebrates, whether external stimuli or internal stimuli and conditions must work through the central nervous system. It follows from this relationship that ethological motivation theories have value primarily through their relevance to actual events in the central nervous system. The terms and concepts employed in such theories are, thus, of critical importance to the understanding of the relationship between the nervous system and behaviour.

Some ethologists may prefer to refrain from making inferences about nervous processes on the basis of their behavioural observations. If so, then to be consistent with neural concepts of motivation the terms "tendency", "drive", and their substitutes would become unnecessary. These terms, which are used interchangeably in this paper, have received two basic types of definition in ethology. One type defines them as reflecting "the state of an animal. . . ." (Tinbergen, 1959: 29); the other type, purposely even less explicit, defines them as "the complex of internal and external factors leading to a given behaviour" (van Iersel & Bol, 1958: 5).

If "tendency" in confined in its meaning to external stimulus factors, then with adequate and accurate description of them the concept of a tendency becomes superfluous and misleading. If "tendency" is defined as including internal factors (both definitions above), then by definition it concerns the state of the central nervous system. If "tendency" is meant to describe explicitly only the behaviour and not to imply anything about the nervous system, then the description would better be made

127

in terms of the units of observation and the correlations between the different measures (these need not be mathematically stated); "tendency" in this case is also superfluous and misleading.

Since confusion is at present inevitable in this subject it would seem advisable either to be explicit in making statements about the theoretical neural state in motivation by using a term such as "neurobehavioural mechanism" (NBM) or not to mention neural states and to restrict the discussion to the units of observation actually used and the correlations between the different observations. In both cases "tendency" and "drive" need not be used at any time and ambiguity would be reduced.

V. Conclusion

Ethologists have been confined primarily to observations of external events. Their form, their quantitative variation, and the correlations between them as analysed by ethologists are some of the phenomena which the neurobehavioural sciences seek ultimately to explain. In the case of agonistic behaviour explanations are sought especially for behavioural relationships concerning (1) temporal proximity and sequence, (2) postural similarities and differences, and (3) general environmental and discrete stimulus contexts. When such phenomena are objectively and clearly described and correlated by ethologists they become more useful to the neurobehavioural sciences. It is hoped that ethologists will provide detailed and quantitative information of these types for the cat, monkey, and other species in common use in the neurobehavioural sciences.

Conversely, ethology may also make use of the contributions of the neurological sciences. Because the motivation and organization of behaviour are mediated through the nervous system, whatever is known about neural mechanisms in behaviour should be utilized by ethologists in the improvement of research hypotheses and theories. The existence in the brain of a multirepresentational system which when activated leads to threat followed sometimes by attack and the existence in close anatomical relationship to it of areas which when activated lead to escape are neurobehavioural facts of direct relevance to ethology.

Summary

1. The methods of ethological investigations are primarily concerned with drawing correlations between externally observable events. Neuroethology concerns additionally direct correlations with activity and structure of the nervous system.

2. Integrated threat behaviour may be elicited from three different brain areas: in the amygdala, hypothalamus, and midbrain of the cat. The neural substrate for the "motivation" of threat behaviour is, therefore, not unitary but multirepresentational.

3. The neural substrate for "motivation" of threat behaviour in the cat overlaps that for escape. This overlap helps to explain the temporal, postural, and situational similarities of threat to escape behaviour.

4. Attacks may be elicited with high intensity stimulation at some places which at the same and lower intensities of stimulation first yield threat. This functional relationship helps to explain the temporal, postural, and situational similarities of threat to attack behaviour.

5. The basic "mood" elicited by stimulation in these motivation areas is primarily determined by the intensity and locus of stimulation. Dummies and other aspects of the external environment (conversations in the case of man) play a role in the orientation of the behaviour and may bring out some external expressions of the elicited motivation which might not otherwise appear.

6. The attack-escape theory and the terms "drive", "tendency", and "neurobehavioural mechanism" are discussed. Non-neural concepts of drive or tendency are considered to be superfluous and misleading.

Acknowledgments

The authors wish to thank the many people with whom they have discussed these problems and to thank particularly Drs. R. A. Hinde, H. E. Rosvold and N. Tinbergen.

REFERENCES

BAERENDS, G. P., R. BROUWER and H. T. J. WATERBOLK, 1955 Ethological studies on *Lebistes reticulatus* (Peters). I. An analysis of the male courtship pattern. *Behaviour*, 8, 249–334.

FERNANDEZ DE MOLINA, A., and R. W. HUNSPERGER, 1959 Central representation of affective reactions in forebrain and brain stem: electrical stimulation of amygdala, stria terminalis, and adjacent structures. *J. Physiol.*, 145, 251–265.

FERNANDEZ DE MOLINA, A., and R. W. HUNSBERGER, 1962 Organization of the subcortical system governing defence and flight reactions in the cat. *J. Physiol.*, 160, 200-213.

HEATH, R. G., and W. A. MICKLE, 1960 Evaluation of seven years experience with depth electrode studies in human patients. In *Electrical Studies on the Unanesthetized Brain*. New York: Hoeber, Inc., 214–247.

HEATH, R. G., R. R. MONROE, and W. A. MICKLE, 1955 Stimulation the amygdaloid nucleus in a schizophrenic patient. *Amer. J. Psychiat.*, 111, 862–863.

HEIMSTRA, N. W. and G. NEWTON, 1961 Effects of prior food competition on the rat's killing response to the white mouse. *Behaviour*, 17, 95–102.

HESS, W. R., 1932 *Beiträge zur Physiologie des Hirnstammes. I. Die Methodik der lokalisierten Reizung und Ausschaltung subkortikaler Hirnabschnitte.* Leipzig: Georg Thieme.

HESS, W. R., 1957 *The Functional Organization of the Diencephalon.* New York, London: Grune and Stratton.

HESS, W. R., and M. BRÜGGER, 1943 Das subkortikale Zentrum der affectiven Abwehrreaktion. *Helv. physiol. Acta*, 1, 33–52.

HINDE, R. A., 1952 The behaviour of the Great Tit (*Parus major*) and some other related species. *Behaviour, Suppl.*, 2, x & pp. 201.

HINDE, R. A., 1959 Unitary drives. *Anim. Behav.*, 7, 130–141.

HUNSPERGER, R. W., 1956 Affektreaktionen auf elektrische Reizung im Hirnstamm der Katze. *Helv. physiol. Acta*, 14, 70–92.

HUNSPERGER, R. W., 1959 Les représentations centrales des réactions affectives dans le cerveau antérieur et dans le tronc cérébral. *Neuro-Chirurgie*, 5, 207–233.

HUNSPERGER, R. W. and J. L. BROWN, 1961 Verfahren zur gleichzeitigen elektrischen Reizung verschiedener subcorticaler Areale an der wachen Katze. Abwehr- und Fluchtreaktion. *Pflügers Arch. ges. Physiol.*, 274, 94.

HUNSPERGER, R. W., and O. A. M. WYOO, 1953 Quantitative Ausschaltung von Nervengewebe durch Hochfrequenzkoagulation. *Helv. physiol. Acta*, 11, 283–304.

HOLST, E. VON, and U. VON SAINT PAUL, 1958 Das Mischen von Trieben (Instinktbewegungen) durch mehrfache Stammhirnreizung beim Huhn. *Naturwissenschaften*, 45, 579.

HOLST, E. VON, and U. VON SAINT PAUL, 1960 Vom Wirkungsgefüge der Triebe. *Naturwissenschaften*, 18, 409–422.

IERSEL, J. J. A. VAN, and A. C. A. BOL, 1958 Preening of two tern species. A study on displacement activities. *Behaviour*, 13, 1–88.

KARLI, P., 1956 The Norway rat's killing response to the white mouse: an experimental analysis. *Behaviour*, 10, 81–103.

LEYHAUSEN, PAUL, 1956 Verhaltensstudien an Katzen. *Z. Tierpsychol.*, Beiheft 2.

MACLEAN, P. D., and J. M. R. DELGADO, 1953 Electrical and chemical stimulation of fronto-temporal portion of limbic system in the waking animal. *Electroenceph. clin. Neurophysiol.*, 5, 91 100.

MOYNIHAN, M., 1955 Some aspects of reproductive behavior in the Black-headed Gull (*Larus ridibundus ridibundus* L.) and related species. *Behaviour, Suppl.* 4, x & pp. 201.

NAKAO, H., 1958 Emotional behavior produced by hypothalamic stimulation. *Amer. J. Physiol.*, 194, 411–418.

OKERMAN, B., B. ANDERSSON, E. FABRICIUS, und L. SVENSSON, 1961 Observations on central regulation of body temperature and of food and water intake in the pigeon (*Columba livia*). *Acta physiol. scand.*, 50, 328–336.

Scott, J. P., 1958 *Aggression.* University of Chicago Press.

Sem-Jacobsen, C. W., and A. Torkildsen, 1960 Depth recording and electrical stimulation in the human brain. In *Electrical Studies on the Unanesthetized Brain.* New York: Hoeber, Inc., 275–287.

Tinbergen, N., 1952 "Derived" activities; their causation, biological significance, origin, and emancipation during evolution. *Quart. Rev. Biol.,* **27,** 1–32.

Tinbergen, N., 1959 Comparative studies of the behaviour of gulls (Laridae): a progress report. *Behaviour,* **15,** 1–70.

Ursin, H., 1960 The temporal lobe substrate of fear and anger. *Acta psychiat. neurol.*

scand., **35,** 278–396.

Wiepkema, P. R., 1961 An ethological analysis of the reproductive behaviour of the Bitterling *(Rhodeus amarus* Bloch). *Arch. néerl. Zool.,* **14,** 103–199.

Wyss, O. A. M., 1945 Ein Hochfrequenz-Koagulationsgerät zur reizlosen Ausschaltung. *Helv. physiol. Acta,* **18,** 18–24.

Wyss, O. A. M., 1950 Beiträge zur elektrophysiologischen Methodik. II. Ein vereinfachtes Reizgerät für unabhängige Veränderung von Frequenz und Dauer der Impulse. *Helv. physiol. Acta,* **8,** 18–24.

Wyss, O. A. M., 1957 Nouveaux appareils éléctrophysiologiques (VII). *Helv. physiol. Acta,* C49–C50.

12

Hypothalamic Norepinephrine: Circadian Rhythms and the Control of Feeding Behavior

Abstract. *The time of day is a decisive determinant of the effects of* l-*norepinephrine on feeding behavior. During the dark, direct application of* l-*norepinephrine to the hypothalamus of rats suppressed feeding behavior. During the light, treatment with the same dose of* l-*norepinephrine facilitated feeding behavior. Thus,* l-*norepinephrine has dual and opposite effects on feeding behavior. A hypothalamic substrate that fluctuates in a circadian rhythm could account for both actions of* l-*norepinephrine.*

The addition of exogenous *l*-norepinephrine (*l*-NE) to the lateral hypothalamus affects feeding behavior. Both stimulant (*1*) and suppressant (*2*) effects have been reported, but the conditions that determine when each of these opposite actions will occur are unknown. This has led to the development of a controversy between proponents of the noradrenergic-feeding and the noradrenergic-satiety theories. The implications of each theory have been reviewed by Hoebel (*3*). We now report data that appear to resolve the controversy. The effects of the addition of exogenous *l*-NE to the hypothalamus appear to be dependent on differences in the internal state of the hypothalamus associated with the environmental cycle of darkness and light. In the dark, this treatment suppressed feeding behavior. In the light, the same dose of *l*-NE applied to the same hypothalamic site facilitated feeding behavior. The type of light-dark cycle and the time of drug administration generally have been omitted in the psychological literature on rodent feeding behavior. Some information, however, is available about these important variables (*1, 2*).

130

Thus, it is possible to evaluate earlier experiments in terms of present results.

Eight male albino rats (Charles River) at a body weight of 300 g were adapted to a reversed light-dark cycle with cool white fluorescent lights turned off automatically at 0740 hours E.S.T. and on at 1940. The rats were housed in automated housing (Environmental Sciences) and given free access to pellets of Purina Laboratory Chow and filtered tap water. Three to four days a week they were placed in Plexiglas chambers in the dark at 1030 hours for a period of 1 hour. The chambers contained a full burette of milk (undiluted Pet condensed milk) and a full burette of filtered tap water. Both liquids were at room temperature. Each burette was covered by a Plexiglas guard and contained a stainless steel wire that was connected to a contact-sensitive relay, designed to record the number of licks made by the rat. We used printout counters and cumulative recorders to monitor the rate of the licks. Spillage of milk and water was collected in petri dishes, which allowed total consumption of milk and water to be calculated at the end of the hour for each rat. In addition, the rats were weighed before and after the feeding test. Stabilization of milk-licking behavior developed within 2 to 3 weeks. Some water-licking behavior also occurred but was negligible in quantity. We then anesthetized the rats with pentobarbital and surgically implanted bilateral cannulas that were aimed for the lateral hypothalmic site, as in (2). Several weeks were allowed for recovery from surgery and stabilization again of milk-licking behavior. l-Norepinephrine hydrochloride was administered by means of duplicate inner cannulas that were weighed on a Cahn electrobalance (G-2) before and after they were loaded with the powdered drug. Immediately prior to the milk-licking test, empty cannulas were removed from the brain and immediately replaced with cannulas that had been loaded with a 25 μg dose of l-norepinephrine hydrochloride (4). All drug treatments were preceded by at least one control day, and the drug treatments did not occur more often than once each week. No drug treatment was made unless the rats were fully stabilized from the prior treatment. Insertions of empty inner cannulas were without effect on feeding behavior.

All tests during week 1 occurred about 3 hours after the onset of darkness at 1030 hours. Treatment with l-NE produced a suppression of mean intake of milk and mean number of licks of milk from the control levels prior to treatment with l-NE (Fig. 1a). Both suppressions were statistically significant ($t = 2.00$, d.f. = 7, $P < .05$ for intake and $t = 3.27$, d.f. = 7, $P < .01$ for number of licks). None of the subsequent control tests prior to drug treatment in the dark differed significantly from the first control or from each other. All of these control results were averaged.

During week 2, tests were made approximately 3 hours into the light at 2230 hours. This and subsequent control tests prior to drug treatment in the light did not differ significantly from each other and also were averaged. The control in the light shows less mean intake of milk and fewer mean licks of milk in comparison to the control in the dark (Fig. 1, a and b). Both decreases were statistically significant ($t = 2.56$, d.f. = 7, $P < .025$ for intake and $t = 3.28$, d.f. = 7, $P < .01$ for number of licks). The suppressant effects of light on rodent feeding behavior are well known (5). Treatment of the hypothalamus with l-NE at 3 hours into the light had no statistically significant effect on mean intake of milk or mean number of licks of milk ($t = 1.06$, d.f. = 7, $P > .05$ for intake and $t = 0.63$, d.f. = 7, $P > .05$ for number of licks). In weeks 3 and 4, treatments with l-NE were made 1

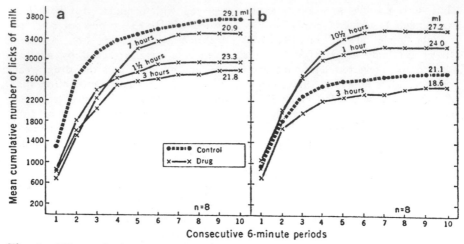

Fig. 1. Effects of direct bilateral application of *l*-norepinephrine (25 μg) to the lateral hypothalamus at various times of the day, on mean cumulative number of licks of milk and mean total intake of milk in milliliters. The number of hours between the beginning of the light (b) or dark (a) period and the start of the drug test is indicated on each drug curve. Control scores were obtained on the day prior to treatment with *l*-norepinephrine.

and 10½ hours into the light. In both cases (Fig. 1b) *l*-NE increased milk intake and milk licking of the rats significantly from the controls in the light ($t = 1.93$, d.f. = 11, $P < .05$ and $t = 2.50$, d.f. = 7, $P < .025$, respectively for intake; $t = 3.59$, d.f. = 11, $P < .005$ and $t = 2.40$, d.f. = 7, $P < .025$, respectively for number of licks). Individual controls prior to drug treatment are in Table 1 for each of the three drug treatments administered in the light.

In weeks 5 and 6, the rats were tested during darkness (Fig. 1a), and *l*-NE once again significantly suppressed both mean intake of milk and mean number of licks of milk (at 1500 hours, about 7 hours into the dark, $t = 2.06$, d.f. = 7, $P < .05$ for intake and $t = 3.06$, d.f. =7, $P < .01$ for number of licks; and at 0900 hours, approximately 1½ hours into the dark, $t = 3.44$, d.f. = 7, $P < .01$ for intake and $t = 1.12$, d.f. = 7, $P > .05$ for number of licks. The reinstatement of the suppressant effects of *l*-NE in the dark indicates that the long-term series of treatments with

NE did not measurably affect the final behavioral response to *l*-NE. Therefore, it is unlikely that drug-induced irritations or lesions are responsible for the opposite effects of *l*-NE on feeding behavior. Individual mean control scores prior to drug treatment are shown in Table 1 for each of the three drug treatments given in the dark.

The role of circadian rhythms in the neurochemical control of feeding behavior can no longer be ignored. Rats are nocturnal animals. Their feeding

Table 1. Mean control scores prior to treatment with *l*-norepinephrine. Number of rats is indicated in parentheses.

Time of test (hours from onset)	Mean licks of milk (No.)	Mean intake (ml)
Dark		
1½ (8)	3812	30.1
3 (8)	3676	30.5
7 (8)	3760	26.7
Light		
1 (8)	2750	20.6
3 (8)	2309	19.3
10½ (12)	3270	23.6

pattern is circadian. When they are maintained on a 24-hour, light-dark cycle, aproximately 80 percent of their total food intake occurs in the dark. The main feeding period begins at the same time each night, with clocklike regularity. The feeding rhythm remains circadian after the animals are blinded or if they are kept in continuous light, but the onset of the feeding period begins to drift by a fixed number of minutes each day. These experiments reveal an internal clock that runs somewhat faster or slower than 24 hours per day. This indicates that the circadian feeding rhythm is endogenous; that is, it is synchronized but not driven by the environmental light-dark cycle (5). Hypothalamic lesions eliminate all manifestations of the endogenous circadian feeding rhythm (6). The hypothalamus may contain parts of an oscillation system that drives the feeding rhythm.

Two circadian rhythms of *l*-NE have been identified in the rat hypothalamus. One of these rhythms, present in the anterior hypothalamus, has a peak in the middle of the daily dark period and low points throughout the light period. The second rhythm, present in the posterior hypothalamus, also reaches a peak in the middle of the dark period and reaches its lowest point at the end of the dark period (7). It is not known if either of these rhythms is endogenous. In cats, however, it has been established that the circadian rhythm of *l*-NE in the anterior hypothalamus is endogenous. It persists in constant light (8). Interestingly, *l*-NE appears to act as a synaptic transmitter in control of circadian rhythms of pineal hydroxyindole-*O*-methyltransferase (9) and liver tyrosine aminotransferase (10). It has been proposed that the liver aminotransferase rhythm may be generated, in part, by the periodic release of *l*-NE in the periphery (10). Perhaps the periodic release of *l*-NE in the anterior hypothalamus serves a

similar function and participates in the generation of the feeding rhythm. However, there is no direct evidence for *l*-NE or any other hypothalamic substance as a part of the circadian oscillator for feeding behavior.

How can the same dose of a synaptic transmitter substance that suppresses feeding behavior during the dark, facilitate it in the light? Different internal states of the hypothalamus produced by dark and light may be responsible for the opposite effects. The concentration of endogenous *l*-NE in the anterior hypothalamus reaches its lowest value in the light (7). Perhaps in this state, the addition of exogenous *l*-NE would increase endogenous concentrations. However, this would not be a large increase, even with the high doses used in our study, because of the extensive catecholamine uptake mechanisms in the hypothalamus (11) and because of the presence of the enzyme monoamine oxidase, which would degrade much of the absorbed *l*-NE (12). We predict that a shift in a hypothalamic substance such as *l*-NE from low to medium concentrations may be responsible, in part, for the shift in feeding behavior from the relatively anorexic condition associated with light to the normally hungry condition produced in the dark. We view the phenomenon of noradrenergic elicited feeding as a transient shift from an anorexic state induced in the light to a higher level of feeding that normally occurs in the dark. Our results suggest that low concentrations of hypothalamic *l*-NE may be an essential prerequisite for the demonstration of the facilitatory effects of *l*-NE on feeding behavior.

The concentration of hypothalamic *l*-NE and other biogenic amines also can be lowered by lesions of the lateral hypothalamus (13). Exogenous *l*-NE enhances the recovery of feeding behavior when administered intraventricularly to anorexic rats recovering

133

from such lesions (*14*). Thus, both lateral hypothalamic anorexia and light-induced anorexia are reversed by treatment with *l*-NE. Both forms of anorexia may be related to the excessively low concentrations of hypothalamic *l*-NE.

The concentration of *l*-NE in the anterior hypothalamus reaches its highest concentration in the dark. In this state, the addition of exogenous *l*-NE would be expected to raise endogenous concentrations, but again the increase would be small. We propose that a shift of a hypothalamic substance, such as *l*-NE, from medium to high concentrations may be responsible, in part, for the shift from normal hunger to satiety. Results with phentolamine, an agent that prevents endogenous *l*-NE from occupying alpha-adrenergic receptors in the periphery, provide support for this proposal. Direct application of phentolamine to the lateral hypothalamus in darkness blocked the effects of satiety on feeding behavior, as indicated by intense overeating (*2*). This suggests that certain forms of obesity may be related to a failure to generate concentrations of *l*-NE high enough to suppress feeding.

In conclusion, we find that treatment of the lateral hypothalamus with *l*-NE has opposite effects on feeding behavior that depend on the time of day the drug was administered. During the dark, treatment with *l*-NE suppressed feeding behavior. During the light, treatment with the same dose of *l*-NE facilitated feeding behavior. Sufficient evidence exists that the neurochemistry of the hypothalamus is not a steady state, but fluctuates according to a variety of rhythms (*15*). Circadian differences in a hypothalamic substance, such as *l*-NE, may provide the basis for the dual and opposite actions of exogenous *l*-NE on feeding behavior. These findings appear to resolve the controversy between the noradrenergic-feeding and the noradrenergic-satiety theories.

D. L. MARGULES
MICHAEL J. LEWIS
JAMES A. DRAGOVICH
ADRIENNE S. MARGULES

Department of Psychology,
Temple University,
Philadelphia, Pennsylvania 19122

References and Notes

1. S. P. Grossman, *J. Physiol. London* **202**, 872 (1962); J. L. Slangen and N. E. Miller, *Physiol. Behav.* **4**, 543 (1969); S. L. Leibowitz, *Nature* **226**, 963 (1970). Leibowitz's experiments were conducted with rats adapted to a normal dark-light cycle (lights on at 0600 and off at 1800 hours E.S.T.), and all drugs were administered in the light (S. L. Leibowitz, personal communication).
2. D. L. Margules, *Life Sci.* **8**, 693 (1969); *J. Comp. Physiol. Psychol.* **73**, 1 (1970). All of these experiments were conducted with rats adapted to a reversed dark-light cycle (lights off at 0800 hours and on at 2000 hours, E.S.T.), and all drugs administered in the dark.
3. B. G. Hoebel, *Annu. Rev. Physiol.* **33**, 533 (1971).
4. The 25-μg dose of *l*-NE may seem high, particularly in comparison to the total amount of *l*-NE found in the hypothalamus (about 0.1 μg). It is possible to suppress or stimulate feeding behavior with doses of *l*-NE as low as 5-μg. Quantities below 5 μg fail to affect feeding behavior. Possibly, these quantities are too small to influence sufficient postsynaptic receptor sites. Uptake mechanisms on the surface of vascular, glial, and neuronal cells remove *l*-NE from the extracellular hypothalamic space. These mechanisms, in conjunction with intracellular monoamine oxidase, may reduce substantially the amount of exogenous *l*-NE that penetrates to the synaptic receptor sites.
5. C. P. Richter, *Comp. Psychol. Monogr.* **1**, 55 (1922); J. LeMagnen, in *Handbook of Physiology*, section 6, "Alimentary canal," C. F. Code, Ed. (American Physiological Society, Washington, D.C., 1967), vol. 1, p. 11; I. Zucker, *Physiol. Behav.* **6**, 115 (1971).
6. C. P. Richter, in *Sleep and Altered States of Consciousness*, S. S. Kety *et al.*, Eds. (Williams & Wilkins, Baltimore, 1967), p. 8; J. W. Kakolewski, E. Deaux, J. Christensen, B. Case, *Amer. J. Physiol.* **221**, 711 (1971); C. M. Brooks, R. A. Lockwood, M. L. Wiggins, *ibid.* **147**, 735 (1946); S. Balagura and L. D. Davenport, *J. Comp. Physiol. Psychol.* **71**, 357 (1970); F. K. Stephen and I. Zucker, *Proc. Nat. Acad. Sci. U.S.A.*, in press.
7. J. Manshardt and R. J. Wurtman, *Nature* **217**, 574 (1968).
8. D. J. Reis, M. Weinbren, A. Corvelli, *J. Pharmacol. Exp. Ther.* **164**, 135 (1968).
9. J. Axelrod, S. H. Snyder, A. Heller, R. Y. Moore, *Science* **154**, 989 (1968).
10. I. B. Black, J. Axelrod, D. J. Reis, *Nature New Biol.* **230**, 185 (1971).

11. A. Philippu, U. Burkat, H. Becke, *Life Sci.* **7**, 1009 (1968).
12. The increase in endogenous *l*-NE that results from administration of *l*-NE in the light is probably quite small. It fails to produce any indication of a behavioral rebound 24 hours after the treatment (*1*). Such rebounds occur 24 hours after administration of the same dose of *l*-NE in the dark, and these have been taken as behavioral evidence of end-product inhibition in the hypothalamus (*2*).
13. A. Heller, J. A. Harvey, R. Y. Moore, *Biochem. Pharmacol.* **11**, 859 (1962).
14. B. D. Berger, C. D. Wise, L. Stein, *Science* **172**, 281 (1971).
15. A. Reinberg and F. Halberg, *Annu. Rev. Pharmacol.* **11**, 455 (1971).
16. We thank J. Wellbrock for technical assistance and Dr. A. Lubin and Dr. S. Roberts for their helpful comments. Supported by grant MH19438 from the National Institute of Mental Health.

9 May 1972: revised 24 August 1972

13

Daily Hormone Rhythms in the White-Throated Sparrow

Albert H. Meier

Reprinted from AMERICAN SCIENTIST, Vol. 61, No. 2, Mar.-Apr. 1973, pp. 184-187
Copyright © 1973 by The Society of the Sigma Xi

Changing temporal relations of the daily rhythms of corticosterone and prolactin control the orderly sequence of migratory and reproductive conditions in the annual cycle of Zonotrichia albicollis

Man has long been intrigued by the migration of birds. He has wondered about the purpose, the stimulus, the means of orientation, and the energetic support for prolonged flight. At least since the time of Job (ca. 1520 B.C.) the inability to fathom these mysteries has taunted him and stimulated his curiosity concerning one of nature's most closely guarded secrets. Only recently have the efforts of many dedicated naturalists and experimentalists coalesced sufficiently so that the bare outlines of the physio-

Albert H. Meier, professor of zoology and physiology at Louisiana State University, Baton Rouge, has been studying the role of daily rhythms in hormone function since 1965. His studies of migration and reproduction in the white-throated sparrow have been supported by the National Science Foundation and the United States Public Health Service. He holds a Research Career Development Award from the U. S. Public Health Service. Address: Department of Zoology and Physiology, Louisiana State University, Baton Rouge, LA 70803.

logical mechanisms controlling the seasonal changes in reproduction and migration may be visualized.

The principal ecological imperative for migration involves breeding and raising young in an area that is highly suitable for the support of life at one time of year and relatively hostile at other times. In the North Temperate Zone, many birds migrate to northern breeding grounds in the spring and return to southern wintering quarters in the fall. The relationship between migration and reproduction is well coordinated throughout the year, suggesting that the control mechanisms are also closely integrated.

The timing of migration and reproduction indicates that birds are receptive to environmental stimuli which change during the annual cycle. In this regard, perhaps the most significant single contribution in this century was made by Rowan (1926; 1929) in a series of experiments about fifty years ago. He demonstrated that an increasing day length in spring, rather than an increasing temperature as was believed at the time, was the principal environmental stimulus for vernal migration and reproductive maturation. The daily addition of several hours of artificial light during midwinter in Edmonton, Canada, induced premigratory fattening, nocturnal migratory restlessness, and evidences of reproductive development in the slate-colored Junco, *Junco hyemalis*.

The importance of day length in setting the times of migration and reproduction has been verified in a large number of birds. However, the mechanisms involved in migration and reproduction are not simply static systems which react directly to seasonal changes in the day length. Although "long days" may stimulate migratory and reproductive conditions in the spring, they do not maintain these conditions indefinitely. Migration ceases, of course, when the birds reach the breeding grounds. After the young are raised in the summer, the reproductive system regresses while the days are still long. Reproductive photorefractoriness to long days generally persists until winter when the days are too short to be stimulatory.

The intervention of short days is important in some birds for reestablishing reproductive photosensitivity. Fall migration occurs when the days are growing shorter; however, a decreasing photoperiod does not appear to exert a major stimulus. Thus, photoperiodic cues are interpreted by the bird in various ways during the year. Although the changes in day length during the year have a critical role in setting the times of migration and reproduction, an important timing mechanism appears to be within the bird itself.

Daily rhythms of prolactin

During the previous seven years, my colleagues and I have been investigating the control mechanisms for migration and reproduction in the white-throated sparrow, *Zonotrichia albicollis* . . . This bird winters in the southeastern United States and breeds in northern North America. Our collections were made from wintering flocks near Baton Rouge, Louisiana, in the vicinity where the great illustrator James Audubon lived and painted the white-throated sparrow and numerous other birds that still abound in the thickly forested, rolling countryside.

Our initial studies involved prolactin, a hormone produced by the pituitary gland in all the major vertebrate classes. Prolactin has long been considered one of the most intriguing hormones. Although it is best known for its stimulatory effect on mammary glands, it is obvious that this effect is a recent evolutionary contribution. I became interested in prolactin in 1964 while studying the white-crowned sparrow, *Zonotrichia leucophrys gambelii*, in the laboratory of D. S. Farner at Washington State University. Prolactin injections stimulated increases in body-fat stores (Meier and Farner 1964) and induced migratory restlessness (Meier, Farner, and King 1965) resembling that observed in caged birds during the migratory seasons. However, when we attempted later to duplicate these experiments with the white-throated sparrow, our first attempts were unsuccessful.

When confronted with conflicting information of this sort, there are two classical conclusions that many scientists have reached. One either places the reason for the inconsistency on the other scientist or attributes it to the time-honored "species differences." In this instance, the first alternative was too painful to accept because I was the "other scientist," and the second alternative seemed to be dodging the issue. The key consideration that eventually led to a new and successful approach was the growing realization that the photoperiodic effect involved endogenous daily, or circadian, rhythms.

The basic theory of reproductive photoperiodism was formulated by Bünning in 1932 (see also Bünning 1960). He postulated that there are two systems in living organisms which are responsive to changes in the daily photoperiod. One system is activated by the first few hours of light, resulting in a photosensitive phase or interval in a second system during the latter portion of the day. If the photoperiod is sufficient, light is present during the sensitive phase and the photostimulation of the second system results in gonadal growth. The applicability of this hypothesis for the photostimulation of the reproductive system in a bird was demonstrated by Hamner (1963), as well as by several other investigators. The identification of the internal mechanisms, however, remained undone.

An enormous amount of research in recent years has demonstrated the existence of daily fluctuations in many physiological parameters, including hormones. Because hormones have strong and pervasive influences in vertebrates, we wondered whether a daily fluctuation of endogenous prolactin might have important effects. We wondered whether it made a difference at what time of day prolactin is released in larger quantities from the pituitary gland. Would injections of prolactin at one time of day elicit different responses from those caused by injections at another time of day?

When we tested whether the time of injection was important in the white-throated sparrow, the results were clear and dramatic (Meier and Davis 1967). In lean birds during the summer, daily injections of prolactin administered during the afternoons of sixteen-hour daily photoperiods caused rapid increases in body-fat stores. Within one week of injections, the fat stores were as high as those found in feral birds during migration. In addition to a stimulation of fattening, the injections of prolactin during the afternoon induced nocturnal restlessness in this nocturnal migrant. On the other hand, daily injections administered during the morning caused losses in fat stores and did not induce nocturnal restlessness. Even huge doses of prolactin in the morning did not reverse the loss of fat stores. Apparently, a physiological time is at least as important as a physiological dose so far as prolactin is concerned.

Since the early work by Bates, Riddle, and Lahr (1937), other workers have discovered that injections of prolactin caused a regression in the gonadal weights of many avian species. In several instances, though, prolactin did not have an antigonadal effect. In at least one instance (Meier and Farner 1964), prolactin had a progonadal effect. It was argued and generally assumed that prolactin inhibited the release of FSH (follicle-stimulating hormone) and did not inhibit the gonadal responses to the gonadotropic hormones. However, when we tested this assumption in the white-throated sparrow, we found that injections of prolactin could block the gonadotropic effects of FSH and LH (luteinizing hormone) injections (Meier 1969). Apparently, prolactin exerts a strong antigonadal effect by inhibiting gonadal sensitivity to the gonadotropic hormones. In addition, the time relations between the daily injections of prolactin and the gonadotropic hormones were critical. Although prolactin was inhibitory in some temporal relations, in other relations with the gonadotropic hormones it was stimulatory.

Temporal synergism of corticosterone and prolactin

The presence of daily variations in responses to prolactin suggested to us that another system was responsible for entraining these rhythms. This system, we felt, must be regulated by the daily photoperiod and must be capable of entraining other rhythms. Taking our cues from studies of mammals, particularly those by Halberg and his collaborators (Halberg 1969), it seemed likely that the adrenal corticoid hormones might be involved. We made the initial tests in fish, lizards, and pigeons and found that daily injections of adrenal corticoids did entrain daily rhythms of fattening response to prolactin in all these vertebrates maintained in continuous light (Meier, Trobec, Joseph, and John 1971).

Similar experiments were performed with both photorefractory white-throated sparrows during the summer and photosensitive white-throated sparrows during the winter (Meier and Martin 1971; Meier, Martin, and MacGregor 1971). In birds maintained in continuous light in order to eliminate photoperiodic entrainment, prolactin injections were administered daily at 0, 4, 8, 12, 16, or 20 hours after an injection of corticosterone, the principal adrenal corticoid in birds. The responses varied depending on the temporal relations of the hormones. The 12-hour relation (prolactin administered 12 hours after corticosterone) caused explosive increases in body-fat stores. The 8-hour relation caused marked decreases in fat stores. Considerable fattening was also elicited by 0-hour and 4-hour rela-

tions. Thus, corticosterone injections entrained a bimodal rhythm of daily fattening responses to prolactin.

In addition to a daily rhythm of fattening responses, corticosterone injections also set the times of reproductive responses to prolactin. Gonadal growth occurred as a result of prolactin injections administered 12 hours after corticosterone, even in the photorefractory birds. On the other hand, gonadal weights were strongly inhibited by the 8-hour relation between corticosterone and prolactin, as evidenced in the photosensitive birds. Other hormonal relations had relatively little effect.

The variable responses to the temporal patterns of corticosterone and prolactin injections indicated that some seasonal conditions of migration and reproduction may be regulated by a temporal synergism of these hormones (see Fig. 1). The 12-hour relation of corticosterone and prolactin induced conditions found during vernal migration, whereas the 8-hour relation induced conditions associated with the summer photorefractory period. This hypothesis was supported by assays of corticosterone (Dusseau and Meier 1971) and prolactin (Meier, Burns, and Dusseau 1969). In May, during the vernal migratory period, there is a 12-hour interval between the daily increase in plasma concentrations of corticosterone and the daily release of pituitary prolactin. In August, during the photorefractory period, the interval between the hormones is about 6 hours.

We have recently obtained further evidence that supports the hypothesis that a temporal synergism of the daily rhythms of corticosterone and prolactin regulates seasonal conditions of migration and reproduction (Martin and Meier, in press). White-throated sparrows were treated with

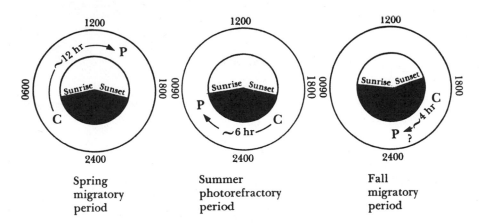

Figure 1. Daily rhythms of endogenous corticosterone (C) and prolactin (P) vary seasonally. The daily increases in concentrations of the hormones in the plasma change with respect to the daily photoperiod and with respect to each other. (From Dusseau and Meier 1971; and Meier, Burns, and Dusseau 1969).

corticosterone and prolactin in various temporal patterns, and the birds were maintained in continuous dim light except for several hours when they were kept in orientation cages under the open night sky. A 4-hour relation and a 12-hour relation of hormone treatment induced nocturnal restlessness, whereas other relations were ineffective. In addition, the restlessness was oriented northward in the birds that received daily injections of prolactin 12 hours after the injections of corticosterone and southward in those that received prolactin 4 hours after corticosterone.

Similar results were obtained when the birds were kept on a 16-hour photoperiod and injected with prolactin only at various times. Injections of prolactin given at 12 hours after the daily rise in concentrations of plasma corticosterone promoted fattening, gonadal growth, and migratory restlessness oriented northward; injections of prolactin 4 hours after the increase in plasma corticosterone promoted fattening and migratory restlessness oriented southward (see Fig. 2). These results were obtained under the natural spring sky as well as under the fall sky. Thus, the major determinant for orientation in the white-throated sparrow does not reside in seasonally changing environmental cues but rather in seasonal changes in the interpretation of the cues.

The control of a diverse group of physiological and behavioral parameters by a temporal synergism of corticosterone and prolactin can account for an orderly sequence of highly integrated events that occurs in the annual cycle of the white-throated sparrow (see Fig. 3). There remain, however, many important questions concerning the regulation of the daily rhythms of the hormones as well as the means by which the temporal synergism organizes and produces the seasonal conditions. Further progress will continue to depend on the combined energies of many biological disciplines, from the physiological to the behavioral, in order to comprehend more fully the complexities involved. A realization that daily rhythms constitute important components in temporal organization should help considerably.

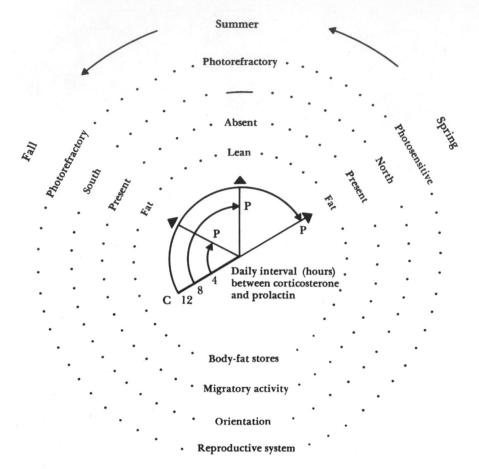

Figure 2. Responses to exogenous corticosterone (C) and prolactin (P) vary depending on the temporal relation of the injections. Daily injections of prolactin administered 12 hours after corticosterone stimulate spring conditions of migration and reproduction. Similarly, an 8-hour pattern of hormone injections induces summer photorefractory conditions, and a 4-hour pattern of injections induces fall conditions.

It is not clear whether the reflections of Job on avian migration and reproduction played a role in his eventual recovery. It seems reasonable, though, that it may have improved his outlook on life. People who work with sparrows are generally more concerned with the wonder of nature than with the immediate practical benefits that may derive from their studies. Nevertheless, we feel that our studies along with those of other biologists may have important practical consequences as well. For example, our laboratory has demonstrated that temporal patterns of adrenal corticoids and prolactin can regulate the amounts of body-fat stores in representative species of fish, frogs, lizards, birds, and mammals (Meier 1972). It is no longer tenable to dismiss daily rhythms as inconsequential anomalies. Instead, the evidence is accumulating from many quarters that daily rhythms constitute the basic units in the temporal organization of living systems.

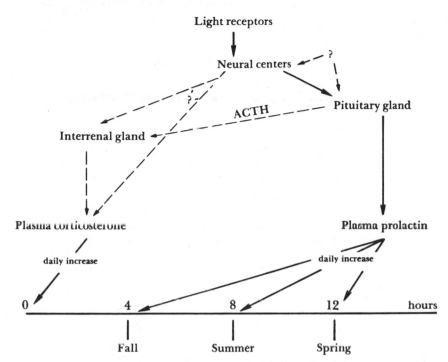

Figure 3. Schema of the mechanisms that control the seasonal events of migration and reproduction. The numbers, locations, and specific functions of the light receptors are imprecisely known. Dashed lines represent processes not yet proven. The neural centers receive photic and other environmental information as well as physiological information. It is not known how all of this information is integrated for the production of specific rhythms of corticosterone and pro-lactin. Although the hypothalamic centers secrete releasing and inhibitory hormones that affect the production of ACTH and prolactin, the assumption that daily rhythms of the hypothalamic hormones synchronize or drive the daily rhythms of corticosterone and prolactin is still open to question. The avian interrenal gland is stimulated by ACTH; however, the daily rhythm of plasma corticosterone may not depend on a daily rhythm of ACTH.

References

Bates, R. W., O. Riddle, and E. L. Lahr. 1937. *Amer. J. Physiol.* 119:610–14.

Bünning, E. 1932. *Ber. lt. bot. Grs.* 54:590–607.

Bünning, E. 1960. Cold Spring Harbor Symp. *Quant. Biol.* 25:249–56.

Dusseau, J. W., and A. H. Meier. 1971. *Gen. Comp. Endocrinol.* 16:399–408.

Halberg, F. 1969. *Ann. Rev. Physiol.* 31:675–725.

Hamner, W. M. 1963. *Science* 142:1294–95.

Job 39:26–27.

Martin, D. D., and A. H. Meier. In press. *Condor.*

Meier, A. H. 1969. *Gen. Comp. Endocrinol.* 13:222–25.

Meier, A. H. 1972. *Gen. Comp. Endocrinol. Suppl.* 3:499–508.

Meier, A. H., J. T. Burns, and J. W. Dusseau. 1969. *Gen. Comp. Endocrinol.* 12:282–89.

Meier, A. H., and K. B. Davis. 1967. *Gen. Comp. Endocrinol.* 8:110–14.

Meier, A. H., and D. S. Farner. 1964. *Gen. Comp. Endocrinol.* 4:584–95.

Meier, A. H., D. S. Farner, and J. R. King. 1965. *Anim. Behav.* 13:453–65.

Meier, A. H., D. D. Martin, and R. Mac-Gregor III. 1971. *Science* 173:1240–42.

Meier, A. H., and D. D. Martin. 1971. *Gen. Comp. Endocrinol.* 17:311–18.

Meier, A. H., T. N. Trobec, M. M. Joseph, and T. M. John. 1971. *Proc. Soc. Exp. Biol. Med.* 137:408–15.

Rowan, W. 1926. *Proc. Boston Soc. Natl. Hist.* 38:147–89.

Rowan, W. 1929. *Proc. Boston Soc. Natl. Hist.* 39:151–208.

14

Sexual Cyclicity in Captive Lowland Gorillas

Reprinted from SCIENCE, 5 September 1975, volume 189, pages 813-814

Abstract. *Oppositely sexed pairs of gorillas exhibit some behavior indicative of higher cognitive functioning, such as individual partner preferences and varied copulatory positions, but also mate in a cyclic manner closely related to the degree of female genital swelling. The latter finding is contrary to predictions based on their advanced position in phylogeny.*

Nearly 30 years ago Beach (*1*) proposed that animals with relatively larger forebrains and, consequently, greater cognitive abilities were less dependent on gonadal hormones for the regulation of their sexual behavior than were animals with smaller forebrains. Evidence in support of this hypothesis has appeared over the years, but relatively little systematic data have been obtained on the more advanced species, in particular the higher primates (*2*). In a study of the sexual behavior of gorillas, I obtained evidence contrary to Beach's hypothesis in that the behavior of these apes, close taxonomic affiliates of man, was rather closely related to the phase of the female's sexual cycle.

My primary objective was to evaluate periodicity in sexual behavior and female genital swelling of gorillas for evidence that the behavior was hormone-dependent. The female gorilla does not possess a sex skin as extensive as that of the chimpanzee or baboon, but it does exhibit tumescence of the perineal labia with a cycle (*3*) comparable to that of these other species and presumably under similar hormonal control (*4*). The subjects were four male and nine female lowland gorillas (*Gorilla gorilla gorilla*), ranging in age from 7 to 10 years at the beginning of the study in 1971. The animals were housed in cages consisting of an inner compartment 2.3 m by 2.3 m by 2.1 m high and an outdoor compartment 2.3 m by 4.1 m by 2.7 m high, interconnected by a guillotine door. When they were not being tested, the females were caged in isosexual pairs and the males

were caged alone. Oppositely sexed pairs were tested approximately daily throughout the sexual cycle of the female. Before each test the male was confined to the outer compartment of his home cage (the test cage) and the female was introduced into the inner compartment. The test was begun after the door separating the compartments was raised and the female had entered and was confined to the outer compartment. The test was terminated after 30 minutes had elapsed without the occurrence of copulation. The labial tumescence (LT) of the female was rated on a scale of 0, 1, or 2, corresponding respectively to detumescence, minimal tumescence, or maximal tumescence, as previously described (*3*).

I paired the 13 gorillas in 20 different combinations for approximately 2000 tests over a 3-year period. All nine females copulated with at least one male partner, but only seven females from nine pairs copulated on a regular basis and were involved in copulation that culminated in ejaculation. For 4 of the 11 pairs that did not copulate or copulated very infrequently, the females failed to exhibit tumescence of the perineal labia, suggesting hormonal inadequacy as the basis for the lack of sexual activity. However, in the seven remaining infrequently copulating pairs, the female did exhibit cyclic fluctuations in LT. Since the females in these latter pairs did copulate when tested with different males, their differential responsiveness may be an example of individual partner preferences. My interpretation of this finding is that some factor other than hormonal, presumably a higher cognitive variable, exerted an important influence on the sexual behavior of these animals (5).

A clearer indication of higher cognitive capabilities in the gorilla was found in the varied copulatory positions taken by the subjects. Of the seven females that copulated regularly, four used the dorsal-ventral position exclusively, one used the ventral-ventral position exclusively, and two used both positions. Of the four males, two used only the dorsal-ventral position and two used both positions. Moreover, a number of variations of the two major positions were also seen. These data indicate that the primary copulatory position for these animals was dorsal-ventral, but that a considerable amount of variability was exhibited within and between positions. This type of variability in copulatory positions contrasts with the relatively invariable, stereotyped performance characteristic of species below the level of the apes and appears to represent a prerogative related to the latter's pronounced degree of encephalization.

Another conspicuous feature of the gorilla's sexual interactions was related to the different roles played by the male and female before copulation. The male was quite unobtrusive in soliciting copulation, his positive responses before copulation consisting primarily of approaches to the female and occasional touches with the back of the hand. The female, on the other hand, was very assertive, backing forceful-ly into the male, frequently pushing him against a wall, and actively rubbing her genitalia against the male by rhythmically raising and lowering her rump while emitting a soft, high-pitched fluttering vocalization. The male also vocalized during copulation and in a pitch similar to the female's, but his vocalization consisted of a short burst rather than the prolonged refrain of the female.

The most significant finding, however, was the relationship between various measures of sexual behavior and the female's labial condition (Table 1). Since the tests were conducted approximately daily throughout the female cycle, the proportion of tests conducted at each degree of LT reflects the proportion of days each degree of LT was represented in the cycle. Based on a cycle of 31 or 32 days (3), LT ratings of 0, 1, and 2 were obtained on 19, 9, and 3 days per cycle, respectively. The females presented themselves sexually to the males during all degrees of LT, but they exhibited significant differences in the percentage of tests on which they presented for the different degrees of LT. All females presented least frequently during their detumescent condition and exhibited successive increases with successively greater degrees of LT. A similar pattern of significant differences was found for the frequency with which females presented per test and for the female success ratio. This is one of the few laboratory studies of primate sexual behavior in which a consistent relationship was found between female

Table 1. Sexual behavior of captive lowland gorillas for different labial tumescence (LT) ratings of the females. The female success ratio is defined as 100 times the number of copulations per test divided by the number of female presentations per test; N is the total number of tests.

| LT rating | N | Presentations | | Female success ratio | Copulations | | Ejaculations | |
		Percent of tests	Per test (median)		Percent of tests	Per test (median)	Percent of tests	Per test (median)
0	705	23	1.6	3	1	0.3	0	0.0
1	342	41	2.1	33	33	1.3	14	1.0
2	102	66	2.8	41	43	1.8	37	1.0
Totals	1149*	34*	2.1†	23*	12‡	1.4	7‡	1.0

*$P < .001$, Friedman two-way analysis of variance (14).   †$P < .05$, Friedman two-way analysis of variance (14).
‡LT = 1 compared with LT = 2, $P < .05$, Wilcoxon matched-pairs signed-ranks test (14).

sexual behavior and phase of the sexual cycle (*6*).

The data on the female success ratio, a measure of female attractiveness (*7*), indicate that the females not only solicited copulation more frequently at successively greater degrees of LT, but were also more successful at those times. Examination of the data on copulation and ejaculation (*8*) supports the relationship between sexual behavior and LT. The males copulated on 12 percent of their tests (approximately 4 days per cycle) and ejaculated on 7 percent of their tests (approximately 2 days per cycle). However, they rarely copulated and never ejaculated with females in the detumescent condition, and exhibited the greatest frequency of these sexual interactions at the maximal degree of LT. However, the frequency of copulations and ejaculations per test did not differ significantly for the different degrees of LT. These data suggest that whether or not copulation and ejaculation occurred on a particular day was importantly influenced by the stage of the female's cycle, but that the frequency with which they occurred on a given day was not so influenced. In fact, although all the males exhibited multiple copulations and ejaculations on sòme tests, their characteristic pattern was to ejaculate on a single copulation and then to interact sexually no further.

These findings on the varied copulatory positions used by gorillas and the relatively passive and assertive roles, respectively, of males and females augment and extend the reports describing gorillas living in zoos (*9*). Those reports, based primarily on single pairs of animals, also indicated that sexual interactions occurred at fairly regular 30-day intervals and were confined to a period of 1 to 4 days. The present study of the largest group of adult gorillas in captivity confirmed those findings and also yielded data on the relationship between sexual behavior and the female's genital swelling.

The parallel changes in tumescence of the female's perineal labia, sexual receptivity, and occurrence of copulation described above represent the first systematic data of this type on gorillas and indicate that sexual behavior of this species is quantitatively different from that of the chimpanzee. Yerkes and co-workers (*10*) and Young and Orbison (*11*) reported that chimpanzees mated throughout the sexual cycle, although most copulated more frequently during the period of maximal genital swelling of the female. They concluded that hormonal influences were relatively unimportant in the sexual behavior of these apes because the expression of estrus could be altered substantially and even completely overwhelmed in any given subject by significant individual differences and social factors associated with the test situation.

Their findings supported Beach's hypothesis (*1*) in that the chimpanzee exhibited a degree of behavioral independence from hormonal control commensurate with its relatively advanced brain capacity. The gorillas of the present study mated less frequently, in general, than did the chimpanzees and their sexual interactions were more closely related to the degree of genital tumescence of the females. These data suggest that sexual behavior of gorillas is under relatively greater hormonal control than that of chimpanzees and is a reflection of the variability in behavioral regulation that exists among these higher primates. No comparable studies have been conducted on the third species of great ape, the orangutan, but limited observations of wild (*12*) and captive (*13*) specimens, as well as my own observations of four captive pairs, suggest that sexual behavior in that species shows less evidence of cyclicity than in either the gorilla or the chimpanzee.

The three great ape species, therefore, appear to comprise a continuum on which animals with similar forebrain development and intelligence differ in the extent to which their sexual behavior is regulated by hormones. Although confirmation of the hormonal mediation of the behavior observed in this study requires further research, these findings suggest that hormonal factors play a more significant role in the regulation of sexual behavior in gorillas than would be predicted on the basis of their advanced position in phylogeny.

RONALD D. NADLER

Yerkes Regional Primate Research Center, Emory University, Atlanta, Georgia 30322

References and Notes

1. F. A. Beach, Psychol. Rev. 54, 297 (1947).
2. B. L. Hart, Psychol. Bull. 81, 383 (1974).
3. C. R. Noback, Anat. Rec. 73, 209 (1939); R. D. Nadler, ibid. 181, 791 (1975).
4. Based on limited evidence from baboons and some macaques, S. Zuckerman [Proc. Zool. Soc. Lond. (1930), p. 691] hypothesized that for all female monkeys and apes that possessed a sexual skin, estrogen would prove to be the hormone responsible for the swelling and coloration changes. Subsequent evidence on pigtail monkeys [G. G. Eaton and J. A. Resko, J. Comp. Physiol. Psychol. 86, 919 (1974)], baboons [J. Gillman and C. Gilbert, S. Afr. J. Med. Sci. 11, 1 (1966)], and chimpanzees [C. E. Graham, D. C. Collins, H. Robinson, J. R. K. Preedy, Endocrinology 91, 13 (1972); C. E. Graham, Folia Primatol. 19, 458 (1973)] has supported Zuckerman's hypothesis.
5. Although my subjective impressions support this interpretation, it is also possible that some hormonal condition not reflected in the perineal labia accounted for the behavioral deficiency.
6. G. G. Eaton, A. Slob, J. A. Resko, Anim. Behav. 21, 309 (1973).
7. R. P. Michael, in Endocrinology and Human Behaviour, R. P. Michael, Ed. (Oxford Univ. Press, London, 1968), pp. 69–93.
8. The behavioral pattern associated with ejaculation was similar to that observed in other nonhuman primates [C. R. Carpenter, J. Comp. Psychol. 33, 113 (1942); R. D. Nadler and L. A. Rosenblum, Brain Behav. Evol. 2, 482 (1969); ibid. 7, 18 (1973)]. Thrusting was terminated by a prolonged period of insertion during which the male's body was rigid and spasmodic muscle contractions were visible. Moreover, when visual examination was possible after the animals separated, a cloudy fluid, apparently semen, was seen emanating from the penis.
9. W. D. Thomas, Zoologica 43, 95 (1958); E. M. Lang, Int. Zoo Yearb. 1, 3 (1959); T. Reed and B. F. Gallagher, Zool. Gart. 27, 279 (1963); G. B. Schaller, The Mountain Gorilla (Univ. of Chicago Press, Chicago, 1963); R. J. Frueh, Int. Zoo Yearb. 8, 128 (1968); J. Tijskens, ibid. 11, 181 (1971); J. P. Hess, in Comparative Ecology and Behavior of Primates, R. P. Michael and J. H. Crook, Eds. (Academic Press, London, 1973), pp. 508–581.
10. R. M. Yerkes and J. H. Elder, J. Comp. Psychol. Monogr. 13 (1936); R. M. Yerkes, Hum. Biol. 11, 78 (1939).
11. W. C. Young and W. D. Orbison, J. Comp. Psychol. 37, 107 (1944).
12. J. Mackinnon, Oryx 11, 141 (1971).
13. H. Fox, J. Mammal. 10, 37 (1929); M. Asano, Int. Zoo Yearb. 7, 95 (1967); W. L. Heinrichs and L. A. Dillingham, Folia Primatol. 13, 150 (1970); P. F. Coffey, Annu. Rep. Jersey Wildl. Preserv. Trust (1972), p. 15.
14. S. Siegel, Nonparametric Statistics for the Behavioral Sciences (McGraw-Hill, New York, 1956).
15. A report on this research was presented by the author at the symposium "Human sexuality as a science" during the AAAS meeting in New York City, January 1975. This work was supported by NSF grant GB-30757 and PHS grant RR-00165. I thank L. D. Byrd, C. E. Graham, and A. A. Perachio for suggestions on an earlier version of the manuscript.

7 April 1975

15

Activation of the Male Rat's Sexual Behavior by Intracerebral Implantation of Androgen

JULIAN M. DAVIDSON

Department of Physiology, Stanford University School of Medicine, Stanford, California

ABSTRACT. The hypothesis that sexual behavior in the male results from the direct action of testosterone on specific brain regions was tested, using intracerebral implantation of crystalline testosterone propionate in castrate male rats. Following establishment of a criterion for the loss of sex behavior in castrates, it was determined that the minimum effective dose of subcutaneously administered testosterone for stimulation of the accessory sex glands was less than that required for restoration of sex behavior. It was found that intracerebral implantation of testosterone could result in reappearance of the complete pattern of male sexual behavior in the absence of any histologically demonstrable stimulation of the seminal vesicles or prostate. These results never followed the implantation of cholesterol, showing that they were not due to lesion production by the implant. Although occasional responses were found following implants in other areas of the brain, all the most effective implants were in the hypothalamic-preoptic region, with the most consistent behavioral reactivation resulting from medial preoptic implants. It is concluded that sexual behavior in the male may be virtually completely independent of androgen-sensitive peripheral mechanisms, and that it is dependent upon activation by testosterone of structures lying within the hypothalamic-preoptic region of the brain. (Endocrinology 79: 783, 1966)

Received April 7, 1966.

The preliminary experiments in this study were carried out in 1963 while the author was a USPHS postdoctoral fellow in the laboratory of Frank A. Beach at the University of California, Berkeley, and were supported by USPHS Grant MH 0400 to Dr. Beach. Subsequent work in 1964–65 was supported by USPHS Grant HD 00778 to the author.

THE CAPACITY to display sexual behavior is dependent, in both sexes, on the present or prior exposure of the animal to gonadal hormones (1, 2). Several possibilities exist as to the locus at which these hormones act to elicit sexual behavior. Thus, (a) their action may be to stimulate specific centers in the brain responsible for the organization of this behavior, or (b) it may be more general, affecting widespread regions of the central nervous system, or (c) it may be extracerebral, possibly affecting organs supplying afferent input to the brain.

Evidence has been presented to the effect that estrogens act on specific hypothalamic centers to activate the mating pattern in female cats (3) and rats (4). In the male it would appear a priori to be less likely that sexual behavior is the result of the direct effect of testosterone on a discrete cerebral center, since successful mating in this sex depends upon a highly co-ordinated series of activities rather than a relatively simple reflex lordosis as in the female. In laboratory rodents, the male's behavior is particularly complex, and involves an intricate pattern of mounts, intromissions and ejaculations. Hypothalamic lesions have been shown to inhibit sexual behavior in male guinea pigs (5) and rats (6, 7) and the effects of the lesions were not reversible by testosterone treatment (5, 6). These experiments, although suggesting that hypothalamic structures may be essential for sexual behavior in the male, do not constitute direct evidence that the action of testosterone on the hypothalamus is a causal factor in activation of this behavior. More relevant, perhaps, is a brief report which appeared in 1956 (8) on manifestations of "exaggerated sexual responses" in a small proportion of normal male rats receiving intracerebral injections of a soluble testosterone salt.

This study deals with the question of whether or not the action of testosterone on specific brain structures is sufficient to elicit the full pattern of normal sexual behavior in male rats which have ceased to display this behavior by virtue of long-term castration. The method used is the stereotaxic implantation of small quantities of crystalline testosterone propionate in specific brain regions. Since any effects of the implanted testosterone could be due either to direct effects on brain tissue or to release of the hormone into the systemic circulation, it was necessary to establish a criterion which would permit elimination of the second alternative. Such a criterion was found to be the absence of any histologically demonstrable stimulation of the prostate and seminal vesicles. The results of this study were presented in part at the 22nd International Congress of Physiological Sciences (9).

Materials and Methods

Sexually mature Long-Evans rats[1] were supplied with illumination from 12 AM to 12 PM each day and tests of sexual behavior were conducted during the dark phase of the daily cycle. Stimulus females were prepared by injecting 0.1 mg estradiol benzoate 48–72 hr before the onset of testing. Only females showing highly estrous responses were used. Tests were conducted in glass-fronted semicircular cages 16 in. high, 30 in. wide and 18 in. deep. After the male had been in the cage for 3 min, the female was introduced through a trap door leading from a special compartment. All mounts, intromissions and ejaculatory patterns[2] were recorded on an Esterline Angus event recorder.

Each test was continued until (a) occurrence of the first intromission after the ejaculatory pattern, or (b) 25 min after the first intromission (in the event that no ejaculatory pattern occurred in that time), or (c) 15 min after commencement of testing (in the event that no intromission occurred in that time). If no intromissions had occurred after approximately $7\frac{1}{2}$ min of testing, females were switched to provide an additional stimulus.

The following behavioral indices were measured or calculated for all tests:
Intromission latency. Time from onset of test (introduction of female) to first intromission.
Intromission frequency. Number of intromissions before first ejaculatory pattern.

---

[1] Simonsen Laboratories, Gilroy, Calif.
[2] The term "ejaculatory pattern" is used to describe the overt behavior (e.g., long intromission and slow dismounting, followed by a period of sexual inactivity) which accompanies seminal emission in the intact male, denoting the successful completion of a sequence of sexual behavior (10).

Ejaculation latency. Time from first intromission to ejaculatory pattern.

Mount frequency. Number of mounts (with pelvic thrusts) before ejaculatory pattern.

Post-ejaculatory interval. Time from occurrence of ejaculatory pattern to first intromission of next series.

Following trans-scrotal castration, tests were conducted weekly or biweekly until a criterion for the loss of sexual behavior was satisfied (see Results). Subsequently, males were either subcutaneously administered testosterone propionate in oil, or crystalline testosterone propionate was implanted intracerebrally. Testing was resumed 1–4 days later, and was continued twice weekly thereafter for 1–3 weeks, with 3- or 4-day intervals between tests. Normal body weight increases were found in almost all cases following implantation.

Intracerebral implantation was performed stereotaxically with the aid of deGroot's atlas (11). In most experiments implants were prepared, as described previously (12), by tamping a pellet of crystalline testosterone propionate or cholesterol into one end of a length of 20-gauge stainless steel tubing. The pellets were ejected into specific brain locations after fixation of the tube to the skull with dental cement. The crystalline material was protected during implantation by a thin film of sucrose which covered the tip, and its weight (approximately 200 μg) was determined by differential weighing on a Mettler M5 microbalance. In experiments in which a lesser surface area of exposure of brain to the hormone was desired, molten testosterone propionate was drawn into the end of the 20-gauge tubing and the outside of the tube was cleaned with solvents so that testosterone remained only at the end and flush with it. The precise rate of absorption from these crystalline

depots of testosterone is unknown. It may, however, be presumed that absorption is very slow, since identically prepared implants (pellets or fused) in the median eminence result in testicular atrophy one month following implantation and the crystalline material can usually be observed, at this time period, as a compact mass at the tip of the implant tube.

On cessation of the experiment, seminal vesicles and ventral prostates were fixed in Bouin's solution, paraffin-embedded and stained with hematoxylin and eosin for histological study. Where required, brains were removed, fixed in formalin, frozen, serially sectioned and stained, to determine the location of the implants.

Results

Criterion for loss of sexual behavior. Failure of the ejaculatory pattern on one behavioral test does not imply irreversible loss of this behavior in castrate rats (10). Before attempting the restoration of sexual behavior it was therefore necessary to establish a criterion on which to base a judgment that the spontaneous reappearance of sexual behavior in animals fulfilling this criterion was reasonably improbable. The absence of ejaculatory patterns on three consecutive weekly tests (*i.e.*, 3 "negative tests") appeared to be a suitable criterion. Forty rats were tested for additional periods of 13 to 70 days following three such negative tests. Fig. 1 shows the results. Of 265 tests in these 40 animals,

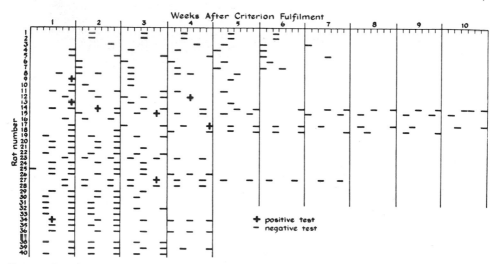

FIG. 1. Mating tests in rats following 3 successive weekly tests in which no ejaculatory pattern occurred ("criterion fulfillment"). Tests scored positive or negative according to whether ejaculatory patterns occurred or not.

only eight tests (3%) were "positive" for completion of the mating pattern. In no case did a rat show more than one positive test during the whole period of testing. Accordingly, this criterion was met before behavioral restoration was attempted by systemic or intracerebral steroid administration, except in a few cases in which the absence of any intromissions on any of five successive daily tests was substituted.

Differential sensitivity to testosterone of morphological and behavioral responses. To compare the minimum effective dose of androgen necessary to restore the ejaculatory pattern with that required to produce histologically demonstrable stimulation of the accessory sex glands, castrates were injected with testosterone propionate in doses of 25, 50, 75 or 100 µg daily for periods of seven or 21 days. Biweekly behavioral tests were conducted in all cases at three- or four-day intervals throughout the period of injections. In seven of the rats, seminal vesicle biopsies were performed after one week's treatment, and injections were then continued for two further weeks.

In 14 castrate control rats sacrificed 21 to 83 days after castration, the follicular epithelium of both seminal vesicles and ventral prostate was invariably squamous. Histological stimulation of these glands was considered to be present only if cuboidal or columnar epithelium was clearly present in a large proportion of follicles. Since the evaluation was qualitative and subjective, all cases in which stimulation appeared to be borderline were regarded as not stimulated. Virtually complete correlation was found between seminal vesicle and prostate stimulation.

Table 1 shows that, with the seven-day injection period, one half of the animals tested manifested accessory sex gland stimulation at the 25 µg dose level; histological stimulation was almost always present at higher dose levels. When injections were continued for 21 days, all animals showed accessory stimulation at all dose levels.

Behavioral restoration (the occurrence of at least one ejaculatory pattern) was obtained with higher dose levels than were necessary in order to stimulate the accessory sex glands, and this was true for all doses and durations of treatment. Restoration of behavior in 100% of animals tested was achieved only with the 100 µg dose level. In no case was behavioral restoration noted in the absence of accessory stimulation. Accessory stimulation in the absence of behavioral restoration was, however, noted in a total of 21 cases. Its occurrence did not appear related to the amount of time elapsing between castration and the onset of injections, which varied in this group from 29 to 151 days.

Testosterone implantation. In *preliminary* experiments testosterone pellets weighing 227 ± 28^3 µg were implanted in 18 rats in locations intended for the midbasal dience-

---

[3] Mean ± standard error of the mean.

TABLE 1. Effects of various subcutaneous doses of testosterone propionate on the histology of the ventral prostates and seminal vesicles and on sexual behavior in castrate rats

Dose (µg/day)	Duration of treatment (days)	No. of rats tested	% of rats responding		Cases of:	
			Accessory sex glands	Sexual behavior (ejaculatory pattern)	Stimulation without behavior	Behavior without stimulation
25	7	10	50	10	4	0
25	21	4	100	50	2	0
50	7	15	87	47	6	0
50	21	7	100	29	5	0
75	7	8	100	75	2	0
75	21	4	100	50	2	0
100	21	4	100	100	0	0
Total: Accessory stimulation without behavior					21	
Total: Behavior without accessory stimulation						0

148

phalic region, but histological study of the brains for verification of implant locations was not performed. On autopsy, four of these showed histological stimulation of the ventral prostate (2 cases) or of both prostates and seminal vesicles. In the remaining 14 rats no signs of accessory sex gland stimulation could be observed. Three to seven tests were performed in the first one to two postoperative weeks at three to four-day intervals. Eight of these animals displayed ejaculatory patterns on at least one of the postoperative tests and four did so on more than one test. Despite the absence of seminal emission, the behavioral pattern was indistinguishable from that occurring in the intact rat. Fig. 2 shows a comparison of the performance of these rats with that of 40 unimplanted animals (discussed above) and of 15 with cholesterol implants. The results suggested that the implantation of testosterone in at least some forebrain areas could result in the reappearance of the full pattern of male sexual behavior, although no conclusions would be drawn with regard to the precise area involved.

In 67 rats, testosterone pellets weighing 220 ± 4^3 µg were implanted in various brain areas, the locations of which were later verified histologically. Five postoperative behavioral tests were performed. In six of these rats, microscopic examination of the accessory sex glands removed at autopsy showed signs of stimulation. Only one of

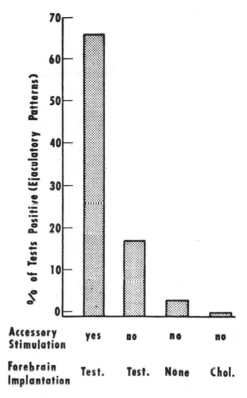

FIG. 2. Percentages of mating tests on which ejaculatory patterns occurred after intracerebral testosterone implantation in preliminary experiments, compared with effects of cholesterol implantation or no treatment.

TABLE 2. Behavioral responses of rats with large testosterone implants (pellet ejected) or cholesterol in various brain regions. Each rat was tested five times following implantation. Tests on which ejaculatory patterns occurred were designated positive.

	No. of rats tested	No. of positive tests						% of tests positive
		0	1	2	3	4	5	
			(figures below are no. of rats)					
Testosterone implants:								
Hypothalamus*	16	3	4	1	4	3	1	44
Medial preoptic†	8			3	4		1	57
Supraoptic and lateral preoptic areas	6	6						0
Thalamus	8	5	2	1				10
Cortex-corpus callosum	7	5	1	1				9
Hippocampus	9	4	3	2				16
Other:	12							20
Posterior colliculus			1	1				
Midbrain		1	2	1				
Zona incerta		1		1				
Caudate-putamen		2	1	1				
Cholesterol implants:								
Hypothalamus (6), Preoptic (6) Thalamus (3)	15	15						0

* Includes area caudal to posterior hypothalamic nucleus but anterior to habenulo-peduncular tract.
† 1.25 mm or less from the midline.

149

these six ejaculated on any of the tests, and it was removed from further consideration. Thirty-nine rats displayed the complete sequence of mating behavior culminating in an ejaculatory pattern on at least one postoperative test, in the presence of completely atrophied accessory sex glands. Mean postoperative latency to first positive test was 7.7 days; the range was two to 20 days. The reactivation of behavioral patterns was particularly striking since, in a number of animals, no mounts (without intromission) were observed on the three pre-implantation tests, i.e., there was no evidence of any of the constituents of sexual behavior in these rats.

Table 2 shows the frequencies of positive tests resulting from implants in different anatomical locations. Although the occurrence of at least one positive test in the series of five performed was found following implants in many different areas of the brain, the frequency of positive tests was much greater for animals with implants in the hypothalamic and medial preoptic than for any other region. Fifty per cent of the hypothalamic and 62% of the medial preoptic rats showed positive responses on more than half the tests performed (i.e., 3 or more positive tests/rat), while none of the animals in any of the other groups responded positively on more than two of the five postoperative tests.

Since one positive test (complete behavioral pattern) in a series is occasionally seen in untreated castrates following "criterion fulfillment" (Fig. 1), the anatomical diagrams (Fig. 3, 4) portray separately animals showing one positive test, and those showing more than one. Behavioral restoration resulted most consistently from implants in a relatively circumscribed area of the medial preoptic region at or close to the plane of juncture of the anterior commissure (Fig. 3B). All of eight animals with implants in this region showed more than one positive test, while no positive tests resulted from implants in the adjacent or immediately anterior areas (Fig. 3A, 3B). The results of hypothalamic implantation were somewhat more variable, and the effective area was more widespread. As shown in Fig. 3C, 3D and 3E, points yielding more than one positive test were dotted throughout this area, and were interspersed with

others yielding one or no positive tests.

Cholesterol implants were placed in 15 rats, in the hypothalamic-preoptic region (n = 12), and in the thalamus. The locations of all of these implants are shown in Fig. 3. None of the 75 postoperative tests performed on these animals was positive.

In order to reduce the surface area of brain exposed to testosterone, thereby decreasing the opportunity for diffusion of testosterone from the implant to distant sites in the brain, smaller implants were used. In these, the hormone was fused to the implantation tube in such a way that the testosterone-brain interface was limited to the bore of the tube. With this procedure, only hypothalamic and preoptic implantation resulted in behavioral restoration, but the percentage of implants in these areas which yielded positive results and the percentage of positive tests were considerably less than in the case of the larger implants (Table 3). However, decreasing the size of the implant did not result in a more precise localization of responsive regions within the hypothalamic-preoptic area, since, as shown in Fig. 4, positive tests were occasionally found to follow posterior hypothalamic and lateral preoptic implantation.

Analysis of restored behavior. Detailed analysis of the behavior records from testosterone-implanted rats showed considerable variability in the latencies and frequencies of the recorded behavior. On many tests in animals with implants in the hypothalamic-preoptic regions, behavioral scores were

150

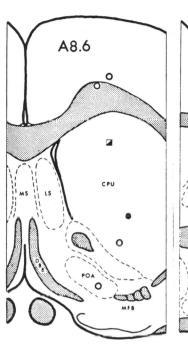

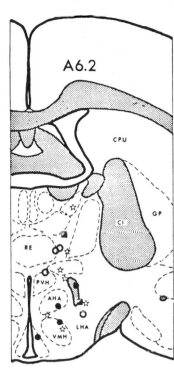

Fig. 3A. Plane of anterior preoptic.   Fig. 3B. Posterior preoptic.   Fig. 3C. Anterior hypothalamus.

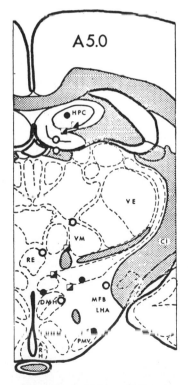

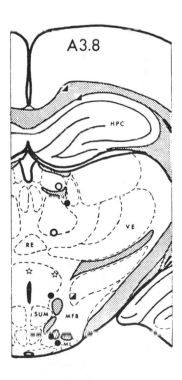

Fig. 3D. Medial hypothalamus.   Fig. 3E. Posterior hypothalamus.

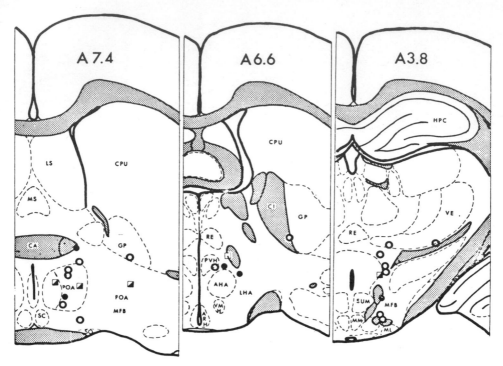

Fig. 4. Intracerebral distribution of small testosterone implants (hormone fused to 20-gauge tubing) at frontal planes passing through posterior preoptic region (*left*); anterior hypothalamus (*center*); and posterior hypothalamus (*right*). See legend to Fig. 3 for abbreviations and other comments.

similar to precastration values in the same animals. On other tests in these groups, and more often in animals with implants in other areas of the brain, the behavior tended to resemble that of long-term castrates with maintenance of sexual behavior (10) in which intromission and ejaculation latencies, intromission and mount frequencies and post-ejaculatory intervals are increased. Fig. 5 shows intromission latencies in rats with implants in the hypothalamic, preoptic and other regions of the brain as well as the values obtained from the same animals on precastration tests. The mean latency was highest in "other" rats, 336 ± 67 seconds, and the difference between this group and the preoptic animals (mean 141 ± 43 sec) was significant at the 5% level (*t* test). The great inter-test variability reduces the significance of the mean differences between groups, but it is clear that low intromission latencies were found most often in rats from both the hypothalamic and preoptic groups, and that the resemblance between the distributions of values from pre-castration and post-implantation tests was much greater than in the case of "other" animals. Similar variability was noted in other behavioral parameters, and no significant differences were found between the mean intromission frequencies, mount frequencies, ejaculation latencies or post-ejaculatory intervals in the three groups. Similarly, the pre-castration scores for these parameters did not differ significantly from the post-implantation scores.

The mean latencies in days from implantation to occurrence of the first positive test for implantation in different brain areas were as follows: medial preoptic area 5.9 ± 1.0 (range 3–11); hypothalamus 6.7 ± 1.2 (range 3–20); other areas 9.3 ± 1.1 (range 2–16).

The comparative latency to first ejaculatory pattern for rats injected with testosterone (see above) was 6.5 ± 0.8. It should be noted that the accuracy of these latency estimates is limited by the fact that tests were conducted only every three or four days. The differences between the latencies of the various groups are not statistically significant.

Discussion

The validity of any attempt to restore sexual behavior after castration by the local action of intracerebrally implanted crystalline material depends upon the demonstration that the observed results are not secondary to 1) the production of lesions or nonspecific irritation at the site of

TABLE 3. Behavioral responses following small testosterone implants (hormone fused to bore of tube). See Table 2 for further details.

	No. of rats tested	No. of positive tests						% of tests positive
		0	1	2	3	4	5	
			(figures below are no. of rats)					
Hypothalamus*	12	7	2	2		1		17
Medial preoptic*	9	5	2		2			17
Anterior and lateral preoptic	3	2	1					7
Other Areas:	8							0
Basal ganglia	2	2						
Thalamus	2	2						
Anterior midbrain	2	2						
N. accumbens	2	2						

* As defined in Table 2.

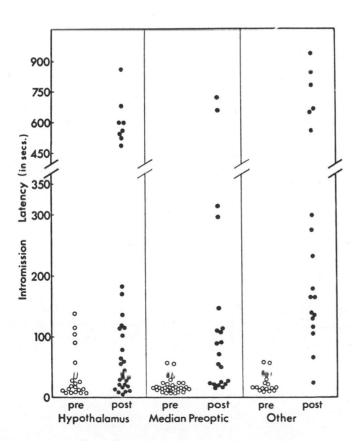

FIG. 5. Intromission latencies in seconds on individual mating tests in animals with large testosterone implants in the hypothalamus, preoptic and other brain regions. Abbreviations: pre, tests before castration; post, tests after implantation of testosterone.

153

implant, 2) "leakage" of the material into the systemic circulation in sufficient quantities to produce the behavior regardless of any local action, or 3) the spontaneous reappearance of behavior in the castrate. These requirements were adequately satisfied in the excellent study of Harris and Michael (3), who showed that intracerebral estrogen implants restore receptivity in the castrate female cat.

In the present investigation on male rats, the complete absence of any behavioral restoration in cholesterol-implanted animals shows that the observed effects of testosterone implantation were not due to destruction or nonspecific stimulation of the brain by the tube and or the crystalline material. Since the minimum effective dose of testosterone to restore behavior was shown here to be higher than that required for histological stimulation of the accessory sex glands, behavioral restoration in the face of completely atrophied seminal vesicles and prostates cannot be explained by release of the hormone from the implant into the systemic circulation. The extreme rarity of ejaculatory patterns in unimplanted rats tested after having fulfilled the criterion of three negative weekly tests, and the fact that this never occurred on more than one test in these animals despite repeated testing, eliminates, too, the factor of spontaneous behavioral restoration as playing a significant role in the results. It may therefore be concluded that the restoration of the full pattern of sexual behavior in the male rat may be achieved by the exposure of brain tissue to testosterone propionate, and that this restoration results from the direct action of the hormone on the brain.

Occasional "positive" tests (*i.e.*, those in which the complete mating pattern occurred) resulted from implants in several different brain areas. However, implants in the hypothalamic-preoptic region were clearly more effective than all others, both with regard to frequency of positive tests and (to a somewhat lesser extent) with regard to quality of the restored behavior. The fact that intromission latency was well above preoperative levels in almost all the relatively few positive tests in rats with extrahypothalamic implants is an indication that sex "drive" or "motivation" was decreased in these animals below their preoperative performance. It was also below that of animals with implants in the hypothalamic-preoptic regions. The latter groups displayed intromission latencies within their preoperative ranges on many of the postoperative tests.

The relatively large extent of the "effective" area constitutes evidence contrary to the concept of a discrete and neuroanatomically well-defined androgen-sensitive sexual behavior "center" in the male. It is of interest that a generally similar intracerebral distribution of estrogen implants effective in restoring female sexual behavior has been described by Michael[1] (13). This widespread distribution may have resulted either from diffusion of hormone to sites distant from the implant, or from the existence of widespread cerebral structures capable of responding to androgen by activating the behavioral pattern. In Michael's studies, autoradiographic evidence was presented which ran counter to the former alternative (14). The present study provides no information as to the actual mobility of the implanted hormone. However, since the best and most consistent results were obtained with implants in the preoptic-hypothalamic region, it is conceivable that the occasional effectiveness of extrahypothalamic implants was due to diffusion of testosterone to this region. The additional possibility that the effectiveness of hypothalamic-preoptic implants was due to diffusion to a smaller "center" within this region (such as the medial preoptic region) was tested by implanting small-surface area, fused implants. If the hypothesis were correct, these implants should have been quite effective but only within a smaller region. They were in fact relatively ineffective at any location, including the medial preoptic, but some positive tests resulted from implants throughout the whole extent of the preoptic-hypothalamic region although none resulted from implants outside it.

---

[1] The resemblance between the anatomical distribution of "effective" implants in this study and Michael's is particularly striking in the plane of the anterior commissure and optic chiasm [compare Fig. 4 (center) with Fig. 3 in ref. 13].

Unlike Michael's conclusions based on estrogen implants in female cats (13), Lisk favors a more discrete estrogen-sensitive sexual behavior center in the anterior hypothalamus of the female rat (4). Although the implants in the present study were relatively large compared to those used by Lisk, testosterone is much less potent than estradiol, and in none of the rats in this study could "leakage" into the systemic circulation be implicated in behavioral restoration. As discussed above, the use of smaller implants reduced the probability of obtaining behavioral activation, but did not increase the anatomical discreteness of the effective area. It appears then that a relatively large area of brain tissue needs to be exposed to the action of the hormone for behavioral activation to occur with reasonable probability. The data are consistent with the following hypothesis: A "network" of testosterone-sensitive neurons exists in the hypothalamic-medial preoptic region of the male rat. Exposure of any significant portion of this "network" to the direct action of the hormone will restore sexual responsiveness to the castrate male. Certain portions of the "network" (e.g., the medial preoptic) appear more responsive than others in terms of the probability of achieving behavioral restoration following testosterone implantation. This may be due to a greater density in these areas of "network" cells or fibers.

Complete behavioral restoration (i.e., all 5 postoperative tests positive) resulted in only two cases. Furthermore, analysis of the behavior of animals with hypothalamic-preoptic implants showed that the restored behavior was not completely normal in many cases. It appears, therefore, that the crystalline testosterone depots were not necessarily capable of the complete restoration of normal male rat behavior. Presumably, the implants could not duplicate precisely the pattern of delivery of testosterone to the brain areas involved which occurs in physiological conditions, especially if a relatively widespread area of brain has to be perfused. On the other hand, the observed discrepancies in behavior might have resulted from the absence of sensory feedback from peripheral receptors, such as the penile papillae, which require androgen stimulation for their maintenance (15). However, since these behavioral discrepancies were often relatively minor or absent, one must conclude that no androgen-dependent extracerebral mechanism appears essential for normal sexual behavior in the male rat.

Acknowledgments

I am most grateful to Dr. F. A. Beach for help in the initial stages of this project, and to Linda Coates, Thornton Ege and Darlene DeManincor for able technical assistance. The testosterone was kindly supplied by Schering Corporation.

References

1. Beach, F. A., Hormones and Behavior, Cooper Square Publisher, Inc., New York, 1961, p. 368.
2. Young, W. C., In Young, W. C. (ed.), Sex and Internal Secretions, vol. II, Williams and Wilkins, Baltimore, 1961, p. 1609.
3. Harris, G. W., and R. P. Michael, J Physiol (London) 171: 275, 1964.
4. Lisk, R. D., Amer J Physiol 203: 493, 1962.
5. Brookhart, J. M., and F. L. Dey, Amer J Physiol 133: 551, 1941.
6. Soulairac, M. L., Ann Endocr (Paris) 24: No. 3 (suppl.), 1963.
7. Phoenix, C. H., J Comp Physiol Psychol 54: 72, 1961.
8. Fisher, A. E., Science 124: 228, 1956.
9. Davidson, J. M., Abstracts of papers, 23rd Intern. Congr. Physiol. Sci., 1965, p. 647 (Abstract).
10. ———, Anim Behav 14: 266, 1966.
11. de Groot, J., Trans Roy Neth Acad Sci 52: No. 4, 1959.
12. Davidson, J. M., and S. Feldman, Endocrinology 72: 936, 1963.
13. Michael, R. P., Brit Med Bull 21: 87, 1965.
14. ———, In Martini, L., and A. Pecile (eds), Hormonal Steroids, vol. 2, Academic Press, New York, 1965, p. 469.
15. Beach, F. A., and G. Levinson, J Exp Zool 114: 159, 1950.

155

16

Suckling stimulation induces aggression in virgin female mice

THE aggression shown by parturient mammals of various species towards strange conspecifics[1,2], referred to as "maternal aggression", ostensibly serves to protect the young. Although much is known about the aetiology of other forms of aggressive behaviour, no information is available about the mechanism responsible for maternal aggression. We found recently that suckling stimulation may be important in the initiation of aggressive behaviour in parturient mice[3]. Animals whose nipples were removed before parturition and that were given foster young each day after delivery did not become aggressive. Moreover, we could not induce aggression in virgin female mice by exposing them continuously to foster young. A reason for this may be that virgin females, because of the lack of nipples, which usually develop during pregnancy, did not receive suckling stimulation from foster young. We hypothesised that if nipple growth were induced in virgin females, they would be able to receive suckling stimulation and thus would exhibit aggressive behaviour comparable with that of parturient animals. The results presented here confirm the hypothesis by showing that suckling stimulation induces aggression.

Eighty-five 60–65-d-old virgin female Rockland–Swiss mice (RS) were divided into four groups. One group of 24 mice, after bilateral overiectomy under ether anaesthesia, received a regimen of hormones that we had found to induce nipple growth. We gave 19 daily subcutaneous injections of 0.02 µg oestradiol benzoate dissolved in 0.02 ml sesame oil and 500 µg progesterone in 0.02 ml oil. (The synergistic action of oestrogen and progesterone in mice is confirmed by the fact that little or no nipple growth occurs in response to injections of either oestrogen or progesterone.) Another group of 20 virgin females was ovariectomised and administered oil for 19 d, while another 19 mice were sham ovariectomised and treated daily for 19 d with oil. A group of 22 mice was left intact and mated with RS males. Each animal was housed singly in an $11 \times 7 \times 5$ inch polypropylene cage, the floor of which was covered with pine shavings. The animals had free access to food and water and were maintained on a 12 h light/12 h dark cycle with lights on between 0600 and 1800. Testing was between 0700 and 0900 and injections were given between 1000 and 1100.

The virgin mice were tested for aggression 24 h after the final injection of oil or hormone. Animals permitted to mate were tested on day 19 of gestation. An adult RS male was placed into the cage of each animal and left for 3 min or until a fight was observed. A fight was defined as 20 s of continuous biting and chasing. Immediately after the aggression test, each animal was presented with five 3–8-d-old RS foster young. (The pregnant animals were allowed to deliver and their young

were removed before presenting them with foster young.) On the next morning each animal was observed for maternal behaviour for 15 min, after which a test for aggression was conducted. Immediately after the aggression test the nipples of each animal were examined (red and distended nipples indicated that the female had received suckling stimulation) and the pups were removed and replaced by five other 3-8-d-olds. This procedure was continued for a further 9 d or until an animal exhibited aggression on two successive tests. The animals were killed at the termination of testing to determine whether milk was present in the mammary tissue.

Every animal exhibited the complete repertoire of maternal behaviour including retrieval of pups, licking of the pups' ano-genital region, and the assumption of the nursing posture. This confirms reports that non-pregnant and non-lactating mice display maternal behaviour[4].

None of the animals fought when tested before exposure to foster young (Table 1). Instead, they sniffed and groomed the male intruder. After exposure to young, 12 of 22 parturient mice and 12 of 24 hormone-treated virgin females were aggressive. The absence of aggression in approximately 50% of the animals in each of these groups is not unusual for we consistently find that 40–50% of parturient RS females are not aggressive[5]. In contrast, only one of 20 oil-treated, ovariectomised and one of 19 oil-treated, sham-ovariectomised virgins fought. Both oil-treated groups differed significantly from the hormone-treated animals and from the parturient animals with respect to the number that were aggressive (χ^2, $P < 0.01$).

Aggression typically occurred on the first test after exposure to pups, although a few animals began to fight on the second or third test. The topography of the fighting behaviour of the hormone-treated virgins was indistinguishable from that of the parturient animals. The initial latency to attack averaged 4 s with the attacks consisting of bites to the neck and flanks of the intruder male. Tail rattling and ano-genital sniffing were observed during fights but not before the initial attack. Intruders, never initiating a fight nor fighting back in response to attack, generally assumed submissive postures.

Daily inspection of the parturient and hormone-treated animals after exposure to pups revealed red and distended nipples. In both groups the pups attached themselves to the nipples such that they could maintain contact with the adult when the latter was lifted by the tail. In contrast, red and distended nipples and attachment were never observed in either of the groups injected with oil. Only the mammary glands of the parturient animals contained milk.

To verify further that the hormone regimen induced nipple

Group	n	No. fighting before presenting foster young	Proportion (and %) of animals fighting after pup exposure	Latency to begin fighting after pup exposure (d)	Average no. of fights[*][†]
Experiment 1					
Parturient	22	0	12/22 (54)	1.8±0.8	—
OVX+EB+P	24	0	12/24 (50)	1.3±0.6	—
OVX+Oil	20	0	1/20 (05)	1.0	—
S–OVX+Oil	19	0	1/19 (05)	2.0	—
Experiment 2					
Parturient–THEL	20	0	4/20 (20)	1.3±0.4	9.8±0.4
Parturient–S–THEL	20	0	13/20 (65)	1.6±1.1	8.6±2.1
OVX+OB+P–THEL	20	0	2/20 (10)	2.5±1.5	8.0±1.0
OVX+OB+P–S–THEL	20	0	12/20 (60)	2.1±2.4	8.2±2.5

Table 1 Aggressive behaviour of different groups of females

[*]Animals in experiment 1 were killed after they fought in two successive tests.
[†]Animals in experiment 2 were given 10 aggression tests. The data are derived from only the animals that fought.
Experiment 1: aggressive behaviour of RS female mice that were either mated (parturient), ovariectomised and primed with oestradiol benzoate and progesterone to induce nipple growth (OVX+OB+P), ovariectomised and treated with oil (OVX+Oil), or sham ovariectomised and given oil (S–OVX+Oil). Aggressive behaviour was scored before exposure to pups (on the last day of hormone or oil treatment or on day 19 of gestation) and on each succeeding day for 10 d after presentation of foster young or until fights were displayed on 2 successive days. Experiment 2: aggressive behaviour of RS females that were mated and either thelectomised (parturient–THEL) or sham thelectomised on day 19 of pregnancy (parturient–S–Thel) and of ovariectomised hormone-treated mice either thelectomised (OVX+OB+P−THEL) or sham thelectomised (OVX+OB+P–S–THEL) on the day after the last hormone injections.

156

growth, six more animals were placed into each treatment group. They were killed either after 19 d of hormone or oil treatment or on day 19 of gestation. The results are summarised in Table 2. On the average, nipple length was 100% and the base diameter was 30% greater for hormone-treated than for oil-treated virgin females. For pregnant animals, the same measures were 200% and 50% greater than those of oil-treated virgins.

Table 2 Average length (mm) and base diameter of a representative nipple of different animals

Group	Average length	Base diameter
Parturient	1.40±0.05*†	0.66±0.06*†
OVX+OB+P	0.99±0.07*	0.55±0.14*
OVX+Oil	0.49±0.08	0.43±0.08
S–OVX+Oil	0.47±0.13	0.41±0.06

Animals were either mated (parturient), ovariectomised and hormonally primed with oestradiol benzoate and progesterone (OVX+OB+P), ovariectomised and treated with oil (OVX+Oil), or sham ovariectomised and given oil (S-OVX+Oil).
*Significantly different (Fisher's least significant difference test, $P < 0.05$) from OVX+Oil and S-OVX+Oil.
†Significantly different from OVX+OB+P.

The findings show that virgin female mice can be induced to show aggression comparable with that of parturient animals by producing nipple growth and presenting them with young. Hormone exposure did not induce fighting directly, for treated animals were never aggressive before exposure to young. Suckling stimulation, therefore, may be essential for the initiation of aggression. To corroborate this we assessed the effect of the suckling stimulus.

Eighty 60–65-d-old virgin female RS mice were divided into four equal groups. Two groups were ovariectomised and given the hormone regimen described above. The nipples of one group were then removed (thelectomy) under ether anaesthesia on the day after the last hormone injection. The other group was sham thelectomised: 1-mm incisions were made along the ventral midline. The two remaining groups were left intact and mated with RS males. One of these groups was thelectomised on day 19 of pregnancy and the other was sham thelectomised. Testing and fostering were as before, except that animals were allowed to complete all 10 aggression tests.

All animals were comparable with respect to maternal behaviour. None was aggressive before exposure to young (Table 1). After exposure to pups, 13 of 20 sham-operated parturient mice and 12 of 20 sham-thelectomised hormone-treated virgins were aggressive compared with four of 20 thelectomised parturient animals and two of 20 thelectomised hormone-treated virgins. The thelectomised groups differed significantly from their sham-operated counterparts with respect to the number that fought ($P < 0.02$). The sham-thelectomised groups did not differ in terms of time taken to become aggressive or the average number of fights exhibited. Daily examination of both sham-operated groups revealed red and distended nipples, indicating that they had received suckling stimulation.

Thus hormone-treated virgins respond to thelectomy in a manner identical to that of parturient mice with thelectomy, significantly reducing the proportion of animals that are aggressive. The data agree with others which demonstrate the attenuating effects of thelectomy on maternal aggression in parturient mice[3].

Finally, 12 hormone-treated virgins and 12 parturient animals that fought the intruder on the third day of exposure to foster young were deprived of their foster young immediately after the fighting test and were retested 24 h later. If fighting did not occur on retesting, five foster young were placed into the cage and, beginning 24 h later, daily aggression tests and daily fostering of young were carried out until fighting recommenced. Animals that fought 24 h after removal of foster young continued daily aggression tests in the absence of the young. As soon as fighting was no longer exhibited, foster young were reintroduced and daily tests were conducted until fighting began.

All animals stopped fighting after pups were removed and resumed fighting when they were replaced. The hormone-treated virgins and the parturient animals did not differ in terms of the number of days after removal required to terminate fighting (range = 1–4 d, median = 1.0 d) or the number of days of pup replacement required to re-establish fighting (all within 1 d). Each animal had red distended nipples and nipple attachment after replacement of foster young. Thus the aggression of both hormone-treated and parturient mice can be eliminated by removing the suckling stimulus and re-established by reintroducing that stimulus.

The fighting behaviour induced in female mice hormonally primed to induce nipple growth could have been a response to chronic stress, produced by chronic suckling by deprived young, rather than a response specific to the suckling stimulus. We do not believe that this is the case, however, because virgin females and lactating mice whose pups had been removed did not attack a male after 24 h of intermittent electric shock to the paws. Moreover, virgins and lactating mice deprived of young did not attack a male after chronic application of painful stimulation to the nipples. Thus fighting in hormone-treated virgin female mice seems to be a response specific to the suckling stimulus.

Our data show that suckling not only supplies nourishment to the young but also affects the relationship between the mother and other adult members of her species. Suckling stimulates the release of hypothalamic and pituitary hormones[6], alters the rate of synthesis of hypothalamic monoamines[7], and changes the firing patterns of hypothalamic neurones[8]. One or more of these changes may induce aggressive behaviour in lactating mice.

This research was supported by grants from the US Public Health Service.

BRUCE SVARE*
RONALD GANDELMAN†

Department of Psychology,
Rutgers University,
New Brunswick, New Jersey 08903

Received November 7, 1975; accepted January 9, 1976.

*Present address: Worcester Foundation for Experimental Biology Shrewsbury, Massachusetts 01545.
†To whom reprint requests should be addressed.

1 Brown, R. Z., Ecol. Monogr., 23, 217 (1953).
2 Scott, J. P., Am. Zool., 6, 683 (1966).
3 Svare, B., and Gandelman, R., Hormone Behav. (in the press).
4 Noirot, E., Anim. Behav., 17, 547 (1969).
5 Svare, B., and Gandelman, R., Hormone Behav., 4, 323 (1973).
6 Grosvenor, C. E., and Mena, F., in Lactation: A Comprehensive Treatise (edit. by Larson, B. L., and Smith, V. R.), 1, 227 (Academic, New York. 1974).
7 Voogt, J. L., and Carr, L. A., Neuroendocrinology, 16, 108 (1974).
8 Brooks, C. M., Ishikawa, T., Korzumi, K., and Lu, H., J. Physiol., Lond., 182, 217 (1966).

17

Journal of Comparative and Physiological Psychology
1970, Vol. 71, No. 2, 210–215

INDUCTION OF OVARIAN ACTIVITY IN FEMALE RING DOVES BY ANDROGEN TREATMENT OF CASTRATED MALES[1]

CARL J. ERICKSON[2]

Institute of Animal Behavior, Rutgers—The State University

Female ring doves were mated with castrated males treated with various amounts of androgen. The oviduct weights and ovulation frequencies of the females reflected the vigor of androgen-induced courtship in their male mates. Correlations of male activity and female response suggest that not all androgen-dependent male courtship patterns are equally effective in promoting female gonadotropin secretion.

In many species of birds successful reproduction is dependent upon the coordination of male and female behavior throughout the stages of courtship, nest-building, incubation, and brooding of young. It is now clear that the behavior patterns which are characteristic of each stage may be at least partially attributed to changes in hormone secretion (Lehrman, 1961). Moreover, the hormone secretion itself may result from behavior performed in an earlier stage of the reproductive cycle or from changes in the environmental stimulus complex. Of particular interest is the fact that stimuli provided by a male are capable of promoting endocrine changes in a female mate which may induce her to build a nest, ovulate, and incubate eggs. Matthews (1939) demonstrated that visual and auditory stimulation alone would induce ovulation in the female

[1] This paper is based on portions of a dissertation submitted in partial fulfillment of the requirements for the PhD degree, Rutgers—The State University, 1965 (Ann Arbor, Mich.: University Microfilms, No. 66-2113). The author wishes to thank Daniel S. Lehrman for his advice and guidance through all stages of this research and preparation of the manuscript. Thanks are also due to Willis Butler for aid in the histological work and to Roy Bruder for his assistance with many of the behavioral observations. The research was supported by a predoctoral fellowship from the National Institute of Mental Health (MH-21,101-02) and by Research Grant MH-02271 (D. S. Lehrman, principal investigator) from the National Institute of Mental Health. Contribution No. 64 from the Institute of Animal Behavior, Rutgers—The State University.

[2] Requests for reprints should be sent to Carl J. Erickson, who is now at the Department of Psychology, Duke University, Durham, North Carolina 27706.

pigeon when tactile contact was prevented by a glass plate placed between members of the pair. Other evidence indicates that the *behavior* of the male is more effective than his mere presence. Erickson and Lehrman (1964) showed that ovarian activity was retarded in female ring doves exposed to noncourting, castrated males as compared to females exposed to active, intact males. In budgerigars male vocalizations stimulate ovulation even when females are kept in darkness (Ficken, van Tienhoven, Ficken, & Sibley, 1960; Vaugien, 1951). Brockway (1965) has presented evidence which indicates that a particular component of the male vocal pattern has special stimulus value in promoting ovarian activity of the female budgerigar. Recent evidence indicates that vocalization alone (Lehrman & Friedman, in press) or in combination with visual stimulation from another bird (Lott & Brody, 1966; Lott, Scholz, & Lehrman, 1967) may also play a significant role in the ovarian activity of the female ring dove. In the ring dove, however, ovulation rarely occurs without some visual access to another bird.

The present study was devised for further understanding of male courtship in the ring dove and its impact on female ovarian activity. Previous studies had raised several issues requiring clarification. First, will testosterone therapy effectively reinstate courtship in castrated males; and, if so, is the vigor of the behavior dose-related or all-or-none? Second, if male courtship is dose-related, does female ovarian activity reflect the vigor of male responses to hormone therapy? And third, do female ovarian responses correlate significantly with the entire spectrum of male hormone-dependent activities or with only particular components of the behavior repertoire?

The basic experimental approach, then, is to treat castrated males with various amounts of androgen and allow them to court females through a glass barrier. Subsequently, the ovarian responses of these females are correlated both with the courtship vigor of their male mates and with the amount of androgen which these males received.

METHOD

Subjects

The subjects were 96 male and 120 female adult ring doves (*Streptopelia risoria*) which had hatched in the laboratory. All were descendants of a mixed stock originally obtained from J. W. Steinbeck of Walnut Creek, California.

Cages and Maintenance

Housing and colony maintenance were as described by Lott et al. (1967).

Experience Prior to Testing

All birds had been hatched in experimental cages and raised by their parents until 21 days of age when they were moved to stock cages and kept in groups of 7–10. Sex and degree of maturity were later determined by exploratory laparotomy. The birds were then placed in experimental cages and allowed to breed, their young being removed when 21 days of age. Several of the birds had been allowed two breeding cycles but never with the same mate. In summary, all birds had had at least one but not more than two previous successful breeding experiences. Immediately prior to the beginning of the experiment all birds were maintained in visual isolation for 4 wk.

Observations

All testing and mating was performed in experimental cages in the colony area. The female's portion of the cage was supplied with a glass nest bowl and a quantity of nest material. A handful of feed was scattered on the floor of both the male's and female's compartments, but food, water, and grit dispensers were not present during testing since the animals often perched on these and terminated interaction for long periods of time. Approximately 15 min. prior to the commencement of the recording period an opaque partition was placed in the testing cage, and the test male and female were placed in their respective compartments on either side. This brief period allowed the animals to recover from the effects of handling and to become accustomed to the new cage conditions. The test was begun when the partition was removed and replaced with a clear glass plate. All observation periods were 30 min. in length, and during this time male behavior was continuously recorded on a keyboard-actuated 20-pen Esterline-Angus operations recorder. Behavior items noted were (a) bowing-coo (frequency and number of bouts[3]); (b) nest-coo (frequency); and (c) wing-flipping (total duration and number of bouts). It should be noted that nest-cooing bouts were not computed since nest-cooing is accompanied by wingflipping and, therefore, the number of wingflipping and nest-cooing bouts is the same except for an occasional wingflipping bout not accompanied by nest-cooing. The *rate* of nest-cooing accompanying wingflipping may vary widely, however, and for this reason nest-coo frequency and wingflipping duration are computed separately.

Experimental Procedure

Presurgery testing. Seven days prior to surgery all males were given behavior tests with females taken from isolation. This test was given for purposes of evaluating the effects of later castration and androgen therapy. Immediately following these observations males and females were returned to visual isolation.

Surgery. Seventy-nine males were gonadectomized in a two-stage operation. At the conclusion of the experiment, all castrates were sacrificed and autopsied, and the abdominal cavity was examined carefully for remnants of testicular tissue. Suspicious fragments were removed for histological examination. No testicular tissue was found. Seventeen males were given a sham operation in which the abdominal wall was opened and the testes manipulated and examined but not removed. Following surgery all birds were allowed a 10-day recovery period in the isolation cages.

Postsurgery treatment. On Day 10 following surgery all males were again tested with females to evaluate the effects of surgery. In order to provide a more uniform base line for the second portion of the experiment, any castrates which showed the bowing-coo, nest-cooing, or wingflipping were eliminated from the study at this point. Following this test the remaining castrates were assigned to five treatment groups. Accordingly, these males received either 5, 15, 25, or 100 μg. of testosterone propionate in .1 ml. of sesame oil. An additional castrated group and an intact sham-operated group received daily injections of the vehicle alone.

Mating. On Day 23 of replacement therapy all males were tested with females in the usual way. At the end of the ½-hr. observation period both male and female remained in the cage. Food, water, and grit dispensers were placed in the areas on both sides of the glass plate, and the male and female remained together for the following 7 days. All males continued to receive injections during this time.

---

[3] The term "bout" refers to a rhythmic repetitive period or occurrences of a behavior item, uninterrupted by other activities. Thus, a bowing-coo bout may consist of one to several hundred bowing-coos. See Miller and Miller (1958) for a description of the behavior repertoire of the ring dove.

Autopsy of females. At the conclusion of the 7-day mating period females were killed for autopsy. The oviduct was removed, cleaned, and weighed. If an egg was present in the shell gland, it was removed and its calcification level noted. In birds which have not ovulated at the time of autopsy, the oviduct weight provides a satisfactory estimation of ovarian activity. Following ovulation, however, oviduct weights fluctuate widely and a more appropriate measure of activity is offered by the following ordinal scale of development presented by Lott et al. (1967): 1—egg high in oviduct; 2—egg in cloaca, soft shell; 3—egg in cloaca, hard shell; 4—egg in nest; 5—egg in nest and egg high in oviduct; 6—egg in nest and egg in cloaca, soft shell; 7—egg in nest and egg in cloaca, hard shell; 8—two eggs in nest. For statistical purposes all females were ranked together whether they had ovulated or not. For example, females exposed to the five groups of castrated males were ranked 1–72 in terms of the degree of their ovarian activity —No. 1 being the bird with the lowest oviduct weight, No. 72 a female with two eggs in the nest.

Use of females as test animals. Since the design of the experiment required the repeated testing of all males, and because the supply of females was limited, all females were rotated as test partners in the preliminary tests. The procedure of rotation ensured that no systematic differences in test experience characterized the females mated with males of the various groups.

Statistical Tests

For comparisons of several groups the Kruskal-Wallis one-way analysis of variance was used.

Correlations are based on Spearman's rank-order method (Siegel, 1956).

RESULTS

Effects of Castration on Male Behavior

Castration resulted in a sharp decline in the performance of certain components of the male repertoire. Ten days following surgery only 1 of the 79 castrated males showed bow-cooing. Six males exhibited wingflipping, and five of these also showed nest-cooing. This is in direct contrast to the sham-operated control group in which 94.1% of the males performed bow-cooing and 88.2% nest-cooed. For purposes of uniformity these seven animals were eliminated from the remaining portion of the experiment.

Effects of Differential Androgen Therapy

Table 1 indicates that the behavior patterns reduced by castration are reinstated through androgen therapy. With the exception of the vigorous group of males given 15 μg., it appears that there is a direct relationship between the dosage of testosterone propionate and activity level. Statistical comparison of the five castrated groups following therapy indicates signifi-

TABLE 1

PERFORMANCE OF SEXUAL BEHAVIOR ITEMS OF CASTRATED MALE DOVES FOLLOWING ANDROGEN REPLACEMENT THERAPY, MEDIAN (RANGE)

Daily amount of treatment	n	Bowing-coo (no.)	Bowing-coo (bouts)	Nest-coo (no.)	Wingflipping (duration in seconds)	Wingflipping (bouts)
Vehicle only	14	0 (0)	0 (0)	0 (0)	0 (0)	0 (0)
5-μg. TP	13	1 (0–16)	1 (0–2)	0 (0)	0 (0)	0 (0)
15-μg. TP	15	22 (0–223)	3 (0–20)	0[a] (0–33)	0[b] (0–234)	0[b] (0–5)
25-μg. TP	15	6 (0–103)	1 (0.7)	0[a] (0–51)	0[b] (0–506)	0[b] (0–5)
100-μg. TP	15	32 (0–221)	3 (0–19)	17 (0–49)	188 (0–519)	2 (0–9)
Intact—vehicle only	17	68 (0–195)	6 (0–13)	58 (0–131)	425 (0–847)	2.5 (0–10)
H[c]		27.88	27.77	24.12	28.60	28.66
p		<.001	<.001	<.001	<.001	<.001

[a] Six positive scores.

[b] Seven positive scores.

[c] Kruskal-Wallis one-way analysis of variance; intact males not included (Siegel, 1956).

cant differences in response on all measures ($p < .001$).

Differences in Female Ovarian Response

Table 2 indicates the extent of ovarian activity in female mates which could see and hear males of the experimental groups. Comparison of the female response reveals highly significant differences in reproductive tract development among the groups ($H = 15.21$, $p < .01$). Moreover, there is a striking peak of activity in the female group which was mated with the highly active group of 15-μg. TP males. Thus, androgen-dependent behavior patterns in the male appear to have a stimulatory effect on ovarian activity in female mates.

Correlation of Specific Male Activities and Subsequent Female Ovarian Response

It has been shown that certain behavior patterns of male ring doves vary with the level of circulating androgen. It has also been demonstrated that levels of development in the female reproductive tract correspond to the activity of male mates. There is a possibility, however, that only particular components of the male behavior repertoire have significance in promoting the release of gonadotropin from the female hypophysis. Such a possibility is suggested by correlations of individual male activity levels on various patterns with responses of the individual female

TABLE 2
MEDIAN RANK OF OVARIAN DEVELOPMENT
(MEDIAN OVIDUCT WEIGHT IN MG.)

Treatment of male mate	n	Mdn	Range	No. of females ovulating
Vehicle only	14	16.5 (682)	2–39 (198 to 2,707)	0
5-μg. TP	13	40.0 (2,363)	1–79 (153 to Level 3)[a]	2
15-ug. TP	15	60.5 (Level 2)	21–71 (616 to Level 3)	9
25-μg. TP	15	44.0 (4,195)	3–88.5 (212 to Level 8)	5
100-μg. TP	15	54.0 (5,187)	4–88.5 (228 to Level 8)	7
Intact—vehicle only	17	60.5 (Level 3)	14–83.5 (511 to Level 7)	11

Note.—Kruskal-Wallis one-way analysis of variance; females with intact males not included (Siegel, 1956).
[a] See text for explanation.

TABLE 3
CORRELATION OF MALE BEHAVIOR ON THE MATING TEST AND DEVELOPMENT OF FEMALE REPRODUCTIVE TRACT ($n = 62$)

Behavior item	Correlation (r_s)	p
Bowing-coo (no.)	.01	>.20
Bowing-coo (bouts)	.13	>.20
Wingflipping (total duration)	.26	<.05
Wingflipping (bouts)	.27	<.001
Nest-coo (no.)	.14	>.20

mates. Table 3 presents rank-order correlations between various male behavior patterns and female ovarian response.[4] Two measures, the time spent wingflipping and the frequency of wingflipping performance (bouts), correlate significantly with female ovarian activity. The bowing-coo (number and bouts) reflects little relationship to the female response. Nest-cooing, an activity which accompanies wingflipping but varies in rate, shows positive but low correlation to reproductive tract development.

DISCUSSION

Previous studies have indicated that castration effectively reduces the expression of several male behavior patterns normally performed in the presence of the female (Erickson, Bruder, Komisaruk, & Lehrman, 1967; Erickson & Lehrman, 1964). The present study confirms these observations and demonstrates that androgen therapy is capable of reinstating these activities. Moreover, male castrates maintained on androgen are effective in stimulating female ovarian activity, and this ovarian activity seems to be a direct reflection of the amount of male activity induced. This correspondence is especially apparent with respect to males receiving 15 μg. of hormone. In terms of the bowing-coo these males were actually more active than those receiving 25 μg. Female mates of these males showed a higher percentage of ovulation than any other group with the exception of the intact controls.

Although this may suggest that males which bow-coo actively are more potent in promoting female gonadotropin secre-

[4] Males and females of the untreated castrate group and the 5-μg. group are excluded from this analysis because of the large percentage of males which failed to respond on any of the measures.

161

tion, correlation of male performance and female response on an individual basis indicates a contrary conclusion. In the latter instance it was clear that ovarian activity was greater in those females that had been mated with males displaying the most wingflipping at introduction. Support for the view that wingflipping may be more effective than bow-cooing in initiating the events leading to oviduct growth and ovulation comes from observations which show that females exposed to their own image in a mirror will ovulate (Lott & Brody, 1966; Matthews, 1939) and female-female pairings frequently result in rapid ovulation by both members of the pair. It should be noted that although females may nest-coo and wingflip vigorously, they almost never exhibit the bowing-coo.

Although it is suggested that certain components of the male repertoire may have special significance with respect to female gonadotropin secretion and consequent ovarian activity, the conditions of the present experiment were inappropriate for a detailed explication of the causal sequence involved. Since the stimulus males could also perceive the females, it is not impossible that females communicate their individual predispositions for ovulation at the outset, and this is reflected in differences in male behavior—in effect, a reversal of the proposed causal relationship. Such an interpretation seems unlikely, however. In previous work (Erickson, 1965) the present author has shown that male performance is highly reliable when different female stimulus animals are used on successive courtship tests. Thus, male performance appears to be relatively independent of female behavior differences, at least under the conditions of observation in these studies.

It should also be noted that during the 7-day interval between the observation of male behavior and the evaluation of female response, striking behavior changes occur in both animals. Therefore, female gonadotropin secretion may be more directly induced by male behavior occurring after the observation period, or even by the characteristic behavior patterns elicited from the female herself. For example, it is not unlikely that ovarian activity is stimulated most directly by the female's active participation in courtship and nest-building. Work currently in progress should provide further evidence on this matter.

REFERENCES

BROCKWAY, B. F. Stimulation of ovarian development and egg laying by male courtship vocalization in budgerigars (*Melopsittacus undulatus*). *Animal Behaviour*, 1965, **13**, 575–578.

ERICKSON, C. J. A study of the courtship behavior of male ring doves and its relationship to ovarian activity of females. Unpublished PhD dissertation, Rutgers—The State University, 1965.

ERICKSON, C. J., BRUDER, R. H., KOMISARUK, B. R., & LEHRMAN, D. S. Selective inhibition by progesterone of androgen-induced behavior in male ring doves (*Streptopelia risoria*). *Endocrinology*, 1967, **81**, 39–44.

ERICKSON, C. J., & LEHRMAN, D. S. Effect of castration of male ring doves upon ovarian activity of females. *Journal of Comparative and Physiological Psychology*, 1964, **58**, 164–166.

FICKEN, R. W., VAN TIENHOVEN, A., FICKEN, M. S., & SIBLEY, F. C. Effects of visual and vocal stimuli on breeding in the budgerigar (*Melopsittacus undulatus*). *Animal Behaviour*, 1960, **8**, 104–106.

LEHRMAN, D. S. Hormonal regulation of parental behavior in birds and infra-human mammals. In W. C. Young (Ed.), *Sex and internal secretions*. Baltimore: Williams & Wilkins, 1961.

LEHRMAN, D. S., & FRIEDMAN, M. Auditory stimulation of ovarian activity in the ring dove. *Animal Behaviour*, in press.

LOTT, D. S., & BRODY, P. N. Support of ovulation in the ring dove by auditory and visual stimuli. *Journal of Comparative and Physiological Psychology*, 1966, **62**, 311–313.

LOTT, D. S., SCHOLZ, S. D., & LEHRMAN, D. S. Exteroceptive stimulation of the reproductive system of the female ring dove (*Streptopelia risoria*) by the mate and by the colony milieu. *Animal Behaviour*, 1967, **15**, 433–437.

MATTHEWS, L. H. Visual stimulation and ovulation in pigeons. *Proceedings of the Royal Society of London*, 1939, **126B**, 557–560.

MILLER, W. J., & MILLER, L. S. Synopsis of behaviour traits of the ring neck dove. *Animal Behaviour*, 1958, **6**, 3–8.

SIEGEL, S. *Non-parametric statistics for the behavioral sciences*. New York: McGraw-Hill, 1956.

VAUGIEN, L. Ponte induite chez la Perruche ondulée maintenue à l'obscurité et dans l'ambiance des volières. *Compte Rendu de l'Académie des Sciences, Paris*, 1951, **232**, 1706–1708.

(Received May 6, 1969)

18

Development and Hormonal Control
of Territorial Marking in the Male
Mongolian Gerbil (*Meriones unguiculatus*)

GARDNER LINDZEY
D. D. THIESSEN
ANN TUCKER

Department of Psychology
University of Texas
Austin, Texas

LINDZEY, GARDNER, THIESSEN, D. D., and TUCKER, ANN (1968). *Development and Hormonal Control of Territorial Marking in the Male Mongolian Gerbil* (Meriones unguiculatus). DEVELOPMENTAL PSYCHOBIOLOGY, 1(2): 97–99. Male Mongolian gerbils were either castrated or sham operated at 30 days of age. Assessment of territorial marking was carried out every 6 days beginning at 52 days of age and extending to 100 days of age. After each marking test, 10 intact males received 640 µg testosterone propionate subcutaneously; 10 other intact animals and 10 castrates received oil injections. The ventral sebaceous scent gland, used by the gerbil to deposit sebum on objects during the marking response, was measured after each test. Behavioral marking and the scent gland were entirely absent in castrates. Relative to controls, marking in hormone-treated animals began earlier and reached higher frequencies. Gland development was also responsive to the hormone, but lagged behind marking activity.

castration   development   gerbil   gland size   territorial marking   testosterone propionate

T HE MONGOLIAN GERBIL (*Meriones unguiculatus*) is rapidly becoming an important subject of research. It is hearty, thrives well on ordinary Purina Laboratory Chow, has little odor, and requires no water other than that derived from the metabolism of foods. Behaviorally, the animals have much to recommend them. They are docile, highly curious, learn well, and show some unusual species-common behaviors. In our laboratory, all animals tested appear deficient in visual cliff behavior, but are not blind (Thiessen, Friend, & Lindzey, in press), and a certain proportion of those tested convulse when placed in a strange environment. The latter effect appears to be regulated by a dominant gene. Further, as this paper shows, gerbils mark territories with a ventral scent gland.

The species *unguiculatus* is of the genus *Meriones*, subfamily *Gerbilinae*, family *Cricatidae*, and order *Rodentia* (Schwentker, 1963). Its close relatives include the Cotton rat of the Eastern United States, the Golden and Chinese hamster, the Deer mouse and the Desert rat of North Africa. As an adult, the gerbil weighs from approximately 70 to 90 g.

The behavioral marking response, discussed in this paper, is particularly interesting. Both males and females rub a midventral sebaceous gland over low-lying prominent objects, leaving a sebum which is oily to the touch and musky in odor. The frequency of marking is about twice as great in the male, as in the

Received for publication 15 February 1968.

Developmental Psychobiology, 1(2): 97–99 (1968)

female, corresponding roughly to the sex difference in gland size. The response is highly discrete, involving a rapid approach to an object, a sniffing of the object, a mounting and press or skimming of the ventral sebaceous gland against the surface followed by a rapid forward dismount. The complete marking response is accomplished in about 1 sec.

The sebaceous gland is a highly organized, fusiform pad which is orange in color. In a well-developed male, it measures approximately 3.0 cm in length, 0.7 cm in width, and 0.2 cm in depth. It is easily seen by parting the ventral hairs enfolding the gland. Marking behavior disappears and the sebaceous gland atrophies following castration of the male, but are easily reinstated with injections of testosterone propionate (Thiessen *et al.*, in press). It seems likely that this marking behavior involves territorial signaling, since a male becomes more hesitant in a field recently occupied by another gerbil of either sex. This paper describes the ontogeny of marking in the male, and associates the development of the gland and the behavior with the presence or absence of androgen.

METHOD

Marking behavior and general activity were assessed in a grey wooden field measuring 1 square meter and lined off into 16 squares of equal size. A roughened Plexiglas peg, measuring 2.6 cm in length, 1.2 cm in width and 0.7 cm in height, was positioned at each of the 9 lined intersections. The field was surrounded by grey wooden walls 46 cm high. On each wall was mounted one 15 w light bulb, shielded at the top, and focused into the interior of the field.

Thirty male gerbils were included in the experiment. At 30 days of age, at least 40 days before maturity, 10 males were castrated under general anesthesia and the additional 20 males received a sham operation. Beginning at 52 days of age, all animals were tested individually for 5 min in the marking field. A test consisted of placing a gerbil in one square of the field and recording the number of peg marks and line crossings during a 5-min period. A mark was recorded whenever an animal skimmed a peg, flattening its abdomen over the surface. A line crossing was recorded whenever an animal stepped over a line with all four feet. The entire apparatus was thoroughly cleaned after each test with a 70% alcohol solution. Testing was carried out during the midpart of the light half of a 12-hr, light–dark cycle, in a darkened room, with only the apparatus lights providing illumination. Immediately after each test, the animal was weighed and the sebaceous gland exposed with a hair depilatory (Sergex) and then measured in cm (length × width). After the gland measurement, each animal was injected subcutaneously with 0.05 ml of testosterone propionate (640 mg/injection), of the vehicle alone (safflower oil). The oil vehicle allowed slow assimilation of the hormone. Ten castrate and 10 intact gerbils received the vehicle only, and 10 intact gerbils received testosterone propionate. Testing, measurement, and injections were continued once every 6 days until the animals were 100 days of age.

RESULTS

Figure 1 illustrates the development of marking, and Figure 2 shows the differential development of gland size. Analyses of variance revealed highly significant variations between groups ($F = 4.61$ and 23.60, respectively; $df = 2$ and 27; $p < .05$ and $p < .01$). Most of the between group variation is accounted for by the castrate animals, but all groups differ among themselves (Ducan Multiple Range Test: $p < .05$) after group separation is apparent on *day 6* for marking and *day 7* for gland size. The mean body weight for the castrate animals exceeded that of the intact and sham animals ($p < .05$), and no differences in activity appeared among the three groups.

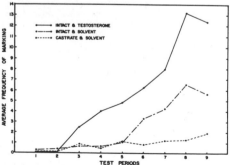

FIG. 1. Development and hormonal control of territorial marking in the Mongolian gerbil. Testing began at 52 and ended at 100 days of age. Frequency of marking recorded for 5-min periods.

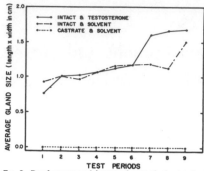

FIG. 2. Development and hormonal control of ventral sebaceous gland size in the Mongolian gerbil. Measurements began at 52 and ended at 100 days of age. Sebaceous pad measurement expressed in cm and converted to a single index of length × width.

DISCUSSION

The development of marking and the concomitant size of the ventral sebaceous gland are clearly related to the androgen status of the male. Marking generally begins around 80 to 90 days of age, but the onset may be either accelerated by large doses of testosterone propionate, or virtually eliminated by pre-pubertal castration. By 100 days of age, when intact gerbils approach their highest marking frequency, gerbils receiving large doses of hormone show significantly greater marking activity. The response of the gland to testosterone propionate is evident and even shows a critical period for hormone-sensitivity (unusual for testosterone-dependent tissues), in that hormone treatment is not effective until the animals are 94 days of age. Marking is evidently more sensitive to the hormone level than is gland size, suggesting, perhaps, a central nervous system effect. This notion is presently being explored by implanting crystalline testosterone into the hypothalamus of castrate males in an attempt to elicit marking behavior independent of a change in the sebaceous gland.

NOTES

This work was supported by Grant MH-1407-01 and Research Development Award MH-11, 174-01 to Delbert D. Thiessen from the National Institute of Mental Health, U.S. Public Health Service. Wayne Kamin kindly helped on all aspects of the problem.

Mailing address: D. D. Thiessen, Department of Psychology, University of Texas, Austin, Texas 78712, U.S.A.

REFERENCES

SCHWENTKER, V. (1963). *Veterinarian, 6:* 5–9.
THIESSEN, D. D., FRIEND, H. C., and LINDZEY, G. (In press). *Science.*

19

Effects of Testosterone Treatment in Adulthood on Sexual Behaviour of Female Pseudohermaphrodite Rhesus Monkeys

(Reprinted from Nature New Biology, Vol. 242, No. 117, pp. 119–120, March 28, 1973)

DANTCHAKOFF[1] observed 35 years ago that prenatal androgen treatment not only masculinized the reproductive anatomy of female guinea-pigs but also seemed to influence the development of their central nervous system since they displayed male like mounting in adulthood. Normal female guinea-pigs also display male-like mounting[2], however, and Dantchakoff made no attempt to quantify the behavioural responses.

Work with guinea-pigs, rats, and hamsters has since demonstrated[3-9] that if androgen is present at some critical period in development, there is an increased probability that the individual (genetic sex is irrelevant) will exhibit male-like sexual behaviour if and when androgen is present in adulthood. The converse is also true; and in rodent species at least, the presence of androgen early in development also inhibits responsiveness to oestrogen in adulthood and thereby diminishes the probability of activating female sexual responses with the usual hormonal treatments.

These changes indicate only an altered sensitivity to gonadal hormones in the rodent brain, not masculine or feminine brain differentiation[10,11]. But the masculinization of micturition patterns in juvenile dogs[12] and of play, threat, and mounting behaviour in juvenile female pseudohermaphrodite rhesus monkeys[13,14] after foetal treatment with testosterone implies functional modification of the central nervous system because masculinization was expressed without the later presence of gonadal hormones.

The experimental production of these female pseudohermaphrodite rhesus monkeys and their juvenile social experience have been reported previously[13,14]. Pregnant females were injected with 5 to 25 mg of testosterone propionate (TP) daily, through the middle trimester of the normal 168-day gestation (Table 1). At birth each female pseudohermaphrodite had a

Table 1 Prenatal Testosterone Propionate (TP) Treatment of Female Pseudohermaphrodite Rhesus Monkeys

Pseudohermaphrodite female No. 1616	TP administered on gestation days	Total TP administered (mg)
828	40 to 69	600
829	40 to 69	600
1239	38 to 66	825
1656	40 to 111	610
836	40 to 89	750
1616	39 to 88	750
1640	39 to 88	750

phallus, empty scrotum, ovaries, and no external vaginal orifice. They were weaned at 90 to 110 days and randomly assigned to social test groups of five to six animals which contained combinations of normal males and females and some prepuberally castrated males. These groups were studied for about 100 days during the first year and for 50 days during the second, third, and fourth years. The frequency with which the juvenile pseudohermaphrodites displayed patterns of threatening, play initiation, rough-and-tumble play, and chasing play was intermediate between that of control males and females, and their juvenile developmental pattern of mounting resembled that of control males and differed from that of control females[13,14]. We here report their adult behaviour and compare their responses after ovariectomy to exogenous testosterone treatment with those of ovariectomized female controls.

For this study, seven female pseudohermaphrodites and five female controls of equivalent age (5–7 yr) and social experience[13,14] were ovariectomized at least 1 month before the tests began. They were then paired weekly with stimulus females which had been primed with 10 μg of oestradiol benzoate per day for 10–18 days before each test. After 6 weeks of testing (six 10-min tests per individual), all experimental and control animals received 0.5 mg TP/kg daily for 24 weeks, followed by 15 mg TP per animal daily for 6 weeks. During the final 10 weeks of TP treatment, they were tested twice weekly.

After ovariectomy and before TP treatment, only one recorded behavioural response of the pseudohermaphrodites was significantly different from that of the female controls (Table 2). They were more frequently aggressive (potentially injurious attacks, predominantly by biting) toward the stimulus females. After the TP treatment, the pseudohermaphrodites remained significantly more aggressive than the controls and also had significantly high "proximity" scores, that is, they more frequently sat next to the stimulus females. Moreover, the pseudohermaphrodites had a significant increase in the frequency of "yawning", "grooming" the stimulus female, and "sexual exploration" (closely examining her genitalia). The female controls also showed a significant increase in "yawning",

Table 2 Mean Frequency per Test of Sexual and Social Behaviours of Ovariectomized Female Pseudohermaphrodite (♀) and Ovariectomized Control Female (♀) Rhesus Monkeys before and after Treatment with Testosterone Propionate (TP)

Behaviour	Before TP		After TP	
	♀	♀	♀	♀
Aggression	0.28	0.07 *	0.28	0.09 *
Proximity	0.12	0.07	0.26	0.09 *
Yawn	0.88	0.27	8.04 †	7.76 ‡
Groom	0.10	0.17	0.18 †	0.07
Sex exploration	0.43	0.33	1.04 †	0.61 ‡
Display	0.26	0	0.57	0.34 ‡
Mount	0.43	0.10	0.91	0.09

 * Significant difference (using the t test, $P < 0.05$) between female pseudohermaphrodites and their controls.
 † Significant difference (using the t test for matched pairs, $P < 0.05$) before and after TP administration to pseudohermaphrodites.
 ‡ Significant difference (using the t test for matched pairs, $P < 0.05$) before and after TP administration to female controls.

166

"sexual exploration", and "display" (a stiff-legged, strutting walk with the tail arched over the back) after TP treatment.

Before and after TP treatment, the pseudohermaphrodites mounted the stimulus females more frequently than did the female controls, but because of the extreme variability between individuals the differences were not statistically significant (Table 2). Before the TP treatment, none of the animals achieved intromission. After 30 weeks of TP, none of the female controls attained intromission, but one pseudohermaphrodite (No. 1640) achieved both intromission and ejaculation during two separate tests. Mean quantitative values for the two ejaculatory tests were as follows: frequency of mounts without intromissions, 9.5; frequency of mounts with intromissions, 5.5; pelvic thrusts per intromission, 4.1; inter-intromission interval, 1.1 min; latency to ejaculation, 8.5 min. Intromission thrusts were shallower than those characteristic of normal males, but the behavioural pattern of ejaculation (a rigid posture with a pause in thrusting) was identical to that of normal males. Ejaculation included seminal emissions that formed vaginal "plugs" when coagulated. This complete ejaculatory behaviour by only one of seven pseudohermaphrodites may have been due to social and experiental factors rather than to the effectiveness of the prenatal treatment because three of them (Nos. 1640, 1616, and 1239) frequently masturbated to ejaculation with seminal emission in their home cages.

The sexual and other types of behaviour that increased in frequency after the TP treatment in adulthood indicated only an increased sensitivity to TP administered prenatally. On the other hand, the aggressiveness of the pseudohermaphrodites before the administration of TP in adulthood, like their masculinized juvenile behaviour, indicated that the adult rhesus brain was functionally modified by prenatal androgen. The foetal rhesus testis secretes high levels of testosterone that fall to non-detectable amounts after birth[15]. Our findings suggest that testosterone supplied by injection to the pregnant rhesus can modify the central nervous system of genetic female foetuses so that they are predisposed to acquire predominantly masculine patterns of behaviour. Neither the mechanism nor the site of action for this effect is known. It is possible that for development of ejaculatory behaviour early hormonal actions on the brain are essential[16], and that the ability of the foetal and neonatal brain tissue to form aromatized metabolites of testosterone[17] may constitute an essential aspect of the mechanism of testosterone action.

We thank Mr Jens Jensen for technical assistance. The work was supported in part by grants from the National Institutes of Health, US Public Health Service. Publication No. 635

G. G. Eaton
R. W. Goy
C. H. Phoenix

Oregon Regional Primate Research Center,
505 NW 185th Avenue,
Beaverton, Oregon 97005,
and Wisconsin Regional Primate Research Center,
1223 *Capitol Court,*
Madison, Wisconsin 53706

[1] Dantchakoff, V., *CR Soc. Biol.*, **127**, 1255 (1938).
[2] Young, W. C., in *Sex and Internal Secretions* (edit. by Young, W. C.), **2**, 1173 (Williams and Wilkins, Baltimore, 1961).
[3] Phoenix, C. H., Goy, R. W., Gerall, A. A., and Young, W. C., *Endocrinology*, **65**, 369 (1959).
[4] Goy, R. W., Bridson, W. E., and Young, W. C., *J. Comp. Physiol. Psychol.*, **57**, 166 (1964).
[5] Grady, K. L., Phoenix, C. H., and Young, W. C., *J. Comp. Physiol. Psychol.*, **59**, 176 (1965).
[6] Barraclough, C. A., and Gorski, R. A., *J. Endocrinol.*, **25**, 175 (1962).
[7] Harris, G. W., and Levine, S., *J. Physiol.*, **163**, 42P (1962).
[8] Crossley, D. A., and Swanson, H. H., *J. Endocrinol.*, **41**, xiii (1968).
[9] Eaton, G., *Endocrinology*, **87**, 934 (1970).
[10] Goy, R. W., in *The Neurosciences: Second Study Program* (edit. by Schmitt, F. O.), 196 (Rockefeller Univ. Press, New York, 1970).
[11] Beach, F. A., in *Proceedings of the Biopsychology of Development* (edit. by Tobach, E.), 249 (Academic Press, New York, 1971).
[12] Beach, F. A. in *Mammalian Reproduction* (edit. by Gibian, H., and Plotz, E. J.), 437 (Springer-Verlag, Berlin, 1970).
[13] Young, W. C., Goy, R. W., and Phoenix, C. H., *Science*, **143**, 212 (1964).
[14] Goy, R. W., in *Endocrinology and Human Behaviour* (edit. by Michael, R. P.), 12 (Oxford University Press, 1968).
[15] Resko, J. A., *Endocrinology*, **87**, 680 (1970).
[16] Hart, B. L., *Physiol. Behav.*, **8**, 841 (1972).
[17] Naftolin, F., Ryan, K. J., and Petro, Z., *Endocrinology*, **90**, 295 (1972).

Printed in Great Britain by Flarepath Printers Ltd., St. Albans, Herts.

20

Aggression in Adult Mice: Modification by Neonatal Injections of Gonadal Hormones

Abstract. *Incidence of spontaneous aggression in adult male mice given a single injection of estradiol benzoate (0.4 milligram) when they were 3 days old was less than that of controls injected with oil. Aggressiveness was increased among adult females injected with either estradiol or testosterone propionate (1 milligram) at the same age. The increased aggressiveness noted among females given androgen was further documented during subsequent mating tests, when these females often attacked, wounded, and, in one case, killed naive males.*

The sexual differentiation of particular behavioral or neuroendocrine control systems may be influenced by the presence of gonadal hormones during infancy in rodents (*1*). For example, neonatal administration of androgens to females results in an acyclic, male-like secretion of gonadotropin during adulthood rather than in the cyclic pattern characteristic of normal adult females (2). Similarly, sexual behavior of female rats may be masculinized to a degree if they are given neonatal injections of androgen, or that of males may be feminized if they are castrated during infancy, provided that appropriate gonadal hormones are administered during adulthood (1, 3). Estrogens, depending upon the time and dose of their injection, may mimic some of these effects of androgens (4). We hypothesized that aggressive behavior could also be modified following treatment with androgens or estrogens during infancy. Our results demonstrate that aggressiveness was increased in adult female mice if they were given either androgen

or estrogen as neonates; aggressiveness in adult males was partially suppressed if they were injected with estrogen during infancy.

Complete litters of 3-day-old C57BL/6J mice of both sexes were injected subcutaneously with 0.05 ml of corn oil containing either 1 mg of testosterone propionate, 0.4 mg of estradiol benzoate, or nothing. Mice were weaned at 21 to 25 days of age and housed singly until tested for aggressiveness at 80 to 90 days of age. Spontaneous aggression (5) was measured in test chambers (12 by 12 by 6 inches) with removable partitions in the middle. Single mice of the same sex and treatment were placed on either side of the partition. It was removed 20 minutes later and the mice were observed until a fight was initiated, or for a maximum of 15 minutes (Table 1). The same pair of mice was tested once daily for three consecutive days, after which vaginal smears were obtained for five consecutive days from all females. All males and 12 females from each group were then autopsied to verify the expected effects of neonatal injections on reproductive tract morphology. Ovaries, uteri, and testes were weighed and examined histologically. Seminal vesicles were homogenized in water and analyzed for fructose (6).

The remaining females from each of the three groups received subcutaneous injections of progesterone (0.3 mg per mouse per day) for 8 days to induce estrous cycles (7). On the afternoon of the 8th day, they were paired with naive males in the females' home cages. Our purpose in this secondary experiment was to verify the lack of mating in females treated neonatally with testosterone or estradiol and to follow a suggestion by Barraclough that changes in aggressiveness might be more obvious in such a situation (8). Incidence of fighting was recorded for the first hour after pairing, and all pairs were in-

Table 1. Number of pairs (of same sex) in which fighting occurred at least once during three encounters and total number of fights occurring during all three encounters.

Neonatal treatment	Fighting at least once (No.)	Fights in three encounters (No.)
	Males	
Oil	23/24	51/72
Testosterone	18/19	46/57
Estradiol	10/20	20/60
	Females	
Oil	1/24	1/72
Testosterone	5/18	10/54
Estradiol	4/14	5/42

Table 2. Number of male-female pairs in which severe fighting occurred within the first hour after pairing and number in which wounding of one member occurred within 18 hours. Females had been previously tested in the primary experiment (Table 1), after which they were given progesterone daily for 8 days and then paired with naive males.

Neonatal treatment of females	Fighting (1st hour)	Wounding (18 hours)
Oil	0/20	0/20
Testosterone	11/13	8/13*
Estradiol	4/14	0/14

* One pair in which female was wounded, three pairs in which male was wounded, and one pair in which male was killed.

169

spected for wounding and presence of vaginal plugs on the following three mornings. Males used in this experiment were about 100 days old, intact, and sexually and experimentally inexperienced; each male had been housed with four or five others since weaning.

The results of the primary experiment, in which mice were given the opportunity to fight only members of the same sex and treatment group, are presented in Table 1. Spontaneous fighting occurred at least once during three encounters in all but one pair of males in each of the two groups that received injections of either oil or testosterone during infancy. Neonatal injections of estradiol reduced the incidence of fighting in adult males to 50 percent ($P<.01$). Only 4 percent of the control females fought, whereas fighting among pairs that had received neonatal injections of either testosterone or estradiol increased to 28 and 29 percent, respectively ($P < .05$ in both cases).

The secondary experiment, in which females were injected with progesterone for 8 days and then paired with normal males, revealed marked aggressiveness on the part of females injected neonatally with testosterone (Table 2); fighting among such pairs was often vicious and usually initiated by the females. Females treated with estradiol also fought with males, but both the incidence and severity of fights were lower. No fighting was noted among pairs in which the female had been injected only with oil in infancy. No vaginal plugs were found in any females receiving steroid neonatally, but 55 percent of the females injected with oil had plugs during the 3 days after pairing.

The effects of neonatal injections of estradiol or testosterone on vaginal cycles and reproductive tracts were similar to previous findings (2, 4) and will be reported here only to an extent necessary for correlation with the behavioral data. Neonatal injections of

Table 3. Body weight, relative (paired) organ weights, and fructose concentrations in seminal vesicles of males treated neonatally with oil, testosterone, or estradiol; body and relative uterine weights of similarly treated females (mean ± standard error).

Neonatal treatment	Males					Females		
	No.	Body wt. (g)	Testes (mg/g body wt.)	Seminal vesicle (mg/g body wt.)	Seminal vesicle fructose (µg)	No.	Body wt. (g)	Uterus (mg/g body wt.)
Oil	48	27.8 ± 0.4	7.53 ± 0.52	2.42 ± 0.27	174.0 ± 6.3	12	22.6 ± 0.7	3.19 ± 0.31
Testosterone	37	27.3 ± 0.5	5.93 ± 0.20*	1.98 ± 0.08*	137.0 ± 5.4*	12	28.2 ± 1.0*	5.23 ± 0.67*
Estradiol	40	24.4 ± 0.4*	4.61 ± 0.41*	1.07 ± 0.15*	48.2 ± 4.7*	12	23.1 ± 0.8	2.37 ± 0.38

* Significantly different from oil controls, as determined by analysis of variance, with a probability of at least $P < .05$.

estradiol in males resulted in decreased body and reproductive organ weights and relative aspermia. Injections of testosterone in infancy also decreased weights of male organs but to a lesser extent than that caused by estradiol (Table 3). All vaginal smears obtained from all females injected neonatally with either steroid contained approximately 80 percent cornified cells and 20 percent leukocytes, and ovaries of such females were polyfollicular and devoid of corpora lutea. Body and uterine weights were increased among females injected neonatally with testosterone.

Androgen is a necessary prerequisite for attack behavior in inexperienced male mice (9), whereas estrogen administered during adulthood has no effect on aggressiveness of males (10). The reduction in spontaneous aggression shown by males injected with estrogen in our study was correlated with large changes in their reproductive tracts, and secretion of testicular androgen was probably considerably reduced. Weights of reproductive organs were also lower in males given neonatal injections of androgen, but they were as aggressive as control males. These facts suggest that those males injected with androgen neonatally probably had sufficient androgen in their circulation during adulthood to permit a high degree of aggressive behavior, whereas those that received estrogen did not. The amount of fructose in seminal vesicles, a good correlate of androgen titers (6), was reduced by 72 percent among males given estradiol in infancy compared to that in controls given oil (Table 3). The comparable figure for males receiving testosterone neonatally was only 21 percent and, hence, the postulate appears reasonably good on this basis.

The low incidence of spontaneous aggression found among control females agrees well with observations of other workers using mice (11). Androgen will not increase aggressiveness in either immature or mature gonadectomized females (12). However, neonatal injections of testosterone, and to a lesser extent estradiol, increase aggressiveness in females after maturity. These effects were significant in both experiments although more dramatic in the uncontrolled secondary experiment where some previously tested females were paired with naive males in the females' home cages after receiving progesterone to induce estrous cycles. Under such conditions mating did not occur, and the females usually attacked and sometimes wounded males. Wounding was sufficiently severe to cause death in one case. The reasons for the dramatic effects observed in this experiment are not readily obvious because of its uncontrolled nature and the data are presented only as an extreme example of a phenomenon observed in the primary experiment. Two investigators have reported that "masculine or aggressive responses" interfered with normal female sexual behavior when rats were treated with estrogen or testosterone in infancy (13) but not to the extent shown in the present study with mice.

A reasonable hypothesis to explain the increased aggressiveness of females treated neonatally with gonadal hormones is the alteration of a neural mechanism whose sexual differentiation is normally regulated by androgen in infancy. Such a concept parallels the conclusions of many studies dealing with either sex behavior or the hypothalamic control of gonadotropin secretion, and some degree of experimental mimicking of androgen by estrogen is well documented in this respect. It does not seem reasonable at this time, however, to suspect the hypothalamus at the expense of other neural structures because the number of brain areas known to function in aggression is relatively large (14). Furthermore, as evidenced by changes in body weight in

both sexes, the effects of early administration of steroids may be widespread.

F. H. BRONSON
CLAUDE DESJARDINS

Jackson Laboratory,
Bar Harbor, Maine 04609

References and Notes

1. S. Levine and R. F. Mullins, *Science* **152**, 1585 (1966).
2. C. A. Barraclough, *Endocrinology* **68**, 62 (1961); R. A. Gorski, *J. Reprod. Fertil. Suppl.* **1**, 67 (1966).
3. R. E. Whalen and D. A. Edwards, *Anat. Rec.* **157**, 173 (1967).
4. G. W. Harris and S. Levine, *J. Physiol. London* **181**, 379 (1965).
5. J. P. Scott, *Amer. Zool.* **6**, 683 (1966).
6. J. S. Davis and J. E. Gander, *Anal. Biochem.* **19**, 72 (1967).
7. C. A. Barraclough, *Fed. Proc.* **15**, 9 (1956).
8. ———, personal communication.
9. E. A. Beeman, *Physiol. Zool.* **20**, 373 (1947); E. B. Sigg, C. Day, C. Colombo, *Endocrinology* **78**, 679 (1966).
10. J. E. Gustafson and G. Winokur, *J. Neuropsychiat.* **1**, 182 (1960).
11. E. Fredericson, *J. Comp. Physiol. Psychol.* **45**, 89 (1952).
12. J. V. Levy, *Proc. West Virginia Acad. Sci.* **26**, 14 (1954); J. Tollman and J. A. King, *Brit. J. Anim. Behav.* **6**, 147 (1956).
13. H. H. Feder, *Anat. Rec.* **157**, 79 (1967); A. A. Gerall, *ibid.*, p. 97.
14. J. M. R. Delgado, *Amer. Zool.* **6**, 669 (1966).
15. This investigation was supported in part by PHS grants FR-05545-05 and HD-00767.

21

BEHAVIORAL BIOLOGY 16, 373-378 (1976), Abstract No. 5310

BRIEF REPORT

Inhibition of the Ejaculatory Reflex in B6D2F$_1$ Mice by Testosterone Propionate[1]

THOMAS E. MCGILL, STEVEN M. ALBELDA,
HENRY H. BIBLE, and CHRISTINA L. WILLIAMS

Department of Psychology, Williams College,
Williamstown, Massachusetts 01267

Testosterone propionate was administered in adulthood to neonatally androgenized female house mice and castrated male house mice of the B6D2F$_1$ genotype. Lengthening of Ejaculation Latency (a measure of mating time) occurred in both groups. Castrated males showed a proportional increase in Ejaculation Latency with increasing dose. This result, which we believe to be the first of its kind, suggests that testosterone propionate inhibits the ejaculatory reflex in this genotype.

Research with guinea pigs (Grunt and Young, 1952), rats (Beach and Holz-Tucker, 1949; Davidson, 1966; Larsson, 1966), hamsters (Whalen and DeBold, 1974), and CD2F$_1$ house mice (Champlin, Blight, and McGill, 1963) indicates that castrated animals on different doses of testosterone propionate (TP) did not differ greatly in Ejaculation Latency (EL: time from first intromission to the ejaculatory reflex). When differences have been found,

[1]This research was supported by Research Grant 07495 from the Institute of General Medical Sciences, U.S. Public Health Service. Some of the data were collected while one of us (C. W.) was a National Science Foundation Research Trainee in the Undergraduate Science Education Program (GY-10612). Hormone preparations were generously supplied by Dr. Rudolf Neri, Schering Corporation, Bloomfield, New Jersey.

they have been in the direction of reduced EL's for those animals receiving higher doses of TP (Beach and Holz-Tucker, 1949; Larsson, 1966; Whalen and DeBold, 1974). The present paper reports experiments showing that TP inhibited the ejaculatory reflex (when "inhibition" was defined as lengthening EL) in three different treatment conditions in B6D2F$_1$ house mice: neonatally androgenized (NA) females and both prepubertally and post-pubertally castrated males. In the latter cases the increase in EL was proportional to the amount of hormone injected daily.

Manning and McGill (1974) observed that female B6D2F$_1$ mice neonatally androgenized by injections with 100 µg of TP on the day of birth exhibited the full range of masculine behavioral responses (including the ejaculatory reflex) when tested with receptive females in adulthood. These behavioral patterns occurred with or without further TP treatment. In the first experiment of the present report, we matched two groups of NA females on the basis of EL in preliminary tests. One group of 18 animals (called Group A) was then injected daily with 100 µg TP in 0.04 ml Arachis oil. A second group of animals (Group B) received the oil vehicle only. Injections continued for 10 weeks with weekly sex tests (Manning and McGill, 1974) occurring during the last nine weeks, or until the animals had exhibited the ejaculatory reflex on three tests. Following these tests, animals were "rested" (no tests; no injections) for six to nine weeks and the treatments were reversed: Group A animals received the oil vehicle and Group B animals received TP. A further nine-week period of behavioral testing followed treatment reversal. Figure 1 shows the results for EL in the two treatment periods. In both cases females receiving TP therapy took significantly longer

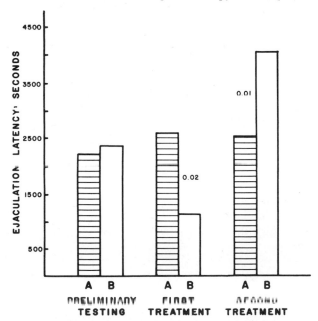

Fig. 1. Response (Ejaculation Latency) of neonatally androgenized B6D2F$_1$ female mice to testosterone propionate (TP) treatment in adulthood. First treatment: Group A received 100 µg of TP. Second treatment: Group B received 100 µg of TP.

TABLE 1

Effect of Daily Testosterone Propionate (TP) Injections on Ejaculation
Latencies (EL) of Prepubertally Castrated Male B6D2F$_1$ Mice

Group	Number of ejaculators	EL (sec)
Intacts	22	833
20 μg TP	12	1235
60 μg TP	10	1837
150 μg TP	9	3080

to reach the ejaculatory threshold. In addition, NA females injected with TP in adulthood had significantly more sexual contacts (mounts and intromission-like responses) than did females receiving the oil vehicle only.

The increase in EL observed in NA B6D2F$_1$ female mice given TP in adulthood suggested that a similar effect might occur in castrated males. We decided to test this possibility and, at the same time, determine the effects of different doses of TP for both pre- and postpubertally castrated males.

To test the response of prepubertal castrates, B6D2F$_1$ males were gonadectomized at 3 1/2-4 weeks of age and at seven weeks daily injections of TP were begun. Groups of 13 animals were given injections of either 20 μg, 60 μg, or 150 μg of the hormone. An intact control group of 30 males received oil injections in the same volume (0.04 ml). Sex tests were conducted once a week for the following 6 weeks with daily injections continuing throughout.

A mean EL was determined for those animals observed to ejaculate, and a mean of these means was determined for each group. These results are presented in Table 1.

A significant treatment effect was confirmed by Kruskal-Wallis analysis ($P < 0.001$). Two tailed Mann-Whitney U-Tests indicated statistically significant differences ($P < 0.05$) between the 150 μg group and both the 20 μg group and the intact males. The difference between the 60 μg group and the intact animals was also statistically significant.

The number of intromissions preceding ejaculation followed the same rank order as EL and statistically significant differences were found for the same comparisons.

Next, we repeated the experiment using groups of 16 B6D2F$_1$ males who were castrated at 10 weeks of age after experiencing three precastration ejaculations. A group injected with 500 μg of TP daily was added. Injections began on the day of castration and continued for nine weeks with weekly sex tests occurring during the last eight weeks. Once again, a mean of means was determined for the 12-16 males that ejaculated in each group. These averages are shown in Fig. 2.

It is clear from Fig. 2 that the general relationship between EL and daily dose of androgen that was found for prepubertally castrated males was also observed when animals were castrated after puberty and sexual experience. An analysis of variance revealed a significant treatment effect (Kruskal-Wallis, $P < 0.001$). Mann-Whitney U-Tests were used to compare the various groups. These tests revealed that the 500 μg group had significantly longer EL's than

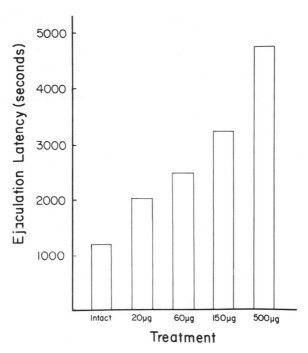

Fig. 2. Response(Ejaculation Latency) of castrated B6D2F$_1$ male mice to different doses of testosterone propionate (TP) injected daily.

all other groups ($P < 0.002$). The 150 μg animals had significantly longer EL's than the 20 μg and intact groups ($P < 0.02$), while the 60 μg males were significantly different from intacts ($P < 0.002$). There was no significant difference between the intact males and the 20 μg males.

The number of intromissions preceding ejaculation followed the same rank order as EL and also was significant by the Kruskal-Wallis Test ($P < 0.001$). Significant differences by Mann-Whitney U-Tests were identical to those for EL, except that 500 μg vs. 150 μg did not reach significance and 60 μg vs. 20 μg did. There were minor differences in certain P values between the two measures. These findings indicate that each group had a similar rate of intromission; that is, approximately the same number of intromissions per unit time. A "fine-grained" analysis (McGill, 1962; 1970) involving several behavioral parameters is required to determine possible differences in such items as length of intromissions, thrusts per intromission, and interintromission intervals.

If androgen, indeed, inhibits the ejaculatory reflex in this genotype, then castrated B6D2F$_1$ males should ejaculate more rapidly than intact animals do. Fortunately, such a comparison is possible since many B6D2F$_1$ males exhibit the complete masculine behavioral pattern, including the ejaculatory reflex long after castration (McGill and Haynes, 1973; McGill and Manning, in press). After gonadectomy castrated B6D2F$_1$ males go through a "difficult period." EL increases greatly over the first five-seven weeks of postcastration testing but then declines to a stable level by about week 16. Unpublished observations show that during weeks 17-26, the castrates ejaculate significantly faster than intact control males.

As far as we have been able to determine, the experiments reported here are the first to demonstrate a lengthening of EL associated with TP treatment in any species. The most directly comparable previous work has been done with the laboratory rat and hamster. Beach and Holz-Tucker (1949) found that increasing doses of TP caused a decrease in EL's of castrated rats. Whalen and DeBold (1974) examined the effect of increasing doses of testosterone, dihydrotestosterone, and androstenedione on the sexual behavior of castrated male hamsters. In general, all of these androgens produced shortened EL's with increases in dose.

In contrast to the results of our study, Champlin, Blight, and McGill (1963) found no statistically significant differences between castrated $CD2F_1$ male house mice (resulting from a cross between inbred BALB/c females and inbred DBA/2 males) receiving 32 μg TP daily and males of the same strain receiving 1024 μg. If the results of that small sample study are valid, then males of different genotypes within a species respond differently to hormone treatment. Whether the phenomenon holds for all genotypes within the house mouse species, or only for certain genotypes, some of the previously observed behavioral differences between strains (McGill, 1962, 1970) may be due to different endogenous androgen levels. For example, BALB/c males have EL's of over one hour while the EL's of $B6D2F_1$ males average 10-20 min. We have recently assayed plasma testosterone levels in BALB/c and $B6D2F_1$ males and found that the plasma testosterone levels in the BALB/c males were two to three times the values found in the $B6D2F_1$ males; the difference was highly significant.

The three experiments of the present report show that TP inhibits the ejaculatory response in both NA $B6D2F_1$ females and in castrated males receiving androgen therapy. For the males, the lengthening of EL was directly proportional to the amount of hormone injected daily. These unexpected findings raise many interesting causal and functional questions regarding the hormonal control of masculine sexual behavior of males and females in this species.

REFERENCES

Beach, F. A., and Holz-Tucker, A. M. (1949). Effects of different concentrations of androgen upon sexual behavior in castrated male rats. *J. Comp. Physiol. Psychol.* **42**, 433-453.

Champlin, A. K., Blight, W. C., and McGill, T. E. (1963). The effects of varying levels of testosterone on the sexual behaviour of the male mouse. *Anim. Behav.* **11**, 244-245.

Davidson, J. M. (1966). Characteristics of sex behaviour in male rats following castration. *Anim. Behav.* **14**, 266-272.

Grunt, J. A., and Young, W. C. (1952). Consistency of sexual behavior patterns in individual male guinea pigs following castration and androgen therapy. *J. Comp. Physiol. Psychol.* **46**, 138-144.

Larsson, K. (1966). Individual differences in reactivity to androgen in male rats. *Physiol. Behav.* **1**, 255-258.

Manning, A., and McGill, T. E. (1974). Neonatal androgen and sexual behavior in female house mice. *Horm. Behav.* 5, 19-31.

McGill, T. E. (1962). Sexual behavior in three inbred strains of mice. *Behaviour* 19, 341-350.

McGill, T. E. (1970). Genetic analysis of male sexual behavior. *In* G. Lindzey and D. D. Thiessen (Eds), "Contributions to Behavior-Genetic Analysis," pp. 57-88. New York: Appleton-Century-Crofts.

McGill, T. E., and Haynes, C. M. (1973). Heterozygosity and retention of the ejaculatory reflex after castration in male mice. *J. Comp. Physiol. Psychol.* 84, 423-429.

McGill, T. E., and Manning, A. (In press). Genotype and retention of the ejaculatory reflex in castrated male mice. *Anim. Behav.*

Whalen, R. E., and DeBold, J. F. (1974). Comparative effectiveness of testosterone, androstenedione and dihydrotestosterone in maintaining mating behavior in the castrated male hamster. *Endocrin.* 95, 1674-1679.

The Development
of Behavior

Part 3 examined material that might be described as part of the "biological background" of animal behavior. We covered several examples showing that hormone treatment early in life can affect the development of an organism and have important consequences for its adult behavior. Now Part 4 deals specifically with the study of behavioral development. However, the independent variables used in the experiments that follow are primarily *experiential* rather than physiological. Under intensive investigation in recent years, this area of research has given rise to major insights into problems of behavioral determination. More such developments are to be expected in the future.

It is in the area of the development of behavior that the seemingly unquenchable nature–nurture controversy most often arises. Reading 22 is an attempt by a distinguished student of the development of behavior to put the problem in a new perspective where behavioral determinants with specific outcomes are sharply divided from those with general effects. P. P. G. Bateson tries to show that differences of opinion about certain aspects of the development of behavior actually stem from different perceptions of the same evidence by the scientists involved in the controversy. An interesting exercise would be thinking of other behavioral examples for Bateson's four-fold classification scheme.

In Reading 23, William R. Thompson illustrates the importance of strict control in behavioral experimentation. This experiment shows that environmental factors affecting later behavior actually begin to operate before the birth of the organism. Note the careful control of possible postnatal maternal effects through the use of cross-fostering.

In Reading 24, R. A. Hinde and Yvette Spencer-Booth provide rather startling evidence on the effects of brief separation from the mother on later behavior in rhesus monkeys. As with prenatal influences, a brief separation has long-term behavioral consequences.

Reading 25 reports interesting experiments concerned with factors affecting sexual maturation in female mice. Note the careful separation of variables that might possibly have been involved in producing accelerated sexual maturation.

Conducting extensive studies in behavioral development for many years, J. P. Scott has been very interested in the concept of critical periods in development. In Reading 26, Scott and co-authors John M. Stewart and Victor De Ghett present the general theory of critical periods as well as an example that should be of interest to almost every reader—the dog and its critical period for primary socialization.

The old saying has it that "birds of a feather flock together." In Reading 27 C. L. Pratt and G. P. Sackett present the fascinating finding that monkeys reared under particular circumstances prefer social interaction with similarly reared monkeys even when the animals are strangers to one another.

Bird-song has been a favorite behavior pattern for those involved in the nature–nurture controversy. By judicious selection of species, one could "prove" that bird-song is completely innate, or completely learned, or partially innate and partially learned. Present investigators find such controversies fruitless. Instead, the song of each individual species is viewed as the result of a complex developmental process that must be painstakingly analyzed by careful experiments. Reading 28 by Peter Marler and Miwako Tamura is a good example of modern work in the area.

The next two papers are primarily concerned with imprinting, a phenomenon generally thought of as occurring in the young of particular species of birds. But imprinting is not restricted to birds, or even to developing organisms, as Peter Klopfer shows in Reading 29. As originally proposed, imprinting had several characteristics, one of which was that the effect was irreversible. In Reading 30, E. A. Salzen and C. C. Meyer provide a test of this hypothesis.

In the final reading of this section, G. P. Sackett reports the surprising, and perhaps disturbing, observation that monkeys reared in isolation from birth to nine months of age respond in species-typical ways to colored slides of infant monkeys or threatening adults.

Reprinted from:
ADVANCES IN THE STUDY OF BEHAVIOR, VOL. 6
© 1976
ACADEMIC PRESS, INC.
New York San Francisco London

Specificity and the Origins of Behavior

P.P.G. BATESON

SUB-DEPARTMENT OF ANIMAL BEHAVIOUR
UNIVERSITY OF CAMBRIDGE
CAMBRIDGE, ENGLAND

I. INTRODUCTION

What factors during development determine the special ways in which an individual animal eventually will behave? What decides the specific form and patterning of its behavior? What gives a behavior pattern its unique character, making it different from other behavior patterns? It would be useless to pretend that the attempts to answer these questions about the ontogeny of behavior bring widespread agreement. Nor is there harmonious consensus among those who study behavior as to the ways these questions should be answered or even about the nature of relevant evidence.

The debate about the best ways to study behavioral development has, of course, been extensive (see Barnett, 1973; Beach, 1955; Dawkins, 1968; Eibl-Eibesfeldt, 1961, 1970; Ewer, 1971; Hailman, 1967; Hebb, 1953; Hinde, 1968, 1970a; Jensen, 1961; Konishi, 1966; Kuo, 1967; Lehrman, 1953, 1970; Lehrman and Rosenblatt, 1971; Lorenz, 1961, 1965; Moltz, 1965; Schneirla, 1956, 1966; Thorpe, 1956, 1963; Tinbergen, 1963). It would be quite wrong to suggest that nothing has been achieved as a result of this debate. In particular, many of the disagreements have been shown to arise from differences in interest and emphasis. Those ethologists influenced by Lorenz have been primarily interested in the origins of behavioral adaptiveness, whereas others studying behavior, particularly those who were influenced by the writings of Kuo, Schneirla, and Lehrman, have been principally concerned with development in the individual animal. Even though this point was clarified many years ago (e.g., Tinbergen, 1963), the controversy has rumbled on. Despite frequent announcements of the death of the nature-nurture dichotomy of behavior, a distinction between activities that are learned and those that are not is still widely used. In part this has been because classifications of the origins of behavior have been frequently muddled with classifications of behavior itself. To state that inheritance and the environment determine the characteristics of behavior is not the same as urging that all behavior patterns can be divided into those that are inborn and those that are environmentally determined. As I shall point out later, a residual confusion between the sources of behavioral distinctiveness and the origins of its adaptiveness is still found in the literature.

I believe, however, that the reasons for the persisting, wide and often bitter differences of viewpoint are much more deeply seated than could be explained by mere errors of logic. In this chapter the possibility is explored that different people perceive the same body of data in different ways. Where some see sharp discontinuities, others see smooth gradations, and, accordingly, classifications differ. In order to develop the argument, I shall first consider factors in development that are responsible for the distinctiveness of behavior. I believe that when these sources of difference are scrutinized, it becomes much easier to understand why classifications of behavior in terms of developmental origins have generated so much heated argument.

II.   INITIAL DETERMINANTS IN THE DEVELOPMENT OF BEHAVIOR

It is customary now to distinguish between the factors that control behavior from moment to moment and those that are responsible for its development (e.g., Hinde, 1970a). The distinction may not always be easily drawn in practice since a factor responsible for the development of a behavior pattern may lie close in time to the occurrence of that behavior. In general, though, sources of behavioral distinctiveness usually lie considerably farther back in time from the behavior they affect than controlling conditions. Developmental determinants are *initiating* agents that lastingly give a behavior pattern its peculiar characteristics differentiating it from other types of behavior; of course, a lasting effect is not necessarily irreversible under all conditions.

Once one starts to trace back through the nexus of events that precede a behavior pattern, there might seem no obvious stopping point. However, what is usually meant by a *developmental determinant* of an individual's behavior is a factor that was responsible for the distinctiveness of the individual's behavior and which operated at some point in the life of that individual. Wherever I refer to "determinant" in this chapter I use it in this special sense.

Few people would disagree nowadays that part of the initial determinants of behavior are already present in latent form within the fertilized egg; some determinants are, perhaps, present as cytoplasmic factors, but most are represented in the nucleus of the zygote—presumably in genetically coded form. An important semantic issue is at what stage a gene is to be treated as a developmental determinant. I believe a gene would generally be regarded as a determinant at the time of its activation. However, to discover the actual moment when gene expression occurs for the first time is an extraordinarily difficult task for embry-

TABLE I
A CLASSIFICATION OF DEVELOPMENTAL DETERMINANTS OF BEHAVIOR

Determinants	Determinants with specific effects	Determinants with general effects
Inherited	A	B
Environmental	C	D

ology, and most statements about inherited determinants will be based on inference rather than evidence.

There is also widespread acceptance that other necessary conditions for the development of any pattern of behavior lie in the environment in which the animal grows up. Difficulties and disagreements arise, however, because both the inherited and the environmental determinants of behavior can be further subdivided into those that exert specific effects and those that have general effects.

1. General and Specific Effects of Determinants

It is important to ask whether it is possible to draw a sharp line across the continuum that runs from those determinants affecting only one pattern of behavior to those having such general effects they are necessary for life itself. In principle, though, the determinants of behavior could be placed somewhere in the matrix shown in Table I. An example of A might be the gene affecting the hygienic behavior in honeybees (*Apis mellifera*) that involves the uncapping of hive cells containing diseased larvae (Rothenbuhler, 1967). A representative of B might be a gene that affects the responsiveness of *Drosophila melanogaster* to light (Benzer, 1967); loss of responsiveness to light, not surprisingly, has a widespread effect on all visually guided activities. An example of C might be the experience of chicks (*Gallus gallus*) that have pecked at small objects painted with bitter-tasting substances; as a result, they develop a selective aversion for pecking at these objects (e.g., Lee-Teng and Sherman, 1966). Finally D might be early experience of crowded conditions by locusts (*Locusta migratoria*) subsequently leading them to become migratory (Ellis, 1964).

The distinction between specific and general effects of determining events poses a number of difficulties. How can we ever be certain that a determining event has only one outcome? Any determinant that seems to have a highly specific effect on behavior is in danger of being reclassified as having more general consequences after further study. For example, further analysis of the honeybee may show that the genes affecting hygienic behavior have pleiotropic effects on other dissimilar behavior patterns. Even after the most convincing demonstration that differences between one animal and another in the way they make nests, say, is dependent on differences in the way they were reared, an experimenter is in no position to claim that other differences in behavior will not subsequently be found. On the other hand, if he finds that the experimental operation is the source of differences in nest-building, aggressive behavior, and feeding, he would probably not even wish to claim that it had highly specific effects. Therefore, it might seem that the categories of determinants with specific outcomes are liable to be eroded by the collection of fresh evidence, and individual cases will tend to move to the right in the matrix shown in Table I.

However, if a determinant affects a number of apparently different types of behavior, does it necessarily mean that its consequences are general? Could not those categories be thought of as having some special feature in common? Perhaps the determinant imposes some constraint on the way the animal's head can be moved and this limitation shows up most noticeably when the animal is making a nest, threatening another individual, or feeding. Alternatively the non-specific effects on behavior may themselves turn out to be consequences of a highly specific behavioral outcome of a developmental process. The point is, then, that the placing of a particular determinant in the matrix shown in Table I is always subject to alteration in either direction as fresh evidence becomes available.

A related point is that a decision on how to classify a determinant may depend critically on the level at which its consequences are assessed. For example, phenylketonuria is a hereditary disease which, among other things, results in rather general disorders of behavior. However, the disease is caused by a specific deficiency of the liver enzyme phenylalanine hydroxylase (Hsia, 1967). Does the classifier utilize this knowledge about the specificity of the genetic determinants of the disease at the biochemical level? Or does he, as seems more logical, consistently apply behavioral criteria throughout and classify the determinants of phenylketonuria as having general effects?

2.   The Problem of Behavioral Units

Another issue impinges crucially on the distinction between specific and general consequences. How should behavior patterns themselves be divided up? Are there obvious units that would provide a basis for the distinction between one behavior pattern being affected by some preceding event and many patterns being affected? It is an important question, but, once again, there is little agreement about the answer to it. The traditional response of many ethologists has been to argue that "natural" units of behavior become apparent to anyone who knows and loves his animals. On this view it is possible to assemble an *ethogram*—a complete inventory of behavior patterns shown by a species. However, thoughtful reviewers of the field have pointed out that selection of evidence is inevitable in the study of animal behavior as in everything else and that any ethogram will reflect the interests and preconceptions of its compiler (see, e.g., Marler and Hamilton, 1966, pp. 711-717; Hinde, 1970a, pp. 10-13).

Furthermore, many difficulties remain even when it is possible to obtain agreement about the ostensive definition of a behavior pattern after pointing it out as it occurs or after detailed description. For instance, the same display given in two different contexts may serve two different biological roles in communicaton; although the message is the same the meaning is different in each case (e.g., Smith, 1968). For purposes of classification, do we have two behavior patterns or one? Another illustration is provided by the great tit (*Parus major*) which hammers with its bill in exactly the same way when it is feeding and when faced with a stimulus that evokes attack. Blurton-Jones (1968) argued that the behavior patterns are different because he found that one increased in frequency after food-deprivation but the other did not. His experiment did not settle the matter, as Andrew (1972) points out, because the motor pattern of hammering may be controlled by the same stimulus in both cases. The food-deprived great tit may hammer more frequently at food because, as a result of its own searching behavior, it sees more food than objects evoking attack. So we are left with the dilemma whether or not we should split or lump bill-hammering in the two situations. Yet another difficulty is that, even with the most unequivocal items of behavior for inclusion in a classic ethogram, the temporal pattern of occurrences may be such that different measures of the behavior yield different results. For instance, the "chink" call given by chaffinches (*Fringilla coelebs*) when mobbing potential predators first increases in frequency and then declines gradually. Now, when Hinde (1960) measured the response of chaffinches to a stuffed owl and a toy dog, he found that on three measures the owl was more effective than the dog; the chaffinches called more at the owl than at the dog during the first 6 minutes of presentation; they responded more rapidly to the owl; and their calling at the owl waned more slowly. However, the time taken to reach the peak rate of calling was shorter when the birds were presented with the dog; the birds' calling in response to the dog apparently warmed up more quickly than was the case with the owl. In order to account for results such as these, it is necessary to postulate a number of underlying processes that interact to produce the temporal pattern of calling (Hinde, 1970b). Where does that leave the treatment of "chinking" as a unitary end product of development? Whatever way one chooses to deal with this particular example, it serves to warn that the types of measure chosen may have a profound effect on the interpretation of how the behavior is controlled and initially determined.

It is easy to lose patience with arguments such as these on the grounds that, despite some imprecision, most people know what they are talking about. But how public are the rules that each of us uses? The difficulties in communication are not trivial and, indeed, present a major problem to philosophers. The issue is stated succinctly by Goldman (1970, p. 1) at the beginning of a book devoted to the topic. He writes:

> Suppose that John does each of the following things (all at the same time): (*1*) he moves his hand, *(2)* he frightens away a fly, *(3)* he moves his queen to king-knight-seven, (*4*) he check mates his opponent, (*5*) he gives his opponent a heart attack, and (*6*) he wins his first chess game ever. Has John here performed *six* acts? Or has he only performed *one* act, of which six different descriptions have been given?

The relevance of this problem to my argument is that the way in which behavior is divided up into units is very much a matter of opinion which, in turn, is a reflection of what questions about behavior are considered to be important. The relative weights given by the classifiers to factors involved in development and control, to context, to consequences of behavior, and to its biological function differ from one school of thought to the next. Classifications of behavior depend very much on the interests of the compiler and what may seem a natural unit from one vantage point may not even be noticed from another (cf. Hinde, 1970a). A decision about how finely behavior should be divided or about what features of behavior are important would obviously have profound effects on the placing of determinants on the specific-general scale. For example, if a gene affects all aspects of migratory behavior in a bird, its effects would be treated as specific if migration is regarded as a single pattern of behavior and nonspecific if the different aspects of migration were regarded as separate activities.

3. A Continuum in Effects of Determinants

A final difficulty that threatens a simple division of determinants into those with specific outcomes and those with general ones is the likelihood of continuity. If one category of conditions affects single patterns of behavior and another category of conditions affects all the behavior patterns in an animal's repertoire, every type of intermediate between these two extremes is possible in principle. In practice, intermediates are posing difficulties for simple dichotomies. For instance, an important criterion used to characterize conditions responsible for learning is that the lasting consequences on behavior of these training conditions are limited in extent. If environmental conditions have diverse effects on behavior persisting for a long time, those effects are not ordinarily attributed to learning. For example, when a rat is handled early in infancy and subsequently its behavior is found to be affected in a whole variety of different ways, it is not thought to have learned anything as a result of the handling. Nevertheless, the line of demarcation is arbitrary. Again, when kittens are exposed to vertical or horizontal lines at a particular stage in development, the kittens are subsequently said to be unresponsive to lines placed at right angles to the familiar orientation (Blakemore and Cooper, 1970; Blakemore, 1973). In some ways these effects are rather similar to those of imprinting in which a bird's social responsiveness is narrowed down to the familiar object. However, the birds have no difficulty in detecting unfamiliar conspicuous objects which they actively avoid, whereas the kittens appear to be unable to detect lines of unfamiliar orientation. Consequently, all behavior patterns dependent on the detection of lines placed at right angles to the familiar orientation would presumably no longer occur in the kittens, and the effect of their early experience would have much more general consequences than that of the young birds. Most people would now want to treat imprinting as an example of learning, but the effects of restricted visual experience on the kittens is much less easily classified. It is worth noting that even the effects of imprinting are relatively nonspecific in as much as the learning process affects the subsequent occurrence of nonsocial behavior such as feeding and grooming by narrowing the range of objects with which the bird associates. In the absence of the mother or her substitute, the birds will generally abandon all other activities while they search for her. Furthermore, imprinting has marked facilitating and constraining effects on what the animal can subsequently learn (Bateson, 1973).

Both Schneirla and Lehrman were concerned about the arbitrary way in which ethologists and experimental psychologists alike have so neatly demarcated the conditions necessary for learning from other types of experience. Lehrman (1970, p. 32) illustrated the conceptual problem facing us by sketching in the stages between environmental conditions having very general effects and those having highly specific effects. He listed the following points on the continuum:

1. Effects on neural development of nonbiological conditions (temperature, light, chemical conditions in the environment).

2. Nonspecific effects of gross stimulus input.

3. Developmental effects of practice passively forced during ontogeny.

4. Developmental effects of practice resulting from spontaneous activity of the nervous system.

5. Links and integrations between behavioral elements resulting from early, nonfunctional partial performances.

6. Interoceptive conditioning resulting from inevitable tissue changes and metabolic activities.

7. Simple conditioning to stimulation resulting from spontaneous movements.

8. Simple instances of conventional conditioning and learning.

Where does all this take us, then? A classification of determinants into those that have specific effects and those that have general effects is likely to be revised as fresh evidence is collected. Furthermore, it assumes a classification of behavioral units or types about which there may not be widespread agreement. Finally, it cuts arbitrarily across a continuum. None of these points render such a classification useless but they do mean that a sharp distinction between determinants with specific and general effects may create conceptual difficulties when attempts are made to unravel the processes involved in development.

III. CLASSIFICATION OF BEHAVIOR IN TERMS OF DEVELOPMENTAL DETERMINANTS

So far I have tried to outline the difficulties inherent in one classification that rests in part on the nature of long-lasting effects on behavior. It is now useful to reverse the procedure and consider a classification of behavior patterns in terms of developmental determinants. Naively it might be supposed that correspondence can be found between the two classifications. Indeed, preformationist views have from time to time slipped into ethological discussions of the origins of behavior. Behavior patterns are sometimes thought of as encapsulated in latent form in the fertilized egg; they are like Japanese flowers that will unfurl under the right environmental conditions. But even the most ardent preformationist does not insist that the blueprint for behavior, to use Lorenz's metaphor, is the same as bricks, mortar, and a work force. In other words, even for the extreme nativist, a host of environmental conditions will obviously be necessary if the behavior pattern is to develop. Therefore it is not necessary to consider a class of behavior patterns that might be determined by a single factor alone. A much more plausible class is one in which the determinants of the behavior patterns are of the type shown in Fig. 1. In this case, a behavior pattern can be determined by one or more factors specifically affecting it as well as by one or more determinants that have general effects. In the example given in Fig. 1, each letter could represent many determinants each of which had the long-term influence indicated by the arrow. Determinant A is necessary for Behavior 1 alone, whereas B is necessary for 1 as well as many other behavior patterns.

Inasmuch as many classifications of behavior have been concerned exclusively with developmental determinants that have specific outcomes, such as A, they have rested on a distinction, which is usually implicit, between determinants

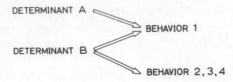

FIG. 1. Determinant A has a specific effect on Behavior 1. Determinant B has a non-specific effect on Behavior 1 and many other patterns. The arrows indicate that the determinants are necessary for the development of the behavior patterns to which they point.

with specific effects and those with general effects. As we have already seen, this distinction raises a number of difficulties; even so it is worth following the logic of this classification. Table II shows the various categories of behavior available if we concentrate on determinants that have specific effects on the development of behavior.

In some terminologies, behavior patterns in category G would be called "innate." For example, Tinbergen (1951, p. 2) represented most ethologists at the time when he wrote: "Innate behaviour is behaviour that has not been changed by learning processes." Tinbergen has changed his views a great deal since then, but some ethologists still cling to the old definition. For example, although admitting a preference for the term "endogenous," Ewer (1971) thought "innate" could be usefully applied to behavior that matures without practices or example. She took this to be Lorenz's position, although he (Lorenz, 1965) had changed his explicit definition of innate and now uses it as a synonym for "phylogenetically adapted." According to this concept, the specific details of the behavior that adapt the animal to its natural environment were selected during the evolution of the animal's species. Now, as has frequently been emphasized, natural selection acts on phenotypic outcomes not on the genotype. So the distinction between phylogenetic and ontogenetic sources of adaptiveness is not the same as the distinction between inherited and environmental determinants. Lorenz (1965) made this point strongly himself and argued that the outcomes of learning processes, such as imprinting, would have been selected during evolution. In other words, the *learned* preferences of birds for members of their own species are innate in the sense of being phylogenetically adapted. Despite this valuable clarification, old habits die hard. A dichotomy of origins of adaptation is all too easily used to justify once again a dichotomy of behavior and, to compound the muddle, also to refute the existence of behavior patterns specifically affected by both inherited and environmental determinants (category H in Table II). The confusion is evident even in Lorenz's (1965, p. 71) book

TABLE II

CLASSIFICATION OF BEHAVIOR IN TERMS OF DEVELOPMENTAL
DETERMINANTS WITH SPECIFIC EFFECTS ON BEHAVIOR

Inherited	Environmental	
	No determinants with specific effects	At least one determinant
No determinants with specific effects	E	F
At least one determinant	G	H

188

in which he wrote: "I strongly doubt that the motor co-ordination of phylo-genetically adapted motor patterns are at all modifiable by learning." One can only suppose that by an unconscious association of ideas, he was using "phylo-genetically adapted" as a synonym for "innate" in the old sense, namely for behavior that is not changed by learning processes.

Behind the inconsistent and inaccurate terminology lies a coherent point which bears directly on the matrix shown in Table II: in the preceding quota-tion, Lorenz was in effect denying the existence of behavior patterns specifically affected by both inherited and environmental determinants (category H). He was still thinking in terms of his old notion of the "intercalation" of inborn and learned components of behavior. This idea of "instinct-learning intercalation" was also pursued energetically by Eibl-Eibesfeldt (1970) who argued strongly against the view that blended intermediates constitute the majority of behavior patterns. Among other examples, he cited his own study of squirrels (*Sciurus vulgaris*) opening nuts in which a complex sequence can be analyzed into com-ponents some of which are learned and some of which are thought to develop without specific opportunities for practice. However, he seems to suggest that because some behavioral sequences can be analyzed in this way, all behavior can be. Is it really possible to break up the fully developed song of an experienced male chaffinch into components, some of which are specifically affected by experience and some of which are not? Even though we know that many factors have been responsible for the detailed specification of the song (Thorpe, 1961), it does not follow that somehow these factors will correspond to constituents of the final behavioral product. Rather than liken the development of such be-havior to the insertion of days into an existing calendar (*intercalare*), I suggest a more appropriate analogy would be the baking of cake. The flour, the eggs, the butter, and all the rest react together to form a product that is different from the sum of the parts. The actions of adding ingredients, preparing the mixture, and baking all contribute to the final effect. The point is that it would be nonsensical to expect anyone to recognize each of the ingredients and each of the actions involved in cooking as separate components in the finished cake. For similar reasons, I think those cases in which a simple relationship can be found between the determinants of behavior and the behavior itself will be exceptional. Behavior patterns that are affected by both inherited and environmental determinants with specific effects will lie in category H in Table II.

On the face of it, category E in Table II should be empty. However it is logically possible and, indeed, rather likely that the necessary conditions for the development of a behavior pattern are frequently those shown in Fig. 2. Be-

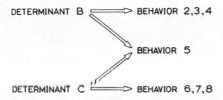

FIG. 2. The special properties of Behavior 5 arise from developmental determinants with many other effects on behavior. The arrows indicate that the determinants are necessary for the development of the behavior patterns to which they point.

havior 5 is determined by B or C both of which have other effects as well. A hypothetical example might be provided by a *Drosophila* mutant whose reduced rate of courtship was known to be due to the general effects of a single gene on its visual system. If this gene only expresses itself when such *Drosophila* are reared at a certain temperature and the environmental condition also affects other patterns of behavior, then the distinctive courtship would, indeed, be an example of behavior falling into category E. Such cases would be particularly interesting because they would lie outside the framework in which the origins of behavior are conventionally treated.

Two other points are worth making about the classification shown in Table II. First, the cell in which a behavior pattern is placed will depend critically on what is meant by a "determinant with a specific effect." If, on the one hand, a liberal view of specificity is taken and the line is drawn toward the general end of the specific-general scale, the major proportion of behavior patterns will, of course, be classified as being affected by both inherited and environmental determinants; if, on the other hand, stringent criteria are used to define specificity, the behavior patterns will be more evenly distributed in the matrix.

The second point is that if we were omniscient and were able to quantify all the determinants exclusively affecting any given behavior pattern occurring at a particular stage of development, it would be possible to build up a scatter diagram such as is shown in Fig. 3. I cannot, of course, justify the relative positions of the four entirely hypothetical dots placed on the scatter diagram which is unsatisfactory, in any event, because it misrepresents the dynamics of behavioral development. Any one diagram can be nothing more than a snapshot of a changing scene. The positions of some behavior patterns would, doubtless, move more during development than others. Many would move to the right on the scatter diagram as the behavior patterns became increasingly enriched and differentiated by experience. Some might move upward or diagonally as fresh genes affecting the details of already established behavior patterns became activated during development. Although lability of behavior is, in general, taken as

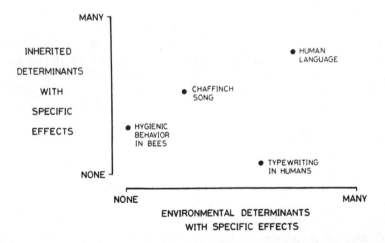

FIG. 3. A scatter diagram showing hypothetical points that might be placed on it if all the developmental determinants with specific effects were known.

evidence for the influence of environmental factors, it would clearly be a mistake to assume that this was always the case. In any event, lability of a behavior pattern means that it might have to be moved around in the matrix shown in Table II.

All of this might be taken to suggest that any kind of classification of behavior based on origins is useless. I think that to adopt such a view would be unduly purist since many people evidently do find it helpful to break up diverse and extensive material into manageable units so that they can think about it more easily. Rapid pigeon-holing of the evidence may frequently be misleading, but it certainly helps to unclutter the mind. Who is to say when it is better to disregard rather than focus on the relations and continuities between conventional categories? I shall consider this question in the next section.

IV. THE NATURE OF "RELEVANT" EXPERIENCE

I have tried to show that differences of opinion about the classification of behavior in terms of origins stems from different perceptions of the evidence. Lorenz has drawn a sharp distinction between factors responsible for the detailed characteristics of behavior (on which its adaptiveness to particular environmental conditions depends) and those factors necessary for continuity in development. Schneirla and Lehrman have objected to this formulation and where Lorenz saw two discrete categories, they perceived a spectrum of determinants. It is common enough in any science for different people to classify the same body of data in totally different ways. But it is worth while to ask whether some type of evidence can be found that would break the apparent impasse. Some progress in this direction may be made by looking more closely at the thinking underlying the experimental strategy proposed by Lorenz (1965).

Lorenz strongly argued for an experimental approach in which it would be possible to identify internal mechanisms responsible for the adaptiveness of behavior by systematically excluding likely sources of environmental "information." The isolation experiment, as it is called, clearly has been of service in eliminating possible explanations for the determination of some behavior patterns. It has suggested hypotheses that are fruitful in the sense that they can be tested. On the other hand, isolation experiments cannot provide direct tests of the hypotheses they propose. In order to demonstrate rigorously that a suspected source of variation does, indeed, have the effect it is supposed to have, that factor must be manipulated directly (see Hinde, 1968). If that cannot be done, progress may still be made by watching what happens when the suspected source of variation fluctuates spontaneously. Either way, the isolation experiment can usefully precede but does not replace direct analysis of behavioral determinants.

As a strategy, Lorenz's approach has the great merit of being positive and directed. Rather than bother about possible unknown sources of variation, the prescription to the experimenter is straightforward: if you consider something as the source of variation, then remove it. However, there are difficulties in this general approach which bring us to the nub of the whole problem. How does the experimenter know when he has excluded everything that is important? As Schneirla and Lehrman frequently asked: Can the experimenter tell the difference between "relevant" and "irrelevant" experience?

Even when considering experience that has a specific effect on behavior, it may be very difficult to know in advance when an animal is likely to generalize the effects of one kind of training to a novel situation. Can we really be so certain that we know what are equivalent types of experience for an animal? The potential importance of this question, which is discussed by Schneirla (1966) and Gottlieb (1973a), is easily underestimated. However another mattter polarizes opinion even more sharply. As we consider experiences with decreasingly specific outcomes at what point do we suddenly say that they are no longer providing relevant information? For Lorenz (1965, p. 37) this was not a problem and he took the following no-nonsense approach in his book:

> No biologist in his right senses will forget that the blueprint contained in the genome requires innumerable environmental factors in order to be realised in the phenogeny of structures and functions. During his individual growth, the male stickleback may need water of sufficient oxygen content, copepods for food, light, detailed pictures on his retina, and millions of other conditions in order to enable him, as an adult, to respond selectively to the red belly of a rival. Whatever wonders phenogeny may perform, however, it cannot extract from these factors information which simply is not contained in them, namely, the information that a rival is red underneath.

Lorenz saw a clear difference between experiences that produce their adaptive effects on behavior through learning and those experiences that are required for normal development and, when witheld, damage the animal in some way. Lorenz may have been led to this position, because many of the early experiments on the effects of sensory deprivation did, indeed, have pathological effects inasmuch as they resulted in degeneration in the deprived sensory modality (see Riesen, 1966).

More recent work had suggested that nonspecific experience can have facilitating effects on development which are not easily predicted in advance. A wide body of evidence indicates that the development of functional connectivity of many neurons in the central nervous system can be markedly changed by stimulation (e.g., Jacobson, 1969; Horn et al., 1973; Riesen, 1975). Examples at the behavioral level of unexpected effects of stimulation are also beginning to appear in the literature. For instance, exposure of domestic chicken eggs to light before hatching had a marked effect on the responsiveness of the chicks to conspicuous objects after hatching (Dimond, 1968; Adam and Dimond, 1971). Similarly, relatively short periods of exposure to constant white light after hatching markedly enhanced the responsiveness of one-day-old domestic chicks to a visually conspicuous object (Bateson et al., 1972; Bateson and Wainwright, 1972; Bateson and Seaburne-May, 1973; Kovach, 1971). After exposure to constant light for as little as 18 minutes, chicks approached a flashing, rotating light more rapidly than those kept in the dark (Fig. 4), and the effects persisted for at least 12 hours and probably much longer (Bateson, unpublished data). The differences between chicks exposed to light and those kept in the dark could not be attributed to difference in handling or differences in the temperature at which the chicks were kept, and the likelihood that the light-exposed chicks

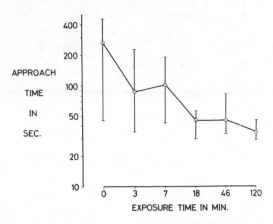

FIG. 4. The effects of varying exposure to a constant light on time taken to approach a flashing light by domestic chicks. Medians and interquartile ranges are given for the time taken to approach. Each group consisted of 8 birds. Both scales are logarithmic. (From Bateson and Seaburne-May, 1973.)

were generally aroused and, therefore, approached rapidly did not appear so attractive after the effects of stimulation in other modalities were examined. Prior exposure to tape recordings of loud peep calls in the dark made the chicks less responsive to a conspicuous visual stimulus (Bateson and Seaburne-May, 1973). Similarly, Graves and Siegel (1968) found that after gentle stroking in the dark domestic chicks took longer to approach a moving object than unstimulated chicks. These results suggest that the birds must be stimulated in the visual modality if visually guided behavior is to be facilitated.

Indeed, one-day-old chicks that have previously been exposed to light for an hour are much more accurate when pecking at millet seed than dark-reared birds (Vauclair and Bateson, 1975). The difference in accuracy was obtained when the chicks were unable to move their heads during the period of exposure to light. The difference might be attributed to deterioration in performance in the dark-reared birds rather than to improvement in the light-exposed ones. However, in a careful study, Cruze (1935) reared and fed chicks in the dark for varying amounts of time before giving them an opportunity to peck at millet seed. He showed that the accuracy of pecking in naive birds continued to improve over the first 5 days after hatching. Although this improvement can probably be attributed, at least in part, to increasing motor coordination (e.g., Bird, 1933), it seems unlikely that visual acuity could have been markedly *declining* over the first 5 days after hatching. Eventually, of course, prolonged rearing in the dark does lead to deterioration of pecking performance and Padilla (1935) had great difficulty in eliciting any pecking from chicks reared in the dark for 14 days from hatching. It may be useful, therefore, to distinguish between the effects of light that influence the initial *development* of visually guided behavior and the effects of light that are necessary for the *maintenance* of the behavior once it is already established. The distinction is illustrated in Fig. 5.

Light seems to have a remarkably similar effect on the development of depth perception in hooded rats. Tees (1974) found that although the performance of dark-reared rats on the visual cliff initially improved with age, the rate of im-

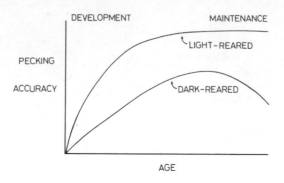

FIG. 5. Schematic diagram of effects of light- and dark-rearing on chicks' pecking accuracy at different ages.

provement was not as rapid as in light-reared rats. Up to around 60 days of age, then, light seemed to have a facilitatory effect on development. However after 80 days of age the performance of the dark-reared rats sharply deteriorated whereas that of the light-reared rats remained stable; in the older animals light appeared to serve a maintenance function.

Returning to chicks, the explanation for the relatively nonspecific effects of light on approach behavior and pecking in chicks may be that activation of the visual pathways by mere use enables visual stimulation to elicit visually guided behavior more readily. The visual systems of young dark-reared birds are not, on this view, damaged or functionally degenerate but are less well developed than the previously stimulated animals.

A similar explanation may account for some remarkable results obtained by Gottlieb (1971). By devocalizing Peking duckling embryos (*Anas platyrhynchos*) between 24 and 25 days after the beginning of development and just before the embryos penetrated the air space, Gottlieb seriously disrupted the preference of ducklings for the maternal call of their species after hatching. If the same operation was done immediately after the ducklings had broken into the air space, when they vocalize more and presumably can hear much better, the operated animals performed just as well as normal animals, strongly preferring the maternal call of their species to the chicken call (Fig. 6). This evidence strongly suggests that sounds the duckling emits itself shortly after it has broken into the air space play a part in the normal development of its auditory preferences.

However generalized the outcome of stimulation in the visual and auditory modalities of young birds, the effects on the development of their social relations with their natural mothers would undoubtedly be adaptive. It would seem, then, that relatively nonspecific stimulation can provide "information" in Lorenz's sense. If that point is accepted, the sharp distinction between determinants of behavior that have specific outcomes and supply relevant "information" and those that have general outcomes and are irrelevant begins to evaporate. It would be absurd, though, to use such evidence as a last-ditch defense of environmentalism, since an identical argument can be mounted in favor of analyzing inherited determinants with relatively nonspecific outcomes.

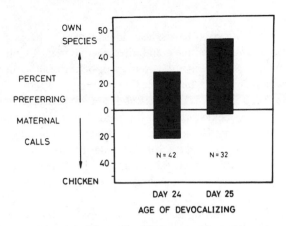

FIG. 6. Auditory preferences of devocalized Peking ducklings. The embryos were devocalized before (day 24) or after (day 25) they had broken into the air space in the egg. At 48 hours after hatching, each duckling was given a 5-minute choice test between the maternal call of its own species and the maternal call of the chicken. (From Gottlieb, 1971.)

The essential point is that if factors with nonspecific effects are disregarded, the chances of unraveling the variety of conditions necessary for the detailed determination of any behavior pattern will be greatly reduced. It may prove helpful to distinguish, as Gottlieb (1973b) has done, between "facilitative precursors" and "determinative precursors." This is a distinction between factors in development that have quantitative effects and those that have qualitative effects. Of course, in the grey area between the two categories, it is probably difficult to decide whether a factor facilitates a process that has already been established or is responsible for the development of a new one. In any event the distinction is not the same as the one between specific and general effects. Facilitative factors may have highly specific effects on development, and, conversely, determinative factors may have general consequences.

V. CONCLUSION

In this chapter, I have argued that a classification of behavior patterns in terms of their developmental determinants depends critically on a sharp division between determinants with specific outcomes and those that have general effects. Where the line is drawn is very much a matter of opinion, and it is hardly surprising that many people have regarded the classificaton of behavior into "innate" and "acquired" as an unwarranted abstraction. Even if the distinction between specific and general is accepted as a matter of convenience, four categories of behavior rather than two are needed. A third category is needed because many behavior patterns are likely to be affected by both inherited and environmental determinants with specific outcomes. The suggestion that such patterns can invariably be unscrambled into intercalating innate and acquired behavioral components is not convincing. The fourth category is needed for cases where the distinctiveness of the behavior arises from the interaction of inherited and environmental determinants both having general effects on a wide variety of behavior patterns.

The four-part classification has its uses inasmuch as it helps many people to think more easily about complex and diverse material. Furthermore it does provide a focus for research into the sources of behavioral distinctiveness. Environmental determinants that have specific outcomes are, almost by definition, mediated by learning processes, and it is undoubtedly a useful tactic in anyone's strategy for studying behavioral development to deny an animal particular opportunities for learning. However, aids to thought at one stage of analysis can become shackles at the next and eventually hinder further understanding.

Certain types of evidence, such as that provided by the development of social preferences in young chicks and ducklings, do not fit easily into a framework in which experience is either "relevant" or "irrelevant." Therefore, when study moves from preliminary sorting of complex material to detailed and comprehensive analysis, it becomes increasingly necessary to recognize the assumptions underlying a classification of behavior patterns in terms of origins. In this chapter I have attempted to uncover these assumptions in order to prepare the way for an integrated approach to the study of behavioral development.

Acknowledgments

The problems discussed in this chapter are steeped in controversy; therefore, I have shown drafts to a large number of friends in different disciplines. Whereas I benefited enormously from their comments, it must not be assumed, of course, that my views are necessarily theirs. In any event, I am greatly indebted to the following for their help: G. Barlow, C. Beer, R. Dawkins, C. Erickson, Ariane Etienne, G. Gottlieb, R.A. Hinde, N.K. Humphrey, P. Leyhausen, A. Manning, R. Rappaport, Amelie Rorty, J.S. Rosenblatt, B.A.O. Williams.

References

Adam, J., and Dimond, S. A. 1971. The effect of visual stimulation at different stages of embryonic development on approach behaviour. *Anim. Behav.* **19**, 51-54.

Andrew, R. J. 1972. The information potentially available in mammal displays. *In* "Nonverbal Communication" (R. A. Hinde, ed.), pp. 179-206. Cambridge Univ. Press, London and New York.

Barnett, S. A. 1973. Animals to man: the epigenetics of behavior. *In* "Ethology and Development" (S. A. Barnett, ed.), pp. 104-124. Spastics Int. Med. Pub., London.

Bateson, P. P. G. 1973. Internal influences on early learning in birds. *In* "Constraints on Learning: Limitations and Predispositions" (R. A. Hinde and J. Stevenson-Hinde, eds.), pp. 101-116. Academic Press, New York.

Bateson, P. P. G., and Seaburne-May, G. 1973. Effects of prior exposure to light on chicks' behaviour in the imprinting situation. *Anim. Behav.* **21**, 720-725.

Bateson, P. P. G., and Wainwright, A. A. P. 1972. The effects of prior exposure to light on the imprinting process in domestic chicks. *Behaviour* **42**, 279-290.

Bateson, P. P. G., Horn, G., and Rose, S. P. R. 1972. Effects of early experience on regional incorporation of precursors into RNA and protein in the chick brain. *Brain Res.* **39**, 449-465.

Beach, F. A. 1955. The descent of instinct. *Psychol. Rev.* **62**, 401-410.

Benzer, S. 1967. Behavioral mutants of *Drosophila* isolated by counter current distribution. *Proc. Nat. Acad. Sci. U.S.* **58**, 1112-1119.

Bird, C. 1933. Maturation and practice: their effects upon the feeding reaction of chicks. *J. Comp. Psychol.* **16**, 343-366.

Blakemore, C. 1973. Environmental constraints on development in the visual system. *In* "Constraints on Learning: Limitations and Predispositions" (R. A. Hinde and J. Stevenson-Hinde, eds.), pp. 51-73. Academic Press, New York.

Blakemore, C. and Cooper, G. F. 1970. Development of the brain depends on the visual environment. *Nature (London)* **228**, 477-478.

Blurton-Jones, N. J. 1968. Observations and experiments on causation of threat displays of the Great Tit (*Parus major*). *Anim. Behav. Monogr.* **1**, 74-158.

Cruze, W. W. 1935. Maturation and learning in chicks. *J. Comp. Psychol.* **19**, 371-408.

Dawkins, R. 1968. The ontogeny of a pecking preference in domestic chicks. *Z. Tierpsychol.* **25**, 170-186.

Dimond, S. J. 1968. Effects of photic stimulation before hatching on the development of fear in chicks. *J. Comp. Physiol. Psychol.* **65**, 320-324.

Eibl-Eibesfeldt, I. 1961. The interactions of unlearned behaviour patterns and learning in mammals. *In* "Brain Mechanisms and Learning" (J. F. Delafresnay, ed.), pp. 53-73. Blackwell, Oxford.

Eibl-Eibesfeldt, I. 1970. "Ethology: The Biology of Behavior." Holt, New York.

Ellis, P. E. 1964. Marching and colour in locust hoppers in relation to social factors. *Behaviour* **23**, 177-192.

Ewer, R. F. 1971. Review of "Animal Behaviour," 2nd Ed., by R. A. Hinde. *Anim. Behav.* **19**, 802-807.

Goldman, A. 1970. "A Theory of Human Action." Academic Press, New York.

Gottlieb, G. 1971. "The Development of Species Identification in Birds." Univ. of Chicago Press, Chicago, Illinois.

Gottlieb, G. 1973a. Neglected developmental variables in the study of species identification in birds. *Psychol. Bull.* **79**, 362-372.

Gottlieb, G. 1973b. Introduction to behavioral embryology. *In* "Studies on the Development of Behavior and the Nervous System. Vol. 2. Behavioral Embryology (G. Gottlieb, ed.), pp. 3-45. Academic Press, New York.

Graves, H. B., and Siegel, P. B. 1968. Prior experience and the approach response in domestic chicks. *Anim. Behav.* **16**, 18-23.

Hailman, J. P. 1967. The ontogeny of an instinct. *Behaviour*, Suppl. XV.

Hebb, D. O. 1953. Heredity and environment in mammalian behaviour. *Brit. J. Anim. Behav.* **1**, 43-47.

Hinde, R. A. 1960. Factors governing the changes in strength of a partially inborn response as shown by the mobbing behaviour of the chaffinch (*Fringilla coelebs*): III The interaction of short-term and long-term incremental and decremental effects. *Proc. Roy. Soc., Ser. B* **153**, 398-420.

Hinde, R. A. 1968. Dichotomies in the study of development. *In* "Genetic and Environmental Influences on Behaviour" (J.M. Thoday and A.S. Parkes, eds.), pp. 3-14. Oliver & Boyd, Edinburgh.

Hinde, R. A. 1970a. "Animal Behaviour: A Synthesis of Ethology and Comparative Psychology," 2nd Ed. McGraw-Hill, New York.

Hinde, R. A. 1970b. Behavioural habituation. *In* "Short-Term Changes in Neural Activity and Behaviour" (G. Horn and R. A. Hinde, eds.), pp. 3-40. Cambridge Univ. Press, London and New York.

Horn, G., Rose, S. P. R., and Bateson, P. P. G. 1973. Experience and plasticity in the central nervous system. *Science* **181**, 506-514.

Hsia, D. Y.-Y. 1967. The hereditary metabolic diseases. *In* "Behavior-Genetic Analysis" (J. Hirsch, ed.), pp. 176-193. McGraw-Hill, New York.

Jacobson, M. 1969. Development of specific neuronal connections. *Science* **163**, 543-547.

Jensen, D. P. 1961. Operationism and the question "Is this behavior learned or innate?" *Behaviour* **17**, 1-8.

Konishi, M. 1966. The attributes of instinct. *Behaviour* **27**, 316-328.

Kovach, J. K. 1971. Interaction of innate and acquired: color preferences and early exposure learning in chicks. *J. Comp. Physiol. Psychol.* **75**, 386-398.

Kuo, Z. 1967. "The Dynamics of Behavioral Development." Random House, New York.

Lee-Teng, E., and Sherman, S.M. 1966. Memory consolidation of one-trial learning in chicks. *Proc. Nat. Acad. Sci. U.S.* **56**, 926-931.

Lehrman, D. S. 1953. A critique of Konrad Lorenz's theory of instinctive behavior. *Quart. Rev. Biol.* **28**, 337-363.

Lehrman, D. S. 1970. Semantic and conceptual issues in the nature-nurture problem. *In* "Development and Evolution of Behavior" (L. R. Aronson, E. Tobach, D. S. Lehrman, and J. S. Rosenblatt, eds.), pp. 17–52. Freeman, San Francisco, California.

Lehrman, D. S., and Rosenblatt, J. S. 1971. The study of behavioral development. *In* "The Ontogeny of Verbebrate Behavior" (H. Moltz, ed.), pp. 1–27. Academic Press, New York.

Lorenz, K. 1961. Phylogenetische Anpassung und adaptive Modifikation des Verhaltens. *Z. Tierpsychol.* **18**, 139–187.

Lorenz, K. 1965. "Evolution and Modification of Behavior." Univ. of Chicago Press, Chicago, Illinois.

Marler, P. R., and Hamilton, W. J. 1966. "Mechanisms of Animal Behavior." Wiley, New York.

Moltz, H. 1965. Contemporary instinct theory and the fixed action pattern. *Psychol. Rev.* **72**, 27–47.

Padilla, S. G. 1935. Further studies on the delayed pecking of chicks. *J. Comp. Psychol.* **20**, 413–443.

Riesen, A. 1966. Sensory deprivation. *Progr. Physiol. Psychol.* **1**, 117–147.

Riesen, A.H. 1975. (Ed.) "The Developmental Neuropsychology of Sensory Deprivation." Academic Press, New York.

Rothenbuhler, W. C. 1967. Genetic and evolutionary considerations of social behavior of honey bees and some related insects. *In* "Behavior-genetic Analysis" (J. Hirsch, ed.), pp. 61–106. McGraw-Hill, New York.

Schneirla, T. C. 1956. Interrelationships of the "innate" and the "acquired" in instinctive behavior. *In* "L'Instinct dans le Comportement des Animaux et de l'Homme" (P.-P. Grasse, ed.), pp. 387–452. Masson, Paris.

Schneirla, T.C. 1966. Behavioral development and comparative psychology. *Quart Rev. Biol.* **41**, 283–302.

Smith, W. J. 1968. Message-meaning analyses. *In* "Animal Communication" (T. Sebeok, ed.), pp. 44–60. Indiana Univ. Press, Bloomington.

Tees, R. C. 1974. Effect of visual deprivation on development of depth perception in the rat. *J. Comp. Physiol. Psychol.* **80**, 300–308.

Thorpe, W. H. 1965. "Learning and Instinct in Animals." Methuen, London.

Thorpe, W. H. 1961. "Bird-Song." Cambridge Univ. Press, London and New York.

Thorpe, W. H. 1963. Ethology and the coding problem in germ cell and brain. *Z. Tierpsychol.* **20**, 529–551.

Tinbergen, N. 1951. "The Study of Instinct." Oxford Univ. Press, London and New York.

Tinbergen, N. 1963. On aims and methods of ethology. *Z. Tierpsychol.* **20**, 410-433.

Vauclair, J., and Bateson, P.P.G. 1975. Prior exposure to light and pecking accuracy in chicks. *Behaviour* **52**, 196–201.

WILLIAM R. THOMPSON

23 Influence of Prenatal Maternal Anxiety on Emotionality in Young Rats

The purpose of the observations reported in this article * was to test the hypothesis that emotional trauma undergone by female rats during pregnancy can affect the emotional characteristics of the offspring. By now, a good deal of evidence favoring this possibility has accumulated from diverse sources, including teratology (1), pediatrics (2), experimental psychology (3), and population biology (4). While none of the studies done has directly confirmed this hypothesis, many of them indicate that such hormones as cortisone, adrenalin, and adrenocorticotropic hormone, injected into the mother during pregnancy, have drastic effects on the fetus via the maternal-fetal blood exchange. Since strong emotion may release such substances into the mother's blood stream, there are grounds for supposing that it may have an important influence on fetal behavioral development. This experiment was the first in a projected series designed to examine this question in detail.

The rationale of the procedure was to create a situation which would predictably

* This research was done at Queens University, Kingston, Ontario, and supported by grants from the Queens Science Research Council and the National Science Foundation. Grateful acknowledgment is made to C. H. Hockman for his invaluable aid in helping to build the apparatus and to test the animals.

arouse strong anxiety in female rats, and to provide them with a standard means of reducing this anxiety; then to expose them to the anxiety-arousing situation during pregnancy, but block the accustomed means of escaping it. The assumption was that strong, free-floating anxiety would be generated in the pregnant females, and that any endocrine changes resulting would be transmitted through the maternal-fetal blood exchange to the fetus. The experiment was done by training five randomly chosen female hooded rats in a double compartment shuttlebox, first to expect strong shock at the sound of a buzzer, and then to avoid the shock by opening a door between the compartments and running through to the safe side. When the rats had learned this, the five experimentals, together with five control females, were mated to five randomly chosen males in a large cage. As soon as the experimentals were found to be pregnant (by vaginal smears), they were exposed to the buzzer three times every day in the shock side of the shuttlebox, but with the shock turned off and the door to the safe side locked. This procedure was terminated by the birth of a litter. The controls were placed in breeding cages during the same time.

Possible postnatal influences were controlled by crossfostering in such a way as to yield a design with six cells, each con-

199

TABLE 1

Comparison of experimental and control animals on two tests of emotionality

ITEM	TEST A		TEST B	
	AMOUNT OF ACTIVITY (DISTANCE)	LATENCY OF ACTIVITY (SECONDS)	LATENCY TO LEAVE CAGE (MINUTES)	LATENCY TO FOOD (MINUTES)
Tests given at age 30 to 40 days				
Experimentals	86.0	146.3	14.9	23.7
Controls	134.5	56.8	5.2	11.8
F values	(15.79, 14.21, 13.57)	(8.51, 7.91, 8.07)	(16.13, 16.46, 15.62)	(31.73, 25.66, 25.87)
p	< .001	< .01	< .001	< .001
Tests given at age 130 to 140 days				
Experimentals	114.5	71.5	4.8	11.6
Controls	162.3	26.8	2.1	6.2
F values	(9.77, 9.12, 8.76)	(4.95, 4.79, 4.57)	(2.39)	(4.48)
p	< .01	< .05	> .05	< .05

taining ten offspring with two main variables—namely, prenatal and postnatal treatment. The data obtained from tests given to the young were examined by means of analysis of variance. In all tests of significance, three error estimates were used: the within-cell variance, the within-plus-interaction variances, and the within-plus-interaction plus between-postnatal-treatment variances. Thus, as shown in Table 1, all tests of significance reported involve three F values.

The emotional characteristics of the 30 control and 30 experimental offspring were compared by two tests given at 30 to 40 and 130 to 140 days of age. In test A, measures of amount and latency of activity in an open field were taken in three daily sessions of 10 minutes each. In test B, emotionality was measured by latency of leaving the home cage, and latency of reaching food at the end of an alley way leading out from the cage after 24 hours' food deprivation. In the second test, the maximum time allowed an animal to reach food was 30 minutes. In the measures used, low activity and high latency were taken as indices of high emotionality.

The results are summarized in Table 1. On test A, striking differences between experimentals and controls were ob-

tained in amount of activity, both at 30 to 40 days and at 130 to 140 days. On the first testing, a significant interaction was obtained which probably represents genetic variation. On the second measure, experimental animals showed a much higher latency of activity than controls at both ages of testing. In neither of these activity measures were there any significant differences due to postnatal treatment or interaction besides the one mentioned.

In test B, experimental animals were slower to leave the home cage than controls at the first age of testing. There was no significant difference between groups in this measure, however, at 130 to 140 days of age. Similarly, experimentals showed a much higher latency than controls in getting to food at the end of the alley way at the first age of testing. The difference was less at the later age of testing. At both ages, significant interaction variances were found As before, both may well be due to genetic variation. On neither of the measures used in test B were any significant differences found between methods of postnatal treatment.

It is clear from this analysis that the experimental and control animals differ strikingly on the measures of emotionality used, and that these differences persist to a great extent into adulthood. While there is no question about the reliability of these differences, there is some ambiguity regarding their cause. Thus, we do not know exactly how the stress used had effects. It is possible that the buzzer was strong enough to act on the fetuses directly rather than indirectly by causing release of hormones in the mother. Only a more careful repetition of the experiment will throw light on this problem.

A more serious objection than this is that, besides the main factor of prenatal stress, genetic variation could also have been responsible for the offspring differences if there had been inadverent selection of nonemotional mothers for the control group and emotional mothers for the experimental group. However, several points argue against this possibility. Choice of female animals for the two groups was carried out randomly, and at least some of the genetic variance was included in the error estimates used to test the main effects. Further, an examination of scores within and between individual litters indicates that interlitter variances tend to be smaller than intralitter differences. This means that, in the population used, genetic variation was relatively slight compared with environmental variation. Consequently, it is improbable that even if accidental selection had occurred it could have resulted in an experimental group genetically very different from the control group.

Accordingly, we may state that there are some grounds for supposing that prenatal maternal anxiety does actually increase the emotionality of offspring. This conclusion is offered tentatively until further experimentation has been completed.

REFERENCES

1. FRASER, F. C., and T. D. FAINSTAT: Am. J. Diseases Children, 82, 593 (1951).
2. SONTAG, L. W.: Am. J. Obstet. Gynecol. 42, 996 (1941).
3. THOMPSON, W. D., and L. W. SONTAG: J. Comp. and Physiol. Psychol. 49, 454 (1956).
4. CHITTY, D.: "Adverse effects of population density upon the viability of later generations," in The Numbers of Man and Animals (Oliver and Boyd, London, 1955).

24 Effects of Brief Separation from Mother on Rhesus Monkeys

Temporary absence of the mother affects behavioral
development in rhesus monkeys (*Macaca mulatta*).

R. A. Hinde and Yvette Spencer-Booth

Reprinted from
9 July 1971, Volume 173, pp. 111-118

That the development of the human
child's personality is profoundly influenced by his social environment
is a commonplace. That a temporary
or permanent disruption of the child's
relationship with the most important
individual in that environment—his
mother—may have far-reaching effects,
is also becoming widely accepted (*1*).
But the severity of the effects, the bases
of the wide individual variation that
they show, and the conditions under
which they can be ameliorated, are
still matters for dispute. Since an experimental approach is not possible
with human subjects, a number of
workers have investigated the possibilities of using monkeys instead (*2–4*).

In the present series of experiments
(*5–9*), rhesus monkeys (*Macaca mulatta*) were used. The animals were
kept in small groups (one male, two
to four females, and their offspring),
each group occupying an outdoor cage
(5.50 by 2.44 by 2.44 meters) which
was connected to an inside room (2.29
by 1.83 by 1.37 meters). These conditions were intended as a compromise
which provided the animals with a reasonably complex (though, of course, not

a natural) social environment, while permitting reasonably precise experimental control and recording. The groups
proved to be social units in the sense
that the responses of individuals to
mildly disturbing situations were reduced by the presence of their companions (*10*).

Course of the Mother-Infant Relationship

It was first necessary to standardize
appropriate recording techniques and
to study the development of the
mother-infant relationship in this situation. This is illustrated in Fig. 1; the
data are based on observations of 16
infants during the first 6 months of
life, and eight infants thereafter. The
data were collected between 0900 and
1300 hours, each infant being watched
for 6 hours a week, a fortnight (2
weeks), or a month, according to age
(*11*). The time off the mother increased to a plateau at the end of the
first year, and then further, to 100
percent at the end of the second year.
This increase was associated with a
decrease in the mother's role in initiating nipple contacts, as measured by
the ratio of the number of nipple contacts initiated by the mother (M) to the
total number successfully initiated by
either mother (M) or infant (A). Con-

Dr. Hinde is a Royal Society Research Profes-
sor and honorary director, Medical Research
Council Unit on the Development and Integra-
tion of Behaviour at Madingley, Cambridge
University, Cambridge, England. Dr. Spencer-
Booth, who died 30 April 1971, was a Scientific
Officer in that unit.

comitantly, the absolute frequency of nipple contacts attempted by the infant that were rejected by the mother (R) increased up to the end of the first year. So did the proportion of nipple contacts or attempted nipple contacts that were rejected by the mother [R/ (A + M + R)]. The subsequent decrease in the absolute frequency of rejections implies lessening demand by the infant.

The proportion of the time off the mother that the infant spent more than 60 centimeters away from her (60 centimeters is roughly the distance a rhesus mother can reach in a hurry) increased to about 60 percent at the end of the first year, and thereafter remained fairly constant. Each time the distance between mother and infant changed from more than 60 centimeters to less, or vice versa, we recorded whether the approach or leaving was due to mother or infant. The difference between the proportions of approaches (Ap) and of leavings (L) due to the infant (% Ap − % L) gives a measure of the infants' role in maintaining proximity (12). This was negative for the first 4 to 5 months, indicating that the mother was then primarily responsible for maintaining proximity to the infant: their roles were subsequently reversed.

Mother-Infant Interaction

All the measures illustrated in Fig. 1 depend on the behavior of both mother and infant. To understand the dynam-ics of changes in mother-infant interaction, it is necessary to tease apart the roles of mother and infant. Some progress toward this can be made by considering the relations between measures (13). Table 1 shows the predicted consequences, on selected measures of mother-infant interaction, of four simple types of change in the behavior of mother or infant. For instance, if the infant changes in such a way that he tends to leave the mother more, the time off the mother and the time at a distance from the mother (> 60 centimeters) will increase, and the relative frequency of rejections [R/ (A + M + R)] and the infant's role in maintaining proximity (% Ap − % L) will decrease. Similar predictions can be made for each of the four types of change shown on the left of Table 1. Of course the implied hypothesis that changes in mother-infant interaction depend on only four types of basic change will prove an oversimplification. However it is useful in showing how correlations between measures provide a means of differentiating between changes in the interaction immediately due to changes in the mothers' behavior and those due to changes in the infants' behavior. Thus Table 1 shows that a change in time off the mother would be positively correlated with the frequency of rejections if immediately due to the mother, and vice versa.

Spearman rank order correlation coefficients between the median values of

Table 1. Predicted direction of changes in various measures of mother-infant interaction produced by four simple types of change in the behavior of either mother or infant.

Change in behavior	Time off mother	Relative frequency of rejections R/(A+M+R)	Time at a distance from mother (> 60 cm)	Infant's role in maintaining proximity (% Ap − % L)
Infant leaves mother more	+	−	+	−
Infant seeks proximity more	−	+	−	+
Mother seeks proximity more	−	−	−	−
Mother leaves infant more	+	+	+	+

some of the measures for successive age periods (weeks, fortnights, or months) are shown in Table 2. They show that the increase, with age, in time off the mother was positively correlated with the relative frequency of rejections, and thus immediately due to changes in the mothers' behavior. Similarly, the positive correlations between the time spent at a distance from the mother and the infants' role in maintaining proximity also indicate a prime role of the mother in the increasing independence of the infant. The positive correlations between the relative frequency of rejections and the infants' role in maintaining proximity can be interpreted on the view that it was those infants that were rejected most which had to play the greatest role in maintaining proximity [for further interpretation of these and other correlations, see (5) and (13)]. Thus, even during the first few months,

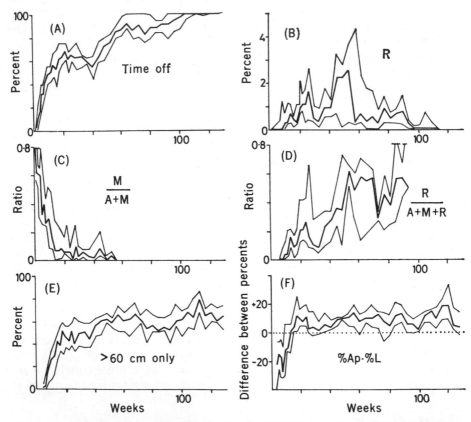

Fig. 1. Age changes in mother-infant interaction. Medians and interquartile ranges for $N = 16$ (weeks 1 to 24) or $N = 8$ (weeks 24 to 130). (A) Number of ½-minute periods in which infant was off mother (as percentage of the number of ½-minute periods observed). (B) Number of times infant attempted to gain nipple and was rejected, per 100 ½-minute periods (R). (C) Maternal initiative in nipple contacts [M (number of times infants were picked up by mothers) divided by A (number of times infants were accepted by mother) + M]. (D) Relative frequency of rejections [R/(A + M + R)]. (E) Number of ½-minute periods that infants spent wholly more than 60 centimeters from their mothers (as percentage of the number of ½-minute periods in which they were off her). (F) Infant initiative in maintaining proximity [percentage of approaches (Ap) due to infant minus percentage of leavings (L) due to infant [% Ap − % L)].

Table 2. Rank order correlation coefficients between median values of measures for all individuals in each age span. Rejections were rare in the early period.

	Relative frequency of rejections R/(A+M+R)	Time at a distance from mother (> 60 cm)	Infant's role in maintaining proximity (% Ap— % L)
Time off			
Week 1–6 (N=6)	No data	+ .98†	+ .96†
Week 7–20 (N=7)	+ .54	+ .96†	+ .94†
Week 21 + (N=17 to 20)	+ .69†	+ .74†	+ .38*
R/(A+M+R)			
Week 1–6 (N=6)		No data	No data
Week 7–20 (N=7)		+ .50	+ .36
Week 21 + (N=17 to 20)		+ .57†	+ .63*
> 60 cm			
Week 1–6 (N=6)			+ .79
Week 7–20 (N=7)			+ .93†
Week 21 + (N=17 to 20)			+ .57†

* $P < .05$.   † $P < .01$.

when it was the mother who was primarily responsible for maintaining proximity between mother and infant, it was changes in the mothers' behavior that were immediately responsible for the decrease in proximity with age.

This conclusion was somewhat surprising, since the growing independence of the infants was correlated with, and appeared to be due to, their increasing physical strength and activity. But the conclusion is in harmony with Hansen's (14) earlier findings that the frequency of mother-infant contacts decreases more slowly with age when the "mother" is an inanimate (and therefore nonrejecting) mother surrogate than when it is a real mother; and also with Harlow's (15) emphasis on the persistence of mutual clinging in infants raised in groups without mothers. Kaufman and Rosenblum (16), by contrast, argue that maternal rejections serve to increase rather than decrease the infants' dependent behavior, but this is probably a short-term effect and most likely to be conspicuous when a mother rejects an infant not yet ready to leave (17). Of course it is not implied that the changes in the mothers' behavior arise endogenously: they may be initiated by changes in the infants' behavior, such

as its demand for milk or locomotor activity. These are consequences of development, which, in turn, depends on the mother. The infants' development involves both increasing locomotor activity and self-initiated exploratory behavior, and increasing readiness to respond to the changing behavior of the mother with greater independence. While the changes in the relationship due to age thus depend on complex interactions between mother and infant, this analysis shows that the mother plays a large role in promoting the infants' independence.

Table 1 can be applied not only to changes with age, but also to individual differences at any given age. The correlations between measures taken from different infants in one age span show that differences among mothers are of prime importance in determining individual differences in mother-infant interaction in the early months, but that differences among infants are important subsequently (5).

Individual differences among the mother-infant relationships tend to remain stable from one age period (6 weeks) to the next over the first 5 to 6 months of life, as indicated by significant rank order correlations in measures of mother-infant interaction from one age period to the next. The correlations are rather less strong for the time spent by the infant off and at a distance from its mother than for measures such as the relative frequency of rejections and the infants' role in maintaining proximity. But the correlations are usually not sufficiently high to permit long-term predictions from one age period to another some months distant (5).

Other Social Companions

So far we have considered mother and infant as though they were unaffected by the other animals present. This is far from the case, for the other animals in the group may interact with both mothers and infants. Females other than the mother interact with infants more than males do; and, among the former, interactions are most common with nulliparous females about 2 years old. The nature of the interaction varies with the age of the infant, partly because of the decreasing protectiveness of the mother. In general, tentative maternal responses from animals other than the mother are commoner with younger infants · than with older ones, whereas grooming or aggressive responses are commoner with older infants (18).

The mothers vary in the extent to which they permit other animals to interact with their infants. They are most likely to tolerate females with which they are frequently in company and subordinate females that they can control (19). Thus, the infants' relations with the other females are affected by the relations between those females and the mother. Furthermore, the infants' relations with its mother are affected by the other females present, for infants living alone with their mothers were off them more, and went to a distance from them more, than did infants with group companions

present. Since the infants living alone with their mothers were rejected more and played a larger role in maintaining proximity with their mothers, the difference was primarily due to the mothers (Table 1). In a group situation, the mothers tend to keep their infants near them more—in part, to protect them from other females (13, 20).

Since the male is not impartial, often interceding in disputes, the mother's relations with the male may affect other females' interactions with the infant. Infants also spend much time in interaction with peers (21); in a social situation, relations with peers are influenced by the relations among the adults. The infant is thus part of a complex nexus of social relations, each link in which may be influenced by many types of change elsewhere. The nature of this nexus varies considerably between species of primate (17): in some, mothers allow much more interaction between their infants and other adults than is the case in rhesus monkeys (22).

Temporary Removal of the Mother

The separation experiments were carried out when the infants were between 21 and 32 weeks old (Table 3). At this age, infants still get some milk from their mothers, but they are capable of feeding themselves. The mother is playing only a small part in initiating nipple contacts, rejections are still increasing in frequency, and the infant has recently become primarily responsible for maintaining proximity (Fig. 1). Whereas female infants before week 20 tend to be more independent of their mothers than males are, males are more independent than females after week 30 (11).

Our experiments involved removing the mother from the group to a distant

Table 3. Scheme of separation experiments. One of the infants in the second group died before its second separation.

Groups		Length of separation (days) of infants at:		
Separations	N	21 weeks	25–26 weeks	30–32 weeks
Single	5			6
Two	5 + 1	6		6
Two	5		6	6
Long	6			13

Table 4. Days on which data were collected for short- and long-separation experiments. Solid lines enclose separation periods; broken lines indicate separation periods within which data were pooled.

Length of separations	Days on which data were collected		
	Pre-separation	Separation	Postseparation
6 days	1 2 3	4 5 . 7 . 9 |10¦11 12 . 14 . 16 ¦	. . . 23 . . 30 . . 37
13 days	1 2 3	4 5 . . . 9 . . 12 . . 15 16 |	17 ¦18 19 . 21 . 23¦ . . 30 . . . 37 . . . 44 .

indoor cage for 6 or 13 days. During the mother's absence, the infant remained with the same animals he had been with before, though the mother's removal from the nexus sometimes led to changes in the relationships among individuals.

Data on mother-infant relations were collected routinely from birth until the infants were 2½ years old, but we shall be concerned here only with the periods immediately before, during, and for a month after the mother's removal. The days on which data on the mother-infant relationship and infa activity were gathered are shown Table 4.

Over the period studied, the age which the experiment was performed made little difference in the response to separation. In Figs. 2 and 3, the data for all first 6-day separations are pooled ($N = 16$, see Table 3) (6). These figures show the median for each measure and the individual data for those two animals with the extreme preseparation scores. When the mother was removed, there was much whoo-calling, an indication of distress. The locomotor and play activity decreased,

and the frequency with which the infant was seen sitting motionless increased. Some infants spent time on, or leaning against, group companions.

When the mother was first returned, most infants spent more time on her, and less of their time off her at a distance from her, than before separation. The latter was associated with an increase in the infants' role in maintaining proximity and so, by an argument similar to that of Table 1, was immediately due to a change in the infants. When the mother first returned, there were frequent tantrums. [In fact, in the days after the mother's return, there were complex interactions between the demands of the infant and the behavior of the mother that are not shown in the pooled data of Figs. 2 and 3 (3).]

More qualitative descriptions of the effects of a period of maternal deprivation are in substantial agreement with these data (2), although it seems that bonnet monkey (Macaca radiata) infants, which receive more care from social companions in the mother's absence, are less affected than rhesus infants are (4).

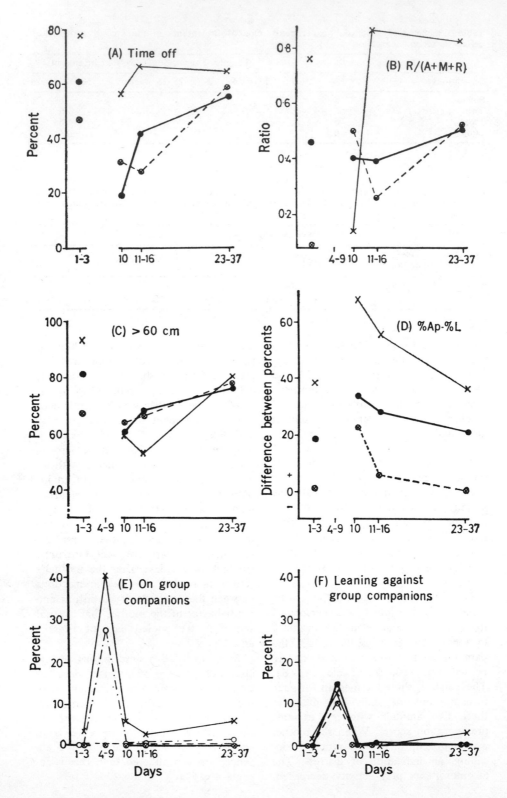

208

Length of the Separation

The effects of removing the mother for a few days are dramatic and last for at least some time after the mother's return. Is the severity of the effects related to the length of the separation period? Data on the separations at 30 to 32 weeks permit comparison between the effects of a single 6-day period of separation, the second of two 6-day periods, and one 13-day period (8). Figure 4 shows that, during the postseparation period, there were no significant differences between the groups in the time spent off and at a distance from the mother. Though the infants that had had a long separation were rejected less by their mothers, this was largely a reflection of a difference between groups before separation. There were, however, marked differences in the behavior of the infants when off their mothers. Those infants that had had only a 6-day separation experience had less depression of ac-

tivity and recovered from it more rapidly, were more active when active, and gave fewer distress calls than did the infants that had had two 6-day separation experiences or one 13-day separation. Thus the relation between the length and the effects of a separation period is quantitative.

Sources of Variation

One of the most conspicuous features of the data from these experiments was the marked individual variability (see Figs. 2 and 3). In order to assess the factors contributing to this, it is convenient to use an index of the distress shown by the infants when the mother is removed. We have seen that in the mothers' absence the infants tend to move about less, to sit more, and to whoo-call more. The 16 infants whose data are shown in Figs. 2 and 3 were ranked on each of these measures and the ranks summed to give a "distress index." The three measures on which the distress index is based are positively, but not always significantly, correlated with each other: some infants respond to the mother's disappearance primarily with depressed locomotor activity, others with frequent crying. Clearly, the distress index is useful only so long as the basis of the correlations between those measures is the focus of interest: it will cease to be useful when individual differences in the pattern of responding are studied.

The distress index was positively correlated between the different phases of the separation experiment (Table 5). This was, in part, a reflection of similar correlations between its constituent measures. Since the rank ordering of individuals on these measures remained similar, even though the absolute values and ranges of the measures increased, the effects of the mother's removal can be seen as an accentuation of preexisting differences (7).

Fig. 2. Effects of a period of separation from mother on relations of infant with mother and group companions. Break in abscissa indicates period of mother's absence. Ordinates: (A) number of ½-minute periods in which infant was off mother, as percentage of ½-minute periods observed; (B) ratio of number of times infants were rejected to total number of times infants attempted to gain nipple and were rejected by their mothers; (C) number of ½-minute periods in which infants were more than 60 centimeters from their mothers, as percentage of ½-minute periods off mother; (D) infant's role in maintaining proximity, as shown by difference between percentage of approaches and percentage of leavings due to infant; (E) and (F) percentage of ½-minute periods watched in which infant was recorded on or leaning against group companions. Thick line gives median for 16 infants; thinner lines, data for those two individuals with the most extreme preseparation scores on the measure in question (except for "On group companions," where median was zero throughout and three individuals are shown). Modified from (8).

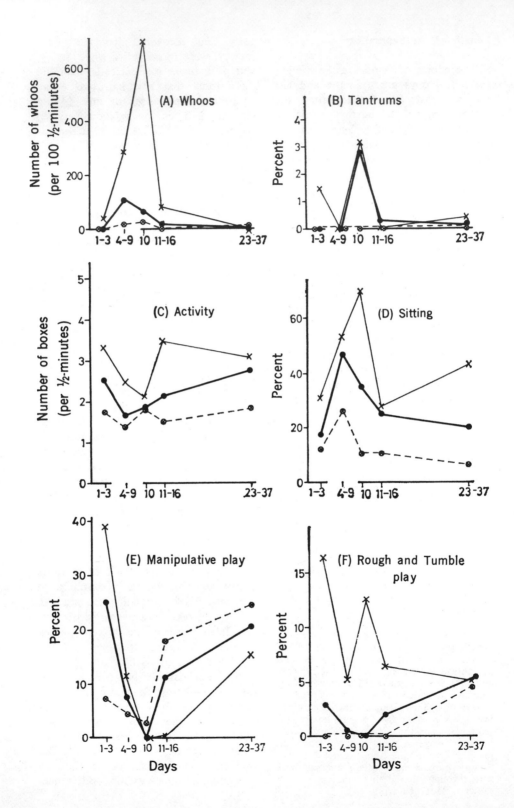

210

Table 5. Distress index. Spearman rank order correlation coefficients between the different stages of the 6-day separation experiments. ("Separation mean"—mean of all days on which data were collected during separation period.)

Stages of 6-day separation experiments	Correlation coefficients
Preseparation	+ .68† ⎤⎤⎤⎤
Separation mean	+ .40] + .53*] + .55*] + .60*
Day 10	+ .50*
Days 11–16	+ .66†
Days 23–37	

* $P < .05$.   † $P < .01$.

Table 6. Spearman rank order correlation coefficients between distress index in three postseparation periods and contemporaneous or preseparation measures of mother-infant relationship.

Measures of mother-infant relationship	Correlation coefficient		
	Day 10	Days 11–16	Days 23–37
(a) Contemporaneous			
Time off	− .09	− .39	− .37
R/(A+M+R)	+ .18	+ .51*	+ .59*
> 60 cm	− .32	− .58*	+ .11
% Ap − % L	+ .22	+ .51*	+ .45*
(b) Preseparation			
Time off	+ .02	− .07	− .13
R/(A+M+R)	+ .41	+ .59*	+ .43*
> 60 cm	− .26	− .28	− .18
% Ap − % L	+ .30	+ .64†	+ .41

* $P < .05$.   † $P < .01$.

When we examined the extent to which the variation among individuals in response to maternal separation could be attributed to various factors (6), the age of the infant (within the range studied) and the extent to which the infant was taken on by group companions proved to be of minor importance at most (23). The parity of the mother and the presence of peers could not be examined, since the mothers were multiparous and the peers were present in nearly all cases. There were, however, some differences related to sex. The mean distress index of male infants was higher throughout the experiment; the constituent measures did not differ significantly between the sexes before separation, but during and after the separation the males sat and whoo-called significantly more than the females did in some periods.

Of greater interest was the finding that the distress index was closely related to certain aspects of the mother-infant relationship (7). Table 6 (a) shows the rank order correlations between the distress index and four measures of the contemporaneous mother-infant relationship after the mothers' return. The correlations were low on the day that the mothers were returned, when most infants spent most of their time on their mothers. Subsequently, there was some tendency for those infants that were most distressed to be off their mothers least, and to spend less of that time at a distance from her. But more conspicuous were the tendencies for those that were most distressed to be (i) those that were rejected most by their mothers and (ii) those that played the greatest relative role in maintaining proximity with her (that is, high % Ap − % L).

Now most of the various measures of the relationship were correlated between pre- and postseparation periods —although the nature of the relationship changed, the individual differences remained similar. The question therefore arises as to whether the distress index after separation was related to the nature of the relationship before separation. Table 6 (b) shows that distress after separation was related, once

Fig. 3. Effects of a period of separation from mother on infants' behavior when off mother. Break in abscissa indicates period of mother's absence. Ordinates: (A) number of whoos per 100 ½-minute periods off mother; (B), (D), (E), and (F) number of ½-minute periods in which behavior indicated was recorded, as percentage of number of ½-minute periods off mother; (C) mean number of sections (out of 16) of the pen entered per ½-minute period off mother. Modified from (8).

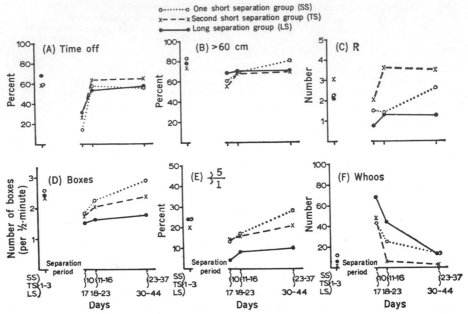

Fig. 4. Effects of removal of the mother (for 6 days, for the second of two periods of 6 days, or for 13 days) on behavior of rhesus infants in the month after the mother's return. Ordinates: (A) number of ½-minute periods as percentage of ½-minute periods observed; (B) number of ½-minute periods as percentage of ½-minutes off mother; (C) number of times infant attempted to gain nipple and was rejected, per 100 ½-minute periods; (D) number of boxes per ½-minute period; (E) number of ½-minute periods in which infants entered > 5 boxes, as percentage of ½-minute periods in which entered > 1 box; (F) number of whoos per 100 ½-minute periods (8).

again, to the frequency with which the mother had rejected the infant and the infant's role in maintaining proximity before separation. Thus the distress index was only loosely related to the more obvious measures of the mother-infant relationship, such as how much time the infant spent on or near its mother, but was clearly related to measures that could colloquially be described as indicators of tension in the relationship. It will be remembered that it is these same measures that are more stable with age.

We must now ask how much the postseparation distress index is related to aspects of the preseparation mother-infant relationship and how much to the postseparation relationship. Calculation of Kendall τ partial rank order correlation coefficients (Table 7) shows that, as might be expected, the pre-separation relationship is the more im-

portant just after the mother's return, but subsequently the postseparation (contemporaneous) relationship becomes paramount.

Thus the response of rhesus monkeys to a period of maternal deprivation is related to the nature of the mother-infant relationship both before and after the separation period. Such correlations do not justify the conclusion that differences in the mother-infant relationship cause the differences in the infant's response to separation, but it seems likely.

Long-Term Consequences

So far, we have been concerned with the consequences of a short period of maternal separation on behavior during the month after the mother's return. But how long do the effects last? Are we dealing with a temporary phenome-

Table 7. Kendall τ partial rank order correlation coefficients between measures of mother-infant interaction and postseparation distress index.

Measures of mother-infant relationship	Correlation coefficient		
	Day 10	Days 11–16	Days 23–37
R/(A+M+R) Preseparation, postseparation constant	.23	.24	.08
Postseparation, preseparation constant	.09	.13	.37
% Ap−% L Postseparation, preseparation constant	.19	.40	.23
Postseparation, preseparation constant	.12	.17	.23

non of no long-term importance? To assess this, mother-infant relations, infant activity, and infant performance in a variety of simple tests (24) were assessed when the infants were 12 and 30 months old. Most of the tests consisted of confronting the infant with mildly disturbing or frustrating situations. Infants that had been given one or two 6-day separation experiences were compared with control infants that had lived with their mothers continuously. (The tests were not devised in time for all of the control infants, which were studied earlier than the separated infants, to be tested, and a few of the separated infants were not available for the tests.) The diversity of the procedures used make it impossible to describe the results succinctly, but two selected examples are given in Tables 8 and 9. Table 8 gives measures of infant activity off the mother at 12 and 30 months. Table 9 contains data, obtained when 12-month-old juveniles had been moved with their mothers to a strange laboratory cage, on the readiness with which the infants would approach objects that were placed in an adjoining cage, from which their mothers were excluded.

The results of these and other tests can be summarized as follows (9). At 12 months, there were few significant differences in measures of the mother-

Table 8. Activity of infants when off mothers: controls (infants with no separation experience) and infants that had had one (SS) and two (TS) 6-day separation experiences when 20 to 30 weeks old (see Table 3). Mean number of sections of the pen (out of 16) entered per ½-minute period (a). A measure of how active when active—the ratio of the number of ½-minute periods in which the infant entered five or more boxes, to the number of periods in which it entered more than one box (b). Percentage of ½-minute periods in which the infant was seen sitting (c). Rough and tumble (d) and approach/withdrawal (e) social play. Manipulative (nonsocial play (f). P values based on Mann-Whitney U test, two-tailed.

Measure of activity	12-month infant			30-month infant		
	Controls (N = 3)	SS (N = 5)	TS (N = 8)	Controls (N = 8)	SS (N = 5)	TS (N = 5)
(a) Sections of pen entered (mean)	3.1	3.2	2.8	3.1	2.4	2.5
(b) $\frac{>5}{>1}$ (%)	36	32	25	42	36	28
(c) Sitting (%)	2	10	35	23	27	35
		P = .012			P = .020	
	P = .036					
(d) Rough and tumble (%)	6	13	5	4	0	0
					P = .02	
(e) Approach/withdrawal (%)	0	1	.5	1	0	0
		P = .05			P = .14	
(f) Manipulative (%)	5	16	17	5	11	13
		P = .032			P = .01	
	P = .024					

Table 9. Tests in strange laboratory cage at 12 months. The test object was placed in an adjoining cage that was connected by an aperture large enough for the infant, but too small for the mother. Test objects were a mirror outside the far wall of cage for 30 minutes; a mirror with grass in front of it, for 15 minutes; and a banana, for 10 minutes. Figures give latency to enter this cage, total time in the cage, and median length of the visit. Mirror and grass tests were on the sixth day after moving to strange cage; banana test on days 1 and 6 (figures give medians of sums of individual scores). P values based on Mann-Whitney U test, two-tailed.

Measure (minutes)	Controls ($N = 6$)	SS ($N = 5$)	TS ($N = 8$)
Mirror (day 6)			
Latency	.2	1.3	2.4
		$P = .042$	
Total time	7.9	6.5	6.7
	$P = .03$		
Median visit length	.4	.3	.3
		$P = .03$	
Mirror and grass (day 6)			
Latency	.2	.1	.4
Total time	9.7	8.7	4.2
		$P = .004$	
Median visit length	.8	.6	.3
		$P = .03$	
Banana (days 1 and 6)			
Latency	1.7	4.7	6.2
Total time	6.7	5.4	5.5
Median visit length	.9	.6	.2
		$P = .008$	

infant relationship between control and previously separated infants, either in the home run or after mother and infant had temporarily been removed to a strange indoor cage. Nor were there any significant differences in measures of the infants' readiness to approach strange objects in the home run. The previously separated infants were, however, less ready than the controls to take vitamin pills from Y.S.-B., who had been involved in catching the mothers. In addition, the previously separated infants still showed some signs of the depression of locomotor and play activity that occurred during the separation period (Table 8). But it was when the infants were confronted with strange objects in the strange cage that

differences between controls and experimentals became most conspicuous. Experimentals were less ready to approach, by a number of measures; and the differences tended to be more marked with the infants that had had two previous separation experiences than with those that had had just one (Table 9). Thus, the effects of a mere 6-day absence from the mother were clearly discernible 5 months later. When tested at 30 months of age (2 years after the separation experience), the differences were much less marked, but the previously separated infants were still less active than the controls (Table 8) and differed significantly from them on a number of other measures in a manner similar to that seen at 12 months.

Summary and Conclusion

To summarize, data on the course of development of mother-infant relations in rhesus monkeys have been presented; a method for teasing apart the relative roles of mother and infant in causing changes or differences in the interaction described; and the complexity of the social nexus, within which the relationship is set, stressed. When the mother is removed for a few days, the infant calls a great deal at first and then shows depressed locomotor and play activity. These symptoms may last for a month after the mother's return. Simple tests given 6 months and even 2 years later strongly suggest that the differences (between infants that have had such a separation experience and infants that have not) are persistent.

Are these data relevant to the human case? The rhesus monkey has no verbal language and a much less complex social development than man. Furthermore, its social environment is quite different from that found in any human culture. Parallels between monkey and man must therefore be scrutinized carefully before being used as a basis for generalization. But the facts show that a brief separation experience produces in rhesus monkey infants symptoms that are very similar (except for the apparent absence of a "phase of detachment" on reunion) to those in human infants (25). While age of separation, within the rather narrow limits used here, was a variable of minor importance, the effects of the separation varied, as in the human case, with the length of the separation experience and the sex of the infant. Differences in the techniques of experimenters, as well as differences in the species, prevent precise comparisons of the roles of the mother-infant relationships; nevertheless, the nature of the relationship appears to be an important variable in both monkey and man. There would seem, therefore, to be strong reasons for thinking that we are dealing with comparable phenomena. If that is the case, the fact that monkeys function at a simpler conceptual level than man limits the complexity of the explanatory hypotheses necessary in the human case. In addition, the finding that such a brief separation experience, involving removal of the mother but no exposure to a strange environment, can produce effects lasting for months or years in rhesus monkeys strengthens the evidence that long-term effects may occur also in man. Finally, this analysis provides bases for attempts to predict individual differences in the effects of a period of separation on rhesus infants, and the parallels with man suggest that examination of the same variables in the human case would be worthwhile.

References and Notes

1. R. A. Spitz, Psychoanal. Study Child 1, 53 (1945); ibid. 2, 313 (1946); J. Bowlby, Maternal Care and Mental Health (monograph series 2, World Health Organization, Geneva, 1952); Attachment & Loss, vol. 1, Attachment (Hogarth, London, 1969); M. D. Ainsworth, in Deprivation of Maternal Care (Public Health Paper No. 14, World Health Organization, Geneva, 1962), p. 97; C. M. Heinicke and I. J. Westheimer, Brief Separations (International Universities Press, New York, 1965).
2. G. D. Jensen and C. W. Tolman, J. Comp. Physiol. Psychol. 55, 131 (1962); B. Seay, E. Hansen, H. F. Harlow, J. Child Psychol. Psychiat. 3, 123 (1962); B. Seay and H. F. Harlow, J. Nerv. Ment. Dis. 140, 434 (1965).
3. Y. Spencer-Booth and R. A. Hinde, J. Child Psychol. Psychiat. 7, 179 (1967).
4. I. C. Kaufman and L. A. Rosenblum, Ann. N.Y. Acad. Sci. 159, 681 (1969); L. A. Rosenblum and I. C. Kaufman, Amer. J. Orthopsychiat. 38, 418 (1968).
5. R. A. Hinde and Y. Spencer-Booth, Anim. Behav., in press.
6. Y. Spencer-Booth and R. A. Hinde, ibid., in press.
7. R. A. Hinde and Y. Spencer-Booth, J. Child Psychol. Psychiat. 11, 159 (1970).
8. Y. Spencer-Booth and R. A. Hinde, Anim. Behav., in press.
9. ———, J. Child Psychol. Psychiat., in press.
10. T. E. Rowell and R. A. Hinde, Anim. Behav. 11, 235 (1963).
11. R. A. Hinde and Y. Spencer-Booth, ibid., 15, 169 (1967).
12. R. A. Hinde and S. Atkinson, ibid., 18, 169 (1970).
13. R. A. Hinde, Ann. N.Y. Acad. Sci. 159, 651 (1969).
14. E. W. Hansen, Behaviour 27, 107 (1966).
15. H. F. Harlow, in Behavior of Non-human Primates, A. M. Schrier, H. F. Harlow, F. Stollnitz, Eds. (Academic Press, New York, 1965), vol. 2.
16. I. C. Kaufman and L. A. Rosenblum, in Determinants of Infant Behaviour, B. M. Foss, Ed. (Methuen, London, 1969), vol. 4.
17. R. A. Hinde, in Behavior of Non-human Primates, A. M. Schrier and F. Stollnitz, Eds. (Academic Press, New York, in press), vol. 3.
18. Y. Spencer-Booth, Anim. Behav. 16, 541 (1968).
19. T. E. Rowell, R. A. Hinde, Y. Spencer-Booth, Anim. Behav. 12, 219 (1964).
20. R. A. Hinde and Y. Spencer-Booth, in Primate Ethology, D. Morris, Ed. (Weidenfeld & Nicolson, London, and Aldine, Chicago, 1967).
21. H. F. Harlow, in Advances in the Study of Behavior, D. Lehrman, R. A. Hinde, E. Shaw, Eds. (Academic Press, New York, 1965).
22. P. Jay, in Primate Behavior, I. DeVore, Ed. (Holt, Rinehart & Winston, New York, 1965); L. A. Rosenblum, I. C. Kaufman, A. J. Stynes, Anim. Behav. 12, 338 (1964).
23. Of course this does not mean that intervention by group companions cannot ameliorate the effects of the mother's absence; indeed, there is evidence from other species that it can (4). The small size of the differences that are found between infants which were and were not taken on by group companions could be due to an interaction between susceptibility to maternal deprivation and probability of receiving care from social companions.
24. Y. Spencer-Booth and R. A. Hinde, Behaviour 33, 179 (1969).
25. C. Heinicke and I. Westheimer, Brief Separations (International Universities Press, New York, 1965).
26. This work was supported in the first instance by the Mental Health Research Fund, and continuously by the Medical Research Council and the Royal Society. We are grateful to T. E. Rowell, M. Bruce, I. Davies, C. Perkis, R. Pollard-Urquhart, S. Richards, and others for help at various times, and to members of the Cambridge University Department of Veterinary Clinical Studies for their advice and help with the husbandry of the animals. We would also like to thank N. Humphrey for his comments on the manuscript.

215

25

BEHAVIORAL BIOLOGY, **12**, 101-110 (1974), Abstract No. 4111

Contact Stimulation, Androgenized Females and Accelerated Sexual Maturation in Female Mice[1]

LEE C. DRICKAMER

Biology Department,
Williams College,
Williamstown, Massachusetts 01267

Adult female mice previously given 100 μg of testosterone propionate on day 1 of age were placed with 21-day-old female mice producing an acceleration of vaginal introitus and first vaginal estrus in the young females. Young test females placed with intact adult males attained first estrus at 27.9 days of age while test females placed with neonatally androgenized females matured at 30.5 days of age and single control females matured at 34.0 days of age. Androgenized females caused an acceleration of maturation in young females, but not by the production of an acceleratory pheromone as in adult intact males. A second experiment demonstrated that contact stimulation with the young females by neonatally androgenized adult females was equal in quantity and similar in quality to the contact by normal adult males. In a final experiment young females were exposed to the continued presence of a neonatally androgenized adult female and a daily sample of male-soiled bedding containing the maturation-accelerating pheromone; these females matured at 28.1 days of age, the same age as young females placed with intact adult males. Contact stimulation may explain the acceleration of sexual development produced by placing young female mice with neonatally androgenized adult females. Such contact could explain a portion of the male-acceleration effect; that portion not accounted for by pheromonal factors.

The presence of an adult male mouse, soiled bedding from a cage containing one or more males, or male urine all lead to an acceleration of sexual maturation in young female mice, where maturation is defined in terms of first vaginal cornification (Vandenbergh, 1969; Vandenbergh, Drickamer, and Colby, 1972; Colby and Vandenbergh, 1974; Fullerton and Cowley, 1971; Kennedy and Brown, 1970). Accelerated sexual maturation (Vandenbergh *et al.*, 1972) produced by the daily addition of male bedding, containing a pheromone, is intermediate (33.0 days of age for first vaginal estrus) when compared with control females housed alone (35.5 days) and with young females placed with adult intact males (28.0 days). The male pheromone,

[1]This research was supported in part by Grant MH 24483-01 from the National Institutes of Mental Health and by the Sloan Foundation Discretionary Fund at Williams College.

216

without the actual presence of the adult male, does not account for all of the accelerated sexual development.

Delays in female mouse sexual development occur when the animals are grouped: a pheromone is at least partially responsible for this phenomenon (Castro, 1967; Cowley and Wise, 1972; Drickamer, 1974). Production of the maturation-delaying pheromone is dependent upon social contact among the grouped females. The studies reported here examined the hypothesis that contact stimulation is also important to the male acceleration-of-maturation phenomenon.

Castrated adult male mice do not produce the maturation-accelerating pheromone (Vandenbergh, 1969) and they also do not show consistent male-like sexual behavior patterns (McGill and Tucker, 1964; McGill and Haynes, 1973). Female mice and rats given injections of androgen at birth exhibit male-like sexual behavior patterns as adults, even without additional androgen in adulthood (Sachs et al., 1973; Barraclough, 1971; Manning and McGill, 1974). Based on these previous findings I hypothesized that adult female mice previously androgenized at birth (called n.a. adult females in this paper) might provide male-like contact stimulation and thus accelerate the sexual development of young females housed in the same cage. Alternatively the adult n.a. females, with or without additional injections of testosterone propionate as adults might produce a male-like pheromone leading to acceleration of maturation in young female mice exposed to the pheromone. Three separate experiments were conducted to test these hypotheses regarding contact stimulation, male-like behavior of adult n.a. females, and their effect on sexual maturation of young female mice.

GENERAL METHODS

All of the mice used in these experiments were from a randomly bred closed-closed colony laboratory strain of *Mus musculus* (derived from the International Cancer Research Strain and supplied by Ward's Scientific Establishment, Rochester, NY). All colony and experimental mice were caged in standard shoebox cages of opaque polypropylene measuring 15 × 28 × 15 cm deep with fitted wire-mesh lids. All mice were supplied with mouse chow (Wayne Lab Blox) and water ad lib. Bedding, consisting of finely ground wood shavings and a piece of nesting cotton, was changed weekly.

Pregnant female mice were caged individually beginning approximately 1 wk before parturition, and each cage was checked daily for the presence of a litter. On the day after birth the number of pups in each litter was counted and each pup was sexed. Litters were reduced to 10 pups with the restriction that at least one male pup remained in each newly constituted litter. Litters of fewer than 10 pups were discarded. Twenty-one days after birth the immature females were weaned and assigned, at random, to a treatment group.

Two rooms were used in these investigations: Both the colony and experimental rooms were maintained at 20-23°C and 30-70% relative humidity. Each room was equipped with overhead fluorescent lighting and a daily light timer set for a 12-hr light/12-hr dark cycle.

Prior to the series of experiments a group of 1-day-old female mice was given 100 μg injections of testosterone propionate contained in 0.02 cc of peanut oil. When mature these adult n.a. females were used in all three experiments and were 60-90 days of age at the time of testing.

EXPERIMENT I

Purpose

Experiment I was designed (1) to determine whether the presence of an adult n.a. female would lead to accelerated sexual development in young females and (2) to test for the possible role of a pheromone in producing these effects.

Methods

At 21 days of age 300 immature female mice were weaned and assigned, at random, to 10 different treatment groups ($n = 30$ mice/treatment): (a) alone as controls; (b) with an intact adult male; (c) alone and receiving a daily 80-cc sample of soiled bedding from a cage containing an adult male; (d) with an adult intact female; (e) alone and receiving a daily 80-cc sample of soiled bedding from a cage containing an adult female; (f) with an adult n.a. female; (g) alone and receiving a daily 80-cc sample of bedding from a cage contining a n.a. female; (h) with an adult n.a. female receiving daily injections of 100 μg of testosterone propionate; (i) alone and receiving a daily 80-cc sample of soiled bedding from a cage containing an adult n.a. female receiving daily injections of 100 μg of testosterone propionate in 0.02 cc of peanut oil; (j) with an adult intact female receiving daily injections of 50 μg of estradiol benzoate in 0.02 cc of peanut oil to produce a condition of constant estrus. Soiled bedding from cages containing adult males or females consisted of finely ground wood shavings, urine, and fecal material. The condition of constant estrus in females receiving injections of estrodial benzoate was confirmed by taking vaginal smears. In cages where bedding was added daily a 320-cc sample was removed at mid-week to prevent the accumulation of bedding in the cage.

Each mature female mouse was checked daily beginning on day 21 for the occurrence of vaginal introitus. Beginning on the day of vaginal introitus a vaginal lavage was taken each day and the cellular contents were examined with a microscope to determine the occurrence of first vaginal estrus. An estrus smear contained at least 95% cornified epithelial cells (see Vandenbergh, 1969 or Vandenbergh *et al.*, 1972, for details). Each mouse was weighed to the nearest 0.1 g at 21, 28, 35, and 42 days of age and on the days of vaginal introitus and first vaginal estrus. The data were analyzed using a one-way analysis-of-variance and Duncan's New Multiple Range Test ($a = .01$).

TABLE 1

Mean Ages and Body Weights (± SE) at First Vaginal Estrus in Female House Mice under 10 Different Treatment Conditions. $N = 30$ Mice/Treatment. Within a Vertical Column Those Means Not Connected by the Same Vertical Line are Significantly Different at the .01 Level

Treatment	First estrus	
	Age (days)	Weight (g)
Adult male present	27.9 (0.4)	19.5 (0.7)
Adult n.a. female present	30.3 (0.4)	20.6 (0.7)
Adult n.a. female (TP) present[a]	30.6 (0.4)	20.4 (0.4)
Adult male bedding added daily	30.7 (0.6)	20.8 (0.6)
Adult female present	32.5 (0.6)	21.7 (0.4)
Adult female bedding added daily	32.6 (0.6)	21.8 (0.6)
Adult female (EB) present[a]	32.6 (0.5)	22.0 (0.4)
Adult n.a. female bedding added daily	33.1 (0.6)	21.8 (0.6)
Adult n.a. female (TP) bedding added daily[a]	33.2 (0.3)	22.1 (0.5)
Control	34.0 (0.8)	22.6 (0.5)
$F (9, 290)$	6.05	2.97
Prob.	$P < .01$	$P < .01$

[a](TP) indicates n.a. females given additional daily injections of testosterone propionate as adults; (EB) indicates adult females given daily injections of estradiol benzoate.

Results

The results of Experiment I (Table 1) revealed that immature females housed with adult n.a. females, with or without additional adult androgen injections, matured significantly earlier than control mice and at the same age as mice receiving a daily sample of male-soiled bedding. Mice caged with males matured first and young females receiving a daily sample of male-soiled bedding were intermediate in an age of first estrus between controls and those females housed with adult males. No other test group mean differed significantly from singly caged control mice. Measurement of the age of vaginal introitus (Table 2) revealed substantially the same picture: vaginal introitus occurred earlier in females caged with adult males or adult n.a. females and in cages receiving a daily dose of male-soiled bedding.

There were no significant differences in body weights of the mice in any treatment group at any of the four weekly age intervals. The patterns of differences in body weight at the occurrence of vaginal introitus and first vaginal estrus were similar. For both measures of maturity there was a significant trend for lighter mean body weights among mice that matured sooner. That is, female mice caged with adult males matured sooner (27.9 days) and weighed less (19.5 g) at maturity than did control females (34.0 days; 22.6 g).

TABLE 2

Mean Age and Body Weight (± 1 SE) at Vaginal Introitus in Female House Mice Under
10 Different Treatment Conditions. N = 30 Mice/Treatment. Within a Vertical
Column Those Means Not Connected by the Same Vertical Line
Are Significantly Different at the .01 Level

Treatment	Vaginal introitus	
	Age (days)	Weight (g)
Adult n.a. female (TP) present	25.4 (0.5)	16.6 (0.2)
Adult male present	25.5 (0.5)	16.9 (0.5)
Adult n.a. female present	26.1 (0.4)	17.5 (0.6)
Adult male bedding added daily	26.2 (0.5)	17.6 (0.4)
Adult female bedding added daily	27.2 (0.5)	18.0 (0.2)
Adult female present	27.2 (0.6)	18.2 (0.6)
Adult n.a. female bedding added daily	27.4 (0.5)	18.0 (0.5)
Adult female (EB) present	27.5 (0.5)	18.3 (0.5)
Adult n.a. female (TP) bedding added daily	27.6 (0.5)	18.3 (0.5)
Control	28.0 (0.6)	19.6 (0.6)
F (9,290)	3.44	3.63
Prob.	$P < .01$	$P < .01$

EXPERIMENT II

Purpose

This experiment assessed the nature and amount of social contact
between immature female mice and (a) adult males, (b) adult females (c) adult
n.a. females, and (d) adult females receiving daily injections of 50 μg of
estradiol benzoate.

Methods

Individual immature (21- to 30-day-old) female mice were paired with
adult mice from each of the four categories listed above. Thirty observations,
each 30 min in duration, were made for each of the categories of adult
stimulus mice. During observations the frequencies of social contact, correctly
oriented sexual mounts, and thrusting while mounting were recorded.
Analyses-of-variance were used to determine whether any differences existed
in the behavior patterns of the stimulus mice.

Results

The data and analyses (Table 3) revealed no significant differences in
total social contact between the stimulus mouse and the immature females
across all four test groups. However, for mounts and thrusts the adult males
and adult n.a. females exhibited significantly higher frequencies. In both

TABLE 3

Mean Frequency (± 1 SE) For Each of Three Behaviors Per One-Half Hour Observations. Young Female Mice, 21–30 Days of Age were Paired with Either an Adult Intact Male, an Adult Intact Female, an Adult Intact Female Receiving a Daily Injection of 50 µg Estradiol Benzoate or an Adult Intact Female Androgenized on the Day After Birth with a 100-µg Injection of Testosterone Propionate. N = 30 Observations/Treatment. Means Within Each Vertical Column Marked with an Asterisk (*) Are Significantly Different from Other Means in the Same Column at .01 Level

	Contact	Mounts	Thrusts
Adult male	63.6 (4.9)	3.6 (1.2)*	3.3 (1.7)*
Adult female	71.7 (5.0)	0 (0)	0 (0)
Adult n.a. female	72.1 (4.9)	5.3 (0.9)*	2.9 (1.1)*
Adult female given daily EB injections	72.3 (6.0)	0 (0)	0 (0)
F (3,116)	0.69	3.64	5.46
Prob.	NS	$P < .025$	$P < .01$

behavior categories the androgenized females had frequencies very similar to those for adult males. Normal adult females and females given estradiol benzoate showed no mounting or thrusting behavior during the observations.

EXPERIMENT III

Purpose

This experiment was designed to test the combined effect of the presence of an adult n.a. female and daily dose of male-soiled bedding on the timing of sexual maturation in young females.

Methods

The procedures for this experiment were identical to those in Experiment I. Five test groups were used: (a) young females caged alone as controls; (b) young females caged with an adult intact male present; (c) young females caged with an adult n.a. female present; (d) young females caged alone and receiving a daily 80-cc sample of male-soiled bedding; and (e) young females caged with an adult n.a. female and also receiving a daily dose of 80 cc of male-soiled bedding. Twenty-five mice were used for each treatment group.

Results

Analysis-of-variance revealed that those female mice housed with males and those caged with adult n.a. females where male-soiled bedding was added had earlier mean ages for both vaginal introitus and first vaginal estrus (Table 4). Young females caged with adult n.a. females or receiving only the

TABLE 4

Mean Age at Vaginal Introitus and First Vaginal Estrus (± 1 SE) for Mice Under Five
Different Treatment Conditions N = 25/Treatment. Within a Vertical Column
Those Means Not Connected by the Same Vertical Line Are
Significantly Different at the .01 Level

Treatment	Age (days)	
	V.I.	1st E
Male present	25.5 (0.2)	28.0 (0.3)
n.a. Female present + male bedding added	25.3 (0.3)	28.1 (0.4)
n.a. Female present	26.1 (0.2)	31.3 (0.4)
Male bedding added	26.2 (0.3)	31.5 (0.6)
Control	28.3 (0.3)	35.9 (0.4)
F (4,120)	4.82	5.63
Prob.	$P < .01$	$P < .01$

male-soiled bedding were intermediate in age at sexual maturity, and control
females housed alone with no additional stimulation matured last.

DISCUSSION

The following conclusions can be made based on these three experiments: (1) The presence of an adult n.a. female mouse leads to earlier sexual maturation in immature female mice. (2) Adult n.a. females do not produce a sexual maturation-accelerating pheromone, even when they are given adult injections of testosterone propionate. (3) The heightened levels of activity associated with constant estrus in adult n.a. females or females given estradiol benzoate does not lead to earlier sexual maturation. (4) A major factor contributing to the acceleration of maturation is the contact stimulation between androgenized females and the immature mice. (5) The full acceleration of maturation produced by the presence of adult male mice may require at least two factors; pheromones and contact stimulation.

Previous investigators have demonstrated that the presence of an adult male mouse leads to accelerated sexual development in young females and that this phenomenon is mediated, in part, by a urinary pheromone (Vandenbergh, 1969; Vandenbergh and Colby, 1974). However, mice receiving daily samples of male-soiled bedding matured at an intermediate age, between control females housed alone and those females caged with an adult intact male. Some additional factor(s) must contribute to the earlier maturation of young females housed with a male.

When adult n.a. females, which show male-like sexual behavior patterns (Manning and McGill, 1974; Experiment II), are caged with young mice these immature females matured significantly earlier than singly caged control females, or immature females caged with normal intact adult females. Some aspect of the behavior of adult n.a. females must be responsible for this acceleration; they do not produce any maturation-accelerating pheromone(s).

Experiment II demonstrated that the amount and nature of contact stimulation, mounting and thrusting, provided by adult n.a. females did not differ significantly from stimulation provided by adult males. The final Experiment (III) demonstrated the additive effect of male pheromone and contact stimulation. Immature females caged with adult n.a. females and given a daily dose of male-soiled bedding matured at the same age as females caged with adult intact males. The presence of an adult male leading to accelerated sexual development of a young female mouse may result from both a pheromone and contact stimulation, or alternatively the presence of the adult male may somehow enhance the effect of the pheromone leading to earlier sexual maturation. It is impossible to differentiate these two hypotheses, because the second alternative cannot be tested, except possibly when the male pheromone has been isolated and identified chemically.

In Experiment I no significant differences were found in the overall body growth rates across all of the treatment groups. There were, however, significant differences in the mean weights of mice at vaginal introitus and first vaginal estrus: Mice that matured earlier weighed less at sexual maturity. Conversely, mice that attained first estrus at older ages also weighed significantly more at the age of maturation. Studies on the separation of sexual and morphological development in rodents and humans have produced conflicting results. Vandenbergh (1967, 1969), Vandenbergh et al. (1972), and Drickamer (1974) with mice, Orbach and Kling (1966) with rats, and Damon et al. (1969) with humans reported data suggesting a separation of physical growth and sexual maturation processes. In contrast Montiero and Falconer (1966) with mice, Kennedy and Mitra (1963) with rats, and Frisch and Revelle (1969) and Frisch, Revelle, and Cook (1970) with humans reported data supporting a critical body weight hypothesis; attainment of a certain body weight triggers metabolic and hormonal changes within the organism leading to sexual maturation (see Ramirez, 1971). Results of the present experiments support the hypothesis that physical body growth processes and the mechanisms triggering sexual maturation are, at least in part, under separate control.

For both humans and laboratory mice social factors and nutrition are important factors influencing the trend toward earlier maturation of females (Tanner, 1962, 1966; Zacharias and Wurtman, 1969; Damon et al., 1969; Vandenbergh, 1967; Kennedy and Brown, 1970; Vandenbergh et al., 1972). Acceleratory and inhibitory influences on sexual maturation in mice are mediated, in part, by pheromones (Vandenbergh, 1969; Drickamer, 1974; Fullerton and Cowley, 1971; Cowley and Wise, 1972). Studies of delayed sexual maturation of females housed in groups and the present study on acceleration of maturation produced by the presence of adult n.a. females indicate that contact stimulation is important to the process of sexual development. Studies of rodent populations have shown that at high effective densities, where females are exposed to a complex set of stimuli, sexual development is retarded. Studies must now focus on the simultaneous effects of contact stimulation and pheromones under conditions combining varied densities of female mice and different levels of acceleratory stimulation.

ACKNOWLEDGMENTS

This research was supported in part by NIMH Grant 24483-01 and by Sloan Discretionary Funds from Williams College. I thank Dr. T. E. McGill for reading the manuscript, James Adams and Roslyn Poznansky for technical assistance, and the Schering Corp., Bloomfield, NJ., for supplying the hormones.

REFERENCES

Barraclough, C. A. (1971). Hormones and the ontogenesis of pituitary regulating mechanisms. *In* C. H. Sawyer and R. A. Gorski (Eds.), "Steroid Hormones and Brain Function," Berkeley, Univ. of California Press.

Castro, B. M. (1967). Age of puberty in female mice: Relationship to population density and presence of adult males. *An. Acad. Brasil Cienc.* 39, 289-302.

Colby, D. R., and Vandenbergh, J. G. (1974). Regulatory effects of urinary pheromones on pheromones on puberty in the mouse. *Biol. Reprod.* (in press).

Cowley, J. J., and Wise, D. R. (1972). Some effects of mouse urine on neonatal growth and reproduction. *Anim. Behav.* 20, 499-506.

Damon, A., Damon, S. T., Reid, R. B., and Valadian, I. (1969). Age at menarche of mothers and daughters, with a note on accuracy of recall. *Hum. Biol.* 41, 161-174.

Drickamer, L. C. (1974). Sexual maturation of female house mice: Social inhibition. *Develop. Psychobiol.* (in press).

Frisch, R. E., and Revelle, R. (1970). Height and weight at menarche and a hypothesis of critical body weights and adolescent events. *Science* 169, 397-398.

Frisch, R. E., Revelle, R., and Cook, S. (1971). Height, weight and age at menarche and the 'critical body weight' hypothesis. *Science* 174, 1148-1149.

Fullerton, C., and Cowley, J. J. (1971). The differential effect of the presence of adult male and female mice on the growth and development of the young. *J. Genet. Psychol.* 119, 89-98.

Kennedy, G. E., and Mitra, J. (1963). Body-weight and food intake as initiating factors for puberty in the rat. *J. Physiol. (London)* 166, 408-413.

Kennedy, J. M., and Brown, K. (1970). Effects of male odor during infancy on the maturation, behavior and reproduction of female mice. *Develop. Psychobiol.* 3, 179-189.

McGill, T. E., and Tucker, G. R. (1964). Genotype and sex drive in intact and in castrated mice. *Science* 145, 514-515.

McGill, T. E., and Haynes, C. M. (1973). Heterozygosity and ejaculatory reflex after castration in male mice. *J. Comp. Physiol. Psychol.* 84, 423-429.

Manning, A., and McGill, T. E. (1974). Early androgen and sexual behavior in female house mice. *Horm. Behav.* 5, 19-31.

Montiero, L. S., and Falconer, D. S. (1966). Compensatory growth and sexual maturity in mice. *Anim. Prod.* 8, 179-193.

Orbach, J., and Kling, A. (1966). Effect of sensory deprivation on onset of puberty, mating, fertility and gonadal weights in rats. *Brain Res.* (Osaka) 3, 141-150.

Ramirez, V. D. (1971). Sex and brain-pituitary function at puberty. *In* C. H. Sawyer and R. A. Gorski (Eds.), "Steroid Hormones and Brain Function," Berkeley, Univ. of California Press.

Sachs, B. D., Poliak, E. L., Krieger, M. S., and Barfield, R. J. (1973). Sexual behavior: Normal male patterning in androgenized female rats. *Science* 181, 770-772.

Tanner, J. M. (1962). "Growth at Adolescence," Oxford, Blackwell.

Tanner, J. M. (1966). The trend towards earlier physical maturation. *In* J. E. Meade and A. S. Parkes (Eds.), "Biological Aspects of Social Problems," Edinburgh, Oliver and Boyd.

Vandenbergh, J. G. (1967). Effect of the presence of a male on the sexual maturation of female mice. *Endocrinology* 81, 345-348.

Vandenbergh, J. G. (1969). Male odor accelerates female sexual maturation in mice. *Endocrinology* 84, 658-660.

Vandenbergh, J. G., Drickamer, L. C., and Colby, D. R. (1972). Social and dietary factors in the sexual maturation of female mice. *J. Reprod. Fert.* 28, 397-405.

Zacharias, L., and Wurtman, R. J. (1969). Age at menarche, Genetic and environmental influences. *N. Engl. J. Med.* 280, 868.

26
Critical Periods in the Organization of Systems

J. P. SCOTT
JOHN M. STEWART
VICTOR J. DE GHETT
Bowling Green State University
Bowling Green, Ohio

The general theory of critical periods applies to organizational processes involved in the development of any living system on any level of organization and states that the time during which an organizational process is proceeding most rapidly is the time when the process may most easily be altered or modified. Complex organizational processes involving 2 or more interdependent subprocesses may show 1 to several critical periods, depending on the time relationships of the subprocesses. The nature of the relationships between interdependent processes operating on different levels is again dependent on time and is a more meaningful formulation than that of the old "innate-acquired" dichotomy. These theoretical considerations lead to the conclusion that understanding a critical-period phenomenon rests on analyzing the nature of the organizational process or processes involved. An example is given in a review of research on the critical period for primary socialization (social attachment) in the dog. Evidence that attachment has taken place consists of discriminative behavior in relation to familiar and unfamiliar objects and rests on a minimum of 3 processes. 1) organization of the separation distress response; 2) visual and auditory sensory capacities; and 3) long-term associative memory capacities. Once these capacities are developed, the overall attachment process proceeds very rapidly (the critical period for such attachment). Thus the critical periods for the organizational subprocesses precede or slightly overlap that for the overall process. Deeper analyses of these processes must rest on neurophysiological research. The theory of critical periods is a general one that should apply to any developmental organizational process which proceeds at grossly different rates at different times.

Received for publication 24 January 1974

Revised for publication 29 April 1974

Developmental Psychobiology, 7(6): 489-513 (1974)

General Theory of Critical Periods

As presently understood, the theory of critical periods is a general theory of organization. Therefore, it is applicable to a wide variety of organizational phenomena, ranging from embryonic development to the formation of organizations between nations. The only exceptions are those based on the nature of specific organizational processes—those that are continuous and proceed at uniform rates—rather than on the entities which are being organized. Such exceptions are relatively rare and, therefore, one may safely assume that any new or unknown organizational process has a high probability of showing a critical period. The theory of critical periods is thus related to and should be compatible with other theories of organization.

Concept of Systems

The theory of organization largely depends on the concept of systems, which is an old idea in biology, and one so familiar and pervasive that most biologists rarely stop to think about it in the abstract but simply employ the concept in most of their work and ideas.

In its simplest definition, a system is a group of interacting entities. These entities may be anything from planets to people, but they must interact in the true sense: the activities of each affect those of the others. They do not interact in the sense of independent action on a common object, nor in the statistical sense of 1 entity affecting the activity of another but not being affected in return.

A group of entities might act on each other without being organized. An example would be a group of gas molecules striking each other within a chamber. The degree to which a group is organized can be measured by the probability that a given activity on the part of 1 entity in the system will be followed by a given response by other entities. Thus systems can range between being organized to a very low degree, with predictability approaching zero, to systems which are highly organized, with predictability approaching one.

A system is different from a mechanism in that while the action of a mechanism may be highly predictable, its definition does not require that the entity that is affected will in turn have an effect on the machine, or upon the entity which set the mechanism in motion. A mechanism may, however, be part of a system. Thus a mechanistic concept of behavior produces quite different results from a systemic one, but the 2 concepts are not incompatible.

Furthermore, the conventional concept of causality becomes difficult to apply to systems as wholes, in that the usual experimental models assume only that A affects B, and not that B will affect A in turn. The latter phenomenon has led to the concept of equilibrium, which has been applied in many different sorts of systems, including nonliving chemical reactions in which reactions are assumed to go in both directions, and on a grander scale in the ideas of homeostasis in individual living systems and the balance of nature with respect to ecosystems.

We, however, are interested primarily in living systems and in the phenomenon of development, which may be defined as a change in the degree of organization of a system. When we relate such organizational change to a time scale, we enter the particular aspect of the organization of systems to which the concept of critical periods applies.

226

Theory of Critical Periods

In its simplest and most general form, the theory of critical periods states that the organization of a system is most easily modified during the time in development when organization is proceeding most rapidly. This is essentially a descriptive statement of a phenomenon that has been empirically established in relation to a wide variety of organizational processes. Whether or not a critical period exists depends upon the nature of the organizational process. If an organizational process proceeds at a uniform rate during the life or existence of the system involved no critical period exists, as the process can be modified as easily at one time as another (Fig. 1A). At the opposite extreme, an organizational process may proceed rapidly at only 1 limited period and is completely halted either before or afterwards (Fig. 1B). This is, of course, the most extreme example of a critical period, i.e., only 1 limited period during which organization may be modified. In between these 2 extremes is what is probably a more common condition, one in which a process proceeds slowly at first, then more rapidly for a time, and then slows down again but never returns to zero (Fig. 1C). An example is the process of social attachment. Still another possibility is that an organizational process may proceed intermittently, with periods of rapid organization, separated by periods in which little or no organization takes place (Fig. 1D). Such a process has many critical periods, and a good example is the process of learning.

Critical Periods in Complex Processes

Because many organizational processes are complex, we need to consider the theoretical implications of combining 2 or more simple processes. The least-complicated

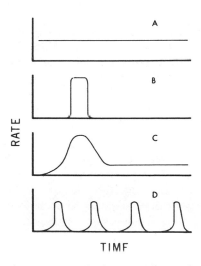

Fig. 1. Organizing processes and critical periods. A: Uniform rate of organization; no critical period. B: High rate of organization during limited time period; precisely defined critical period. C: Process operates at a high rate at one period; at a much reduced level later. D: Intermittent operation of process; repeated critical periods.

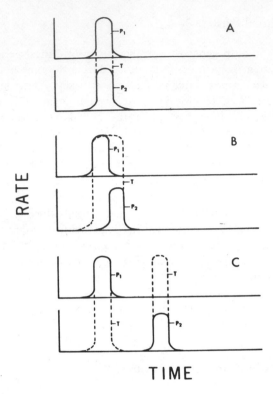

Fig. 2. Critical periods in relation to complex organizing processes. A: Processes P_1 and P_2 coincide in time. T, the critical period for combined processes, coincides with both. B: Processes P_1 and P_2 are immediately consecutive, extending T. C: Processes P_1 and P_2 operate at 2 different times producing 2 critical periods for combined process T.

model is that of 2 interdependent processes, each having a single critical period. Figure 2 shows how these may be related with respect to time. The assumption is that both component processes are necessary to the completion of the total process.

In Figure 2A, processes P_1 and P_2 have critical periods occurring at the same time. Modification of either process during the critical period produces modification of the total process, and the critical period for the total process will of course be identical with that of both subprocesses. Depending on their natures, each subprocess might be affected by different factors. In such a case, a combination of these factors acting at the same time will produce more severe effects. In general, when critical periods coincide, to demonstrate experimentally that 2 subprocesses do, indeed, exist is difficult, as any result could be explained equally well as an additive effect of 2 factors on a single process. The demonstration of 2 processes therefore depends on some form of independent evidence, such as that provided by genetic variation in 1 process but not the other.

Figure 2B represents the case of interdependent processes whose critical periods immediately succeed each other in time. Such is the case where the completion of 1 process is necessary before the 2nd can begin. The critical period for the entire complex process therefore includes both the critical periods of the component processes. If the 2 processes are affected by similar modifying factors, the main effect is to extend the critical period, and simple experimental techniques will not reveal the existence of the

2 processes. If, on the other hand, the 2 are affected by different factors, 2 critical periods will appear experimentally, 1 immediately succeeding the other.

Finally, in Figure 2C, if processes P_1 and P_2 are separated in time, the net effect is 2 separate critical periods. If the subprocesses are affected by different modifying factors, the situation is not changed. Of the 3 cases, this is obviously the one in which the presence of 2 subprocesses can be most easily demonstrated by the technique of experimental modification. Note, however, that Figure 2C superficially resembles Figure 1D, representing repeated critical periods in a single intermittent process.

We can extend the above reasoning to complex processes involving 3, 4, or more processes, anticipating nothing more than greater complexity of results. The larger the number of processes, the more likely the critical periods of each will overlap, and if a sufficient number of such processes exist, the probability is good that all periods in development will be critical for something. This essentially is the case if we consider the development of an organism as a single process, the view that Schneirla and Rosenblatt (1963) advanced in their critique of the theory of critical periods. It also approaches the situation depicted in Figure 1A, although it is unlikely that even the inclusive process of all development proceeds at a uniform rate. This is to say that the theory of critical periods is not inconsistent with the inclusive or holistic approach to development, but rather that each has its own validity. On a scale of complexity of organizational processes, the critical-period theory applies best to simple processes.

It is therefore apparent that the usefulness of the *theory* of critical periods (apart from the *phenomenon* of critical periods, which has enormous practical utility) depends on discovering the nature of organizational processes, particularly in the case of complex organizational phenomena which can only be understood in terms of their component parts. Hypotheses are the lifeblood of research, and we hope that the present theoretical formulation will inject new life into the field of early experience.

Destruction of Organization in Relation to Critical Periods

In most cases, the interest in critical periods is derived from the potential modification of organization. However, we may also be interested in the complete destruction of organization, if only to be able to explain the naturally occurring cases of this phenomenon. Figure 3 illustrates the effect of applying a strongly destructive factor at 3 points in time: before, during, and after a critical period. Supposing that this destructive factor affects only the particular organization studied, we predict that prior to the critical period, when the organization does not yet exist, there will be no effect; that during the critical period there will be a partial effect; and that at any time afterward there will be a complete effect. For the result of destruction, then, there is no critical period (unless one wishes to consider the entire life following the cessation of the organizational process as such), but rather a *critical point* (C) beyond which the effect is absolute.

This theoretical result is consistent with the empirical experimental results of destructive lesions of the nervous system. Done early enough in development, such lesions have little or no effect, somewhat later there may be partial recovery, and still later there may be no recovery.

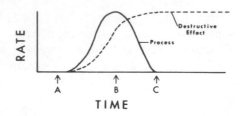

Fig. 3. Destructive effect in relation to timing of an organizational process. The same destructive agent is ineffective at time "A," partially effective at "B," and completely effective at "C."

A complicating factor in the above formulation is differential resistance to destructive agents. A general property of living systems is the tendency to maintain organization under a great variety of potentially disrupting conditions, with the result that they are most susceptible to destructive agents during the process of organization. A mildly destructive agent therefore demonstrates the critical-period phenomenon in a way no different from that of modifying factors.

From a general viewpoint, the young developing organism should therefore be more or less resilient with respect to traumatic injury, depending on the timing of the organizing process or processes involved. Contrary to some of the early theories as to the effects of early experience, one would anticipate that in certain respects an infantile organism is less susceptible to traumatic injury than an adult organism because organizing processes are not yet completed. In particular cases, this depends on the nature of the organizational process involved and whether or not it has gone to completion and so was in fact destroyed. These considerations bring up the further point that any modifying factor (as opposed to a destructive one) should be more effective if applied throughout the entire organizational process (critical period), rather than at only 1 point during the period.

Relationship Between Higher and Lower Levels of Organizational Processes

Until now we have considered only those complex processes which operate on the same level, but any complete consideration of such processes must consider the possible interrelationships between processes operating on different levels of organization. Table 1 is a simplified representation of the principal organizational processes that take place at different levels, and lists processes acting, respectively, on the intraorganismal level, the

TABLE 1. *Relationships Between Organizational Processes On Different Levels.*

Level of Organization	Organizational Process
Social	Formation of Social Relationships (including Attachment Process)
Organismal	Learning
Intraorganismal	Growth and Differentiation

organismal one, and on the social level. Each level is in part determined by that below it, but not completely so.

Of particular interest is the relationship between intraorganismal factors that are ordinarily classed as physiological with those above it, since any aspect of behavior must have a physiological basis. Such physiological processes fall into at least 3 classes: (a) those which maintain organization but are not themselves organizational, as for example metabolism, nutrition, and blood circulation; (b) those which form the physiological bases for organizational processes on higher levels, e.g., the physiological basis of learning; and, (c) those which are themselves organizational, such as growth and differentiation. Considered from the viewpoint of development of an organism, organizational physiological processes (c) precede maintenance processes (a), which in turn precede physiological processes basic to higher organizational processes (b). However, there may be considerable overlap in time in all 3 of these processes. Just how extensive this overlap may be is illustrated in Figure 4, which strikingly illustrates the artificiality of dichotomizing development into "innate" and "acquired" organizational processes on the basis of a single event of birth or hatching, and suggests that this old problem should be abandoned in favor of a more meaningful approach.

Because organizational processes on different levels obviously proceed simultaneously (Fig. 4), a fascinating and fundamental problem arises: what are the relationships between organizing processes at different levels? To begin with, it is well known that organizational processes on a higher level may influence physiological maintenance processes on a lower level, an example being the influence of learning on the process of salivation, as was originally shown by Pavlov. The extent to which higher level processes can influence physiological organizational processes is an open one, but it is at least possible for learning to influence growth through the organization of feeding behavior. Conversely, it is obvious that the outcome of lower level processes can influence the operation of upper level processes, if in no other way than by affecting the capacity to exercise higher level processes.

The effect of a lower level process on an upper one is related to time. If there is an overlap in time, both lower and upper level processes modifying the lower level may have more effect at an early stage than at a time close to the final stage of organization. Once a lower level process has achieved its final stage of stable organization, the effect is to limit processes on the higher level. We may therefore conclude that the relationship between lower and higher level processes depends on the stability of organization achieved in the lower level. If it is unstable, it will produce corresponding instability in the higher processes. If higher processes can influence the lower ones, they could theoretically produce instability in them, thus leading to a circular situation in which organization never proceeded to a final stage of stability.

Further, if a lower level process is unstable, this creates the possibility that it may be in the process of being organized by a higher level process. If this in turn proceeded to completion, the lower level organizational process might thus achieve stability in a different fashion.

It follows also that higher level processes can affect organization at lower levels only if these organizational processes overlap in time. There is now abundant evidence that this actually does occur in the development of learning processes in the higher organisms, i.e., the capacity for learning appears well before final physiological organization of the nervous system takes place, and long before physiological growth and maturation are

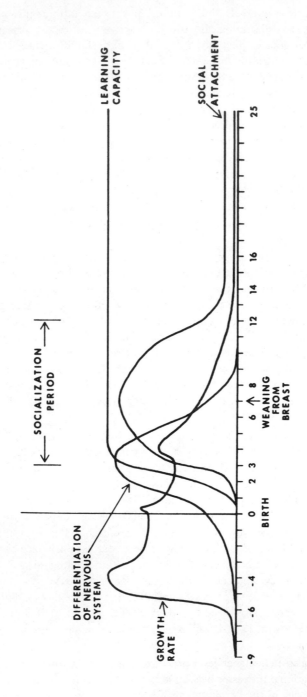

Fig. 4. Time relationships of certain developmental organizational processes in the dog. Processes are indicated semidiagram-matically in terms of weeks prior to and after birth.

completed (see Fig. 4). Further, if a lower level process does not run to completion, a higher level process may impose a continuous effect on organizing processes at lower levels.

In attempting to summarize these relationships, we can say that in general there is a hierarchical relationship between higher and lower level processes, in the sense that higher level processes can impose organization on the lower ones but the latter can only influence higher level organization by providing basic capacities, but not by controlling the organization itself.

Incidentally, one of the consequences of these theoretical findings is that they bring about a reconsideration of the process of adaptation. Organizing processes tend to proceed in the direction of stability, but effective adaptation to a changing environment demands flexibility. We would therefore expect that processes might be evolved that would tend to inhibit the completion of organization to a final steady state. For example, in the process of learning there is a tendency to organize behavior in fixed and stereotyped habit patterns. At the same, time, there appears to be in higher organisms a counterprocess which leads to variation in behavior, and in any particular case the end result is a balance between these processes. Fixed and completely stable organization of behavior is adaptive only in cases where the environment is completely stable.

Analysis of the Critical Period for Primary Socialization in the Dog: an Example

From our theoretical introduction it follows that a basic empirical problem in the analysis of any critical-period phenomenon is determining the nature of the organizational process or processes involved. As a general strategy, a 1st approach is to try to discover the process or processes that appear at the beginning and disappear at the end of the period. In order to do this, the boundaries of the period need to be determined as precisely as possible by experimentally varying a factor which is known to modify the organization of behavior during the critical period. Once the relevenat processes are determined, they can be experimentally analyzed on an individual basis.

Discovery of the Critical Period for Primary Socialization in the Dog

Ordinarily, the discovery of such a period can take place in 2 ways: either by detailed observation of behavioral development or by accidental or deliberate administration of an experimental factor that modifies organization. An example of the latter sort of technique was the discovery of the critical period for early stimulation in rats, which came about as the result of systematic study of the effects of handling on later behavior (Denenberg & Zarrow, 1971; Levine & Lewis, 1959; Schaefer, 1968).

In the case of the dogs, both methods contributed to the discovery. Originally, we did a detailed observational study of behavioral development involving some 500 puppies from 5 pure breeds and hybrids between 2 of them (Scott & Fuller, 1965; Scott & Marston, 1950). As a result of classifying this material, we concluded that the early development of the puppy naturally fell into 3 periods marked by different organizational processes. The *neonatal* period, extending from birth until the opening of the eyes at approximately 2 weeks of age, is chiefly marked by the establishment of the process of neonatal nutrition. All behavior at this time is adaptive to an existence in which the mother takes complete care of the pups. The 2nd or *transition* period, extending from approximately 2-3 weeks (more precisely, from 13 to 19 days on the average), extends from the opening of the eyes to the appearance of the startle response to sound. This period is marked by several transition processes: that from neonatal to adult forms of locomotion (crawling to walking and running), from neonatal nutrition to the adult form (nursing to eating of solid food), from relatively slow-learning capacities to the rapid capacities of adults, the acquisition of adult sensory capacities of sight and vision, and finally the acquisition of the capacity to rapidly form attachments to strange individuals, whether they be dogs, people, or other living animals. We later realized that the puppies also develop a capacity to become attached to particular places, paralleling that of attachment to living things.

The 3rd period, known as the period of *primary socialization*, is characterized by the process of forming attachments to conspecifics, other animals including humans, and to particular places. On the basis of observation, we set the end of this period at the time of final weaning from the breast, which takes place on the average at 7-10 weeks, on the logical grounds that weaning marked a strong change in social relationships.

Separation experiments provided a method of modifying the normal processes of social attachment and site attachment. Results from these tended to verify conclusions from observational studies. For example, in 1 early experiment (Scott, Fuller, & Fredericson, 1951) a puppy taken from the mother at 2 weeks (1 week prior to the beginning of the socialization period) readily adjusted to human care, but 1 taken at 4 weeks showed violent emotional reactions upon separation. On the basis of this and other preliminary experiments, a more extensive experiment was designed.

Experiments Determining the End of the Critical Period

Carrying out this design, Freedman, King, and Elliot (1961) performed the most elegant and extensive experimental test of the critical-period hypothesis that has so far been carried out. They placed pregnant mothers in 1-acre fields with high board fences and allowed them to give birth to their pups and rear them without any direct human contact, except when the experimenters entered briefly to remove 1 of the pups. In a given litter 1 pup would be taken out when 2 weeks old and given opportunity for socialization with human beings during the following week. Subsequent puppies were removed at 3, 5, 7, and 9 weeks and given similar treatment. Finally, the whole litter was brought in and tested for individual responses to human beings at 14 weeks of age.

The results were unequivocal (Fig. 5). Puppies taken at 2 and 3 weeks were relatively unresponsive to human contacts and performed little better than untreated controls at 14 weeks. Those taken at 5 and 7 weeks were highly responsive and also showed maximum levels of performance at 14 weeks. Those taken at 9 weeks were intermediate in both respects. Puppies that were not brought in until 14 weeks showed severe and permanent deficits in their behavior, being extremely fearful in initial contacts, never forming close attachments to human beings, and always showing preferences for dogs. The results with the 9-week-old puppies were unexpected in that they adjusted to human contact effectively but more slowly than younger ones.

It was obvious that the end of the critical period could not be set at 8 weeks, and that it should be placed somewhere between 9 and 14 weeks. Had we anticipated this result we could have tested additional groups of puppies at later ages. As it was, other

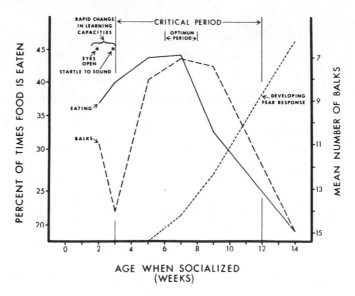

Fig. 5. Nature of the critical period for primary socialization (primary attachment) in the dog. Scores for eating and balks are inversely related, as indicated by the reversed scale on the right.

experiments gave us additional clues as to the duration of the critical period. In an entirely different sort of experiment, Fuller (1967) raised puppies in complete isolation from 3 weeks on, i.e., from the onset of the critical period. Puppies left in isolation until 16 weeks showed drastic disturbances of behavior on being removed, again of such severity that recovery was never complete. Puppies removed as late as 7 weeks, however, appeared to be completely normal in their later behavior. Those taken out at later ages showed increasing distortions of behavior. Scott and Fuller (1965) concluded that the peak of the socialization process occurred at approximately 7 weeks, and that its end should be placed considerably later.

Stanley (1965) did a somewhat similar experiment, except that he raised puppies apart from human beings in small rooms. Before 9 weeks, such puppies immediately approached human experimenters without hesitation. At 9 weeks they showed obvious fearful reactions when exposed to people for the 1st time. Data on guide dogs retained in kennels until 12 weeks or longer before being placed in homes indicated that the probability of failure in later training increased after 12 weeks (Pfaffenberger & Scott, 1959). Somewhat arbitrarily, then, Scott (1970) placed the end of the critical period at approximately 12 weeks and concluded that a major limiting factor was the development of a fear response to the strange, which would have the effect of limiting any contacts with new individuals to very brief periods and thus interfere with the process of attachment.

Waller and Fuller (1961) did the only experiment in which the fear response was manipulated. They kept puppies in isolation from 3 weeks onward under constant treatment with chlorpromazine, which has the effect of decreasing anxiety in human beings and therefore might be presumed to control fear responses. The assumption was that this would prevent the development of fear responses. However, such puppies, when taken off drugs and removed from isolation at 16 weeks, were no different from controls, leading to the conclusion that the capacities for the fear response develop even when it is not evoked. Fuller (1967) later found that if he gave puppies chlorpromazine at the time of emergence it did have some beneficial effect, although it did not produce completely normal puppies. Thus the developing fear response is 1 factor that brings about the end of the critical period, but not the only one.

Subsequent observations on puppies that had undergone isolation showed that they could, given sufficient time, develop attachments at later periods. Woolpy and Ginsburg's (1967) experiments on socializing wild-caught adult wolves lead to a similar conclusion. If daily contact with a human experimenter is enforced on such a wolf over a period of several weeks, its initial reaction is one of extreme fear, persisting for many contacts. Eventually, the wolf will overcome this, make positive contacts, and form an attachment, albeit not one of the same kind that is formed by a young wolf cub during the critical period. The speed of attachment is much slower, covering a period of weeks rather than hours.

We now believe that the canine attachment process which is, of course, a process of organizing a social relationship, can take place at any time in life subsequent to 3 weeks, provided there is not too much interference. There are at least 4 interfering factors. The 1st of these is the developing capacity for fear responses to the strange, which becomes increasingly important after 7 weeks and reaches a maximum by 14 weeks. A 2nd factor is the separation reaction itself. An animal that has already developed a strong attachment to other individuals or places will show an acute and long-lasting emotional distress reaction when separated from them. Under these conditions formation of new attachments is impossible, unless this separation reaction can be relieved in some way. A 3rd factor is interference from previously formed attachments. If an animal is attached to particular individuals or places he tends to stay with these, and thus not to have extensive contacts with new individuals and places unless forcibly kept under such conditions. A 4th factor is the development at sexual maturity of territorial defense reactions, including a tendency to attack strange individuals. This, of course, inhibits both the process of attachment between adult dogs and that between adult dogs and strange people. The first 3 of these interfering or limiting factors become increasingly strong after 7 weeks and have parallel effects that may in combination strongly limit the process of social attachment after 12 weeks.

Optimum Period for Transfer of Relationships

One of the outcomes of this research is that we have been able to define with considerable exactitude the critical period of primary socialization as extending from 3 until 12 weeks of age, and to recommend to prospective dog owners that the optimum period for achieving attachment of a puppy to a human master is somewhere between 6 and 8 weeks of age. A puppy will still do reasonably well up until 12 weeks, but as it grows older it becomes increasingly difficult to control and train, as well as not becoming attached as rapidly.

An interesting additional bit of evidence is the discovery by Theberge and Pimlott (1969) that wolves normally move their litters from the den to another location at about 2 months of age, the same age that we have defined as being an optimal period for making the change from dog to human society. This new area, called a rendezvous site, is usually near a small lake. Here, the litter is left to explore the environment and rest while the pack goes off to hunt.

Also, Woolpy and Ginsberg (1967) have found that wolf cubs that socialized during the critical period and later returned to a zoo pack acted like wild animals as adults. This suggests that there may be a 2nd critical period when the attachment becomes "nailed down," so to speak. This possibility should also be explored in dogs, but does not invalidate the existence of an early critical period when attachment is easy to initiate.

With respect to the original problem of determining the nature of the organizing processes in social attachment, these experiments were somewhat disappointing. Rather than finding a process that ceased with the end of the critical period, we found only a set of interfering factors that limited it. We therefore concluded that more definite clues might be found near the beginning of the critical period.

Experimental Evidence Regarding the Onset of the Critical Period

We may now consider the organizational processes accompanying the onset of the critical period. First, there is the organization of the separation distress response. Fredericson [see Scott et al. (1951)] first studied the phenomenon and discovered that a puppy 2 weeks of age made a rapid adjustment to being removed from a litter and reared in a home; a similar puppy, taken at 4 weeks of age, vocalized for 24 hr without stopping, showing that strong attachments had already been formed. Later, Elliot and Scott (1961) showed that puppies separated from their familiar social companions and placed alone either in the home room or in a strange room otherwise identical with their own would vocalize at high rates as early as 3 weeks of age.

Second, there are gross changes in learning capacity. In studying the development of learning, Fuller, Easler, and Banks (1950) found that pronounced qualitative changes in the ability to respond to classical conditioning techniques took place around 18-19 days of age. After this age, a puppy could form associations as rapidly as an adult. Scott and Marston (1950) then hypothesized that this change in learning capacities accounted for the onset of the process of attachment, i.e., that the younger puppy simply could not remember that he had seen or had contact with a particular individual before. Scott and Marston also stated, on the bases of Fuller's experiment and certain observational data, that puppies appeared to learn little or nothing before this point in development, and hypothesized that the capacities for simple associative learning developed suddenly at this point. Fuller and Christake (1959) challenged this statement and showed that puppies could form associations prior to this age, albeit at a slower rate. Stanley (1970) also challenged the hypothesis of nonlearning in the neonatal puppy and, in a brilliant and careful series of experiments, was able to show that the puppies could be conditioned even very early in the neonatal period. Stanley further showed that the model of operant conditioning was more appropriate than classical techniques in studying sucking, which is 1 of the few kinds of behavior that the neonatal puppy emits, and hence that can be modified by operant learning.

However, it is still true that gross quantitative changes in the capacity for learning do take place during the transition period. Russian experimenters (Klyavina, Kobakova, Stelmakh, & Troshikhin, 1958) have shown that the kind of behavior involved determines the rate at which classical conditioning takes place. The number of pairings to criterion is much lower with respect to feeding behavior than escape behavior. In addition, there are certain qualitative changes which take place. Bacon (1971) has shown that although neonatal puppies can learn to make discriminations, their behavior lacks 1 of the elements of such behavior in adults. The puppies form positive associations in respect to discriminated stimuli, but not negative ones. Finally, 1 result obtained by Fuller and Christake (1959) has never been followed up. They hypothesized that conditioned responses formed through the autonomic system might develop earlier than those developed through voluntary responses (a reasonable assumption based on the commonly held belief that learned emotional responses are important in early development), but obtained quite the opposite results, being unable to condition heart rates until 5 weeks or so, long after good conditioned responses in overt behavior can be obtained.

None of these experiments cover 1 of the most basic factual issues in this field. We still do not have good studies of developmental change in various learning capacities from birth to 3 weeks of age. These results, however, raise some very interesting questions with respect to the effects of experience on infantile animals. The data indicate that young puppies are protected against the consequences of traumatic experiences in early life, and that they do not have the ability to develop inhibitions as the result of such experiences. Furthermore, it has not been established how long memories acquired in the neonatal period last, and whether the information so stored can be retrieved at later ages. This is an extremely interesting theoretical and practical problem now being studied by Misanin, Nagy, Keiser, and Bowen (1971), who have found in another species (rats) that the memory capacity for retaining associations more than 1 hr does not develop until several days after birth. What the situation is in the dog is now being tested by Z. M. Nagy (unpublished data).

Nature of the Organizational Process

The organizational changes that take place in the nervous system as an animal becomes attached to another individual or place are, of course, physiological in nature. At the present time nothing is known directly about such a process or processes, and they must be inferred from behavioral evidence. As indicated above, 2 organizational processes may be inferred to accompany the onset of social attachment: organization of the separation response and that of learning. An obvious hypothesis is therefore that social attachment is the outcome of a process of learning, and various theories have been proposed in the past to account for it on this basis. We can now cite the following evidence concerning the relationship between the attachment and learning processes.

As to the nature of social attachment itself, we can conclude that all that is necessary for the process to take place is some form of perceptual contact. Early experiments by Brodbeck (1954) and W. C. Stanley (private communication) showed that feeding was unnecessary, as attachment would take place even if puppies were fed mechanically, provided they had contact with human beings. Cairns and Werboff (1967) showed that during the early part of the critical period, puppies would begin to show separation distress after 1-2 hr of contact with new animals and that direct interaction with physical contact produced a more rapid effect than mere visual contact. Stanley and Elliot (1962) showed that puppies would learn to run and show improvement in performance when the goal was a completely passive individual. We can conclude that although the attachment process may be slightly modified by external circumstances, it must be primarily an internal one. Even when puppies are punished for making contact with human experimenters, they still become attached (Fisher, 1955). Thus we have a learning process that is not dependent upon either positive or negative external reinforcement. What is it?

To set up the simplest possible hypothesis that will account for the results, the animal must have developed 2 capacities. One is the ability to discriminate between familiar and unfamiliar objects, which in turn depends on the development of adequate sensory capacities. The 2nd is the capacity for memory, i.e., perceiving that an object is familiar would depend on memory of the fact that it had been previously seen or sensed in some other way. Following this reasoning, we conclude that associative learning is an essential part of the attachment process. Repeated exposure to an object renders the association stronger, in accordance with the laws of learning.

It is also possible that the perception of a familiar object evokes a pleasurable emotion that might have a reinforcing function, but this has not been established. What is definite is that separation from a familiar individual or place will evoke distress. From the viewpoint of reinforcement, such distressing experiences should result in the puppy being internally punished for separation and rewarded for reunion. Repeated separations, then, should result in a strengthening of attachment. At the same time, repeated exposures to the familiar object, even when no separation is involved, should result in increasingly strong associations and memories of the familiar object. Thus learning processes should strengthen attachment in 2 ways, but do not themselves comprise the whole process.

Returning to the theoretical problem of the relationships between interdependent organizing processes, we see that the recognizable processes involved include development of: (1) distress vocalization (present at birth but not yet organized with respect to separation); (2) sensory capacities (tactile and olfactory capacities present at birth, visual capacities increasingly organized from 13 to 28 days after birth, and auditory capacities appearing at approximately 19 days); and (3) learning capacities (associative processes becoming increasingly organized from approximately Day 6 to Day 21).

There is obviously a great deal of overlap in the time of rapid organization of sensory and learning capacities, and their critical periods sould consequently overlap during the transition period. However, until both kinds of capacities become well organized, the overall process of social attachment cannot proceed rapidly. We therefore predict that social attachment takes place slowly prior to 3 weeks, but at a gradually rising rate. Thus the critical periods for the component processes take place immediately prior to that of the overall process. This is similar to the situation in Figure 2B, except that in this case the organization of component processes can be altered by quite different modifying factors from that of the overall attachment process, and hence be separately recognizable.

New Evidence from the Onset of the Critical Period

With these concepts of the nature of the organization process, it is possible to return to the problem of the factors which determine the onset of the critical period, and obtain more definite empirical evidence.

A newborn puppy can vocalize at fairly high rates, but does so in response to a variety of stimuli such as pain, cold, and hunger. The work of Jeddi (1970), confirming that of Igel and Calvin (1960), indicates that newborn puppies are attracted both by soft furry surfaces and by heat. The problem then is to devise an experiment which will test the factor of separation apart from physical discomfort. With this in mind, J. M. Stewart and V. J. DeGhett (unpublished data) performed the following experiment.

Beginning at 5 days of age, the separation responses of 4 litters of puppies of diverse genotypes were measured for 10 min under 2 conditions. In the 1st the mother and entire litter were removed from the nest box and home room, and 1 puppy was returned to the nest box. In the 2nd, a puppy was placed in another room in a duplicate nest box whose bottom was covered with a freshly laundered turkish towel and which was kept warm by an overhead lamp providing a temperature between 82° and 85°F (27.8° and 29.4°C). As Figure 6 shows, the responses in the home-room nest box were small and irregular at first; 3 out of 4 litters showed responses close to zero as late as 12 days of age. From 13 to 16 days the responses remained on a plateau and then gradually rose until 21 days, with considerable daily fluctuations. Because Day 13 is the average age at which puppies have their eyes completely open, and Day 19 is the average time at which puppies begin to respond to sounds with a startle response, we may conclude either: (1) that the onset of the capacity to form attachments is dependent upon the development of these 2 capacities, i.e., without vision and hearing the puppy does not become familiar with its litter mates, or more probably, (2) that attachment has taken place, but other sensory modalities do not supply reliable cues that separation has taken place. Vocalization in the strange area, which the puppies should have been able to discriminate from their home nest box by both olfactory and tactile senses from birth onward, indicated change as early as Day 8. Although 1 litter responded at a fairly high rate at 6 days, none of the litters showed more than very small scores on Day 7. Following this, the response rate showed a general trend upward through Day 21.

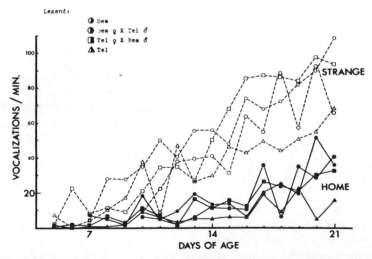

Fig. 6. Distress vocalization rates of puppies isolated under 2 conditions comparing beagles; Telomians, and their hybrids. Note that rates do not consistently rise until after 7 days of age.

One explanation of the above data is that certain puppies mature much more rapidly than others. Taking 50 vocalizations in a 10-min period as a barely minimum response that might indicate something more than accidental reactions to discomfort, we find that only 25-30% of the animals were so responding on Days 5, 6, and 7. This figure jumped to 60% on Day 8 and then gradually rose toward 100% on Day 21 (Fig. 7). The trend indicates that some maturational change has taken place at about Day 8.

If, on the other hand, we take the percentage of animals whose vocalizations are in excess of 800, which is approximately the average at 21 days, we find virtually no animals responding at this rate even as late as the 13th day. From 14 to 17 days the response rate is on a plateau from which it sharply rises through Day 21. The trend again supports the conclusion that the maturation of the visual and auditory senses greatly increases the capacity of the animal to discriminate between familiar and unfamiliar surroundings. The maturational change at 8 days coincides with changes in conditioning ability in Fuller and Christake's (1959) data, and thus may reflect an important change in learning capacities. It could also represent the 1st appearance of the separation distress response as an organized reflex.

However, other performance of individual animals was so inconsistent that we suspected that the data were confounded by accidental stimulation from other sources than separation, i.e., discomfort or lasting effects of previous stimulation. We have therefore attempted to discover a strange situation which will reliably keep neonatal puppies completely quiet. After considerable experimentation we have discovered that, in addition to temperature and contact with a soft surface, as described by Jeddi (1970), a 3rd factor is confinement. A puppy in the neonatal period placed in an incubator at 85°F (29.4°C), resting on a fresh plastic-covered baby's diaper and enclosed by a short length of galvanized stove pipe, will apparently be perfectly content, although it should be possible for it to discriminate by touch and odor the

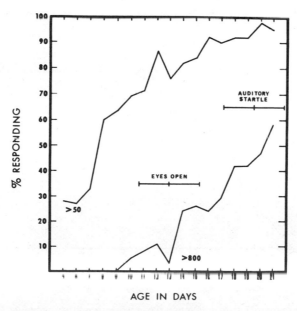

Fig. 7. Developmental changes in percentages of animals responding at minimum and average rates of distress vocalizations when isolated in a strange situation.

difference between this situation and being in the home pen with the mother and littermates. This technique should make it possible to determine more exactly when the puppy begins to be distressed by separation from familiar objects. We anticipate that the distress vocalization rates will in any case be rather low, as similar stimuli will somewhat alleviate the separation reactions of older puppies in the middle of the critical period.

Despite the above qualifications, we can conclude that the results fit the prediction based on our knowledge of the rates of organization of sensory and learning capacities: namely, gradually increasing attachment is evident prior to 3 weeks. Of the various capacities necessary to produce evidence of attachment, either the reflex emotional responses of distress vocalization to a separation or an improved capacity for associative learning, or both, may appear about Day 8. The appearance of visual and auditory capacities may increase the probability that the separation response will be evoked. Further refinement of these results must await further experiments.

One thing is clear: prior to 21 days of age, repeated daily 10-min exposures to strange situations is accompanied by rising rates of distress vocalization. After 3 weeks the same procedure results in a falling rate (Fig. 8). Whatever may be the explanation (and we have not been able to clearly establish that this latter effect is due to habituation to stimulation or to familiarization and the resulting attachment to the strange place), a marked change in behavior in this situation does occur at approximately 21 days of age. This adds further evidence to our original conclusion that 3 weeks should be considered the time of onset of the critical period.

We should remember, of course, that this is the approximate mean of the population, and that individual animals might vary in their development as much as 3 or 4 days on either side of this time. The experimental results of Freedman, King, and

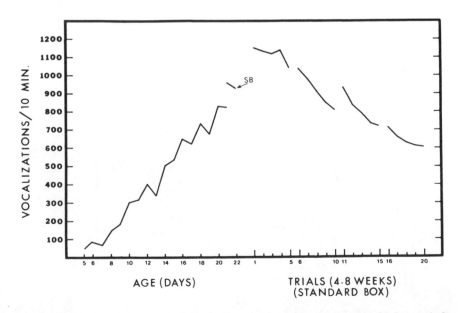

Fig. 8. Effects of repeated daily isolation in the same strange situation before 21 days and after 28 days of age (trials on weekends were omitted on older groups). (Note that the trend is upward in younger animals and downward in older ones.)

Elliot (1961) showed that puppies exposed to strange situations and contact with humans at 2 or 3 weeks of age revealed almost no gain in developing adequate relationships with humans at 14 weeks as compared to controls having no such experience. We conclude that some profoundly important changes in the organization of the brain of the dog take place at approximately 3 weeks. What the nature of these changes may be and how they can be modified is still to be discovered, and this should be a time for a neurophysiologist to study the developing brain of the dog.

Theoretical Controversies

Although we do not wish to emphasize history in this paper, and though many of the past controversies have died down spontaneously, we should not neglect history entirely. We shall therefore discuss some of the most important issues.

One Critical Period or Many?

On the senior author's rereading his early papers, he finds that he assumed that there was only 1 critical period in behavioral development, occurring quite early and being relatively short. Denenberg (1962) questioned this assumption and stated that there might be as many critical periods as there were combinations of independent and dependent variables.

However, at that time (1962), Denenberg did not consider the case of interdependent organizational processes. If the theoretical case illustrated in either Figure 2B or Figure 1C involved separate subprocesses that were differentially affected by different levels of stimulation, this would explain the sort of results obtained by Kline and Denenberg (quoted in Denenberg, 1964), who found that the timing of critical periods was apparently shifted by using different levels of electrical stimulation in their investigation of the effects of early stimulation on developing rat pups. In fact, Denenberg's (1964) Figure 5, showing the effect of 2 levels of shock on subsequent avoidance learning, is essentially identical with our Figure 2C.

Later, Denenberg (1968) made a theoretical analysis based on the assumption of a single unknown organizational process whose critical period is discovered by the systematic application of a single modifying factor. This model is essentially that employed in the discovery of a critical period for the effects of early handling in the rat. His model describes the various ways in which a critical period might be manifested in data and gives results essentially similar to those in our Figure 1, except that he did not consider the possibility of repeated critical periods (Fig. 1D).

Denenberg and Zarrow (1971) advanced the hypothesis that corticosterone modifies the organization of the neuroendocrine system of the developing rat in a critical period shortly after birth in much the same way that the male sex hormone modifies the eventual expression of fighting behavior in the same period in the mouse. In support of this hypothesis, Denenberg, Brumaghim, Haltmeyer, and Zarrow (1967) demonstrated that adrenal cortical activity can be markedly increased by stimulation at this age, but not a few days later. Thus, they made an important step toward identifying the physiological basis of the presumed organizing process, one which is consistent with the earlier work of Levine and Lewis (1959) on the effects of stress.

This hypothesis does not eliminate the possibility that other organizing processes may be affected; in fact, the term "infantile stimulation" covers a great variety of potential modifying factors, including handling, cold stress, shaking, and electric shock. These affect such diverse phenomena as physical growth and emotional reactions in the open-field situation; thus, many organizational processes must be involved.

The multiprocess explanation is not related to the problem of the effect of intensity of a modifying factor upon an organizing process. Modification should be proportional to intensity, but only within limits. Below a lower limit there should be no effect and, at least in certain cases, beyond an upper limit great intensity should produce destructive rather than modifying effects. As explained above, the time relationships should also vary. Actually, this problem is more important theoretically than practically, since the principal characteristic of a critical-period phenomenon is the ease with which effects can be produced during that period.

Further, from the theoretical formulations in the current paper, it is obvious that there can be as many critical periods as there are organizational processes, unless all of these processes are concentrated on one period. In relatively short-lived animals such as dogs, rats, and mice, organizational processes do tend to be bunched together early in development. Thus there was some justification for the first overenthusiastic hypothesis of "one critical period" (Scott & Marston, 1950), which followed a cardinal rule of research strategy, namely, to investigate the simplest possible hypothesis first. However, we can now state that it is an error to speak of *the* critical period, except in relation to a specific organizational process.

Relationship Between Learning and Critical Periods

Early in the history of the study of social attachments, a controversy arose because Lorenz (1937) stated that the process of imprinting in young birds was not the same as Pavlovian conditioning. In the work of Scott and Marston (1950), one of the reasons for setting the onset of the critical period for social attachment in dogs at 3 weeks was the association with a change in conditioning ability that occurred at about this time. They also hypothesized, on theoretical grounds and with no idea of what other organizational processes might be involved, that the time immediately after the onset of learning ability should be critical for a large number of phenomena that are organized by learning.

From the viewpoint of current theory of critical periods, learning itself is a group of organizational processes. In a typical learning sequence a young animal confronted with a problem will at first attempt many different solutions. Finding one that is successful, he quickly reduces his behavior to a stereotyped habit, and thereafter shows little or no improvement. Once the behavior is organized it is difficult to change. Any process of problem-solving behavior thus necessarily has a critical period in its early stages that determines whether or not there will be a successful solution, and if so, what its quality may be. Thus there is a strong theoretical reason for assuming that the time when a young animal first becomes capable of solving problems should be a critical period in development.

The process of social attachment is obviously a special case with respect to learning. There is no problem to be solved; in fact, the animal has to do nothing

244

except to notice its environment in order to form an attachment. As we have hypothesized the attachment process, learning should enter into it in 2 ways. First, the primary process of familiarization appears to be pure associative learning with no rewards involved. Second, once the separation distress reaction occurs (and its development is dependent on organizational processes of growth and differentiation in the nervous system) the behavior of the animal is negatively reinforced for separation, but this occurs internally rather than externally. Thus the overall process of organization in attachment is different from simple Pavlovian conditioning, but learning processes nevertheless form a part of the total organizational process. Primary socialization, or social attachment, is thus a complex organizational process.

Terminology

Certain minor issues have arisen concerning the use of the term critical period. The word "critical" comes from the same root as crisis, derived from a Greek word meaning to separate. Hence crisis and critical have as their most general meaning a turning point. In mathematics and physics a critical point is a point at which a change occurs, as for example the temperature at which a substance changes from a gas to a liquid. As pointed out above, this concept has some utility in behavioral development but in a quite different sense, in that time is involved.

In biology, the term *critical period* introduces the concept of time, which is not present in the concept as used in physics. The biological concept, however, is compatible: a critical period is still a time at which a change occurs. It is also a turning point or choice point between 2 or more paths of development, and hence is associated with the concept of canalization, i.e., a period in development during which the cell or other unit of organization may proceed along any of several paths, and beyond which it is more or less confined to 1 channel.

Any confusion concerning the definition of the concept of critical periods disappears when we apply it to the general theory of organization. A critical period is a time during an organizational process when a change in organization is easily produced. A tendency in some recent literature (Hinde, 1970, p. 564) has been to substitute the term sensitive period for critical period, meaning a time during which an environmental factor will easily produce a change in the organizational process of learning. We see no logical contradiction between these 2 terms. A critical period is also a sensitive period. We do, however, prefer the original term as a more general, sharper, and less fuzzy concept, and one which also emphasizes the practical importance of the phenomenon.

Moltz (1973) has raised the question of whether the term optimum period should be used synonymously with critical period. Again, from the viewpoint of the general theory of organization, the time at which an organizational process is proceeding most rapidly should be the optimum time for developing an organization most completely, identical with the critical period as defined above. It seems desirable, however, to keep the terminology concerning the outcome of an organizing process distinct from the concept of the process itself and its timing.

A Complex or Simple Theory?

It has been alleged that the critical period theory is overly simple and, consequently, that it contributes little to our understanding of behavior and leads to no deeper research. Although the observational description of a critical period is quite simple (and this is one of the practical strengths of the concept), it leads immediately into the study of the nature of organizing processes and eventually to the discovery of how such processes may be manipulated. As this is done, the theory becomes more and more complex, especially where several different organizing processes are integrated with each other, or in cases where related organizing processes overlap in time.

Conclusion

The theory of critical periods is an easy concept to use, both theoretically and practically. It is a tremendously useful practical tool that can be applied to organizational phenomena on any level, from the genetic through to the ecological. Organizational processes are ordinarily easy to identify by observation, and any such process that proceeds at differential rates in time should have a critical period or periods. As a matter of research strategy, anyone confronted with a practical or theoretical problem should first ask whether or not this involves an organizational process. If the answer is "yes," he should consider the possible application of the theory of critical periods.

The theory can be used in a predictive sense also, provided one has some knowledge, even of a rough descriptive nature, of the organizational process involved. For example, in human development the time immediately after birth is a critical period in several ways. For example, the process of neonatal nutrition must be successfully organized as a matter of survival. Perhaps equally important (and this is often overlooked) is the organizational process of the mother's attachment to the infant. Later on, much descriptive and experimental data have established the period from approximately 5 or 6 weeks to 7 months as a critical period for the process of primary socialization (Scott, 1963).

A much longer critical period is that for the process of language organization, which runs from approximately 2 yr until Age 13 or so. It is a continuing scandal of our educational system that this period is not utilized for the teaching of foreign languages. This failure is not unconnected with the fact that the critical period of organization of our educational system is long since past. Like any other well-organized system, it is extremely resistant to change.

Another obvious critical period for human development is that of adolescence, in which sexual behavior is organized either successfully or unsuccessfully. Even more important, this is a period in which an individual first organizes his adult social relationships.

The declining years of life also involve organizational processes, although not in ways as clearly related to time as those of early development. About Age 40 an individual must reorganize his life in relation to declining rates of metabolism and loss of physical strength. For women, the menopause, marked by a more or less abrupt change of hormonal secretions, is traditionally a critical period, involving reorganization of behavior, although this may be as much a social phenomenon—that of children leaving home, with the resulting breakdown of family organization—as one of physiological change. For men, and increasingly often for women, the reorganization made necessary by retirement from active work may also be a critical period.

This brings up the problem of *disorganizational* processes. Throughout this paper we have spoken as if organization is a 1-way process. Actually, disorganizational processes are as much a part of life as the organizational ones, and they might provide a fertile field for the application of the critical-period concept. For example, a functional psychosis is obviously a disorganizational process. An individual faced with an acute problem, with which his behavior as organized cannot cope, may respond by disorganizing his behavior. In some cases this is followed by a reorganization of behavior which does permit coping with the situation. This is, of course, the goal of psychotherapy. We all know, and especially the practitioners of psychotherapy, that efforts to help an individual to reorganize his behavior are in many cases failures and completely frustrating. Considered as a process, any point at which the individual attempts to reorganize his behavior should be a critical period—a time when outside help would be of real value—whereas a similar attempt at times when disorganizing processes are going on would lead to failure. It is possible that a disorganizing process must run its course before any effective reorganization is possible.

Finally, let us say a few words about critical periods in social organization. As many people have discovered through practical experience, it is tremendously difficult to modify a well-established organization, whether it be a local club, General Motors, the government of the United States, or even a university. It is much easier to modify and improve organization by setting up an entirely new group and working with it during the critical period when organization is proceeding most rapidly.

This raises the question of why living systems tend to develop toward stability. An obvious answer is that stability permits a more efficient operation of highly specialized entities that make up the system. The human brain, for example, operates well only within a very limited range of physiological conditions and ceases to function entirely if these are seriously upset more than a few minutes. As human social organization develops, the same tendency is seen. Modern human societies permit the survival and efficient operation of such highly specialized individuals as, let us say, research scientists. However, we must remember that any human individual is fundamentally not highly specialized, but is biologically primitive and generalized, that human social organization has in the past evolved in the direction of one individual operating in many social roles rather than only one, as in the case of ants and termites. It follows that any social organization that results in complete fixity of roles and stereotyped behavior will be unsatisfactory for most of its member. This is, possibly, a new way of defining the intuitive longing for freedom.

Notes

This paper is a revised version of the address given by the senior author as President of the International Society for Developmental Psychobiology at San Diego, California, on November 6, 1973. The research reported was supported in part by the National Institute of Child Health and Human Development under Grant HD-3778.

247

John M. Stewart is now at the Jackson Laboratory, Bar Harbor, Maine.

Victor J. DeGhett is now at the Department of Psychology, State University of New York at Potsdam, Potsdam, New York.

Requests reprints from: Dr. J. P. Scott, Department of Psychology, Center for the Study of Social Behavior, Bowling Green University Bowling Green, Ohio 43403, U.S.A.

References

Bacon, W. E. (1971). Stimulus control of discriminated behavior in neonatal dogs. *J. Comp. Physiol. Psychol.*, 76: 424-433.

Brodbeck, A. J. (1954). An exploratory study on the acquisition of dependency behavior in puppies. *Bull. Ecol. Soc. Am.*, 35: 73.

Cairns, R. B., and Werboff, J. (1967). Behavior development in the dog: An interspecific analysis. *Science, 158*: 1070-1072.

Denenberg, V. H. (1962). An attempt to isolate critical periods in development in the rat. *J. Comp. Physiol. Psychol.*, 55: 813-815.

Denenberg, V. H. (1964). Critical periods, stimulus input, and emotional reactivity: A theory of infantile stimulation. *Psychol. Rev.*, 71: 335-351.

Denenberg, V. H. (1968). A consideration of the usefulness of the critical period hypothesis as applied to the stimulation of rodents in infancy. In G. Newton and S. Levine (Eds.), *Early Experience and Behavior*. Springfield: Thomas. Pp. 142-167.

Denenberg, V. H., Brumaghim, J. T., Haltmeyer, G. C., and Zarrow, M. X. (1967) Increased adrenocortical activity in the neonatal rat following handling. *Endocrinology, 81*: 1047-1052.

Denenberg, V. H., and Zarrow, M. X. (1971). Effects of handling in infancy upon adult behavior and adrenocortical activity: Suggestions for a neuroendocrine mechanism. In D. N. Walcher and D. N. Peters (Eds.), *Early Childhood*. New York: Academic Press. Pp. 39-71.

Elliot, O., and Scott, J. P. (1961). The development of emotional distress reactions to separation, in puppies. *J. Genet. Psychol.*, 99: 3-22.

Fisher, A. E. (1955). The effects of differential early treatment on the social and exploratory behavior of puppies. Unpublished doctoral thesis, Pennsylvania State University, University Park, Pa.

Freedman, D. G., King, J. A., and Elliot, O. (1961). Critical period in the social development of dogs. *Science, 133*: 1016-1017.

Fuller, J. L. (1967). Experiential deprivation and later behavior. *Science, 158:* 1645-1652.

Fuller, J. L., and Christake, A. (1959). Conditioning of leg flexion and cardio-acceleration in the puppy. *Fed. Proc.*, 18: 98.

Fuller, J. L., Easler, C. A., and Banks, E. M. (1950). Formation of conditioned avoidance responses in young puppies. *Am. J. Physiol., 160*: 462-466.

Hinde, R. A. (1970). *Animal Behaviour; A Synthesis of Ethology and Comparative Psychology*, 2nd Ed. New York: McGraw-Hill.

Igel, G. J., and Calvin, A. D. (1960). The development of affectional responses in infant dogs. *J. Comp. Physiol. Psychol.*, 53: 302-305.

Jeddi, E. (1970). Confort du contact et thermoregulation comportementale. *Physiol. Behav.*, 5: 1487-1493.

Klyavina, M. P., Kobakova, E. M., Stelmakh, L. N., and Troshikhin, V. A. (1958). On the speed of formation of conditioned reflexes in dogs in ontogenesis. *J. Higher Nerv. Activ.*, 8: 929-936.

Levine, S., and Lewis, G. W. (1959). Critical periods and the effects of infantile experience on maturation of stress response. *Science, 129*: 42-43.

Lorenz, K. Z. (1937). The companion in the birds' world. *Auk, 54*: 245-273.

Misanin, J. R., Nagy, Z. M., Keiser, E. F., and Bowen, W. (1971). Emergence of long-term memory in the neonatal rat. *J. Comp. Physiol. Psychol.*, 77: 188-199.

Moltz, H. (1973). Some implications of the critical period hypothesis. *Ann. N. Y. Acad. Sci., 223*: 144-146.

Pfaffenberger, C. J., and Scott, J. P. (1959). The relationship between delayed socialization and trainability in guide dogs. *J. Genet. Psychol., 95*: 145-155.

Schaefer, T. (1968). Some methodological implications of the research on "early handling" in the rat. In G. Newton and S. Levine (Eds.), *Early Experience and Behavior*. Springfield: Thomas. Pp. 102-141.

Schneirla, T. C., and Rosenblatt, J. S. (1963). "Critical periods" in the development of behavior. *Science, 139*: 1110-1116.

Scott, J. P. (1963). The process of primary socialization in canine and human infants. *Monogr. Soc. Res. Child Dev. 28*: 1-47.

Scott, J. P. (1970). Critical periods for the development of social behavior in dogs. In S. Kazda and V. Denenberg (Eds.), *The Postnatal Development of Phenotype*. Prague: Academia. Pp. 21-33.

Scott, J. P., and Fuller, J. L. (1965) *Genetics and the Social Behavior of the Dog*. Chicago: University of Chicago Press.

Scott, J. P., Fuller, J. L., and Fredericson, E. (1951). Experimental exploration of the critical period hypothesis. *Personality, 1:* 162-183.

Scott, J. P., and Marston, M. V. (1950). Critical periods affecting the development of normal and maladjustive social behavior in puppies. *J. Genet. Psychol., 77*: 25-60.

Stanley, W. C. (1965). The passive person as a reinforcer in isolated beagle puppies. *Psychon. Sci., 2*: 21-22.

Stanley, W. C. (1970). Feeding behavior and learning in neonatal dogs. In J. F. Bosma (Ed.), *The Second Symposium on Oral Sensation and Perception*. Springfield: Thomas. Pp. 242-290.

Stanley, W. C., and Elliot, O. (1962). Differential human handling as reinforcing events and as treatments influencing later social behavior in basenji puppies. *Psychol. Rep., 10*: 775-788.

Theberge, J. B., and Pimlott, D. H. (1969). Observations of wolves at a rendezvous site in Algonquin Park. *Can. Field Naturalist, 83*: 122-128.

Waller, M. B., and Fuller, J. L. (1961). Preliminary observations on early experience as related to social motivation. *Am. J. Orthopsychiatry, 31*: 254-266.

Woolpy, J. H., and Ginsburg, B. E. (1967). Wolf socialization: A study of temperament in a wild social species. *Am. Zool., 7*: 357-363.

27

Selection of Social Partners as a Function of Peer Contact during Rearing

Abstract. *Three groups of monkeys were raised with different degrees of contact with their peers. The first group was allowed no contact, the second only visual and auditory contact, and the third was allowed complete and normal contact with their peers. Animals of all three groups were allowed to interact socially; they were then tested for their preference for monkeys raised under the same conditions or for monkeys raised under different conditions. Monkeys raised under the same conditions preferred each other, even if the stimulus animals were completely strange to the test monkey.*

The early experiences of primates often have profound consequences on later behavior. In rhesus monkeys exploratory, maternal, sexual, and social behaviors appear extremely vulnerable to early social and sensory restric-tion (*1*). Monkeys reared in isolation tend to withdraw from other animals and huddle by themselves in social situations. If such animals prefer each other over more normal monkeys, they may not be effectively exposed

249

to the stimuli which lead to some degree of social adjustment. The fact that socially normal monkeys may avoid contact with monkeys reared in isolation further retards rehabilitation. We varied the amount of peer contact during rearing and investigated its effect on physical approach to a social partner, in order to determine whether monkeys reared under identical conditions prefer each other to monkeys reared under different conditions.

Three groups of rhesus monkeys were reared from birth in the laboratory without mothers. Each group contained four males and four females. Sets of three animals were matched across groups for age, sex, and test experiences after rearing was complete. The first group (A) was reared from birth to 9 months in individual closed cages. On the first 5 to 7 days they experienced physical, but minimal visual, contact with a human during feeding. No other physical or visual contact with humans or live monkeys occurred during rearing. Changing visual experiences throughout rearing were limited to presentation of pictures of monkeys engaged in various behaviors and pictures of people and inanimate objects (2). From months 9 through 18 the monkeys in group A were housed individually in bare wire cages from which they could see and hear other isolates and humans, but physical contacts were unavailable.

Subjects in the second group (B) were reared individually in a large nursery room in bare wire cages from birth to 9 months. Other monkeys and humans could be seen and heard, but physical contact was not available. From month 9 through 18 the monkeys in group B were housed in the same room as the monkeys in group A; they were in wire cages where they could see and hear, but not touch, one another.

The third group (C) lived in wire cages in peer groups of varying sizes during the first 18 months of life. Rearing conditions and social behavior tests provided physical peer contact during this period. In summary, group A had no early contact with live peers, group B had visual and auditory but no physical contact with peers, and group C had complete peer contact during the rearing period.

When they were 18 months old, sets of monkeys from all groups interacted during social behavior tests in a large playroom (3). Each animal was tested weekly for 12 weeks in three 30-minute sessions. In one weekly session a constant set of one group A, one group B, and one group C monkey of the same sex interacted together; the same animals were always tested together. On the two other weekly sessions constant pairs of groups A and B, A and C, and B and C subjects interacted in groups of four monkeys. After social testing, each subject had received equal playroom exposure to one monkey from its own rearing condition and to two monkeys from each of the other rearing conditions. After playroom testing was completed, the monkeys were tested for their preference for other monkeys reared under the same conditions or for those reared under different conditions.

Testing was done in the "selection circus" (Fig. 1), which consists of a central start compartment that bounds the entrances to six adjoining choice compartments. Wire-mesh cages for the stimulus animals were attached to the outside of appropriate choice compartments. The front walls of the stimulus cages, the outside walls of the choice compartments, and the guillotine doors separating choice compartments from the start compartment were all made of clear plexiglas.

For the testing, the subject was placed in the center start compartment with the plexiglas guillotine doors down for a 5-minute exposure period. The subject could see and hear the stimulus

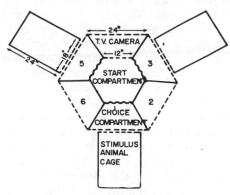

TOP VIEW OF SELF- SELECTION CIRCUS

Fig. 1. Scheme of the "circus" which is constructed of aluminum channels containing plexiglas walls (dotted line), plywood walls (solid line), and plexiglas guillotine doors (wavy line). Wire-mesh stimulus cages with a single plexiglas wall are attached outside choice compartments. In testing, the subject is first placed in the start compartment. It can look into and through the choice compartments, but cannot enter them until the plexiglas guillotine doors are raised by a vacuum lift. Plywood walls block physical and visual access to choice compartments that are not used in the experiment.

Table 1. Mean number of seconds spent with each type of stimulus animal for each rearing condition, averaged over the two test trials.

Rearing condition of experimental animal	Rearing condition of stimulus animal		
	A (totally deprived)	B (partially deprived)	C (peer-raised)
A (totally deprived)	156	35	29
B (partially deprived)	104	214	103
C (peer-raised)	94	114	260

animals, but could not enter the choice compartments near them. Unused choice compartments were blocked off by plywood walls inserted in place of the plexiglas guillotine doors. After the exposure period, a 10-minute choice trial was given. The plexiglas guillotine

doors were raised by a vacuum system; this procedure allowed the subject to enter and reenter choice compartments or to remain in the start compartment. The total time spent in each choice compartment during the test trial was recorded over a closed-circuit TV system.

The monkey's entry into different choice compartments served as our index of social preference. This measure of preference involves visual orientation, but, more importantly, it also involves locomotion toward a specific social object. It may be argued that a measure of viewing time, such as that used by Butler (4) in which monkeys inspected various objects through a small window, is not a proper index of social preference. Although actual physical contact was not available to our subjects, a great deal of nontactile social interaction was possible. Thus, our measure of preference based on physical approach toward a social object seems to be more analogous to an actual social situation than would be a simple viewing response.

Two types of trials were given. In the first, the stranger trial, one stimulus animal from each of the rearing groups was randomly positioned in a stimulus animal cage outside choice compartments 1, 3, or 5. These stimulus animals had received no previous social contact with the test subject but they were the same age and the same sex. A second test was identical with the stranger trial except that the three stimulus animals had received extensive social experience with the test subject during the playroom tests. Before the start of these tests, all 24 subjects had been adapted to the circus during nonsocial exploration tests. The order of serving first as a stimulus animal or as a test subject was randomized across groups.

Analysis of variance of the total time spent in the choice compartment had rearing condition as an uncorrelated variable, and type of stimulus animal and degree of familiarity as correlated variables. Familiarity did not have a significant main effect, and it did not interact with the other variables (all $P > .20$). Rearing condition had a significant effect ($P < .001$), which indicated that total choice time in all compartments differed as a function of early peer contact. Group A subjects spent half as much time (average = 220 seconds) in choice compartments as either group B (average = 422 seconds) or group C (average = 468 seconds) monkeys.

The interaction of rearing condition with type of stimulus animal was also significant ($P < .001$). Table 1 shows this effect, with choice times averaged over the trials with strange and familiar stimuli. These data show that like prefers like—each rearing condition produced maximum choice time for the type of stimulus animal reared under that condition. The data for individual subjects supports this averaged effect. In the group A, two of the eight monkeys did not enter choice compartments. Of the six remaining monkeys, five spent more time in the group A choice compartment than in the other two compartments (two-tailed binomial, $P = .038$, with $p = \frac{1}{3}$, $q = \frac{2}{3}$). In the groups B and C all subjects entered choice compartments, and seven out of eight in each group spent more time with the animal reared like themselves than with the other animals (both $P = .0038$, two-tailed binomial).

The data indicate that social preferences are influenced by rearing conditions. In playroom testing the group C monkeys were the most active and socially advanced groups studied. Therefore, it was not surprising that they discriminated and showed large preferences for both strange and familiar group C animals. The group A monkeys, however, were highly retarded in their playroom behavior, and they did not show much progress over the 12 weeks of social interaction. As expected, these animals did exhibit a low degree of choice time in this study. We also thought that group A monkeys would be least likely to show preferences for a particular type of animal. It was, therefore, surprising to find that they did prefer each other to animals reared under other conditions. The group B animals, which were intermediate in social adequacy in playroom testing, also preferred each other. This result seems to strengthen the idea that animals of equal social capability, whether or not they are familiar with each other, can discriminate themselves from others, and not only discriminate but approach each other.

These results have important implications for studies designed to rehabilitate primates from the devastating effects of social isolation. The fact that socially abnormal monkeys prefer each other poses difficulties in the design of social environments which contain experiences appropriate for the development of normal social responses. Further, the finding that socially normal monkeys do not choose to approach more abnormal ones compounds the problem of providing therapy for abnormal animals.

These data also have implications for attachment behavior in mammals. Cairnes (5) suggests a learning theory approach to the formation of attachments in which the subject will approach a social object as a function of having made many previous responses while the social object was part of the general stimulus situation. Thus, indices of social attachment toward an object are expected to be higher with increases in the probability that this object occurs as part of the stimulus field in the subject's overall repertoire of responses. Although this seems a reasonable approach, the present data present some difficulties for this view. During rearing, the monkeys in group A did not have the same opportunity to

252

learn the characteristics of other monkeys as did the monkeys in groups B and C. Yet, the monkeys in group A did prefer each other to the alternative choices available. Thus, it is possible that the preference shown by group A monkeys was not based on the conditioning of approach behavior to specific social cues, as is suggested by the stimulus-sampling theory of attachment. It is possible that the behavior of group A was motivated by avoidance of cues contained in the social behavior or countenance of the other two types of monkeys. Thus, there may be at least two distinct kinds of processes in the choice of a social stimulus. The conditioning of specific social cues to the response systems of an animal may be one factor, and the avoidance of nonconditioned cues may be a second important factor in the formation of social attachments.

The specific cues used by the monkeys studied here are not known. Neither do we yet know how our animals differentiated between the stimuli. The discrimination may be based solely on differences in the gross activity of the stimulus animals, or on more subtle and specific social cues. Analysis of the specific stimulus components operating in this situation may clarify the nature of the social cues involved. The important question to be answered is whether the types of cues used in selecting a partner are qualitatively different for different rearing conditions, or whether the same aspects of stimulation are simply weighted differently as a function of an animal's rearing history.

CHARLES L. PRATT
GENE P. SACKETT

Primate Laboratory, University
of Wisconsin, Madison 53706

References and Notes

1. H. F. Harlow and M. K. Harlow, Sci. Amer. 207, 136 (1962); G. P. Sackett, Child Develop. 36, 855 (1965).
2. The rearing conditions are described fully by G. P. Sackett, Science 154, 1468 (1966).
3. The playroom situation is described by H. F. Harlow, G. L. Rowland, G. A. Griffin, Psychiat. Res. Rep. 19, 116 (1964).
4. R. A. Butler, J. Comp. Physiol. Psychol. 50, 177 (1957).
5. R. B. Cairns, Psychol. Rev. 73, 409 (1966).
6. Supported by NIMH grant MH-11894.

10 January 1967

28 Culturally Transmitted Patterns of Vocal Behavior in Sparrows

Abstract. Male white-crowned sparrows have song "dialects," acquired in about the first 100 days of life by learning from older males. In the laboratory an alien white-crowned sparrow dialect can be taught. Once the song is established further acoustical experience does not change the pattern. White-crowned sparrows do not copy recorded songs of other sparrow species presented under similar conditions.

The white-crowned sparrow, Zonotrichia leucophrys, is a small song bird with an extensive breeding distribution in all but the southern and eastern parts of North America (1). Ornithologists have long remarked upon the geographical variability of its song. Physical analysis of field recordings of the several vocalizations of the Pacific Coast subspecies Z. l. nuttalli re-

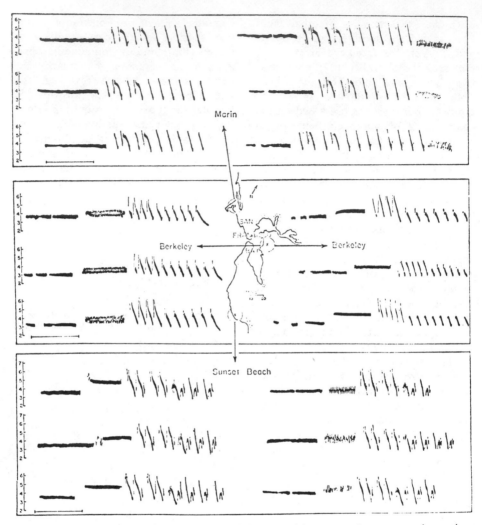

Fig. 1. Sound spectrograms of songs of 18 male white-crowned sparrows from three localities in the San Francisco Bay area. The detailed syllabic structure of the second part of the song varies little within an area but is consistently different between populations. The introductory or terminal whistles and vibrati show more individual variability. The time marker indicates 0.5 second and the vertical scale is marked in kilocycles per second.

veals that while most of the seven or so sounds which make up the adult repertoire vary little from one population to another, the song patterns of the male show striking variation (see 2).

Each adult male has a single basic song pattern which, with minor variations of omission or repetition, is repeated throughout the season. Within a population small differences separate the songs of individual males but they all share certain salient characteristics of the song. In each discrete population there is one predominant pattern which differs in certain consistent respects from the patterns found in neighboring populations (Fig. 1). The term "dialect" seems appropriate for the properties of the song patterns that characterize each separate population of breeding birds. The detailed structure of syllables in the second part of the song is the most reliable indicator. Such dialects are known in other song birds (3).

The white-crowned sparrow is remarkable for the homogeneity of song patterns in one area. As a result the differences in song patterns between populations are ideal subjects for study of the developmental basis of behavior. If young male birds are taken from a given area, an accurate prediction can be made about several properties of the songs that would have developed if they had been left in their natural environment. Thus there is a firm frame of reference with which to compare vocal patterns developing under experimental conditions. Since 1959 we have raised some 88 white-crowned sparrows in various types of acoustical environments and observed the effects upon their vocal behavior. Here we report on the adult song patterns of 35 such experimental male birds. The several types of acoustical chamber in which they were raised will be described elsewhere.

In nature a young male white-crown hears abundant singing from its father and neighbors from 20 to about 100 days after fledging. Then the adults stop singing during the summer molt and during the fall. Singing is resumed again in late winter and early spring, when the young males of the previous year begin to participate. Young males captured between the ages of 30 and 100 days, and raised in pairs in divided acoustical chambers, developed song patterns in the following spring which matched the dialect of their home area closely. If males were taken as nestlings or fledglings when 3 to 14 days of age and kept as a group in a large soundproof room, the process of song development was very different. Figure 2 shows sound spectrograms of the songs of nine males taken from three different areas and raised as a group.

The patterns lack the characteristics of the home dialect. Moreover, some birds from different areas have strikingly similar patterns (A3, B2, and C4 in Fig. 2).

Males taken at the same age and individually isolated also developed songs which lacked the dialect characteristics (Fig. 3). Although the dialect properties are absent in such birds isolated in groups or individually, the songs do have some of the species-specific characteristics. The sustained tone in the introduction is generally, though not always, followed by a repetitive series of shorter sounds, with or without a sustained tone at the end. An ornithologist would identify such songs as utterances of a *Zonotrichia* species.

Males of different ages were exposed to recorded sounds played into the acoustical chambers through loudspeakers. One male given an alien dialect (8 minutes of singing per day) from the 3rd to 8th day after hatching, and individually isolated, showed no effects of the training. Thus the early experience as a nestling probably has little specific effect. One of the group-raised isolates was removed at about 1 year of age and given 10 weeks of daily training with an alien dialect in an open cage in the laboratory. His song pattern was unaffected. In general, acoustical experience seems to have no effect on the song pattern after males reach adulthood. Birds taken as fledglings aged from 30 to 100 days were given an alien dialect for a 3-week period, some at about 100 days of age, some at 200, and some at 300 days of age. Only the training at the age of 100 days had a slight effect upon the adult song. The other groups developed accurate versions of the home dialect. Attention is thus focused on the effects of training between the ages of about 10 and 100 days. Two males were placed in individual isolation at 5 and 10 days of age, respectively, and were exposed alternately to the songs of a

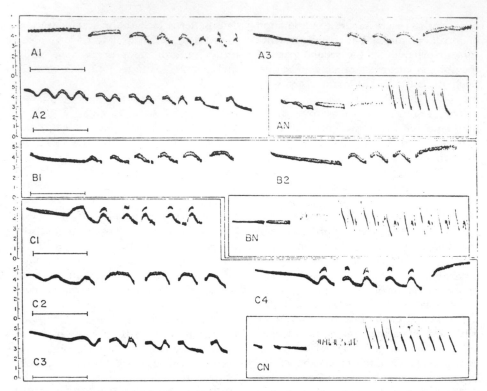

Fig. 2. Songs of nine males from three areas raised together in group isolation. *A1* to *A3*, Songs of individuals born at Inspiration Point, 3 km northeast of Berkeley. *B1* and *B2*, Songs of individuals born at Sunset Beach. *C1* to *C4*, Songs of individuals born in Berkeley. The inserts (*AN, BN,* and *CN*) show the home dialect of each group.

normal white-crowned sparrow and a bird of a different species. One male was exposed at 6 to 28 days, the other at 35 to 56 days. Both developed fair copies of the training song which was the home dialect for one and an alien dialect for the other. Although the rendering of the training song is not perfect, it establishes that the dialect patterns of the male song develop through learning from older birds in the first month or two of life. Experiments are in progress to determine whether longer training periods are necessary for perfect copying of the training pattern.

The training song of the white-crowned sparrow was alternated in one case with the song of a song sparrow, *Melospiza melodia*, a common bird in the areas where the white-crowns were

taken, and in the other case with a song of a Harris's sparrow, *Zonotrichia querula*. Neither song seemed to have any effect on the adult patterns of the experimental birds. To pursue this issue further, three males were individually isolated at 5 days of age and trained with song-sparrow song alone from about the 9th to 30th days. The adult songs of these birds bore no resemblance to the training patterns and resembled those of naive birds (Fig. 3). There is thus a predisposition to learn white-crowned sparrow songs in preference to those of other species.

The songs of white-crowned sparrows raised in isolation have some normal characteristics. Recent work by Konishi (*4*) has shown that a young male must be able to hear his own voice if these properties are to appear.

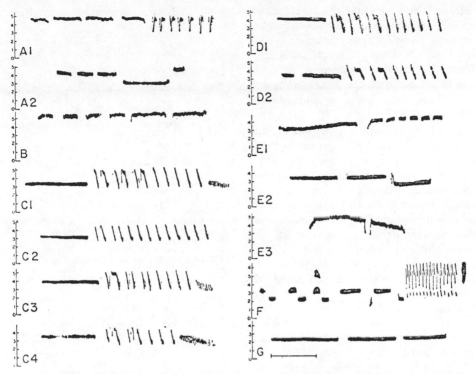

Fig. 3. Songs of 12 males raised under various experimental conditions. *A1* and *A2*, Birds raised in individual isolation. *B*, Male from Sunset Beach trained with Marin song (see Fig. 1) from the 3rd to the 8th day of age. *C1* to *C4*, Marin birds brought into the laboratory at the age of 30 to 100 days. *C1*, Untrained. *C2* to *C4*, Trained with Sunset Beach songs; *C2* at about 100 days of age, *C3* at 200 days, *C4* at 300 days. *D1*, Bird from Sunset Beach trained with Marin white-crowned sparrow song and a Harris's sparrow song (see *G*) from the age of 35 to 56 days. *D2*, Marin bird trained with Marin white-crowned sparrow song and a song-sparrow song (see *F*) from the age of 6 to 28 days. *E1* to *E3*, Two birds from Sunset Beach and one from Berkeley trained with song-sparrow song from the age of 7 to 28 days. *F*, A song-sparrow training song for *D2* and *E1* to *E3*. *G*, A Harris's sparrow training song for *D1*.

Deafening in youth by removal of the cochlea causes development of quite different songs, with a variable broken pattern and a sibilant tone, lacking the pure whistles of the intact, isolated birds. Furthermore, there is a resemblance between the songs of male white-crowned sparrows deafened in youth and those of another species, *Junco oreganus*, subjected to similar treatment. The songs of intact juncos and white-crowns are quite different. Konishi also finds that males which have been exposed to the dialect of their birthplace during the sensitive period need to hear themselves before the memory trace can be translated into motor activity. Males deafened after exposure to their home dialects during the sensitive period, but before they start to sing themselves, develop songs like those of a deafened naive bird. However, once the adult pattern of singing has become established then deafening has little or no effect upon it. Konishi infers that in the course of crystallization of the motor pattern some control mechanism other than auditory feedback takes over and becomes adequate to maintain its organization. There are thus several pathways impinging upon the development of song patterns in the white-crowned sparrow, including acoustical influences from the external environment, acoustical feedback from the bird's own vocalizations, and perhaps nonauditory feedback as well.

Cultural transmission is known to play a role in the development of several types of animal behavior (5). However, most examples consist of the reorientation through experience of motor patterns, the basic organization of which remains little changed. In the development of vocal behavior in the white-crowned sparrow and certain other species of song birds, we find a rare case of drastic reorganization of whole patterns of motor activity through cultural influence (6). The process of acquisition in the white-crowned sparrow is interesting in that, unlike that of some birds (7), it requires no social bond between the young bird and the emitter of the copied sound, such as is postulated as a prerequisite for speech learning in human children (8). The reinforcement process underlying the acquisition of

sound patterns transmitted through a loudspeaker is obscure.

PETER MARLER
MIWAKO TAMURA
Department of Zoology,
University of California, Berkeley

References and Notes

1. R. C. Banks, *Univ. Calif. Berkeley Publ. Zool.* 70, 1 (1964).
2. P. Marler and M. Tamura, *Condor* 64, 368 (1962).
3. E. A. Armstrong, *A Study of Bird Song* (Oxford Univ. Press, London, 1963).
4. M. Konishi, in preparation.
5. W. Etkin, *Social Behavior and Organization Among Vertebrates* (Univ. of Chicago Press, Chicago, 1964).
6. W. Lanyon, in *Animal Sounds and Communication, AIBS Publ. No. 7,* W. Lanyon and W. Tavolga, Eds. (American Institute of Biological Sciences, Washington, D.C., 1960), p. 321; W. H. Thorpe, *Bird Song. The Biology of Vocal Communication and Expression in Birds* (Cambridge Univ. Press, London, 1961); G. Thielcke, *J. Ornithol.* 102, 285 (1961); P. Marler, in *Acoustic Behaviour of Animals,* R. G. Busnel, Ed. (Elsevier, Amsterdam, 1964), p. 228.
7. J. Nicolai, *Z. Tierpsychol.* 160, 93 (1959).
8. O. H. Mowrer, *J. Speech Hearing Disorders* 23, 143 (1958).
9. M. Konishi, M. Kreith, and J. Mulligan cooperated generously in locating and raising the birds and conducting the experiments. W. Fish and J. Hartshorne gave invaluable aid in design and construction of soundproof boxes. We thank Dr. M. Konishi and Dr. Alden H. Miller for reading and criticizing this manuscript. The work was supported by a grant from the National Science Foundation.

14 September 1964

29

Mother Love: What Turns It On?

Studies of maternal arousal and attachment in ungulates may have implications for man

Peter H. Klopfer

Reprinted from AMERICAN SCIENTIST, Vol. 59, No. 4, July-August 1971, pp. 404-407
Copyright © 1971 by The Society of the Sigma Xi

A female that has recently given birth may succor infants other than her own. Then again she may not. Among roe deer (*Capriolus c.*), for instance, females will apparently adopt alien fawns, while the fallow deer females (*Dama d.*) apparently do not. Moose (*Alces a.*) mothers, particularly when they have lost their own young, are very prone to adopt aliens, even to kidnapping young belonging to other females. In goats, on the other hand, alien young are generally excluded from the family circle. The solicitude a female goat extends to kids is limited to her own. Even within this species there may be further variation in the degree to which a female cares for her own. These variations presumably reflect experiential differences, while the differences between goats and moose represent phylogenetic adaptations.

The differences in adaptations are probably related to the different ecological and physiological demands made on the various species. Moose give birth at some distance from their fellows, in contrast to goats, which rarely withdraw far or for long from their herd (Altman 1958). When feeding habits require a high degree of dispersion, there may be less occasion to exclude alien individuals than when the social group is compact and the animals are compelled to maintain a great deal of contact with one another. Under these circumstances, the benefits of care being directed exclusively toward one's own young may outweigh the disadvantages.

Dr. Peter H. Klopfer is professor of zoology at Duke University, Director of the Field Station for Animal Behavior Studies, and Associate Director of the Primate Facility. His interests include ecological problems relating to community structure and faunal diversity, and psychological problems of maternal-filial attachments and early learning. Among his major publications are: Behavioral Aspects of Ecology and An Introduction to Animal Behavior: Ethology's First Century (with J. P. Hailman). The work discussed in this article is supported by a grant from the National Institute of Mental Health and a Research Scientist award. The author gratefully acknowledges the help of his assistants, Catherine Dewey, Diane Chepko, Peter Haas, and John Gamble, and of his ungulate-loving associates, Richard Hemmes, Barrie Gilbert, Donald Adams, and Martha Klopfer. Address: Department of Zoology, Duke University, Durham, NC 27706.

In order to get a clear understanding of how ecological (or historical) variables interact with physiological demands, we need to study an array of related species that show a high degree of ecological diversity. Such diversity is provided by the ungulates and the lemurs, both of which are groups with a great many rather different species. Our work has focused upon the patterns of maternal care in Toggenburg goats and fallow deer and on three species of lemurs (*Lemur catta, fulvous,* and *variegatus*) (Klopfer and Klopfer 1968; Gilbert 1964 and MS; Hemmes MS; Klopfer and Gamble 1967; Klopfer, Adams, and Klopfer 1964; Klopfer and Klopfer 1970).

In the pages that follow, I shall limit myself to an analysis of the mechanisms that underlie maternal care and recognition of young in goats. Understanding these in one species permits predictions that can be tested in the others. Our interest in goats, in short, derives from the part they play in the larger picture sketched above.

Mother-young bonds

A goat is one of the species in which there are stable, specific, and rapidly formed bonds between the mother and the young. As was intimated, this is not universally true of mammals. Ties need be neither very specific nor rapidly formed, nor yet stable.

Our first task, then, is to confirm that in our goats the bond between mother and young is indeed stable, specific, and rapidly formed. The procedure for studying this question entailed allowing our goats—purebred Toggenburgs—to select their own birth sites within the quarter-acre enclosure where they had previously been living. This was an area familiar to them, as were the observers who were present. At the time of birth, the does were assigned to one of two treatment groups. In one of these, the kids were removed immediately at parturition and wrapped in toweling as they emerged into the world. In the other, the first kid born was allowed five minutes of contact with its mother. This interval was measured from the time the doe first began sniffing the kid, usually a few seconds after its birth. Kids born subsequently were quickly covered with towels and removed. If one was born during the five minute "contact" period, the clock was stopped until the completion of parturition and then allowed to resume as soon as the mother once again turned her attention back to her first-born.

Upon removal, the kids were isolated from their mothers for periods of 1, 2, or 3 hours. At the end of this time, they were singly introduced to their mothers for a 10-minute test period. The introduction of "own" kids was alternated in a random fashion with the introduction of littermates or alien kids. The females were scored according to whether they allowed the kid to nurse or rejected it. During this ten-minute test period the behavior of the doe was usually clear-cut, with few instances of ambivalence. Either the doe would reject the kid by abruptly withdrawing from it when it approached and butting it vigorously if it persisted in seeking to nurse or, alternatively, the doe would lick the kid and assume a nursing stance.

In the few instances of ambivalent behavior, cascading effects very quickly brought the mother to one of the two extreme patterns designated acceptance or rejection. Any initial hesitancy on the part of the mother toward an advancing kid might make the kid more cautious in its approach. A hesitation on the part of the kid would then prompt a more severe withdrawal on the part of the doe. In turn, this amplified the hesitation of

the kid, which fed back to increase the rejecting behavior of the female. Thus, through feedback, an animal that was initially ambivalent quickly became an accepting or rejecting mother.

A brief note on the reason for limiting the period of separation to 1, 2, or 3 hours is in order. After 3 hours, an unfed kid is often too weak to make a satisfactory attempt to nurse. If it has been fed, however, it may become so vigorous that, in its attempts to nurse, it "overshoots" the teats. Normally, the teats are found by progressive movements made by the kid along the contours of the mother's ventral body surface. When the lowest extremity, the teat, is reached, the kid opens its mouth, lunges, and, sometimes after several failures, succeeds in making contact and sucking. Especially vigorous kids, however, make these movements so rapidly that they move from one side of the female, under her, to the other, allowing themselves insufficient contact to find the teat.

As for shorter separations, less than an hour did not allow (in preliminary trials) the emergence of a clear difference between the "contact" and "noncontact" groups. With 30-minute separations, about one-third of the noncontact kids would still be accepted. We were obviously dealing with a time-dependent decay process and for pragmatic reasons selected an interval sufficiently long to assure complete "decay." There was no difference in the effects of separations of 1, 2, and 3 hours, so these subgroups were combined in the analysis.

The results of these experiments are clear. Of fifteen mothers allowed 5 minutes' contact with their kids (five in each of the three groups where the kids were separated for 1, 2, or 3 hours), fourteen immediately reaccepted their kids. They also accepted those of their kids with which they had had no contact, but vigorously rejected all aliens, including those of similar age. Of the fifteen mothers that had been immediately separated from their kids at the time of parturition, only two allowed their kids to nurse. The alien kids were vigorously rejected by all fifteen does. There were no differences between experienced and primiparous (giving birth for the first time) does. Littermates of the test kids (experienced does usually produce twins, often triplets) were accepted or rejected according to whether or not the kid with which contact was allowed was accepted or rejected.

In short, the answer to our first question is yes, the response of the mother to her young is formed very rapidly (in as little as five minutes), and it is at least specific to the animal with which contact is made and its siblings. It is also stable, for once nursing has occurred it can be easily reestablished later, although these particular experiments do not demonstrate this fact.

Sensory basis of recognition

The next question we need to ask concerns the cues that are involved in the recognition and acceptance of young. We must answer this question before we can postulate a mechanism whereby recognition and discrimination takes place. We began the inquiry by assuming that the sensory modalities of importance would most likely be vision, sound, or smell. Thus, we repeated the first experiment with animals that were blindfolded shortly before parturition commenced. These animals were accustomed to wearing opaque hoods for some weeks ahead of time. We then selected from all our subjects those pairs of animals (doe and kid) which had not vocalized during parturition, the 5-minute contact period, or during the test.

The results were identical with those obtained in the first instance. This seeming unimportance of sound and vision led us to focus on olfactory cues, leading to the following hypothesis of olfactory imprinting: a female responds to labile and highly attractive elements associated with the birth fluids. These could be attractive at all times, but more probably are only attractive to parturient females as a consequence of the physiological changes associated with parturition. Before this labile substance decomposes, the female's response becomes generalized to include those unique to her kid and its effluents. This model covers the fact that the period during which attachment can occur is limited to a few minutes after parturition, the young thereafter losing their attractive characteristics. It also covers the fact that a mother accepts all of her litter, though no aliens, so long as she has had contact with one of her kids immediately upon birth. Presumably all the kids of her litter share a litter-specific odor.

The supposition that there is some kind of labile attractant associated with the birth fluids is not an unreasonable one. First of all, we know that many changes in olfactory sensitivity are hormone-dependent. For example, estrogen levels in females are known to affect a variety of olfactory thresholds, and estrogen levels change at birth. We also know of labile products associated with the events at birth. In sheep, for instance, the blood fructose of the lamb is very high at the time of birth, but within minutes falls to adult levels. (Let me hastily add that there is no evidence that blood fructose is actually the attractant in question.)

Olfactory imprinting?

The third question must be whether the hypothesis of olfactory imprinting is correct. One way to test it is by predicting that a goat with an impaired sense of smell, or anosmia, at parturition will not accept her young. (Actually, we cannot distinguish between smell and taste in these tests, so anosmia as here used must not be taken literally.) An animal that has been allowed to establish a relationship with her young, on the other hand, should be unable to recognize it subsequently if anosmic at the time of the test. Therefore, the experiments described above were repeated yet again with the animals this time being not only blind and silent but with their olfactory mucosa cocainized either at the time of parturition or at the time of the test. In this instance, of course, all the animals were allowed 5 minutes of contact with one of their own kids. A 10 percent solution of cocaine hydrochloride was sprayed into their nostrils 20 minutes before birth commenced, or 20 minutes before the acceptance test. We had previously shown this to be an effective way of temporarily impairing olfactory ability, though of course it cannot be claimed that this procedure absolutely barred all olfactory cues. The results nevertheless support the view that olfaction was impaired by the administration of cocaine into the olfactory epithelium.

Of nine goats that were allowed 5 minutes of contact with their young, though blindfolded, silent, and cocainized, eight of the nine accepted their young subsequently, and, of the nine, *six also accepted alien kids*. The animals that were not impaired at the time of parturition but were cocainized at the time of the test subsequently rejected their own young as often as they accepted them. Clearly, the hypothesis is not correct. We must distinguish between the elicitation of maternal behavior and the discrimination of own from alien young. The interference with olfaction produced

by the cocaine apparently prevents females from making the discrimination between their own young and the alien young. It does not, however, interfere with maternal solicitude. The events that transform a female from a rejecting to a motherly animal, events which must be exploited within five or so minutes after parturition, must presumably be sought elsewhere than in some kind of external "releasor."

An endogenous basis for maternal care?

The fourth question we must then ask is this: Is maternal solicitude perhaps primarily dependent on endogenous factors rather than stimuli associated with birth fluids? If so, alien animals, even those who are some weeks of age, should also be accepted if presented during the critical 5-minute period after parturition.

Therefore the same experiments were repeated once again, with this difference: as soon as the first kid was born, it was removed and an alien kid several to 28 days old was substituted. These kids had been hand-fed or had been with their own mothers, and thus they were not particularly eager to attach themselves to foreign mothers. For 5 minutes they were held before the experimental mothers, who generally licked them. It should be noted that the aliens were kept from contact with any birth fluids. After 5 minutes' contact they were separated for the standard period of time, to be re-presented 1, 2, or 3 hours later, with the following results.

Of five animals presented with an alien kid for 5 minutes directly after parturition, all five accepted that alien. These five animals, however, rejected all aliens other than the one with which they had contact. Interestingly enough, four of the five accepted their own kids too. It must be remembered that in this instance there had been no contact between any of the does and their own kids.

In an additional six cases, both an alien kid and the doe's own kids were presented simultaneously for 5 minutes directly on parturition. Five of these six animals accepted their own kids and the alien, though other aliens were rejected. Even when the alien was allowed to remain continuously with the mother for 2 full hours after parturition in the total absence of their own kids, in two cases out of four the own kid was subsequently accepted.

What this means is that the response to the alien is indeed specific: that alien with which contact is made during the 5 minutes after parturition is accepted but no others, whether or not the alien be neonatal. But there is a curious lack of symmetry in the response of the mothers. If an adoption has occurred, her own kids continue to be attractive. We can conclude that parturition itself somehow makes the female maternally solicitous. The presence of the birth fluids is apparently unimportant, as shown by the fact that older aliens are readily adopted, provided only that they are present immediately upon parturition. These presentations, incidentally, always occurred at some meters distant from the birthing site, to make certain that the alien would not make contact with the birth fluids.

Some attractive element must nonetheless be associated with own young, though the attraction is provoked only as a consequence of parturition. It is clear that in this respect aliens and own young are not equivalent. The aliens become attractive only if presented at parturition, while own young retain their potential for being attractive for at least some hours.

Since the discrimination between own young and alien young is abolished by application of cocaine to the olfactory mucosa, it appears that the important cues are olfactory.

Basis of individual differences

How are kids individually coded? It is of interest to note the work of Brower, Brower, and Corvino (1964) on the sequestration of molecules by butterflies. They showed that the palatability of butterflies to birds was a function of what the butterflies had eaten. A similar mechanism may be operating here, assuming that kids and their mothers share a common scent. Nonetheless, despite this role of olfactory stimuli provided by the kids, the primary factor is clearly a change internal to the doe. Something happens in the space of a few minutes after parturition which makes her ready, then and only then, to attach herself to a kid. Once she is attached, she displays many of the human signs of distress on the removal of the kid. Spared an attachment, the removal leaves her as nonchalant as any virgin, despite the fact that she may be lactating heavily. What could be happening?

A model that we recently suggested is based on the observation that at the final stage of labor (in sheep), upon the presentation of the head, there is a substantial increase in the circulating level of oxytocin (Folley and Knaggs 1965). This posterior hypophyseal hormone is actually secreted in the hypothalamus, a structure intimately related to much of the less plastic behavior of animals. The release of oxytocin can also be induced by manual dilatation of the cervix (Peeters et al. 1965). Within minutes after its release, however, the oxytocin returns to its normal level, as a consequence of the actions of blood-borne enzymes. Our model suggests that the cervical dilatation accompanying birth induces the release of a substance with a time course similar to that of oxytocin. This may either stimulate specific central ganglia or alter the sensitivity of peripheral receptors so as to make the animal receive stimuli to which it is normally unresponsive. In either case, a short-term sensitivity or responsiveness would be effected during which the basis for a longer-term bond could be forged.

This model then leads to another question: Does oxytocin itself initiate maternal care? While our model does not require that it actually be oxytocin which serves as the maternal hormone, it does present an attractive possibility. Richard Hemmes (MS) has recently begun to investigate this point, first by the administration of exogenous oxytocin. Up to four International Units administered through the external jugular vein 2 hours after parturition appear to produce no particular response. Because the administration of exogenous hormone is beset with so many complications, a negative result does not reveal much.

The next step taken by Hemmes, therefore, was to simulate the endogenous production of oxytocin by cervical dilatation. This was achieved by inserting a rubber balloon into the cervix and inflating it to such a degree as to induce back-arching by the doe. Each animal was stimulated to arch fifty times in 15 minutes. The does' responsiveness toward kids was then tested. For a variety of reasons the results were inconclusive; this experiment has yet to be repeated under conditions similar to those of the tests described earlier. However, the indications are strong that the dilatation of the cervix may suffice to induce at least some features of the maternal response. This is a question on which

we can surely expect more information soon.

Finally, Hemmes is exploiting a technique for transfusing blood from one goat to another. By exchanging blood between a doe in labor with one that is not, he should be able to determine whether blood-borne factors suffice to initiate maternal care.

Finally . . .

It is unlikely that our other subjects, the lemurs, whose young are relatively more helpless than kids, will depend on similar mechanisms of attachment. Yet, the questions they raise are similar, and the differences in underlying mechanisms will be instructive. For instance, *Lemur catta* mothers apparently will allow others in their troop to fondle their infants within the first several days after birth and are keen to handle the infants of other mothers too. In the rather similar *L. fulvous*, however, mothers retain almost exclusive control over their own infants for most of their first month. Finally, in *L. variegatus*, if our lab colony is giving us an accurate indication of normal patterns, the young are left in a nest, to amuse themselves alone except for sporadic nursing interludes (Klopfer and Klopfer 1970). The infant, after it can locomote, must take the initiative in establishing contact with its mother.

We tend to extrapolate from one species to another with great facility. So much of the current discussion on aggression or territoriality in man, for instance, stems from the assumption that, because some animals show territorial or aggressive behavior, similar mechanisms underlie the functionally equivalent behavior in man. We overlook the multiplicity of mechanisms that may be employed to a common end. For example, the bright blue hues of many birds and insects, identical to our eyes, may depend on mechanisms as varied as absorbing pigments or refracting scales. These studies of maternal care should help us to determine the rules of the game, whether and when functional convergence implies uniformity in mechanism.

References

Altman, M. 1958. Social integration of the moose calf. *Animal Behaviour* 6:155–59.

Brower, L. P., Z.u.Z. Brower, and J. Corvino. 1967. Plant poisons in a terrestrial food chain. *Proceedings*, National Academy of Science 57:893–98.

Folley, S. J., and G. S. Knaggs. 1965. Levels of oxytocin in the jugular vein blood of goats during parturition. *J. Endocrinology* 33:301–15.

Gilbert, B. K. 1964. Development of social behavior in the fallow deer (*Dama dama*). *Zeitschrift fur Tierpsychologie* 25:867–76.

Gilbert B. K. The influence of foster rearing on adult social behavior in fallow deer. Duke Univ., Ph.D. thesis.

Hemmes, R. B. The ontogeny of the maternal-filial bond in the domestic goat. Duke Univ., Ph.D. thesis.

Klopfer, P. H., and M. S. Klopfer. 1968. Maternal "imprinting" in goats: Fostering of alien young. *Zeitschrift fur Tierpsychologie* 25:862–66.

Klopfer, P. H., and J. Gamble. 1967. Maternal "imprinting" in goats: The role of chemical sense. *Zeitschrift fur Tierpsychologie* 23:588–92.

Klopfer, P. H., D. K. Adams, and M. S. Klopfer. 1964. Maternal imprinting in goats. *Proceedings*, National Academy of Science 52:911–14.

Klopfer, P. H., and M. S. Klopfer. 1970. Patterns of maternal care in Lemurs: 1. Normative description. *Zeitschrift fur Tierpsychologie* 27:984–96.

Klopfer, P. H. Pattern of maternal care in three species of *Lemur*: Effects of early separation. Unpublished MS.

Peeters, G., M. Debackere, M. Lauryssens, and E. Kuhn. 1965. Studies on the release of oxytocin in domestic animals. In *Symposium on Advances in Oxytocin Research, 1964*, J. H. M. Pinkerton, ed. Pergamon Press.

30

Imprinting: Reversal of a Preference established during the Critical Period

Two distinctive features of "imprinting" originally emphasized by Lorenz[1] were that there was a critical period in which the preference for a particular species was established and that the preference established during this period was permanent and could not be changed by subsequent experience. These two features have been reiterated by Hess[2], who has added a primacy-recency feature, claiming that the first imprinting experience has priority over a subsequent one. These features of imprinting have been questioned by Sluckin and Salzen[3], who treated imprinting as a perceptual learning phenomenon in which the sensitive period is experience dependent and the stability of an imprinted preference is dependent on the amount of experience. More recent reviews by Sluckin[5] and Bateson[6] have supported this view. In particular the perceptual learning view of imprinting has been developed into a neuronal model hypothesis of imprinting by Salzen[4] and it predicts that object preferences established by the imprinting process should be subject to reversal given sufficient exclusive and enforced experience of new objects after the end of the so-called critical period. The present experiment demonstrates a reversal of this kind.

The experiment used Cornish × White Rock chicks hatched in separate boxes and transferred when 12–18 h old to isolation rearing cages. In the centre of each cage separate from the sources of food, water and heat, there was either a dark blue or a green cloth covered paper ball about 5 cm in diameter and hanging approximately 3 cm above the floor. The chicks quickly became strongly attached to the balls, spending much time beside them, and interacted with the balls by pushing, pulling, and pecking them. After three days (12 h light/dark cycle) the chicks were tested for their preference between blue and green balls. The balls were hung midway on opposite long sides of a box (45 × 30 cm). The chick under test was placed in the dark at the mid-point of a short side, a lamp was switched on above the box, and the chick was given 2 min in which to go to and stay with one of the balls. A preference was recorded if the chick after reaching a ball either contacted it, pecked and pulled it, and/or gave pleasure calls, or stayed close (1 in.) beside

265

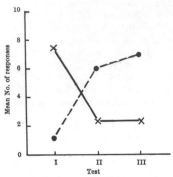

Fig. 1. Mean numbers of responses to the first-learned (×) and second-learned (●) coloured balls in the two-choice discrimination tests. Each of the twenty-two chicks had ten 2 min trials in each of three tests which were given at intervals of three days.

the ball and was silent or pleasure calling. The trial was ended when either of these criteria was reached or at the end of 2 min. The light was switched off, the chick was returned to a holding box and the positions of the objects were reversed ready for the next trial. Ten trials were made with each chick. After testing the chicks were returned to their cages and the balls exchanged so that each chick now had a new and different coloured ball. The chicks began to respond to the new balls on the same day. Three days later the chicks were again tested for their preference between green and blue balls. Then they were returned to their cages for a further three days but this time without any balls present. Finally, a third preference test was given. In this way twelve chicks were imprinted with a blue ball and then given reversal training with a green ball, while ten chicks were tested in the opposite manner.

The results are shown in Fig. 1 in terms of the mean number of responses made by the chicks to each ball in each preference test. The two colour treatments are combined because they were nearly balanced and gave similar results. The results show that at the end of the first 3 days the chicks had an almost exclusive preference for their familiar coloured ball ($p < 0.01$, Wilcoxon test). After the second three day period, this preference had been reversed ($p < 0.01$, Wilcoxon test). At the third test the new preference was maintained or even slightly increased because the responding to the preferred ball had increased significantly since the second test ($p < 0.02$, Wilcoxon test). Twenty-one of the twenty-two chicks made more responses to their familiar ball on the first test. On the second and third tests eighteen and seventeen of these twenty-one chicks made more responses to the newly experienced ball.

There can be no doubt that the period in which the first ball was experienced included any possible critical period for imprinting in chicks. The experience began 18 h after hatching, and Hess[5] has claimed that this period reaches its peak at this time for chicks. Furthermore, the second ball was experienced well after the end of the critical period of 24 h determined by Hess. Many studies[3] have shown that domestic chicks imprint with objects when exposed to them within the first 3 days after hatching. Thus the experiment shows that a strong preference established during this imprinting period can be completely reversed by subsequent experience. It also shows that this reversal was maintained intact after a period in which forgetting of both objects could have occurred. It cannot be said therefore that learning which takes place after the imprinting period is more rapidly forgotten than the learning during the imprinting period. Thus it would seem that the most recently learned preference predominated contrary to Hess's[2] claim of a primacy effect in imprinting. The recent test by Kaye[6] of a primacy effect gave an equivocal result, probably because of the very short training periods involved. The present test used stimulus objects that differed only in colours that were known from pilot studies to be equally easily learned and discriminated by chicks. The design was balanced for colour and similar results obtained with either colour. The training periods were long and the resulting preferences very strongly developed. Finally, the behaviour in the tests involved patterns of social interaction as well as the simple approach response. This study therefore represents the first closely controlled laboratory demonstration of reversibility of imprinting in precocial birds. It confirms the predictions of the neuronal model hypothesis of imprinting and agrees with the field observation of Steven[7]. In the field study by Schein[8] there appeared to be evidence for an irreversible preference for humans among human imprinted turkeys. It should be noted that these turkeys were always able to see humans as well as their later flock companions. Under these circumstances the neuronal model hypothesis would not necessarily predict a change in preference. The present demonstration of reversibility of imprinting in chicks agrees with the demonstration of a reversible social attachment in lambs[9]. Similar results have been obtained with a shape discrimination and a full account will be published elsewhere.

This work was supported by the National Research Council of Canada.

Eric A. Salzen
Cornelius C. Meyer

Department of Psychology,
University of Waterloo,
Waterloo, Ontario.

Received April 21; revised May 22, 1967.

[1] Lorenz, K. Z., *Auk*, **54**, 245 (1937).
[2] Hess, E. H., *Science*, **130**, 133 (1959).
[3] Sluckin, W., and Salzen, E. A., *Quart. J. Exp. Psychol.*, **13**, 65 (1961). Sluckin, W., *Imprinting and Early Learning* (Methuen, London, 1964), Bateson, P. P. G., *Biol. Rev.*, **41**, 177 (1966).
[4] Salzen, E. A., *Symp. Zool. Soc. Lond.*, **8**, 199 (1962).
[5] Hess, E. H., *J. Comp. Physiol. Psychol.*, **52**, 515 (1959).
[6] Kaye, S. M., *Psychonom. Sci.*, **3**, 271 (1965).
[7] Steven, D. M., *Brit. J. Anim. Behav.*, **3**, 14 (1955).
[8] Schein, M. W., *Zeit. Tierpsychol.*, **20**, 462 (1963).
[9] Cairns, R. B., and Johnson, D. L., *Psychonom. Sci.*, **2**, 337 (1965).

31

Monkeys Reared in Isolation with Pictures as Visual Input: Evidence for an Innate Releasing Mechanism

Abstract. *Monkeys reared in isolation from birth to 9 months received varied visual input solely from colored slides of monkeys in various activities and from nonmonkey pictures. Exploration, play, vocalization, and disturbance occurred most frequently with pictures of monkeys threatening and pictures of infants. From 2.5 to 4 months threat pictures yielded a high frequency of disturbance. Lever-touching to turn threat pictures on was very low during this period. Pictures of infants and of threat thus appear to have prepotent general activating properties, while pictures of threat appear to release a developmentally determined, inborn fear response.*

Research on a wide variety of animals has shown that early experiences can be important determinants of later social and nonsocial behavior (*1*). Some of these experiences apparently must occur during a limited developmental period if the animal is to exhibit behavior patterns normal for its species. One important, but relatively neglected, area of study in primate behavior involves determination of the developmental importance of different types of sensory input early in life. The experiment reported here is part of a study examining the effects of visual social and nonsocial stimulation presented to monkeys otherwise reared in total social isolation. The present study asks if totally naive infant monkeys will show differential behaviors toward specific types of visual stimulation, and whether such differential behaviors mature at specific periods during the monkey's development.

Four male and four female rhesus monkeys (*Macaca mulatta*) were reared in individual wire cages (61 by 71 by 71 cm) from birth to 9 months. Three walls, the ceiling, and the area below the wire floor of each cage were covered by Masonite or aluminum panels blocking all visual access to the world outside of the cage. The rear wall of each cage was a rear projection screen, which also blocked visual access to

the outside. The screen had a ground surface, preventing a mirror effect on the inner surface. Thus, reflections of the monkey's own image on the screen were minimized, although slight shadows did appear when the screen was brightly illuminated by the projector. A nonmovable brass lever, 0.6 cm in diameter, projected 7.6 cm into the cage at the bottom right corner of the screen. During rearing, the subjects never saw another monkey, and saw no humans after a hand-feeding period during the first 5 to 9 days of life. Sounds were not controlled, so the monkeys did hear other animals and humans.

On day 14, two types of visual stimulus presentations were begun, as follows.

Experimenter-controlled slides. In this procedure pictures of monkeys engaged in different activities and nonmonkey control pictures were projected on the screen. Each individual picture was available for 2 minutes. Six monkey and two control pictures, randomly selected from a large pool, were projected in a daily test session. In these tests the subject had no direct control over the onset or duration of the picture.

Animal-controlled slides. In this procedure the subject could expose itself to pictures by touching the brass lever, which operated an electronic contact relay circuit. Lever-touching opened a shutter located in front of a Kodak Carousel slide projector, and also operated a printing counter that recorded the touching behavior. The standard schedule of six randomly selected monkey and two control pictures was used in each test session. Each picture was potentially available for 5 minutes and 15 seconds. During the first 15 seconds the shutter automatically opened, exposing the subject to the picture that would be available for the next 5 minutes. During the next 5 minutes each lever touch turned the picture on for 15 seconds. If the subject continued to touch the lever, the picture still went off at the end of 15 seconds, and did not come on again until contact was broken and then the lever was touched again.

Up to the fourth month of life all subjects received a minimum of two experimenter- and two animal-controlled tests each week. At 4 months, four subjects received motion picture stimuli, but these monkeys were still exposed to at least one trial of experimenter-controlled slides each week. The motion picture data parallel the data for slides, and will not be treated here.

The 2- by 2-inch (5- by 5-cm) colored slides used as stimuli were grouped into ten categories (four of which are illustrated in Fig 1). They include pictures of (i) threatening, (ii) playing, (iii) fearful, (iv) withdrawing, (v) exploring, and (vi) sexing monkeys, as well as pictures of (vii) infants, (viii) mother and infant together, and (ix) monkeys doing "nothing." Labels were applied, and stimuli were included for study, only when the pictures received a unanimous title by a panel of eight experienced monkey testers. The final category, (x) control pictures, included a living room, a red sunset, an outdoor scene with trees, a pretty adult female human, and various geometric patterns. Each of the ten categories had at least four different examples. Only one randomly chosen slide from each category was used in any individual test session. Complete randomness was restricted such that no single slide appeared more than once in a 7-day period. The monkey pictures were projected with an approximately life-size image.

In animal-controlled tests, the basic data were the number of lever touches producing shutter openings out of the 20 possible shutter openings in each 5-minute stimulus period. In experimenter-controlled tests data were collected on a checklist by an observer looking through a nonreflecting one-way viewing screen in one wall of the cage. These checklist data were frequencies per 2-minute stimulus period of (i) vocalization, (ii) disturbance behaviors including rocking, huddling, self-clasping, fear, and withdrawal, (iii) playing with the picture, (iv) visual and manual exploration of the picture, and (v) climbing on the walls of the

Fig. 1. Examples of picture stimuli from the ten categories. The four categories selected for illustration here are (top left) fear, (top right) threat, (bottom left) infant, and (bottom right) control. The actual pictures were in natural color.

cage. Sex, threat, and aggressive behaviors were also scored, but sexual responses never occurred, and only three threatening or aggressive responses were observed during the 9 months of testing.

Evidence for behavioral differentiation of the ten types of stimuli during experimenter-controlled tests is shown in Table 1. These results can be summarized as follows. First, threat and infant pictures produced the greatest frequency of response on all measures of behavior. All differences between these two types of pictures and the other pictures taken individually were significant beyond the .01 level. Second, there was more vocalization and disturbance with threat than with infant pictures (both $p < .02$). Third, there were no significant differences between the remaining seven monkey pictures (fear, withdrawal, explore, sex, mother-infant, and nothing) on any measure (all $p > .10$). Tests of overall differences between any of these seven monkey pictures *for each month* during rearing also failed to reveal significant variation (all $p > .05$), indicating that these seven pictures were responded to similarly throughout rearing. Fourth, the control pictures had significantly lower frequencies than these remaining seven monkey pictures pooled together on the vocalization, disturbance, and play measures (all

$p < .04$), but no significant differences appeared on exploration and climbing activity (both $p > .10$).

The development of responding during experimenter-controlled tests is summarized in Fig. 2. In these data the seven monkey pictures that did not differ significantly from one another in overall frequency of eliciting responses are pooled together to form an "other-monkey" category. Disturbance behavior occurred at a uniformly low level throughout the 9-month period for all pictures except threat. Beginning at 2 to 2.5 months, and peaking at 2.5 to 3 months, disturbance behavior consisting primarily of fear, withdrawal, rocking, and huddling occurred at high levels whenever pictures of monkeys threatening appeared on the screen. At 3.5 months this apparently innate fear response to threat stimuli declined. The vocalization measure, a response that reflects disturbance but may also reflect contentment, showed a course of development similar to disturbance, except that vocalizations were relatively high with infant pictures. After the first month of life, pictures of threat and of infants received more exploration and play than did other-monkey or control pictures. Interestingly, the first stimulus pictures to receive a relatively high degree of play were threat, which were played with even during periods when

Table 1. Overall responsiveness to pictures. Mean frequency of the five behavioral measures for all 2-minute periods, during the 9 months of testing.

Stimulus picture	Behaviors				
	Vocalization	Disturbance	Play	Exploration	Activity (climbing)
Threat	0.61	0.83	0.46	5.0	1.7
Infant	.48	.37	.54	4.8	1.8
Withdraw	.27	.20	.17	2.4	0.8
Fear	.27	.15	.26	2.3	.9
Play	.17	.11	.11	2.1	.8
Explore	.22	.19	.19	2.2	.8
Sex	.17	.24	.18	2.2	.8
Mother-infant	.29	.25	.19	2.5	.7
Nothing	.22	.21	.31	2.7	.9
Control	.14	.12	.16	2.6	.7

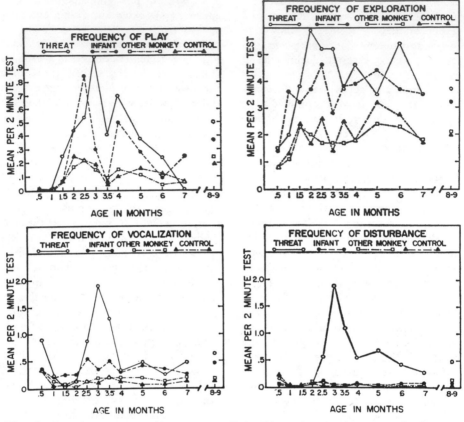

Fig. 2. Reactions to experimenter-controlled slides: the development of play, exploration, vocalization, and disturbance behaviors in response to pictures of threatening monkeys, infants, all other monkey pictures pooled, and control pictures.

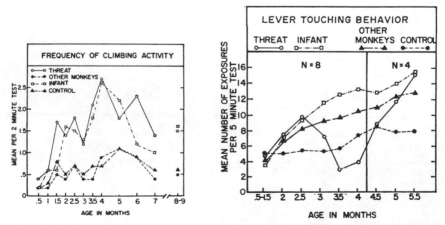

Fig. 3 (left). The development of climbing behavior during stimulation, taken as an index of general activity level. Fig. 4 (right). Frequency of self-exposures to threat, infant, other monkey, and control pictures for the first 6 months of animal-controlled tests.

disturbance responses to threat pictures were high. Climbing responses (Fig. 3), taken as an index of general activity, exhibited a large increase after the first month. Climbing was most frequent in response to threat and infant pictures from this time until the end of testing. Also illustrated in Figs. 2 and 3 is the finding that all measures for all stimuli generally leveled off at about 6 months and remained constant thereafter.

Figure 4 presents the data for animal-controlled pictures for all stimuli through the first 6 months of life. Lever-touching was equal for all pictures during month 1. During months 2 and 3 the subjects began exposing themselves more to pictures containing monkeys than to control pictures. About month 3, when the subjects were beginning to show disturbance to threat pictures in experimenter-controlled tests, lever-touching for threat declined markedly. However, responding for infant and other-monkey pictures continued to increase. During month 4 responding to threat pictures was depressed below the control-picture level, but lever-touching for threat did increase again during months 5 and 6. Lever-touching was generally higher for pictures of infants than for other-monkey pictures from month 2 until the end of testing. Although lever-touching may have been influenced by differential general activity with the various stimuli, it is unlikely that the decrease in response for threat pictures was the result of such a factor. As shown in Fig. 3, climbing activity was similar for both threat and infant pictures from months 2 to 5, yet infant pictures elicited a high level of lever-touching while touching for threat pictures declined markedly.

In general, the shapes of the curves in Figs. 2–4 were characteristic of the behavior of individual subjects. The individual curves tended to follow the major inflections shown in the averaged data, although there were discrepancies of up to 1 month in the exact age at which a given monkey might show a large increase or decrease in a given behavior. The most important case concerns disturbance behavior in response to threat pictures. Two monkeys showed a large increase in disturbance at 2 months of age, four animals showed increases at 2.5 months, and the remaining two subjects showed increased disturbance at 3 months. Kendall's Coefficient of Concordance was calculated to measure the degree of consistency in disturbance with threat pictures between subjects over age blocks. This measure revealed a very high degree of association between subjects ($W = .808$; $p < .001$), indicating that individual animals behaved in a very similar manner toward the threat pictures.

These data lead to several important conclusions. First, at least two kinds of socially meaningful visual stimuli, pictures of monkeys threatening and pictures of infants, appear to have unlearned, prepotent, activating properties for socially naive infant monkeys. From the second month of life these stimuli produced generally higher levels of all behaviors in all subjects. Second, the visual stimulation involved in threat behavior appears to function as an "innate releasing stimulus" for fearful behavior. This innate mechanism appears maturational in nature. Thus, at 60 to 80 days threat pictures release disturbance behavior, although they fail to do so before this age. These fear responses waned about 110 days after birth. This could be due to habituation, occurring because no consequences follow the fear behavior released by threat pictures—consequences that would certainly appear in a situation with a real threatening monkey.

One important implication of these results concerns the ontogeny of responses to complex social communication in primates. These data suggest that at least certain aspects of such communication may lie in innate recognition mechanisms, rather than in acquisition through social learning processes during interactions with other animals. Although the maintenance of responses to socially communicated stimuli may well depend on learning and some type of reinforcement process, the initial evocation of such complex responses may have an inherited, species-specific structure. Thus, these data suggest that innate releasing mechanisms such as those identified by ethologists (2) for insect and avian species may also exist in some of the more complex behaviors present in the response systems of primates.

GENE P. SACKETT

Primate Laboratory, University of Wisconsin, Madison 53715

References and Notes

1. See, for example, R. Melzack, in *Pathology and Perception* (Grune and Stratton, New York, 1965); A. H. Riesen, in *Functions of Varied Experience*, D. W. Fiske and S. R. Maddi, Eds. (Dorsey Press, Homewood, Ill., 1961); M. R. Rosenzweig, *Amer. Psychol.* **21**, 321 (1966); G. P. Sackett, *Child Develop.* **36**, 855 (1965).
2. W. H. Thorpe, *Learning and Instinct in Animals* (Harvard Univ. Press, Cambridge, 1963).
3. Supported by grant MH-4528 from the National Institute of Mental Health. Portions of these data were reported at the 10th Interamerican Congress of Psychology, Lima, Peru, April 1966. I thank C. Pratt and L. Link for assistance with this research.

3 October 1966

PART 5

Sensory Processes, Communication, and Orientation

Human beings have long been curious about the mental life of the creatures who share the earth with us, and we have posed many questions regarding the existence of "consciousness" and similar processes in animals. Some of these questions are forever unanswerable, but modern experimentation has conclusively demonstrated that many animals possess senses and means of communication and orientation that we do not have. Although experimentation on the topics included in this section is difficult and demanding, it can be most rewarding since many of the most interesting problems in animal behavior are found in these areas.

We begin with a paper that is similar to Reading 8, but not because Reading 8 concerns crickets and Reading 32 cricket frogs. Rather because both studies show that the sensitivity of the auditory system has evolved in relationship to the calls of the species. In both species such specificity in acoustic production and reception can act as a reproductive isolating mechanism. In certain other species, however, the auditory systems have evolved in response to different selective pressures, including the presence of enemies. Thus, when a predator develops a system for the capture of certain prey, it should not be surprising that the prey, in turn, evolves mechanisms for the detection and evasion of the predator. For example, it is known that bats are capable of capturing insects by means of echolation. Reading 33 presents one evolutionary "answer" to the bat as it has evolved in several families of moths.

Since the auditory system of certain moth families is tuned to the cry of particular predators, it is "logical" (considering the simplicity of their nervous system) that they do not depend upon hearing to locate a mate. The olfactory sense serves this function. Yet in other species the situation is partially reversed, and olfaction is used to detect and select prey. For example, different species of snakes prefer different kinds of prey. In Reading 34 Gordon Burghardt shows that newborn, unfed snakes exhibit the same species differences in their responses to water extracts of the skin substances of various prey.

In Reading 35 we are introduced to a sense modality that man apparently does not possess: the capacity to produce and/or detect electric currents. Note that this sense is used both to detect objects in the environment and to communicate with other members of the species. The article also shows that knowledge from other disciplines, in this case physics, can be of great aid in the study of animal behavior.

Recent years have witnessed a sometimes bitter controversy as to whether honey bees use the information contained in their dances. This literature is reviewed in Reading 36, and then critical experiments that seemingly resolve the issue are reported. Only time will tell if the case of the "dance-language controversy" has been finally closed.

While Reading 33 provided an example of one species "tuning-in" on the cries of another in order to avoid being eaten, Reading 37 presents the reverse occurring: female fireflies of a particular species mimic the flash responses of females of other species. Males attracted to the mimics do not receive the response that they might have anticipated from females of their own species. In short, they are "seized and devoured." As Aubrey Manning, a long-time student of insect behavior, said when he read this paper, "Leave it to the insects . . . !"

Readings 38 and 39 are concerned with orientation. In the first, Stephen T. Emlen reports on the effects of celestial rotation on the development of migratory orientation in indigo buntings. In the second, C. Walcot and R. P. Green describe how the orientation of pigeons can be altered by the induction of magnetic fields around their heads.

32

Encoding of Geographic Dialects in the Auditory System of the Cricket Frog

Robert R. Capranica, Lawrence S. Frishkopf and Eviatar Nevo

Reprinted from
21 December 1973, Volume 182, pp. 1272-1275

Abstract. *The frequency sensitivity of the auditory nervous system of cricket frogs* (Acris) *varies geographically. This variation is closely matched to the spectral energy in their mating calls, thus enabling them to respond preferentially to the calls of their local dialect.*

The cricket frog gets its name from the cricket-like sound of the male's mating call. The two species that comprise this genus (*Acris*) are both found in the United States: *A. gryllus* occurs in the southeastern part, while *A. crepitans* is found throughout most of the country east of the Rockies. The two species are sympatric in the southeastern states from Virginia to Louisiana (*1*). Behavioral studies in the field have shown that females of each species respond preferentially to the mating calls of their own species (*2*). Thus the species-specific mating call of the male and the selective response of the female provide a reproductive isolating mechanism between these two species in their zone of sympatry. Furthermore, not only are their mating calls species-specific, but they are also geographically specific. When the mating call of a local male *crepitans* and the mating call of a male *crepitans* from a different geographical locality are presented simul-

taneously through separate loudspeakers, a female *crepitans* will respond preferentially to the mating call from her local population. By studying the response of female *crepitans* to synthetic calls, Capranica and Nevo (*2*) have identified the signal characteristics in the male's call that permit the female's recognition of calls from her own local population. This report presents evidence that the auditory nervous system of the cricket frog is "tuned" to the local dialect and thereby provides a major basis for the female's selective response to the calls of males of her own population. Our evidence is based on electrophysiological recordings from cells in the medullary auditory nuclei (*3*) of 17 cricket frogs obtained from New Jersey, South Dakota, and eastern Kansas.

Adult cricket frogs, about 1 inch (2.5 cm) in length and 1.5 to 2.0 g in weight, were anesthetized with Dialurethane (Ciba) (3 $\mu l/g$); exposure of

275

the medulla was made by a ventral approach through the roof of the mouth. Temperature of the animals was maintained between 22° and 25°C. Acoustic stimuli were delivered by a PDR-10 earphone enclosed in a sealed housing which was fitted around the animal's eardrum to provide a closed stimulus system. Stimuli consisted of tone bursts controlled by an electronic switch (Grason-Stadler 820E); sound intensity was monitored by a condenser microphone (B & K 4134) sealed in the earphone housing. Indium-filled pipettes with platinum-black tips (4) were used in recording the spike activity of single cells.

Isolation of single nerve cells in these small frogs was difficult. We typically observed multi-unit activity as the electrode was advanced through the auditory nucleus in the medulla. We succeeded in clear isolation of 66 units: the frequency and threshold sensitivity of the background multi-unit activity closely matched the response properties of the single cells that were isolated. We therefore are confident that the response characteristics of the single units which we present are representative of the auditory sensitivity of cricket frogs from the different geographical populations.

Two distinct types of single units were found in the medullary nuclei of cricket frogs (5). One type was sensitive to low-frequency tones and the other type was sensitive to high-frequency tones. From previous studies of the bullfrog's auditory system (6), we believe that the units tuned to low frequencies derive their input from the auditory nerve fibers that innervate the amphibian papilla in the inner ear, while the units tuned to high frequencies derive their input from fibers from the basilar papilla. In this respect, amphibians are exceptional among vertebrates in having two separate and anatomically distinct auditory receptor organs. Each organ is tuned to a separate frequency region which presumably provides the basis for the bimodal frequency sensitivity of anurans in general (7).

Figure 1A shows representative tuning curves (plot of neuron's threshold versus tonal frequency) for units of each type in *A. crepitans* from New Jersey. The low-frequency unit has its best frequency at 550 hz, whereas the high-frequency unit has its best frequency at 3550 hz (8). Note that the tuning curves are disjoint and do not overlap even at high sound intensities. The disjoint nature of their frequency sensitivities permitted ready classification of different units as either of low-frequency type or of high-frequency type. We found that the best frequencies of the low-frequency units were distributed over the range 200 to 1000 hz, while *all* of the high-frequency units had their best

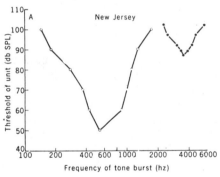

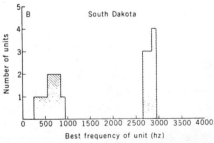

Fig. 1. (A) Tuning curves of a low-frequency unit, best frequency 550 hz, and of a high-frequency unit, best frequency 3550 hz, in *Acris crepitans* from New Jersey (0 db SPL = 0.0002 dyne/cm²). (B) Histogram of best frequencies for 20 low- and high-frequency units in *A. crepitans* from South Dakota.

frequencies narrowly clustered around 3500 to 3550 hz. Furthermore, while the low-frequency units had moderately sensitive thresholds of 35 to 60 db SPL (sound pressure level), *all* of the high-frequency units had relatively poor thresholds around 75 to 80 db SPL. These conclusions are based on the single units that we could isolate as well as the multi-unit background activity that we routinely monitored: all of the auditory (single and multi-unit) activity that we observed in the New Jersey animals was tuned to these two frequency and threshold ranges.

The signal characteristics of a typical mating call from an adult male *crepitans* recorded in New Jersey consists of a sequence of clicks having a stereotyped temporal pattern. The energy in each click has a distinct spectral peak centered around 3550 hz (Fig. 2A). This frequency is a characteristic feature of the mating calls of male *crepitans* in New Jersey: the spectral peaks in the calls of different males, or of the same male at different times, usually show a standard deviation of less than 100 hz from the mean of 3550 hz. A comparison with the sensitivity of the high-frequency units indicates that these units are all selectively tuned to the spectral peak in the male's calls.

Suppose we consider a different population of *Acris crepitans*. Figure 1B shows a histogram of best frequencies for the medullary units that we were able to isolate in *crepitans* from South Dakota. The histogram is bimodal and emphasizes the disjoint frequency sensitivity of the two types of units. Those units sensitive to low frequencies, tuned to the range of 200 to 1000 hz, had thresholds distributed between 35 and 60 db SPL. These units were very similar to the low-frequency units in *crepitans* from New Jersey. The high-frequency (both single and multi) units were all narrowly tuned, not to 3550 hz as in New Jersey *crepitans*, but rather to 2900 hz. All of these high-frequency units, as in the New Jersey

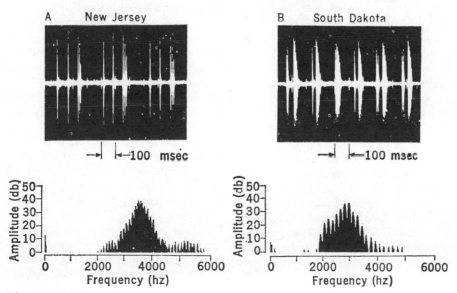

Fig 2. (A) The mating call of an adult male *A. crepitans* from New Jersey consists of a repeated sequence of clicks. The spectrographic section shows the distribution of energy in a single sequence; this distribution remains the same throughout the call. (B) Mating call of an adult male *A. crepitans* from South Dakota. Details are the same as in (A).

animals, were fairly insensitive, with thresholds at their best frequencies around 75 to 80 db SPL.

A representative mating call of an adult male crepitans from South Dakota (Fig. 2B) consists of a sequence of clicks with a temporal pattern that is characteristic of the males from this geographic population. The energy in each click of this call has a spectral peak, not at 3550 hz as in New Jersey crepitans, but instead at 2900 hz. The location of the spectral peak in the mating calls of male crepitans from South Dakota matches the narrow tuning of the high-frequency unit activity in this population of cricket frogs.

These results suggest that, where a shift in spectral energy in the mating call occurs throughout the geographical range, the tuning of the high-frequency units follow this shift. We pursued this conclusion with studies of crepitans from eastern Kansas. The high-frequency units from this population were all narrowly tuned around 3750 hz. The mating calls of males in eastern Kansas have a spectral peak narrowly centered around this same frequency.

Cricket frogs breed in very dense populations in which the distance between calling males is but 1 to 2 m. Behavioral studies in the field reveal that the acoustic intensity of a calling male at a distance of 1 m is typically about 95 db SPL. By employing synthetic mating calls in auditory discrimination tests in the field, Capranica and Nevo (2) found that female crepitans would no longer respond to a stimulus of this intensity if the spectral peak was shifted by as little as 500 hz either above or below the mean for the local population, even though the temporal pattern of clicks was appropriate for that population. We suggest, therefore, that the matched frequency sensitivity of the high-frequency units—and of the basilar papilla from which they presumably derive—provides a basis of such selective response behavior. Our finding that the

high-frequency units all have rather high threshold may therefore have a special significance (9). The relative insensitivity of the high-frequency units permits spectral peak detection, narrowly tuned to the dominant frequency in the local call but not stimulated by neighboring frequencies of lower amplitude nor by the dominant frequencies of other dialects.

The sensitivity of low-frequency units in crepitans does not show any apparent variation with geographic population. Discrimination studies in the field involving filtered mating calls and synthetic calls indicate that the presence of high-frequency energy alone enabled the female to recognize the male's call. This result contrasts with earlier studies of the bullfrog (Rana catesbeiana). The bullfrog requires simultaneous presence of low- and high-frequency sound energy, exciting both basilar and amphibian papillae, to evoke his mating call (10).

The role of the low-frequency units in the cricket frog's auditory behavior requires further clarification (11). In studies of auditory discrimination involving synthetic calls, the addition of low-frequency energy in the range of 200 to 1000 hz to an effective stimulus inhibited the female's response. When this low-frequency energy was removed by filtering, the female readily approached the sound source. Since large species of frogs and toads typically produce low-frequency energy in their mating calls, and since these species are potential predators for the cricket frog, the low-frequency sensitivity—presumably from the amphibian papilla—may function to alert the cricket frog of nearby danger. This function would be consistent with the moderately sensitive thresholds of the low-frequency units.

The frequency sensitivities of the receptor organs in the inner ear of the cricket frog (12) provide a mechanism for discrimination of local dialects based on the spectral energy in the mating call. Only those signals which pass this peripheral filter can affect neural activity in higher auditory centers. Another feature of the dialect is the stereotyped temporal pattern of clicks within the call; this pattern varies with geographic population. The field studies of Capranica and Nevo (2) indicate that the pattern of clicks is a very important feature for call recognition among these animals. The manner in which this temporal information is extracted by the central auditory system is unknown.

One of the interesting aspects of this study is the question of how geographical dialects are maintained, namely, are the distinctive characteristics of a male's mating call and the corresponding matched auditory sensitivity of members of his local population determined by genetic mechanisms, or by environmental influence, or are they learned? Further experiments hopefully will reveal the factors that can lead to the maintenance of geographical variation in communicatory signals. Such studies could also provide insight into how vocal dialects might originate.

ROBERT R. CAPRANICA

*Section of Neurobiology and Behavior,
Cornell University,
Ithaca, New York 14850*

LAWRENCE S. FRISHKOPF

*Department of Electrical Engineering,
Massachusetts Institute of Technology,
Cambridge 02139*

EVIATAR NEVO

*Department of Biology,
Haifa University, Haifa, Israel*

References and Notes

1. R. Conant, *A Field Guide to Reptiles and Amphibians* (Houghton Mifflin, Boston, 1958).
2. E. Nevo, in *Systematic Biology* (National Academy of Sciences, Washington, D.C., 1969), pp. 485–489; R. R. Capranica, *Neurosci.*
   Res. Program Bull. **10**, 16 (1972); ——— and E. Nevo, in preparation.
3. These medullary neurons receive their input from auditory nerve fibers that innervate the receptor organs in the inner ear. Because of the very small size of these animals, it was not possible to record from single primary fibers in the auditory nerve.
4. R. M. Dowben and J. E. Rose, *Science* **118**, 22 (1953).
5. R. R. Capranica and L. S. Frishkopf, *J. Acoust. Soc. Am.* **40**, 1263 (1966).
6. L. S. Frishkopf and M. H. Goldstein, Jr., *ibid.* **35**, 1219 (1963); L. S. Frishkopf and C. D. Geisler, *ibid.* **40**, 469 (1966); L. S. Frishkopf, R. R. Capranica, M. H. Goldstein, Jr., *Proc. IEEE* **56**, 969 (1968).
7. R. S. Schmidt, *Behaviour* **23**, 280 (1964); M. Sachs, *J. Acoust. Soc. Am.* **36**, 1956 (1964); T. Hotta, *J. Physiol. Soc. Jap.* **30**, 779 (1968); R. E. Greenblatt, MIT (Mass. Inst. Technol.) *Q. Prog. Rep.* **92**, 440 (1969); H. Liff, *J. Acoust. Soc. Am.* **45**, 512 (1969); J. J. Loftus-Hills and B. M. Johnstone, *ibid.* **47**, 1131 (1970); J. J. Loftus-Hills, *Z. Vgl. Physiol.* **74**, 140 (1971).
8. The frequency to which a unit is most sensitive is called its best frequency.
9. Anuran species whose mating calls possess spectral peaks at high frequencies generally have poor auditory sensitivity. It has been suggested that this decrease in auditory sensitivity is due to a fall-off in the frequency response of the middle ear [J. J. Loftus-Hills and B. M. Johnstone, *J. Acoust. Soc. Am.* **47**, 1131 (1970)].
10. R. R. Capranica, *The Evoked Vocal Response of the Bullfrog: A Study of Communication by Sound* (Research Monograph 33, MIT Press, Cambridge, Mass., 1965); *J. Acoust. Soc. Am.* **40**, 1131 (1966).
11. ———, *Neurosci. Res. Program Bull.* **10**, 65 (1972).
12. Matched sensitivity of the high-frequency units to the spectral peak in the local male's mating call was also found in *Acris gryllus*. For example, high-frequency units in the medulla of *gryllus* from Georgia are all narrowly tuned around 3600 hz. The mating calls of male *gryllus* in Georgia have a *distinct* spectral peak near this same frequency. The sensitivities of the low-frequency units in *gryllus* are similar to those of *crepitans*, namely, distributed over the range of 200 to 1000 hz. This sensitivity is consistent with the hypothesis that the amphibian papilla provides a warning function for these tiny species. The mating calls of sympatric *crepitans* in Georgia have their spectral peaks centered around 4050 hz. Furthermore, the temporal patterns of clicks in the calls of the two species are very different, so that species-specific recognition involves both spectral and temporal signal characteristics (2). We did not explore the possibility of geographical variation in the auditory system of *A. gryllus*, although the characteristics of their calls also vary geographically (2).
13. The electrophysiological studies were conducted while R.R.C. and L.S.F. were members of the technical staff, Bell Telephone Laboratories, Murray Hill, New Jersey. Recording and analysis of mating calls were supported, in part, by grants GB-18836 to R.R.C. and GB-3167 to E.N. from the National Science Foundation.

20 June 1973; revised 28 August 1973

KENNETH D. ROEDER
and ASHER E. TREAT

33   The Detection and Evasion of Bats by Moths*

A central objective of a large segment of biological and psychological research is to provide a physiological basis for behavior. The first step toward this objective is analytic, and consists of determining the structure and function of neural components after they have been isolated from their connections with the rest of the nervous system. There has been much progress in this direction, and it is now possible to describe in terms of input and output performance the operation of many isolated sense cells, neurons, and muscle fibers, even though the principles of their internal operation are mostly not understood.

The next step, the synthetic process of assembling this information on isolated neural components and relating it to the behavior of the intact animal, is hampered by two kinds of difficulty. The first appears to be methodological, but is somewhat hard to define. When one regards the ever-growing literature on the unit performance of sense cells, nerve cells, and muscle fibers, it is to experience that sense of dismay first

* Much of the experimental work reported in this paper was made possible by Grant E-947 from the U. S. Public Health Service.

encountered at a tender age when the springs, gears, and screws of one's first watch were strewn upon the table. The modus operandi of analysis or taking apart seems to come naturally, and the problems encountered are essentially technical in nature. Synthesis or the derivation of a system from its components seems to lack the a priori logic of analysis.

The second general difficulty is technical, and stems from the fact that even the simplest behavior of the higher animals and man is accompanied by the simultaneous activity of millions of sense cells, nerve cells, muscle fibers, and glands. Even if it were possible to register the traffic of nervous and chemical information generated and received by each and all of these neural elements during the behavior, it is doubtful whether the record would provide a meaningful description of the action.

Even though these problems cannot be solved directly at the present time, they become less formidable if the behavior selected for study is simple and stereotyped, and only a small number of nerve cells are concerned in its execution. These conditions are partly fulfilled by the sensory mechanisms whereby certain nocturnal moths detect the approach of insectivorous bats.

Echolocation and Countermeasure

Bats detect obstacles in complete darkness by emitting a sequence of high-pitched cries or chirps and locating the source of the echoes. As Griffin (1958) and others have shown, this form of sonar is unbelievably precise. By means of it, insectivorous bats locate and track flying moths, mosquitoes, and small flies (Griffin et al., 1960). North American bats, such as *Myotis lucifugus* and *Eptesicus fuscus,* emit chirps about 10 times a second when they are cruising in the open. Each chirp lasts from 10 to 15 milliseconds (msec.) with an initial frequency of 80 kilocycles (kc.) dropping about one octave in pitch toward its end. . . .

The frequencies in these chirps are ultrasonic, that is, inaudible to human ears, which cannot detect tones much above 15 to 18 kc. The higher frequencies used by bats make possible more discrete echoes from smaller objects. The chirps can be rendered audible by detecting them with a special microphone and rectifying the ultrasonic component. They then can be heard through headphones as a series of clicks. These clicks fuse into what Griffin has called a "buzz" when the bat is chasing an insect or avoiding an obstacle.

Several families of moths (in particular the owlet moths or Noctuidae) have evolved countermeasures enabling them to detect the chirps of bats. A pair of ultrasonic ears is found near the "waist" of the moth between thorax and abdomen. . . . An extremely thin eardrum or tympanic membrane is directed obliquely backward and outward into the recess (dark area) found at this point. . . .

Internal to the eardrum is an air-filled cavity that is spanned by a thin strand of tissue running from the center of the eardrum to a skeletal support. . . . This tissue contains the sound-detecting apparatus,

consisting of two acoustic sense cells (A cells). A single nerve fiber arises from each A cell and passes close to the skeletal support, where the pair is joined by a third nerve fiber arising from a large cell (B cell) in the membranes covering the support. The three fibers continue their course to the central nervous system of the moth as the tympanic nerve.

The traffic of nerve impulses passing over the three fibers from A cells and B cell to the nervous system of the moth can be followed if a fine metal electrode is placed under the tympanic nerve. Another electrode is placed in inactive tissue nearby. As each impulse passes the site of the active electrode it can be detected as a small action potential lasting about 1 msec. Since the tympanic nerve contains only three nerve fibers, it is not difficult to distinguish and to read out the respective reports to the nervous system from the pair of A cells and the B cell. A similar experiment in a mammal is practically meaningless since the auditory nerve contains about 50,000 nerve fibers.

This method of detection shows that the A cells transmit organized patterns of impulses over their fibers only when the ear is exposed to sound (Roeder and Treat, 1957). The B cell transmits a regular and continuous succession of impulses that can usually be distinguished from the A impulses by their greater height. The B impulses are completely unaffected by acoustic stimulation, and change in frequency only when the skeletal framework and membranes lining the ear are subjected to steady mechanical distortion (Treat and Roeder, 1959). The B cell behaves in a manner similar to receptors found in other parts of the body that convey information about mechanical stress on joints, muscles, and skeleton. The role of such a receptor in the ear of a moth is unknown.

In the absence of sound, the A cells discharge irregularly spaced and relatively infrequent impulses. . . . A continuous pure tone of low intensity elicits a more regular succession of more frequent impulses in one of the A fibers. . . . The other fiber is not yet affected. Any slight increase in the intensity of the tone causes a corresponding increase in the impulse frequency of the active fiber. When the intensity of the tone is increased to about tenfold that producing a detectable response in the more sensitive A fiber, the second A fiber begins to respond in like manner. Its action potentials are superimposed on those of the first . . . by the method of recording, but actually reach the central nervous system over their own pathway. This experiment reveals two of the ways in which the moth ear codes sound intensity. It is like an instrument having a graded fine adjustment

(the intensity-frequency relation) and a coarse adjustment of two steps (the pair of A cells). Other ways of coding intensity will appear later.

The moth ear responds in this manner to tones from 3 kc. to well over 100 kc., but there is no evidence that it is capable of discriminating between tones of different frequency. It is most sensitive near the middle of its range, that is, to frequencies such as those contained in bat chirps.

In Plate 2, Figure 2 [omitted, ed.], it will be noticed that, in each of the recordings, the intervals between the successive impulses increase as the pure tone stimulus continues. In terms of the nerve code out-

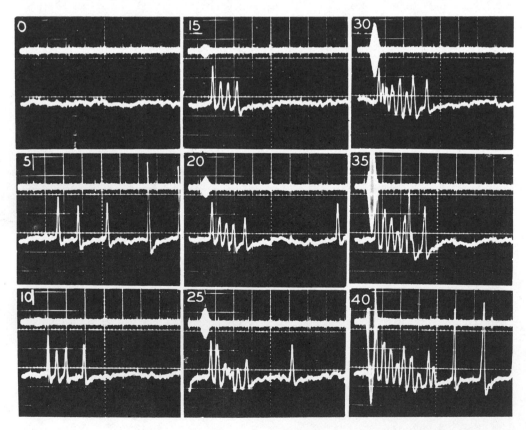

FIG. 1. Tympanic nerve responses (lower traces) of *Noctua* (= *Amathes*) *c-nigrum* to a 70 kc. sound pulse recorded simultaneously by a Granith microphone (upper traces). The numbers indicate the intensity of the sound pulse in decibels above a reference level (0). The threshold of the sensitive A cell lies between 0 and 5 db. The large spikes appearing in some of the records are from the B cell. The less sensitive A cell responds in the 25 db recording. Vertical lines, 4 msec. apart.

lined above, the A cells report that the sound is declining in intensity with time, although in fact it was kept constant. This adaptation to a constant stimulus occurs in most receptors registering changes in the outside world. In terms of our own experience, the impact of our surroundings would be shocking and unbearable if it were not distorted in this manner by sense organs. The brilliance of a lighted room entered after dark would continue to be blinding and the noise of a jet engine would remain unbearable. However, the A cells of the moth's ear adapt very rapidly to a continuous tone, and their full effectiveness as pulse detectors is revealed only when they are exposed to short tone pulses similar to bat chirps.

In the experiment illustrated in Figure 1, a tone pulse of 3 msec. duration was generated at regular intervals. It is similar to a bat chirp except for its regularity and the absence of frequency modulation. A microphone (upper trace) and moth ear (lower trace) were placed within range, and the intensity of the stimulus pulse was adjusted so that it just produced a detectable response in the most sensitive A fiber (0 db). The intensity was then increased by 5 decibel† (db) steps as each recording was made. It will be seen that the microphone begins to detect the sound pulse when it is about 10 db above the threshold of the most sensitive A cell in the moth's ear. As before, the increase in frequency of A impulses is evident if the 5 and 10 db records are compared, and a response of the less sensitive A cell appears first in the 25 db record where the extra peaks of its action potentials overlap those of the more sensitive A unit. In addition to these two ways of coding intensity, two more can now be recognized. If the interval between detection of the sound by the microphone and by the moth ear is compared at dif-

† The decibel (db) notation expresses relative sound pressures. An intensity of 20 db is tenfold that of the reference level (0 db), a 40-db sound is a hundredfold the reference level.

ferent sound intensities, it will be noticed that the tympanic nerve response occurs earlier and earlier on the horizontal time axis. In other words, the latency of the response decreases with increasing loudness. Also, the sense cells are seen to discharge impulses for some time after the sound has ceased, and this after-discharge becomes longer with increasing sound intensity.

The Detection of Bats

These experiments with artificial sounds suggest how the moth ear might be expected to respond to a bat cry. A few laboratory observations were made with captured bats. In one of these experiments, in collaboration with Dr. Fred Webster, the cries of a flying bat were picked up simultaneously by a moth ear and a microphone, and recorded on high-speed magnetic tape. . . . Interesting though they were, these experiments served mainly to show that the full potentialities of the moth ear as a bat detector could not be realized within the confines of a laboratory, and efforts were made to transport the necessary equipment to a spot where bats were flying and feeding under natural conditions.

Finally, about 300 pounds of equipment was uprooted from the laboratory and reassembled at dusk of a July evening on a quiet hillside in the Berkshires of western Massachusetts. Moths attracted to a light provided experimental material. The insect subject was pinned on cork so that one of its ears had an unrestricted sound field, and with the help of a microscope its tympanic nerve was exposed and placed on electrodes. After amplification, the action potentials were displayed on an oscilloscope. They were made audible as a series of clicks by means of headphones connected to the amplifier and were stored on magnetic tape for later study.

It was dark before all was ready, but bats immediately revealed their presence to the moth ear by short trains of nerve impulses that recurred about 10 times a second (Figure 2, A). The approach of a

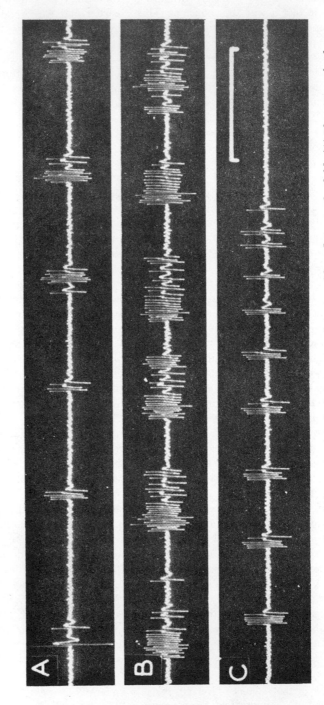

FIG. 2. Tympanic responses of *Noctua* (= *Amathes*) *c-nigrum* to the cries of bats flying in the field. (*A*) The approach of a cruising bat emitting pulses at about 10 per second. (*B*) Tympanic response to the original cry and its echo made by a bat cruising nearby. (*C*) A "buzz." Time line, 100 msec. (From Roeder and Treat, 1961.)

cruising bat from maximum range was coded as a progressive increase in the number and frequency of impulses in each train, first from one and then from both A fibers. It was not long before we learned to read something of the movements of the bats from these neural signals. Long trains, sometimes with two frequency peaks, suggested the chirps of nearby bats that echoed from the wall of a neighboring house (Figure 2, B). An increase in the repetition rate of the trains coupled with a decrease in the number of impulses in each train signified a "buzz" as the bat attacked some flying insect in the darkness (Figure 2, C).

All this was inaudible and invisible to our unaided senses. With a powerful floodlight near the nerve preparation we were able to see bats flying within a radius of 20 feet, and some attacks on flying insects could then be both seen and also "heard" through the "buzz" as coded by the moth's tympanic nerve. However, most of the sounds detected by the moth ear were made by bats maneuvering well out of range of the light. A rough measure of the sensitivity of the moth ear to bat chirps was obtained at dusk on another occasion when the bats could still be seen. The A cells first detected an approaching bat flying at an altitude of more than 20 feet and at a horizontal distance of over 100 feet from the moth—a performance that betters that of the most sensitive microphones.

Direction

Since differences in sound intensity are coded by the tympanic nerve in at least four different ways, the horizontal bearing of a bat might be derived from a comparison of the nerve responses to the same chirp in the right and left ears. A difference in right and left responses might be expected only if each ear had directional properties, that is, a lower threshold to sounds coming from a particular direction relative to the moth's axis.

Directional sensitivity was measured in an open area where echoes were minimal.

A source of clicks of constant intensity was placed on radii to the moth at 45° intervals. The source was moved in and out on each radius until a standard tympanic nerve response was obtained, and the distance from the moth noted. Horizontal distances along eight radii were combined to make a polar plot of sensitivity (Roeder and Treat, 1961). The plot showed that, although there was little difference in sensitivity fore and aft, a click on the side nearest the ear at about 90° relative to the moth's longitudinal axis was audible at about twice the distance of a similarly placed click on the far side.

This led to further field experiments in the presence of flying bats. The tympanic nerve responses from both ears of a moth were recorded simultaneously on separate tracks of a stereophonic magnetic tape. The tape was subsequently replayed into a two-channel oscilloscope and the traces photographed (Figure 3). In the upper record (A) the increase in number of impulses in each succeeding train suggests the approach of a bat. When the signals from right and left ears are compared, it is evident that the greatest difference exists when the signal is faintest, the first response of the series occurring in one ear only. When both ears respond, the differential nature of the binaural response can be seen first as a difference in the number of spikes generated in right and left ears, second in the differential spike frequency, and third in the latency of the response, which is greater on that side generating fewer spikes. It is also evident that, as the sound intensity increases (presumably due to the approach of the bat), the differential becomes less until the responses of right and left ears become almost identical. In another experiment, it was found that the tympanic nerve response saturates, i.e., becomes maximal, when the sound intensity is about 40 db (hundredfold) above threshold. From this it can be concluded that the moth's nervous system receives information that would enable it to determine whether a distant bat was to the right or left, but if the bat was at close quarters

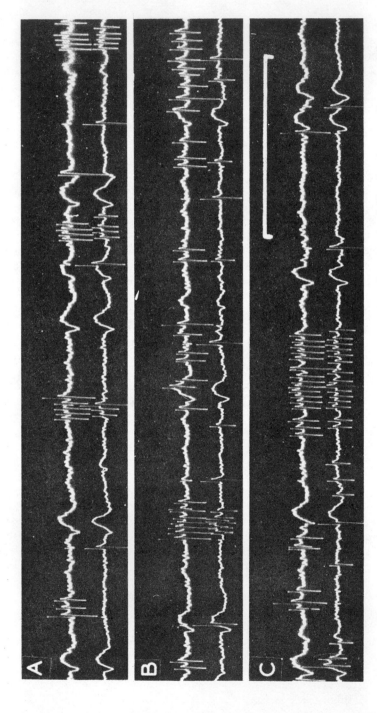

FIG. 3. Binaural tympanic responses of *Feltia* sp. to the cries of red bats flying in the field. The electrocardiogram of the moth also appears on both channels as slow waves. B impulses (large spikes) appear regularly in the records from both tympanic nerves. (A) An approaching bat. Differential response is marked at first (response latency, number of spikes) but has practically disappeared in the final train. (B) A "buzz" registered mainly by one ear. (C) A "buzz" registered by both ears. Time line, 100 msec.

this information would not be available. In Figure 3C, the "buzz" was picked up by one ear only, presumably because during this part of its performance the chirps of a bat are much less intense.

It is tempting to estimate just how close the bat must be before the moth fails to get information on its location. If it is assumed that a bat is first detected at 100 feet and approaches on a straight path at right angles to the moth's course while making chirps of constant loudness, the differential tympanic nerve response would diminish throughout the approach and disappear completely when the bat was 15 to 20 feet away. However, we have not yet determined how much of the information that we are able to read out of its auditory mechanism is actually utilized by the moth in its normal behavior.

The Evasive Behavior of Moths

Although the evasive behavior of moths in the presence of bats must have been witnessed hundreds of times, it is hard to find an adequate account of the maneuvers of either party. The contest normally takes place in darkness, and, even when it is illuminated by a floodlight, the action is too fast and complex to be appreciated by the eye. The flight path of the bat and its ability to intercept and capture its prey have been studied by Griffin (1958) and his students. More recently, Webster (in press) has shown by means of high-speed sound motion pictures that bats become adept at using echoes to plot an interception course with an object moving in a simple ballistic trajectory. Many people have noted the seemingly erratic dives and turns made by moths when bats are near, and similar behavior has been described when moths are exposed to artificial sources of ultrasound (Schaller and Timm, 1950; Treat, 1955).

In an effort to learn more about the behavior of moths under field conditions their flight was tracked photographically as they reacted to a series of ultrasonic pulses simulating bat cries. The sounds were generated by the equipment used in the experiment shown in Figure 1. The pulses were similar in form to those shown, although longer in duration (6 msec.). Each pulse ranged from 50 to 70 kc. with a rise and fall time of about 1 msec. Pulse sequences up to 50 per second could be released on closure of a switch. The sounds were emitted by a plane-surfaced condenser loudspeaker mounted so as to project a fairly directional beam over an open area of lawn and shrubs illuminated by a 250-watt floodlight.

The observer sat behind the sound generator and floodlight, holding in one hand the cable release of a 35 mm. camera set on "bulb," and in the other the switch controlling the onset of the sound-pulse sequence. Many moths and other insects flew out of the darkness into this floodlight arena. A number were attracted directly to the light and were disregarded. Many others moved across the arena at various angles but without marked deviation toward the light. When one of these appeared to be in line with the loudspeaker the camera shutter was opened and the sound pulses turned on.

Two of the tracks registered by the camera as the illuminated moths moved against the night sky are shown in Figure 4. Many insects, including some moths, showed no change in flight pattern when they encountered the sound. In others, the changes in flight path were dramatic in their abruptness and bewildering in their variety. The simplest, and also one of the commonest reactions was a sharp power dive into the grass (Figure 4, left). Sometimes the dive was not completed and the insect flew off at high speed close to the ground. Almost as frequently the dive was prefaced or combined with a series of tight turns, climbs, and loops (Figure 4, right).

It is not known whether these maneuvers are selected in some random manner from the repertoire of individual moths, or whether they are characteristics of different species. However, Webster (in press)

FIG. 4. Flight tracks registered by various moths just before, and immediately following, exposure to a series of simulated bat cries. The dotted appearance of the tracks is due to the individual wingbeats of the moth. The beginning of each track appears in each photograph, and the moth finally flies out of the field.

has shown that bats soon learn to plot an interception course with food propelled through the air in a simple ballistic trajectory. The random behavior elicited by simulated bat cries in the natural moth population seems to be a natural answer to this predictive ability in bats, while the sharpness of the turns must certainly tax the maneuverability of the heavier predator.

The reacting moths shown in Figure 4 were mostly within 25 feet of the camera and sound source, and were exposed to an unknown but probably high sound intensity. Under these circumstances, the evasive behavior appeared to be completely unorientated relative to the sound source, as might be predicted from the binaural tympanic nerve recordings. In some instances, moths flying at a greater distance or only on the edge of the sound beam appeared to turn away from the area and fly off at high speed. This must be checked in future experiments.

The Survival Value of Evasion

In spite of the evidence that the moth ear is an excellent bat detector, and that acoustic stimulation releases erratic flight patterns, one may well ask whether this behavior really protects moths from attack by bats.

This question has been answered (Roeder and Treat, 1960) by observing with a floodlight 402 field encounters between moths and feeding bats. In each encounter we recorded the presence or absence of evasive maneuvers by the moth, and the outcome, that is, whether it was captured by the bat or managed to escape. From the pooled data we determined the ratio of the percentage of nonreactors surviving attack to the ratio of reactors surviving attack. Thus computed, the selective advantage of evasive action was 40 percent, meaning that for every 100 reacting moths that survived, there were only 60 surviving nonreactors.

This figure is very high when compared with similar estimates of survival value for other biological characteristics. It seems more than adequate to account for the evolution of the moth's ear through natural selection even if the detection of bats turns out to be its only function.

Conclusion

As with most investigations, this work raises more questions than it has answered. The role of the B cell remains completely obscure. There is no evidence to connect it with the auditory function even though it is located in the ear, and its regular impulse discharge is a characteristic feature of the tympanic nerve activity of many species of moth (Treat and Roeder, 1959. See also Figure 3). The manner in which the A cells transduce sound waves recurring 100,000 times a second into the much slower succession of nerve impulses remains a mystery, and the synaptic mechanisms whereby information from the A fibers is translated into action by the nervous system of the moth, await investigation.

During the field experiments it was noticed that many other natural sounds initiated impulses in the A fibers. These included the rustling of leaves, the chirp of tree and field crickets, and, in one instance, ultrasonic components in the wingbeat sounds made by another moth. Occasionally, the A fibers discharged regularly as if detecting a rhythmic sound, though none was audible to the observers and its source (if any) remains a mystery. There is no evidence that these identified and unidentified sounds are important in the life of a moth, yet it must be said that a moth can detect them, and a careful study of moth behavior in their presence would be of value.

Several families of moths lack ears and show no response to ultrasonic stimuli. Some of these, such as the sphinx or hawk moths and the larger saturniid moths, are probably too much of a mouthful for the average bat, and might find no survival advantage in a warning device. Others are of the same size and general habits as the noctuids and might be expected to suffer attacks by bats. Included in this group are some common pests such as the tent caterpillar. It will be interesting to learn whether these forms owe their success in survival to some structural or behavioral countermeasure that compensates for the lack of a tympanic organ.

In spite of these unanswered questions, we believe that some progress has been made in putting together the sensory information received by an animal, and relating this to what the animal does. That this has been possible in moths is only because of the small number of channels through which acoustic information reaches the nervous system in these insects. Further examples of this favorable situation have been described in other insects, and still others are waiting to be explored.

REFERENCES

GRIFFIN, DONALD R., 1958 *Listening in the dark.* Yale University Press, New Haven.

GRIFFIN, D. R., F. A. WEBSTER, and C. R. MICHAEL, 1960 The echolocation of flying insects by bats. *Animal Behaviour,* vol. 8, pp. 141–154.

ROEDER, K. D., 1959 A physiological approach to the relation between prey and predator. In: *Studies in Invertebrate Morphology,* Smithsonian Misc. Coll., vol. 137, pp. 287–306.

ROEDER, K. D., and A. E. TREAT, 1957 Ultrasonic reception by the tympanic organ of noctuid moths. *Journ. Exp. Zool.,* vol. 134, pp. 127–158.

ROEDER, K. D., and A. E. TREAT, 1961 The detection of bat cries by moths. In: *Sensory Communication,* ed. by W. Rosenblith. M. I. T. Technology Press, Cambridge, Mass.

ROEDER, K. D., and A. E. TREAT, 1960 The acoustic detection of bats by moths. Proc. XI Internat. Entomol. Congr.

SCHALLER, F., and C. TIMM, 1950 Das Hörvermögen der Nachtschmetterlinge. *Zeitschr. Vergl. Physiol.,* vol. 32, pp. 468–481.

TREAT, A. E., 1955 The response to sound of certain Lepidoptera. *Ann. Entomol. Soc. America,* vol. 48, pp. 272–284.

TREAT, A. E., and K. D. ROEDER, 1959 A nervous element of unknown function in the tympanic organs of moths. *Journ. Insect Physiol.,* vol. 3, pp. 262–270.

34

Chemical-Cue Preferences of Inexperienced Snakes: Comparative Aspects

Abstract. *Different species of new-born, previously unfed snakes will respond with tongue flicking and prey-attack behavior to water extracts of the skin substances of various small animals. However, there are clear species differences in the type of extract responded to by previously unfed snakes, even within the same genus. These differences correspond to the normal feeding preferences shown by the various species.*

It has often been noted that animals can selectively respond to certain highly specific perceptual cues without the benefit of previous experience with those cues (*1*). The stimuli involved usually represent but a small fraction of the entire stimulus situation and are termed sign stimuli or releasers. In many instances the resulting response is also quite specific and stereotyped. For instance, newborn, previously unfed garter snakes (*Thamnophis s. sirtalis*) will respond with prey-attack behavior to extracts of the surface substances of normally eaten prey when these extracts are presented on cotton swabs (*2*). Similar specificity to chemical cues has been demonstrated in many forms of invertebrates and, to a lesser extent, in vertebrates (*1*).

Beyond the existence and analysis of such stimulus–response relations in a particular species looms the broader evolutionary implications. I here report the chemical perception aspects of feeding behavior in a number of species of neonate colubrid snakes.

I presented a variety of extracts from the surface substances of small animals to litters of individually isolated newborn snakes. The animals used in preparing extracts for the testing were: nightcrawler (*Lumbricus terrestris*), leafworm (*Lumbricus rubellus*), redworm (*Eisenia foetida*), turtle leech (*Placobdella parasitica*), slug (*Deroceras*

gracile), cricket (*Acheta domestica*), minnow (*Notropis atherinoides acutus*), guppy (*Lebistes reticulatus*), goldfish (*Carassius auratus*), larval salamander (*Ambystoma jeffersonium*), metamorphosed salamander (*Ambystoma jeffersonium*), cricket frog (*Acris crepitans blanchardi*), and newborn mouse (*Mus musculus*). An extract was made by placing one or more of the intact animals in distilled water (10 ml of water per 1.5 g of body weight) at 50°C for 1 minute and stirring the water gently. The animal was then removed and the resulting liquid centrifuged and refrigerated until use. Extracts for any one experiment were always prepared on the same day.

The snakes were from litters or eggs borne by gravid females captured in the field and maintained in captivity until parturition or egg-laying. Shortly after birth or hatching the snakes were weighed, measured, and then isolated in glass tanks measuring 23 by 14 by 17 cm. Each tank was placed on white shelf paper and the four outside walls were covered with white partitions. The floor of the tank was bare except for a small plastic petri dish containing water. Except for testing periods, the aquaria were covered with glass tops. The temperature of the room in which the snakes were housed never varied more than between 22° to 26°C; during testing the temperature was maintained at

24° to 25°C.

Each member of a given litter of newborn snakes was tested only once on a series of extracts of the surface substances of potential prey. Usually twelve or thirteen different extracts were used. Distilled water was the control. Each subject received a different ordering of the test extracts, systematically balanced insofar as possible for each litter. Testing was carried out over 2 or 3 successive days beginning on the 3rd or 4th day of life and always before any previous feeding or exposure to the extracts (3). The testing procedure consisted of dipping a 15-cm cotton swab into the extract or control, slowly introducing it into the tank, and bringing it within about 2 cm of the snake's snout. If the swab was not attacked within 30 seconds, it was moved closer until it touched the snout gently three times, as actual contact with the lips of the snake is sometimes necessary to elicit an attack. If no attack was made at the end of 1 minute, the swab was removed and the total number of tongue flicks emitted in the 1-minute interval recorded. If the swab was attacked, the elapsed time, measured to the nearest 0.1 second, was recorded (4).

An extract, when not actually eliciting a prey-attack response, would often elicit a large number of tongue flicks over and above that elicited by distilled water. It appeared as though the frequency of tongue flicking was correlated with the intensity of arousal by or interest in the swab. Since previous experiments indicated that the prey-attack response in snakes is mediated by the tongue–Jacobson's organ system (3, 5), tongue-flick data can reasonably be considered along with the attack data in assessing the relative releasing value of various extracts on different species of snakes. The control swab (distilled water) never elicited a prey-attack response in an inexperienced snake, although sometimes a large number of flicks would occur. In most of the species studied here, aggressive behavior was never shown by the newborn snakes; and, indeed, it was impossible to provoke such behavior. A conservative scoring system was used to score the extract given to each snake. The scoring system was based on the assumptions that an actual attack is more definitive than any number of tongue flicks and that a more potent stimulus will lead to an attack with a shorter latency than will a weaker stimulus. The base unit was the maximum number of tongue flicks given by any individual of the litter tested to any of the test stimuli (the maximum was invariably given to a swab containing an extract). A snake which did not attack was given a score identical with the number of tongue flicks it emitted in the 1-minute test period. If the snake did attack, it was given a score identical with the base unit (for that litter) plus one point or fraction for every second or fraction less than 1 minute that it responded. The score for an attacking subject can be represented by base unit + (60 − response latency), measured in seconds.

Figure 1 shows an example of the type of profile obtained when a litter is presented with a series of extracts. The species represented is the eastern plains garter snake, *Thamnophis r. radix*. The 22 living young from a litter of 24 born to a female captured at the Palos Forest Preserve were tested. Each of the previously unfed snakes was presented once with extracts prepared from earthworms, leeches, slugs, crickets, fish, amphibians, and baby mice. Responses to all extracts were significantly higher than those to the control except for those to the baby mouse, slug, cricket, and metamorphosed salamander extracts ($P < .01$, *t*-test). Although no extensive ecological studies have been done on this species,

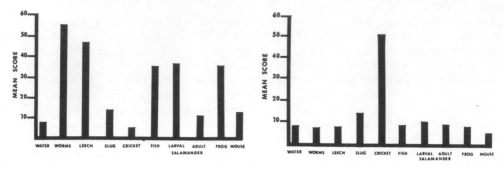

Fig. 1 (left). Response profile of 22 newborn, previously unfed eastern plains garter snakes (*Thamnophis r. radix*) to water extracts from the surface substances of various small animals. The results for the three species of earthworms and the three species of fish have been averaged together.   Fig. 2 (right). Response profile of five newly hatched, previously unfed western smooth green snakes (*Opheodrys vernalis blanchardi*). The snakes were the same age as those in Fig. 1 and were tested at the same time and with the same extracts.

it appears that earthworms, amphibians, fish, and leeches are readily eaten, with worms being probably most common in the natural diet (6). The present results with inexperienced newborn snakes on isolated chemical cues are in remarkably close agreement. Worms as a class were more effective than fish as a class, as there was no overlap in the mean scores for the various worm and fish species. The increased releasing value of the larval salamander over the metamorphosed form (*P* < .0005, *t*-test) is a relationship that has been found frequently in most species of newborn snakes tested that include amphibians in their normal diet. A chemical change in the skin during metamorphosis is probably responsible for the difference.

In sharp contrast to these results were results obtained from the western smooth green snake, *Opheodrys vernalis blanchardi*. This species is oviparous, and five young from a clutch of seven eggs laid by a female captured on the Palos Forest Preserve were tested. The eggs hatched on the same day that the plains garter snakes were born. The green snakes were tested along with the plains garter snakes on the same extracts at the same time. Although each green snake received a different ordering of the extracts, each

sequence was identical to one used with a plains garter snake. In Fig. 2 the results for the green snake are presented for the same extracts as shown for the plains garter snake. The cricket extract was the most potent; indeed, it was the only extract to which actual attacks were made and the only one with a score significantly higher than that of the control (*P* < .004, Mann-Whitney U test, one-tailed). This result becomes more meaningful when it is realized that the cricket extract is the only one which represents an organism eaten by the green snake. In fact, this species apparently will eat only insects, spiders, and perhaps small soft-bodied arthropods.

With literally every procedural detail controlled, clear differences were found in the chemical perception of food objects in the two species of inexperienced snakes. In contrast to the green snake, the plains garter snake does not show any interest in insects as food and the cricket extract received the lowest score of all the extracts. Likewise the green snake was uninterested in those extracts which elicited significant responses in the plains garter snake.

To test these results further, three more closely related forms were studied in the same manner, although only differentiating results for the three earth-worm, three fish, and slug extracts will be presented. The eight surviving young of a litter of ten born to a midland brown snake (*Storeria dekayi wrightorum*) captured on the Palos Forest Preserve were tested. Of the above three classes of prey—earthworms, fish, and slugs—this species is known to eat only worms and slugs. The second species tested was Butler's garter snake (*Thamnophis butleri*). A litter of 15 was obtained from a female caught in southern Michigan. In captivity, Butler's garter snake readily eats worms and fish but not slugs. The third species was the aquatic garter snake, *Thamnophis elegans aquaticus*. A litter of nine was tested which were born to a female found in southern California. Of the three classes of prey, this species is known to eat only fish.

The inexperienced young of the three species were tested on the seven extracts representing worms, fish, and slugs (Fig. 3). All scores for the midland brown snake (*Storeria*) were doubled so as to bring them up to the same scale as the two species of *Thamnophis*. For a given species of snake, all scores above 25 are significantly higher than scores below 25 ($P < .05$, t-test). The responses of the different species of inexperienced snakes to skin extracts parallels the feeding habits of specimens captured in the wild. The generally lower scores shown by the midland brown snake are due mainly to the fact that the frequency of tongue flicking was lower than for the two *Thamnophis* species. This may be an important difference also.

These results clearly indicate that chemical perception in newborn young

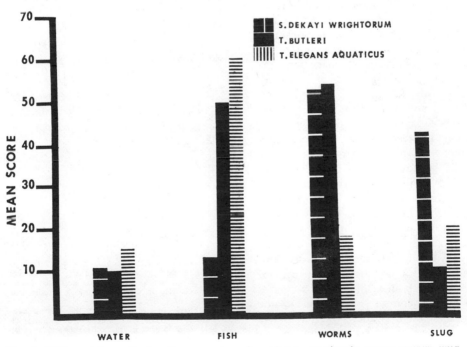

Fig. 3. Differential results for three species of newborn, previously unfed snakes with extracts from the surface substances of three classes of potential prey. Three species of worms and three species of fish were used. All scores for the midland brown snake (*Storeria dekayi wrightorum*) have been doubled to bring them up to the same scale as the two species of *Thamnophis*.

is species-specific. That these are related to the natural feeding ecology of the species is equally clear. But an inexperienced snake will respond to chemical cues that cannot or do not figure in the normal feeding behavior of the species. For instance, the aquatic garter snake (*T. elegans aquaticus*) rarely, if ever, encounters the guppy in nature, yet the inexperienced young readily responded to the guppy extract. Since the aquatic garter snake normally eats fish, however, it is probable that the guppy possesses chemical cues similar or identical to those found in fish which the snake normally eats. In Butler's garter snake (*T. butleri*), the situation is a little more complex, for fish do not constitute any part of the species' normal diet (6). Yet specimens readily eat fish in captivity and newborn young, as shown here, respond significantly to fish extracts. It is, therefore, apparent that the normal feeding habits and ecology of a species are not sufficient to explain the response to chemical cues in newborn young. In this case Butler's garter snake may retain the perceptual side of releasing mechanism that appears to be of no selective advantage in its present mode of life. Of course, retention of the potential to respond to chemical cues from fish by inexperienced snakes would be advantageous if a change in the environment occurred such that fish became a necessary or more easily obtainable food source. The same situation is found with amphibians in this species. Amphibian extracts are responded to by newborn snakes but amphibians also do not form a part of the normal diet of Butler's garter snakes. Therefore, in relation to the extracts used, Butler's garter snake would seem to possess more innate perceptual responsivity than does the aquatic garter snake, which did not respond to any of the worm extracts. Butler's garter snake is generally considered as having evolved from the plains garter snake which not only innately responds to extracts from fish and amphibians, but also normally eats them (7).

The results are open to an evolutionary interpretation. Highly specific stimulus-response information is probably genetically coded in the organism and must in part, at least, be expressed by an innate filtering mechanism at the level of Jacobson's organ or even within the central nervous system itself. This does in no way, of course, rule out the possibility that subsequent feeding experiences (or perhaps even maternal feeding) can influence the feeding preferences of newborn snakes. Indeed, something akin to food imprinting, such as already demonstrated in turtles, may take place (8). In any event, the present data show that innate perceptual differences can be useful in the study of closely related as well as more distantly related forms, and that the analysis of the chemical perceptual mechanisms involved should consider evolution and ecology.

Gordon M. Burghardt
W. C. Allee Laboratory of Animal Behavior, University of Chicago, Chicago, Illinois 60637

References and Notes

1. P. R. Marler and W. J. Hamilton, *Mechanisms of Animal Behavior* (Wiley, New York, 1966), pp. 228–315.
2. G. M. Burghardt, *Psychon. Sci.* 4, 37 (1966).
3. More complete details concerning the subjects, the experimental procedures, and the results are recorded in G. M. Burghardt, thesis, University of Chicago (1966).
4. To check the reliability of the testing procedure, 20 trials (10 water and 10 nightcrawler) were later run on different snakes with a second observer independently timing attack latencies and counting tongue flicks. This second observer did not know which extract was being presented. The average tongue-flick count discrepancy for a given trial was less than 1 and the average latency discrepancy less than 0.5 second. The rank correlation of the 20 trials was highly significant ($r_s = .997$).

5. G. Naulleau, thesis (P. Fanlac, Paris, 1966);
W. S. Wilde, *J. Exp. Zool.* **77**, 445 (1938).
6. Information concerning feeding habits in the
field and in captivity was gleaned from the
following, in addition to personal experience:
C. C. Carpenter, *Ecol. Monogr.* **22**, 235 (1952);
R. L. Ditmars, *The Reptiles of North America* (Doubleday, New York, 1936); P. W.
Smith, *Ill. Nat. Hist. Surv. Bull,* **28**, Art. 1
(1961); R. C. Stebbins, *Amphibians and Reptiles of Western North America* (McGraw-Hill,
New York, 1954); H. W. Wright and A. A.
Wright, *Handbook of Snakes of the United
States and Canada* (Comstock, Ithaca, 1957),

2 vols.
7. K. P. Schmidt, *Ecology* **19**, 396 (1938).
8. G. M. Burghardt and E. H. Hess, *Science*
151, 108 (1966); G. M. Burghardt, *Psychon.
Sci.* **7**, 383 (1967).
9. Assisted by NIH grants MH 776 and MH
13375. I thank E. Lace and other naturalists
at the Palos Division, Cook County Forest
Preserve, P. Allen, and H. Campbell for providing the gravid females and newborn young;
E. H. Hess, E. Klinghammer, G. S. Reynolds,
T. Uzzell, and D. Wake for assistance.

13 April 1967

35 Theodore Holmes Bullock

Seeing the World through a New Sense: Electroreception in Fish

*Sharks, catfish, and electric fish use low- or high-frequency
electroreceptors, actively and passively,
in object detection and social communication*

Reprinted from AMERICAN SCIENTIST, Vol. 61, No. 3, May-June 1973, pp. 316-325
Copyright © 1973 by The Society of the Sigma Xi

*Theodore H. Bullock is a professor in the Department of Neurosciences of the School of Medicine and head of the Neurobiology Unit of the
Scripps Institution of Oceanography at the University of California, San Diego. As a comparative
neurologist he has studied the brain, synapses,
giant fibers, and sense organs of animals from
porpoises to jellyfish. With G. A. Horridge he
wrote a two-volume treatise,* Structure and Function in the Nervous Systems of Invertebrates
*(W. H. Freeman Co., San Francisco).
Born in China, Dr. Bullock did his graduate
work in zoology at Berkeley and then worked
successively at Yale, the University of Missouri,
and UCLA before going to La Jolla, with summers at Woods Hole or other marine stations or on
expeditions. He is a member of the National
Academy of Sciences and the American Philosophical Society and served as president of the
American Society of Zoologists in 1965. He
began to search for electroreceptors in 1959 and
has had many collaborators in that work, including
S. Hagiwara, N. Suga, T. Szabo, P. S. Enger,
S. Chichibu, A. J. Kalmijn, and H. Scheich,
with support from the NSF, ONR, and NIH.
Address: Department of Neurosciences, UCSD,
La Jolla, CA 92037.*

Surely no other branch of physiology
has provided more novelty than sensory physiology, if we may use as
evidence the continued discovery of
new organs and of functions for hitherto unexplained organs. We are
inured to surprises year after year,
such as the demonstration of infrared
receptors, aerodynamic or wind speed
receptors, polarized light detection,
and fantastic abilities in chemoreception, hydrostatic pressure, magnetic-field and sonar reception, and
others.

How should we name new sense organs? This is far from a trivial question and will repay our brief consideration. Given that a stimulus
has been found which elicits impulses
in a sensory nerve, how should the
receptor be designated? An innocent-

sounding issue, it conceals pitfalls that have trapped more than a few physiologists and requires a real understanding of the biological meaning of the organ.

In general we cannot depend on the good fortune we have in the case of eyes and ears, in which the function is more or less obvious from accessory structures or exclusive sensitivity to only one mode of stimulation. Nor can we simply say that the function is determined by the stimulus to which an organ is most sensitive. How many degrees of temperature stimulus are equal to one gram of mechanical stimulus? High sensitivity to one form of stimulus may be incidental to a normal use in a less familiar modality.

Many receptors are ambiguous. They respond, for example, to a certain temperature *and* a certain mechanical event or other input. Are they ambiguous only to us or also to their possessors? How can we be satisfied that we have tried all the relevant forms of stimulation? When can we name an organ by function? An example will help and will lead us into the topic of the title.

The best case illustrating the basic issues and principles involved in this question is that of certain classical structures widespread in the skin of all sharks and rays, called ampullae of Lorenzini. These are tiny bladders, richly innervated and connected to a surface pore by a long canal. After decades of speculation centered on mechanoreceptive functions, electrophysiological techniques in 1938 revealed an unprecedented sensitivity to temperature, and for years the ampullae were the first physiologically studied thermoreceptors (Sand 1938).

Doubts crept in, however, as to whether the ampullae really function as thermoreceptors for the shark.

If so, why the long canal? New studies reactivated the old idea of mechanical sensitivity. Excitation by electric current was also noted, but neither form of energy seemed likely to be the natural "adequate stimulus." Even the eye can be stimulated by pressure and electric shock. Much more interesting was the later discovery that these organs will alter their nerve impulse traffic to the brain when slightly diluted or concentrated seawater is encountered, and we wondered whether their function is detection of variations in salinity of the environment.

Finally—and I believe that word is justified—a Dutch biophysicist named A. J. Kalmijn, in the laboratory of Sven Dijkgraaf, showed that sharks use these ampullae of Lorenzini to detect prey buried in the sand (Fig. 1). He found that the ampullae are excited by the feeble, steady electric field that leaks out of the buried flatfish prey. The shark detects and aims an attack at the flatfish even when it is covered with an agar plate that prevents mechanical, visual, and chemical cues, but is transparent to electric current. It also attacks buried electrodes that carry current equivalent to that of a flatfish. Therefore, we must call the ampullae electroreceptors, by the definitive criterion of behavior.

I have recounted this piece of scientific drama to emphasize the conceptual and tactical problem of defending the proposed function(s) of a receptor (Bullock 1973; Kalmijn and Bullock 1973). Physiological methods alone cannot establish the proper functional designation of a sense organ. There must be evidence of the normal availability and biological significance of the presumptive adequate stimulus, preferably by behavioral response appropriate to that form of information.

Plaice (*Pleuronectes platessa*) under the sand elicits the shark's accurate, directed attack.

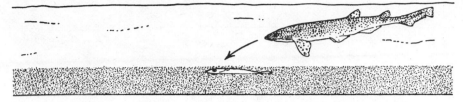

The shark also attacks accurately when the plaice is in an agar chamber, "transparent" to electric current, with a resistivity about equal to that of the medium.

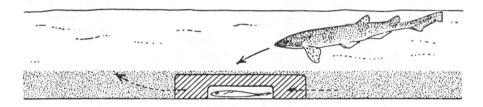

Only if the plaice has been chopped up and frozen for a long time, exaggerating the olfactory stimulus and fragmenting the d.c. field, will the shark diffusely search in the downstream area.

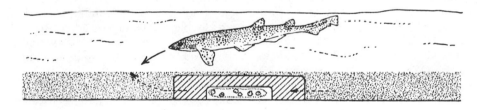

If the agar is covered by an insulating plastic film the response is lost

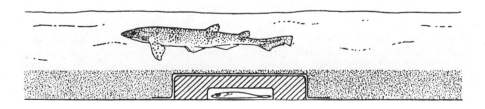

Figure 1. Feeding responses of the shark, *Scylliorhinus canicula*. The agar chamber is not to scale. Solid arrows indicate movements of the shark; dashed arrows, the flow of seawater through the chamber. (From Kalmijn 1971, by permission of the *Journal of Experimental Biology*.)

Electrodes producing a d.c. field like that of the plaice elicit the response.

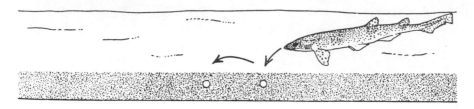

Even in the presence of a piece of food the response is still directed toward the dipole source (only one electrode shown).

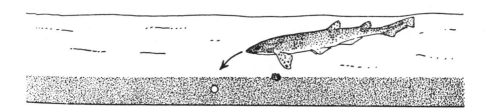

Natural electric stimuli

What does the world look like to a creature with the amazing sensitivity to electrical fields in water of 0.01 $\mu v/cm$ (= 1 $\mu v/m$, or 1 flashlight cell per 1,500 km)? What are the kinds and sources of voltage fields in nature? Which signals are useful and which are noise?

Surveying sixty species of animals from eight major branches of the animal kingdom, Kalmijn, now at the Scripps Institution of Oceanography, found that nearly all, including man, emit into seawater d.c. fields which even at tens of centimeters distance are well above the shark threshold. A.c. fields due to muscle and heart activity are generally much smaller. The d.c. fields are presumably incidental voltages of diverse origin, including millivolt potentials between body fluids and ocean and between different parts of the body. Although quite variable

they may signal the position, orientation, type, and physiological condition of the animal. A wound, for instance, markedly increases the voltage gradient. Even a minor scratch can double the voltage gradient from a man in seawater (to one that would be detectable by a shark at more than 1 meter).

The class of electrical signals arising from living organisms can be called bioelectric signals. Very likely there are biological meanings other than the localization of prey—social signals between conspecifics, for example. This is especially true for the a.c. bioelectric fields emitted by the numerous species of electric fish, as we will see below. The field is just opening up.

In addition to the world of bioelectric signals there is a world of inanimate fields. We are now testing whether large homogeneous fields such as those induced by the motion of water masses, like the great ocean currents and rivers, through the earth's magnetic field can be used as navigational clues. The signals are known to be present; the sharks are amply sensitive to detect them. Similarly, the fields caused by waves and tides and by magnetic variations as well as by the sharks' own swimming are strong enough to be detected by sharks, though we do not know that they use them.

There may conceivably be signals from local fields straying out from ore bodies of certain kinds, possibly useful in the way topographic features of the landscape are. There are fields caused by earthquakes and other episodic events such as eruptions and crustal movements. In shallow water complex activity results from atmospheric disturbances such as lightning and ionosphere reverberations, including what used to be called "static," and also from slow changes like day-night alternations. There are chemical potentials at sharp discontinuities like river bottoms and at gradual boundaries—perhaps less usable for that reason—like river mouths and thermoclines.

Many of the available fields may represent background noise out of which signals must be extracted. Our understanding of the uses made of inanimate fields is quite embryonic, but it is as though a door has been opened. Surprises can be expected—such as the well-documented hypersensitivity to mechanical stimuli in

---

Table 1. Modes and roles of receptor function

MODE
Passive (animal detects fields from
    external sources)
    Animate (membrane potentials of
      other organism)
      d.c. (e.g. gill potentials)
      a.c.-slow (d.c. modulated by
        slow movements)
      a.c.-fast (muscle, heart, electric
        organ action potentials)
    Inanimate sources
      motion of water mass in earth's
        magnetic field
      electrochemical, atmospheric,
        geological processes

ROLE
Electrolocation
    Close range
      passive detection of other fish
      active detection of objects and
        spaces
    Long range
      directional navigation

Electrocommunication
    Constant signals
      related to place, kind, sex,
        individual
    Changing signals
      related to food, threat, attack,
        submission, mating

Active (animal detects fields caused
    by its own activity)
    Animate (membrane potentials from
      own body)
      d.c., a.c.-slow, a.c.-fast (same as
        for passive mode above)
    Inanimate sources
      induced potentials from swimming in earth's magnetic field

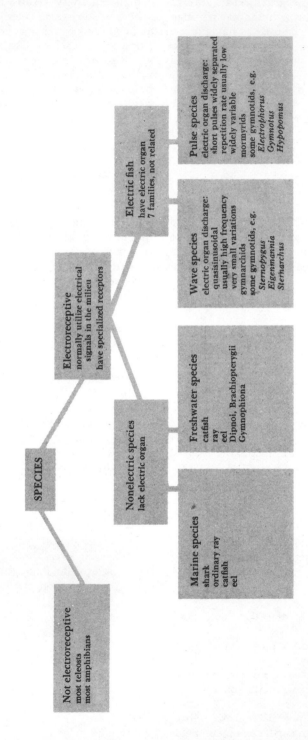

Figure 2. Relationships and characteristics of electroreceptive fish.

catfish at a certain place in Japan during and for some hours before earthquakes, behavior which is evidently dependent on changes in the electrical currents in the earth, picked up by the freshwater stream (Hatai and Abe 1932; Kokubo 1934; Kalmijn and Bullock 1973).

The world of natural waters, being a more or less good conductor, is busy with electrical currents, on a millimeter and a kilometer scale, from d.c. to hundreds of kHz, from fractions of microvolts to several millivolts per meter and locally even higher. Man is "polluting" the waters and the ground with far higher currents in return paths from power lines, leakage, and high-power broadcasting. We are beginning to learn that many fish normally detect and are influenced by feeble currents (Fig. 2).

In what follows we shall recognize two modes of detection—passive and active; two roles in the biology of the species—"electrolocation" of objects and "electrocommunication" of social signals (Table 1); and two classes of sense organs—receptors tuned to low frequency and those tuned to high frequency (Fig. 3). The three dichotomies are not equivalent, and many of the permutations are known or suspected—e.g. passive detection of objects by low-frequency receptors (the shark and flatfish case of Fig. 1), active electrolocation by low-frequency receptors (the case of navigation with the aid of currents induced by swimming in the earth's magnetic field), and passive detection of social signals from conspecifics by high-frequency receptors.

Passive electroreception

We have already seen one example of the passive mode in Figure 1 and suggested others, e.g. the navigation by electric currents induced by water masses such as the Gulf Stream moving

in the earth's magnetic field (see Table 1). How does the shark sample the electric field, since he has to look at two points and complete a circuit between them? Figure 4 shows the relevant features of the animal's interior and skin in relation to a large homogeneous field in the milieu (Kalmijn and Bullock 1973).

In seawater the body fluids of vertebrates have a higher resistivity than the milieu, and the skin is only a moderately good insulator. The ampullary organs are remarkable in two ways: the jelly in the canals has a very low resistivity, whereas the resistivity of the canal walls is very high. The result is that the sense cells in the bottom of the ampullae face outward to a jelly virtually at the same potential as that in the sea at the pore, while the inner face is bathed by body fluid virtually at the same potential as that of the seawater just outside the skin at that point (see Fig. 4). The long canal therefore enables the sense cell to sample the field at two widely separated points. Sharks and rays have hundreds of these canals, radiating in various directions and reaching lengths, in rays, of up to one-third of the body length. The remarkable saltwater catfish *Plotosus* also has these organs and the electroreceptive ability.

What about freshwater species? Figure 4 shows that the situation is drastically different in electrosensitive freshwater species. The body fluids are nearly isopotential throughout because of the high skin resistance, and thus the inner side of a receptor cell is at the same potential as the external field near the middle of the animal. Therefore there is no need for long tubes. We made a study of the unusual freshwater-adapted elasmobranch, the Amazon stingray *Potamotrygon*, and found exactly this—microscopic ampullae and tubes just reaching through

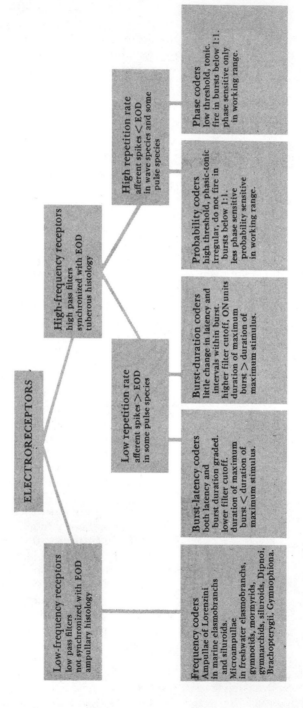

Figure 3. Present knowledge permits recognizing two broad classes and five coding types of electroreceptors, with the characteristics and distributions shown here. EOD = electric organ discharge. ON refers to responses to stimulus onset. See Fig. 7 for explanation of coders.

the skin (Szabo et al. 1972). Virtually the same situation is true for the freshwater teleosts (gymnotids, mormyrids, gymnarchids, and silurids at least) that have ampullary sense organs and electroreception of low-frequency fields.

One further indication of the adaptation of the ampullary receptors is the frequency range of electrical events to which they are sensitive. This is strictly confined to very slow events, between 0.1 and 10 Hz at least in the shark. Sensitivity extending down

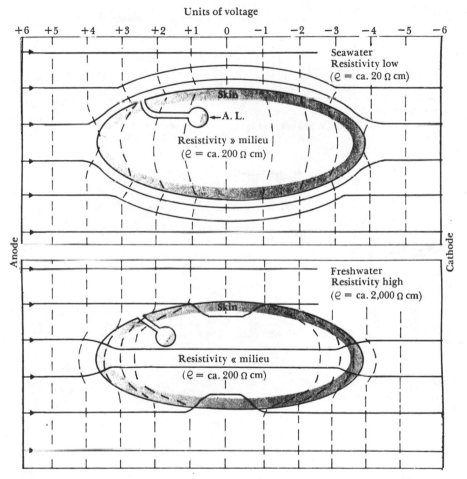

Figure 4. Schematic representation of marine elasmobranch or teleost in seawater (*above*) and elasmobranch or teleost adapted for freshwater (*below*). Note that the mural resistance of the skin in ohm/cm² is relatively low in the marine and high in the freshwater forms shown, which are probably typical for elasmobranchs evolutionarily adapted to the respective habitats and for electric fish. A.L. = ampulla of Lorenzini, with sensory membrane at the bottom and canal with very low resistance jelly and very high resistance walls. The interior represents mean tissue fluid resistivity (ρ). The marine elasmobranch employs a long canal to put a considerable part of the external voltage gradient across its sensory membrane; the well-adapted freshwater form (*Potamotrygon* or electric fish) needs no long canal to achieve the same result away from the electrical equator because the whole interior is virtually isopotential with the equator. The marine form does not load the external field because its tissue fluids are higher in resistivity; the freshwater form loads the external field only moderately because its skin resistance is so high.

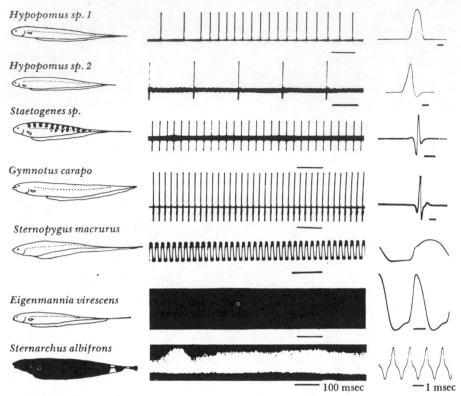

Figure 5. Electric organ discharges from several species of the South American family Gymnotidae. Each discharge is shown beside the fish it comes from and on both slow (*left*) and fast (*right*) time bases. Upward deflection means positivity of the head end. The last two are wave species; all the others are pulse species. (From Hagiwara and Morita 1963, by permission of the American Physiological Society.)

to 0.1 Hz gives the animal access to d.c. fields in the water, because of its own swimming and swerving. The fall in sensitivity above 10 Hz cuts out higher-frequency noise and corresponds with the surprising fact that the fast electrocardiographic, locomotor, and respiratory muscle action potentials of the fish upon which the shark *Scyliorhinus* preys are much weaker than the d.c. stray fields from their gills and buccal membranes.

Passive reception of low-frequency, feeble voltage gradients is now known to occur in elasmobranchs, catfish, and common eels (Anguillidae), and electric fish of the families Gymnotidae, Mormyridae, and Gymnarchidae; it can be suspected in several other groups. It is usually attributable to structures of the type called ampullary organs, but we must not be surprised if some other types turn up in addition. Passive reception by high-frequency receptors will be dealt with below under communication.

Active electroreception

Active electroreception is best known in electric fish in the object-detecting mode, using high-frequency receptors. Even before Kalmijn's discovery of the normal role of ampullae as electroreceptors, Lissmann (1958) had reopened an old question—namely, whether the electric organs in so-called pseudoelectric fish could be functional. It had been concluded that these small electric organs were useless because they were too puny to be offensive or defensive. But Lissmann found that they were discharging in a pulsatile way all the time, night and day (Fig. 5), and he proposed an electrolocating or object-detecting function (Fig. 6).

Here, then, is a proposal for an elec trosensitive system that is *active* in the same sense that a bat's echolocation is. It measures not the time of return of a signal, since that is virtually instantaneous, but the shape of the field. Receptors must be widely distributed over the body, while the brain must do a spatial computation and could, in principle, infer size, distance, shape, and quality of an object, with certain ambiguities (Bennett 1970; Scheich and Bullock,

forthcoming). This proposed function has now been amply confirmed by behavioral experiments.

Largely through the work of my colleagues S. Hagiwara, S. Chichibu, H. Scheich, and several others, as well as important work by M.V.L. Bennett, in New York, and A. Fessard and T. Szabo, in Paris, the required new class of receptors has been found and characterized. These were the first electroreceptors discovered and shown to satisfy the criteria of response to natural stimuli mentioned above. The excitement of discovering receptors for a new modality or view of the world was then enhanced by the finding that there are several kinds of such receptors (see Fig. 3). These differ in their mode of coding intensity, as illustrated in Figure 7.

Unraveling the codes in the sensory nerve fibers occupied several years and is not really completed yet. The fundamental problem illuminated by these exotic electric fish is: How is

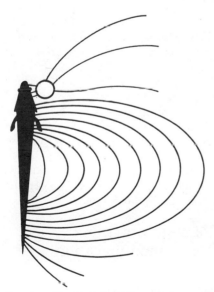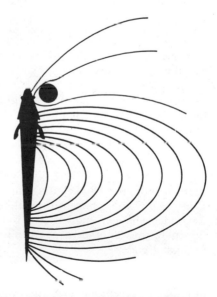

Figure 6. The electric field on one side of an electric fish, shown by current lines, and the distortions due to objects of low conductivity (*right*) and high conductivity (*left*).

(After Lissmann and Machin 1958, by permission of the *Journal of Experimental Biology*.)

305

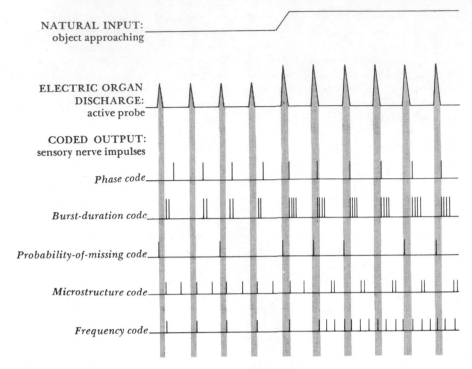

NATURAL INPUT:
object approaching

ELECTRIC ORGAN
DISCHARGE:
active probe

CODED OUTPUT:
sensory nerve impulses

Phase code

Burst-duration code

Probability-of-missing code

Microstructure code

Frequency code

Figure 7. Diagram of several hypothetical types of afferent nerve fibers from electroreceptors, each encoding in a different way the step of intensity of the electric field caused by introduction of a natural stimulating object. All types shown have been found in electric fish except the microstructure code, which is known from other animals. (After Bullock 1968, by permission of the National Academy of Sciences.)

information represented in streams of nerve impulses? Like many basic questions of neurobiology, answers or clues from lower species may help us to understand the brains of higher forms. It had been thought that mean frequency was *the* code of the nervous system, but these fish happen to be favorable for revealing several other codes (see Fig. 7) among the theoretical candidates (Perkel and Bullock 1968). One is a phase shift without frequency change, another is a burst duration, and another a probability-of-missing code.

The frequency code is also used—in ampullary receptors like those of the shark, which "listen" only to very slow potentials, mainly passively, and do not even "hear" the electric

organ discharges because they are too fast or too brief. We worked out some of the codes quantitatively (Scheich et al. 1973), and it seems reasonable to expect to see some of them turn up in respectable higher animals! The general principle we induce is that brains do not operate with just one code, waiting to be broken, but at least several, perhaps a good many, working in parallel (Perkel and Bullock 1968).

Electrolocation and electrocommunication

We have referred to some known and some suspected uses of electroreception in what may be called the electrolocating role (see Table 1), detecting objects or prey in the near field, directions relative to earth or water current in the far field, employing both the active and the passive modes. Little is known of the skill or resolution in electrolocation; however, two methods have been developed for quantitative estimation of performance by Lissmann and Machin and by Heiligenberg.

The first named authors trained *Gymnarchus* to discriminate between two porous pots 50 mm in diameter, both containing aquarium water, but one containing a 1 mm glass rod and the other a 6 mm glass rod. Heiligenberg (manuscript) measured the gain and phase of a response which might be called the electrokinetic response in *Eigenmannia*. A blinded fish, even without special training, tends to hover between vertically suspended strips of plastic just as normal fish do among water plants, roots, and rocks. When the strips are slowly moved left and right relative to the fish, it moves similarly, maintaining a position centered between the strips to an accuracy limited by the frequency of the sinusoidal strip movement and the size and distance between strips. Controls with electrically transparent agar strips show no such following.

These methods indicate that the near-field object location of good insulators is useful to some centimeters, doubtfully to decimeters, varying with water conductivity, object size, container size, and other factors. Passive electrolocation like the sharks' detection of hidden flatfish is useful to sizable fractions of a meter in seawater. These or improved methods should permit future analyses of the parameters of discrimination, the biophysical bases of discrimination, and the brain mechanisms involved in electrolocation.

Another role of electroreception, which is particularly developed in the electric fish—the families that possess electric organs in addition to electroreceptors—is the social signaling or electrocommunicating role. Even if there were no special alterations of the electric organ discharge (EOD) correlated with behavior, the presence of the EOD of the species' characteristic shape and frequency range and of the repetition rate characteristic of the individual provides information on the location, species, individual identity, and in at least one species (*Sternopygus macrurus*, Hopkins 1972) the sex, for any receiver equipped to understand this information. But by adding alterations in EOD many more signals of social significance are possible.

Before looking further at social communication it will be helpful to make still another distinction. Among the electric fish there are "wave" species and "pulse" species, the distinction being based on the form of the electric organ discharge. Wave species emit a continuous, nearly sinusoidal voltage. Different species discharge within characteristic ranges, *Sternopygus macrurus* at 50 to 150 Hz, *Eigenmannia virescens* at 250 to 600, *Sternarchus albifrons* at 750 to 1,250, and *Sternarchorhamphus sp.* at up to more than 2,000 Hz (all at 27°C). The discharge frequency of each individual is ordinarily extremely regular and does not change for long periods, or changes less than 2 percent, with state of arousal, feeding, touch, vibration, light, or conspecific fish.

The pulse species normally discharge at a few pulses per second, up to 100 per sec, usually quite irregularly. The pulses are 0.2 to 2 msec long, and the intervals between them from 10 msec to a second or more. They continually change rate by large percentages, with arousal, food, touch, vibration, light, or other fish, in a highly labile manner.

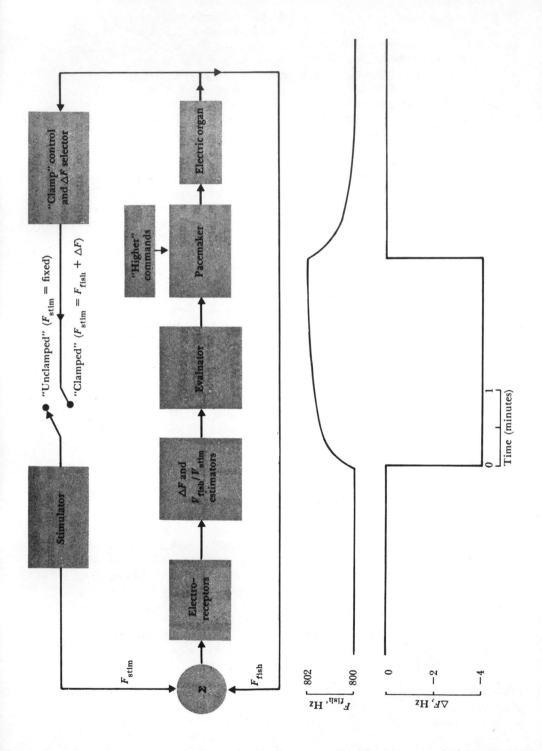

Social signals

We are learning that both wave and pulse species make social signals of some variety by modifications of EOD, and thus a busy electrical world of electrocommunicating signals is added to the cacophony of inanimate electrical events in the microvolt range. Without detailing them and their social contexts, still quite incompletely known, we can safely list several kinds of discharge alterations as correlates of social situations, perhaps having significance as threat, warning, submission, and announcement of food or other object of special interest (Möhres 1957; Black-Cleworth 1970; Hopkins, dissertation; Moller and Bauer, in press). Alterations include frequency rises of different form, amount, and duration; brief chirps or beeps; shorter or longer silent periods (ca. one-quarter to several seconds); and in a few cases amplitude modulations and chords (two frequencies at once).

One form of response deserves special attention because of its accessibility to analysis in terms of sensory input, central evaluation, and EOD control. In species with a steady hum, like *Gymnarchus* and *Eigenmannia*, at 250 to 500 per sec, and *Sternarchus*, at 750 to 1,250 per sec, there is a small but definite frequency shift whenever another fish, or a stimulus simulating one, with a frequency within 1 or 2 percent comes into range (Figure 8). If, for example, a fish is firing along at 800 EOD's per sec and a stimulus appears at 796 per sec (a 0.5 percent difference, undetectable to the human ear if converted into sound), the EOD will glide slowly up to 801 or 802 or 803, depending on the stimulus voltage. If we flip the stimulus frequency every 30 sec from 4 Hz below to 4 Hz above the fish's frequency, the EOD will slide up or down, respectively, again and again without fatigue (see Fig. 9). I call this the JAR, for jamming avoidance response (Bullock et al. 1972a, b). It is a dependable quantifiable, reflexlike, normal social response, inviting study as an example of behavior. We already know something of its afferent and efferent bases from previous analyses of the receptor types, the electric organ, and its control and command system in the brain (Fig. 10).

Eigenmannia and *Sternarchus* give best responses to a difference of 3 or 4 Hz (Fig. 11), respond less to 10 or 15 Hz, and ignore a 30 Hz difference. They respond definitely though weakly to a difference of 0.1 Hz but ignore a stimulus of exactly their own frequency. The fish always responds in the right direction, without hunting, even when the fraction-of-a-second latent period gives it less than a tenth of a beat cycle. It reacts to the second, third, fourth, or fifth harmonic if we add or subtract a few Hz, and the best difference frequency is still 3 Hz; therefore the stimulus is a beat frequency, not the percentage difference.

The electric organ is paced by a command unit in the brain, cycle by cycle, and thus the central nervous system has a reference against which it might measure the strange rhythm. But if the command output is interrupted and the electric organ silenced, it can be shown that the fish does not in fact use the command rhythm, which goes merrily on in the pacemaker center.

The fish, it seems, must get both rhythms through the sense organs.

Figure 8. Schematic diagram of the system causing the JAR (jamming avoidance response) of high-frequency weakly electric fish. ΔF is the stimulus frequency minus the fish frequency. (After Bullock et al. 1972b, by permission of Springer-Verlag.)

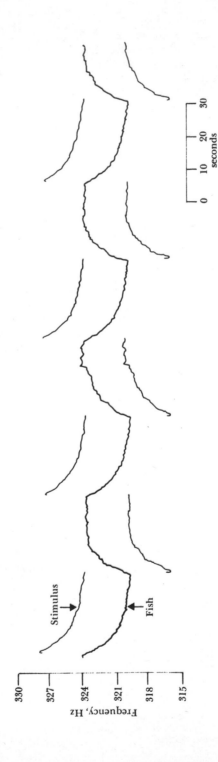

Figure 9. The JAR of *Eigenmannia* stimulated with alternately plus and minus $\Delta F = 4$ Hz, "clamped" so that the stimulator follows any changes in fish frequency to maintain the ΔF. The reversal of stimulus sign occurs automatically every 25 sec. The fish tries to escape by shifting its frequency in the correct direction to increase the ΔF. The two frequencies are continuously measured by triggering a 10 MHz precision clock with zero crossings and counting the clock cycles, without dead time. (From Bullock et al. 1972a, by permission of Springer-Verlag.)

310

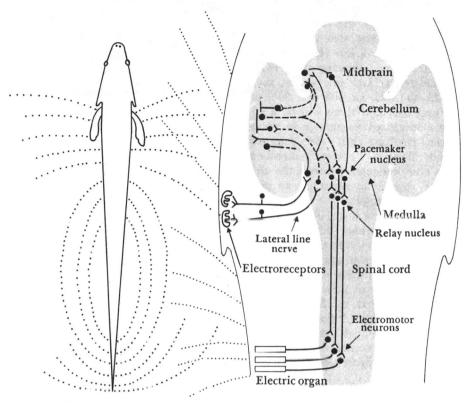

Figure 10. Diagram of the anatomy of the JAR system. Dashed lines represent hypothetical connections. Dotted lines are current lines in the fish's electric organ field. (After Bullock et al. 1972a, by permission of Springer-Verlag.)

They can be any two rhythms, as long as they are a few Hz apart. For example, after silencing the EOD we can deliver two sine waves, such as 323 and 326, or 502 and 505, or 241 and 244, and there will be a response measured as change in pacemaker rhythm; but the response cannot have a "correct" direction if no clue indicates whether the difference should be regarded as plus or minus. Normally, the asymmetry of the EOD wave form, like a clipped sine wave, provides this clue, and then the fish analyzes the small frequency differences in the time domain. The range of effective stimuli can be described in terms of sidebands and can be generated by AM, FM, or phase modulation. Figure 12 shows a recent diagram of the components of the system, considered as an example of input processing and output control.

The smooth and continuously gradable control of frequency of EOD in these wave fish is correlated with a high precision of the regularity of the discharge frequency—at least 100 times better than familiar rhythmic nerve cells and biological clocks. The standard deviation of single intervals (= periods), over thousands of cycles, without averaging, is 0.01 percent or 0.1 μsec in *Sternarchus!*

This regularity allows the fish to use their object-detecting ability in the presence of neighbors with frequencies only a few Hz away and to use receptors that code electric-field strength with high-precision phase shifts. These features of high temporal resolution

311

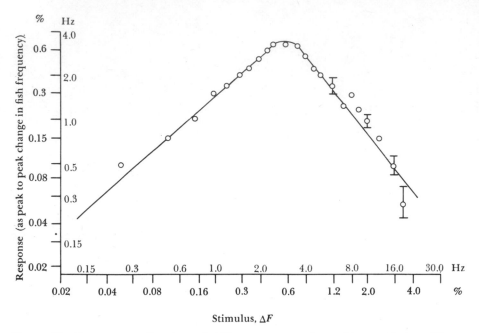

Figure 11. Response as function of ΔF (neglecting its sign). JAR elicited as in Figure 9 with different values of ΔF. Vertical bars are estimated confidence limits. (From Bullock et al. 1972b, by permission of Springer-Verlag.)

require the stable detection of very small frequency differences both in respect to magnitude and sign. We have found higher-order neurons in the midbrain (torus semicircularis) that can do this, apparently by integrating lower-order phase and probability-coded impulses. Other units in the lobus lateralis posterior and cerebellum seem less specialized for microsecond temporal analysis and may be better adapted to object detection.

Wave vs pulse species

We are struggling even to speculate on the adaptive value of the two general types of species, the high-frequency, high-regularity wave species and the low-frequency, irregular pulse species. The former should be more sensitive to object detection, since they sample the world so often and can average many independent samples. They should also be able to detect each other at greater distances,

if the brain can process input as the human brain does in hearing a musical tone against loud noise—that is, they should be less troubled by interference from "static" than the pulse species are. The low-frequency, variable-interval pulse species should be less bothered by coincident discharge from other fish, since simultaneous pulses will be rare, especially a whole series in succession such as normally occurs during every beat cycle of two or more wave fish.

Either type can be expected to detect not only ohmic discontinuities in the milieu but also capacitative impedance because of the regular frequency in the wave species and the uniform, brief spikes in the pulse species. Experiment has confirmed this for single afferent fibers in *Eigenmannia* (Scheich et al. 1973). Here is yet another new look at the world. It may permit discriminating on a new dimension among objects such as

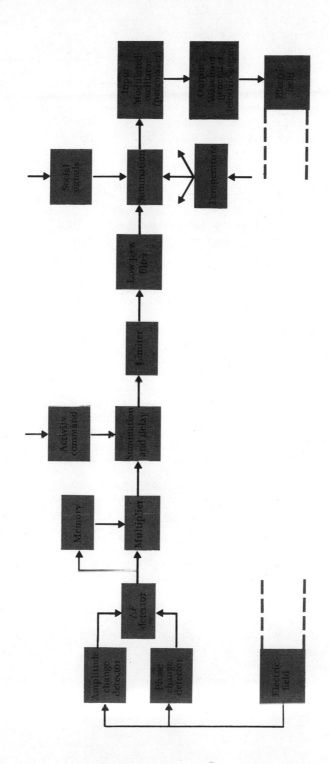

Figure 12. Block diagram of the formal equivalent components inferred in the JAR system. Each block has input-output properties that are still only partly explained. (After Bullock et al. 1972b, by permission of Springer-Verlag.)

other organisms. At present we know little about the capacity of natural objects. The capacity of a cubic centimeter of muscle tissue is about 0.1 to $0.01\mu F$, whereas the single electroreceptor of *Eigenmannia* can detect $0.0004\ \mu F$ in a certain geometry— that is, the receptor is far more sensitive than necessary to detect such an object.

Even for object detection by local conductance discontinuity—objects like rocks or fish—we know little about the parameters used by fish or the limits of performance in respect to discrimination of size, shape, distance, conductance, or motion. Since the electric field from the fish's discharge falls off with the cube of distance and an object distorting the field can be regarded as a dipole source whose effect also falls off with the cube of distance, we may expect object detection to be effective only at close range. However, the consequences of many factors are not yet fully assessed: the high sensitivity of receptors, especially to differences in electric field intensity, and the large number of receptors; the large size of the relevant brain regions; the special geometry of the electric field; the pattern of resistances in the fish's skin and its sense organs; to say nothing of the electrical parameters of objects normally important to these fish.

In spite of many unknowns in the evaluation of their roles in the animal's life, it has recently become clear that several families of fish have developed low- and high-frequency electroreceptors useful in active and passive electrolocation and electrocommunication, and that their sensitivities give access to a busy world of both signals and noise.

References

Bennett, M. V. L. 1970. Comparative physiology: Electric organs. *Annu. Rev. Physiol.* 32:471–528.

Black-Cleworth, P. 1970. The role of electrical discharges in the non-reproductive social behavior of *Gymnotus carapo* L. (Gymnotidae Pisces). *Anim. Behav. Monog.* 31:1–77.

Bullock, T. H. 1968. The representation of information in neurons and sites for molecular participation. *Proc. Nat. Acad. of Sci.* 60(4):1058–68.

Bullock, T. H. Forthcoming. An essay on the discovery of sensory receptors and the assignment of their functions together with an introduction to electroreceptors. In *Handbook of Sensory Physiology III*, A. Fessard, Ed. New York: Springer-Verlag.

Bullock, T. H., R. H. Hamstra, Jr., and H. Scheich. 1972a. The jamming avoidance response of high frequency electric fish. I. General features. *J. Comp. Physiol.* 77:1–22.

Bullock, T. H., R. H. Hamstra, Jr., and H. Scheich. 1972b. The jamming avoidance response of high frequency electric fish. II. Quantitative aspects. *J. Comp. Physiol.* 77:23–48.

Hagiwara, S., and H. Morita. 1963. Coding mechanisms of electroreceptor fibers in some electric fish. *J. Neurophysiol.* 26:551–67.

Hatai, S., and N. Abe. 1932. Responses of the catfish, *Parasilurus asotus*, to earthquakes. *Proc. Imp. Acad.* 8:375–78.

Heiligenberg, W. In press. The electromotor response in the electric fish *Eigenmannia*. *Nature*.

Hopkins, C. D. 1972. Sex differences in electric signalling in an electric fish. *Science* 176:1035–37.

Hopkins, C. D. Patterns of electrical communication among gymnotid fish. Ph.D. dissertation, 1972, Rockefeller University, New York.

Kalmijn, A. J. 1971. The electric sense of sharks and rays. *J. Exp. Biol.* 55:371–83.

Kalmijn, A. J., and T. H. Bullock. Forthcoming. The role of electroreceptors in the animal's life. I. The detection of electric fields from inanimate and animate sources other than electric organs. In *Handbook of Sensory Physiology III*, A. Fessard, Ed. New York: Springer-Verlag.

Kokubo, S. 1934. On the behavior of catfish in response to galvanic stimuli. *Sci. Rep. Tohoku Univ.* (D)9:87–96.

Lissmann, H. W. 1958. On the function and evolution of electric organs in fish. *J. Exp. Biol.* 35:156–91.

Lissmann, H. W., and K. E. Machin. 1958. The mechanism of object location in *Gymnarchus niloticus* and similar fish. *J. Exp. Biol.* 35:451–86.

Möhres, F. P. 1957. Elektrische Entladungen im Dienste der Revierabgrenzung bei Fischen. *Naturwissenschaften* 44:431–32.

Moller, P., and R. Bauer. In press. "Communication" in weakly electric fish, *Gnathonemus petersii* (Mormyridae). II. Interaction of electric organ discharge activities of two fish. *Anim. Behav.*

Perkel, D. H., and T. H. Bullock. 1968. Neural coding. *Neurosciences Research Program Bulletin* 6:221–348.

Sand, A. 1938. The function of the ampullae of Lorenzini with some observations on the effect of temperature on sensory

rhythms. *Proc. Roy. Soc. B* 125:524–53.

Scheich, H., and T. H. Bullock. Forthcoming. The role of electroreceptors in the animal's life. II. The detection of electric fields from electric organs. In *Handbook of Sensory Physiology III*, A. Fessard, Ed. New York: Springer-Verlag.

Scheich, H., T. H. Bullock, and R. H. Hamstra, Jr. 1973. Coding properties of two classes of afferent nerve fibers: High frequency electroreceptors in the electric fish, *Eigenmannia*. *J. Neurophysiol.* 36:39–60.

Szabo, T., A. J. Kalmijn, P. S. Enger, and T. H. Bullock. 1972. Microampullary organs and a submandibular sense organ in the freshwater ray, *Potamotrygon*. *J. Comp. Physiol.* 79:15–27.

36

Honey Bee Recruitment: The Dance-Language Controversy

Unambiguous experiments show that honey bees use an abstract language for communication.

James L. Gould

Reprinted from
29 August 1975, Volume 189, pp. 685-693

More than 2000 years ago, Aristotle pondered the phenomenon of honey bee recruitment. He observed that, although a source of food placed near a hive might remain undiscovered for hours or even days, once a single bee had located the food, many new bees soon appeared (*1*). Aristotle supposed that the bee which found the

The author is assistant professor, Department of Biology, Princeton University, Princeton, New Jersey 08540. This article is based on work done at Rockefeller University, New York.

food and learned its location—the "forager"—led the new bees—"recruits"—to it. It was later shown that if the forager were to be captured on its way back to the food, some recruits still were able to locate the source (*2*).

By 1920, von Frisch had discovered how to train bees to forage at artificial feeding stations, a technical breakthrough which opened the door to the experimental analysis of honey bee behavior. When von Frisch set out an array of plates near the hive containing various scents, he found that recruits approached only those which contained the odor of the food upon which the forager had been feeding. In an elegant series of experiments, von Frisch demonstrated that odors carried back on the waxy hairs of the forager provide recruits with the information necessary to locate the food (*3*). Recruits obtain this information by following "dances" which the forager may perform upon its return to the hive.

In the 1940's, however, while working with food sources at greater distances, von Frisch found that recruits arrived only in the vicinity of the forager's station, and not at the scent plates containing the same food odor set out at very different distances and directions (*4*). At first he supposed that recruits were attracted by the sight, unique hive odor, or the assembly pheromone of the foragers (or all three). This latter odor, produced by a specialized scent gland in the abdomen, had long been known to attract bees. Because recruits continued to arrive preferentially at or near the forager station even at distances of 300 meters, and when the wind carried odors away from the hive, von Frisch concluded that recruits must be provided with other cues, either in the hive or in the field. As a control, he sealed the pheromone-producing scent gland of each of his foragers with shellac, but still most recruits arrived at or near the forager station.

When he observed the behavior of foragers in the hive, von Frisch discovered that the tempo of the dances decreased as the food was moved farther and farther away from the hive, while the orientation of the "waggle run" phase of the dance (relative to gravity in the absence of light on the vertical combs) corresponded to the direction of the food source with respect to the sun. When the food source was moved very close to the hive, on the other hand, a "round" dance was observed, containing neither distance nor direction correlations (*4*).

The dance-language hypothesis which grew out of these observations proposes that recruit bees use the symbolic information in the dances as a guide to the approximate location of the food, and then use the odor information—that on the body of the dancing forager as well as odors left by foragers in the field—and visual information to locate the source exactly.

Objections to the Hypothesis

In 1967, Wenner and Johnson challenged the evidence supporting the dance-language hypothesis, and proposed that recruitment is accomplished solely on the basis of odor information. They repeated von Frisch's experiments with certain modifications, but found that when forager odor was provided at all scent plates, the preference for the vicinity of the actual forager station disappeared (*5*). They challenged many of the more unguardedly enthusiastic statements about recruitment [for example, that recruits fly "rapidly and with certainty" to the food (*6*, p. 57)] as being without experimental foundation.

In response to the challenge, von Frisch reiterated his opinion that the olfaction hypothesis was unlikely and cited two of his own experiments as further evidence (*7*). In one series of "detour" experiments (*6*, pp. 173–182), recruits had successfully located a food source on the other side of an obstacle (in one case a 12-story building) which foragers had been trained to fly around. Because some recruits were observed flying over the center of the obstacle, von Frisch concluded that they must have "known" where to go.

In another experiment (*6*, pp. 153–156) foragers had been trained to a station in one direction, while scent plates were set out at various distances in four directions. The hive was laid flat in the morning so that the dances were performed on a horizontal surface. Under these circumstances the dances were disoriented (although the distance correlation was preserved). Recruits in the morning displayed no preference for either the correct distance or the correct direction. The hive was restored to its normal position in the afternoon, and the dances became oriented. After a time, recruits began to arrive preferentially at scent plates near the forager station. As is discussed below, neither approach is immune to criticism (8).

Experiments to Control for Forager Odor

The work of Wenner and his colleagues stimulated new research designed to investigate recruit behavior more closely (9). Gould et al. (10, 11) and Lindauer (12) devised similar techniques to control for forager odors. They trained groups of foragers in two different directions, then fed one group a concentrated sugar solution and the other a dilute solution. Both solutions contained the same odor. In the hive, only foragers collecting the more concentrated food danced. Since the food odors and bee odors were the same at both stations, only the dance information might be thought to distinguish them. If recruits used the symbolic information, they would be expected to show a preference for the station with the more concentrated sucrose. In fact, they did just that (13).

Esch and Bastian (14), Gould et al. (11), and Mautz (15) sought to follow the behavior of individual recruits. Each found that recruits locate the food neither quickly nor reliably. Instead, recruits typically observe a single dance for about six cycles, fly out for about 6 minutes, return and attend another dance, fly out again, and so on. In the end, only about half of the recruits located the single food sources (situated 120 to 400 m away—in theory only 19 to 57 seconds flying time away).

Locale Odor

About the same time, Wenner and his colleagues began to gather evidence that odors specific to the location of the food (as well as the odors of the food itself) are important (16). Hence, if recruits know the "olfactory landscape," site-specific odors would provide information about the location of the food. Needless to say, distinguishing between recruits that are supposed to be using site-odor information rather than dance-language information would be difficult. The predictions of the two theories are virtually identical. Of course, they are not mutually exclusive— recruits could, in theory at least, use both sorts of information.

Excepting the experiments reported here, the locale-odor hypothesis can effectively account for all the results achieved to date without recourse to the dance-language theory. Even the clever experiment discussed above, in which the hive was first placed on its side and then returned to its normal orientation, can be disposed of in this way (17). Indeed, in similar experiments by New and New and by Wells and Wenner, recruits found the food source whether the dances were correctly oriented or not (18). Presumably bees could have been using site odor in these cases.

Evolutionary Argument

It has been argued that the dance correlations *must* be useful, or they would not exist; that is, evolution would not have selected for a nonfunctional behavior (12, 19). This argument depends on our belief that all behavior that seems somehow "special" is functional to the animal and is open to direct selection. It further supposes that we can correctly guess the function the behavior serves. This same proposition was put forward in a preevolutionary context by Leibniz, and effectively and entertainingly countered by Voltaire in *Candide*. In the present circumstances it seems reasonable to remain cautious, lest we glibly explain away phenomena and inhibit research. There can be no doubt that evolutionary considerations can suggest ingenious theories, but those theories must be tested if ethology is to remain an experimental science.

In the case of the dance language, the evolutionary argument appears particularly persuasive. But if every special behavior must have a "purpose," what can be the purpose of the oriented "dances" of flies?

What is the function of the distance correlations of the lateral movements of moths, or of the buzzing runs of stingless bees? What is the adaptive value of the transformation of angles flown or walked with respect to the sun into angles with respect to gravity, as is observed in ants, bees, beetles, spiders, and so on? In each case there is no evidence that the function is communication (20). If there is a "purpose," we must not yet know it; and that same unknown function could presumably explain the dance correlations as well. If there is *no* adaptive value to these behaviors, then the dance correlations need have none either. In the search for certainty, evolutionary arguments unaccompanied by critical experiments leave much to be desired.

Misdirection Experiments

The results not only of Wenner and his colleagues, but those of experimenters stimulated by the controversy, suggest that, at the very least, odors may affect the speed and accuracy of recruits. An unambiguous experiment might be one in which the symbolic information in the dance would indicate a specific location well away from the place actually being visited by the dancer. One way in which to accomplish this might be to construct a model bee whose dance parameters would be under the control of the experimenter. Progress on this front has been substantial, but this important technique has yet to be perfected (10, 21). Another way would be to cause the dancer to "lie"—to orient its dances to one parameter while recruits would be orienting their "interpretations" to another.

Under some circumstances, honey bees seem to interpret a bright light as the sun, and will orient their dances to it rather than to gravity (6, pp. 135–137, 196–203). (Under certain natural conditions such as swarming, dancing bees regularly substitute sun orientation for gravity orientation.) Since both dancers and dance attenders are reoriented, no misdirection of

recruits occurs (6, pp. 203–204). When the ocelli, the three simple eyes located between the compound eyes, are covered, bees become six times less sensitive to light. They require higher light levels to fly or display phototaxis, but are still capable of foraging and dancing (22). When a light is made bright enough to reorient the dances of normal bees, the dances of ocelli-covered foragers are unaffected (Fig. 1). Only when the light is made still brighter do these treated bees begin to reorient their dances from gravity to light.

Using a light intensity just bright enough to reorient the dances (and dance interpretations) of normal bees, but not sufficient to affect the ocelli-covered bees, it should be possible, assuming that recruits use the dance correlations, to "misdirect" untreated recruits. For example, if

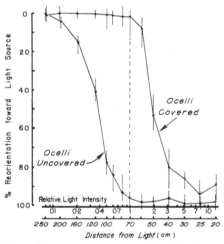

Fig. 1. Dance reorientation to light. The dance orientations of foragers were unaffected by a distant light source. When the source was moved closer, thereby increasing its apparent brightness (illumination), the dance directions shifted until they were oriented completely with respect to the light rather than to gravity. Further increases in brightness were without effect. Bees whose ocelli had been covered also reoriented their dances, but only when the light was much brighter. At an intermediate level (dashed line), the dances of normal bees were completely reoriented, while those of the ocelli-painted foragers were unaffected. Virtually all bees in each group were reoriented to the same degree (the bars indicate the scatter about the average). About 20 dances were measured for each point.

the food were in the direction of the sun, forager dances would normally be pointed up on the comb. A bright light 90° to the left of vertical, on the other hand, would cause normal dances to be directed 90° to the left, toward the artifical sun. Ocelli-covered foragers, however, would not be reoriented, and would continue to perform dances pointing up. If the dance information is used, untreated recruits attending these dances would be expected to interpret the dance directions as being 90° to the right of the artificial sun, and hence 90° to the right of the sun in the field. If, on the other hand, recruits rely solely on odor cues, they would be expected to ignore the dance direction and proceed to the forager station on the basis of odor cues.

Experimental Procedures

Two groups of 10 to 15 foragers were trained in each experiment on orange-scented $0.5M$ sucrose with the use of the Gary and Witherell improvement (23) of von Frisch's technique (6, pp. 17–20). The first group served as a control to measure the degree of reorientation of the dances to light. The ocelli of the foragers in the second group were covered with flat-black enamel paint. Dances of foragers in this second group were monitored, and any that showed some degree of reorientation to the light were eliminated. The light was left on and in the vertical ("up") position for 30 minutes before the experiment began, to adapt the bees to its presence. (Since, in the absence of light, bees orient with respect to vertical, placing a bright light in that same direction does not affect the dances.)

The light was a 650-watt quartz movie light with reflector. To prevent the hive from overheating, an infrared-absorbing filter mounted in an asbestos mask was placed just in front of the light for shielding. The Plexiglas covering the hive acted as a sharp ultraviolet filter for wavelengths shorter than 340 nanometers. The light was directed onto the only dance area easily accessible to returning foragers. Since the light was a finite distance away, danc-

ers on opposite edges of this area saw the light at slightly different angles. The distribution of 100 dances was plotted, and from that the mean parallax error was calculated to be 1.14°.

Another minor source of error arose from the phenomenon known as "residual misdirection." Foragers returning from the field regularly make slight errors in transposing the flight angle into a dance angle (6, pp. 204–217). Oddly enough, this effect disappears when the earth's magnetic field is canceled in the vicinity of the hive (24). In the 400-m direction experiments reported here, the angle of the light was corrected periodically to allow for this effect. (The metal shed in which the hive was located provides substantial shielding from the earth's magnetic field.) Later measurements from video tapes of the dancing revealed that the mean uncorrected error from this source was about 0.8°. The mean total error from both parallax and residual misdirection was of the order of 1.4°.

The observation hive was one frame (comb) thick and three frames high. It contained approximately 8000 Italian bees. The colony was not allowed to accumulate stores of food—a procedure which served to facilitate training and to enhance the tendency to dance.

When the experiments began, the solution at the control station was changed to anise-scented $1M$ sucrose (25). Hence the foragers at this station began to dance occasionally. The solution at the ocelli-treated forager station was switched to $2M$ sucrose containing the experimental scent. These bees began dancing vigorously. The light was moved to the appropriate position, and six recruit stations nearly identical to the forager station were opened.

The recruit stations were conceptually similar to ones designed by Renner (26). They were constructed from airtight plastic pails (Fig. 2). Entering recruits passed through a beam of red light shining on an array of photodiodes. (Bees do not react to red light.) The resulting signal was amplified and transmitted to an event recorder. Thus, the arrival time and location of each

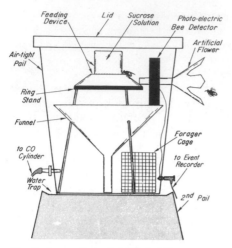

Fig. 2. A recruit station. The station was constructed from an airtight plastic pail. A decoy bee was placed in the artificial flower to induce landing. The "flower," fashioned from a plastic funnel, was painted to appear dark in the center and light on the edge under ultraviolet light. A wire cage of anesthetized foragers located in the station served as a source of bee odors. When a recruit entered a station, it interrupted a beam of red light, causing a photoelectric circuit to signal an event recorder, thereby registering the arrival. As the bee continued in, it came upon a feeding device offering a sucrose solution with the experimental scent. The station was filled with carbon monoxide so that while the recruit fed, it became anesthetized and tumbled off the inclined feeding device into a funnel. The stations stood on the water-filled bottoms of larger pails. The water helped to prevent ants from entering the stations. The pails were about 40 centimeters tall.

arriving bee was automatically registered for later analysis.

In order to attract recruits, it was necessary to bait the stations with food containing the experimental scent, and with a cage of six anesthetized, ocelli-treated foragers to provide bee odors. To provide visual landing encouragement, the station was equipped with a "flower" fashioned from a plastic funnel which was cut and painted to present the appropriate pattern when viewed under ultraviolet light (6, pp. 481–491; 27) and onto which was glued the body of a sun-dried, alcohol-extracted bee. To prevent recruits from returning to the hive where they might dance and indicate the locations of the stations, each recruit station was filled with carbon monoxide.

Entering recruits generally fed for about 30 seconds before becoming completely anesthetized, tumbling off the inclined feeding platform into the funnel below. The order of arrival of recruits at each station was preserved in the order of anesthetized bees in the neck of the funnel. This made it possible to edit the event recorder data accurately on the rare occasions when a fly or small bumble bee was found in the station. At the end of each experiment, the number of anesthetized recruits in each station was compared with the number of "events" registered by the recorder for that station. The numbers always corresponded.

Recruits arriving at the ocelli-treated forager station were captured manually by a technique which precluded the release of alarm odor (11). The dancing in the hive was recorded on video tape for later analysis. Wind speed and direction were recorded continuously with a calibrated Taylor "Windscope" interfaced to two Rustrak recorders. Wet- and dry-bulb temperatures were continuously recorded in the corner of the video field by a Heath two-channel digital thermometer. A Heath digital clock recorded the time in the same manner. Sky conditions and barometric pressure were recorded manually at 10-minute intervals.

At the end of each experiment the forager food was removed and the light was returned to the "up" position. After about 10 minutes, the recruit stations were closed.

The experiments were performed on a relatively flat, mowed field at the Rockefeller University Center for Field Research, Millbrook, New York. The field (Fig. 3A) was bordered on most sides by forest, but beyond the boundary line marked to the northeast, it fell away into an unmowed, grassy valley. Experiments were generally performed in the early morning while the heavy dew typical of this region remained on the ground. Good results could be obtained while the vegetation remained wet, presumably because competition from natural food sources was minimal. The hive was located in a small metal shed in the southwest corner of the field.

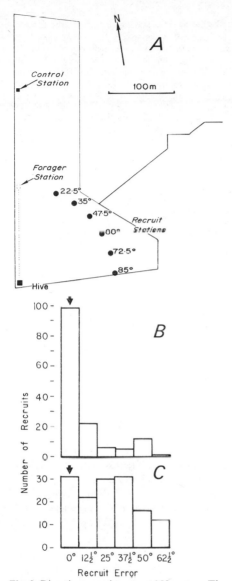

Fig. 3. Direction experiments at 150 meters. The array (A) consisted of recruit stations set out at 12½° intervals as shown. Ocelli-treated foragers were trained to the forager station, while untreated foragers were trained to the control station. The results of two misdirection experiments (Table 1, experiments A and B) have been combined to summarize the recruit distribution (B). The left bar—indicated by the arrow—represents the station specified by the dancing, while the bars to its right represent the averages for single stations at various distances. When the experiment was repeated using Wenner's training technique, quite a different distribution was obtained (C). The data for two experiments (Table 1, experiments C and D) are summarized.

Is Direction Communicated?

Control foragers were trained to the control station 300 m north of the hive. Ocelli-painted foragers were trained to the forager station 150 m north of the hive. Six recruit stations were set out 150 m from the hive at 12½° intervals as indicated in Fig. 3A. In the first experiment, the light shining on the hive was moved every 30 minutes, first to 85° left of vertical, then to 35° left, and finally to 60° left. In each case, recruits favored the recruit station in the direction indicated by the dance in preference to recruit stations located in other directions (as well as to the forager station itself) (Table 1, experiment A). The same results were obtained when the experiment was repeated (Table 1, experiment B). In this case, the light was again moved every 30 minutes, first to 72½° left, then to 35° left (Fig. 3B).

When recruits were presented with odor information for one location—the forager station—and abstract dance information for another—a particular station in the recruit array—they flew to the station specified by the dance correlations. Hence, recruits must be able to use the symbolic direction information in the dance. Analogous results were obtained using distance arrays (28).

Recruit Accuracy

In the 150-m direction experiments, the mean recruit error was 11.9° (31 m). In order to judge the accuracy of recruits at greater distances, the recruit stations were set out 400 m from the hive at 3° intervals (Fig. 4A). The ocelli-treated forager station and control station were established along a line to the northeast at 500 and 400 m, respectively. In the first experiment the light was moved to 42° to the right of vertical for 40 minutes, and then to 33° to the right for another 40 minutes (Table 2, experiment A). In the second experiment the procedure was reversed, with the light first at 33° right for 40 minutes, and then at 42° right for the same period (Table 2, experi-

Table 1. Direction experiments at 150 meters. Asterisks indicate the station signaled by the dancing. Experiments A and B are summarized in Fig. 3B. Experiment A was performed on 28 June 1974. The experimental scent was peppermint. The wind was from 001° at 4.2 miles per hour. The sky was cloudy, the barometric pressure was 771.6 mm-Hg, the relative humidity was 63 percent, and the temperature was 21°C. Experiment B was performed on 16 July with geranium scent. The wind was from 032° at 2.9 miles per hour. The sky was clear, the pressure was 770.8 mm-Hg, the humidity was 66 percent, and the temperature was 22°C. Experiments C and D (Wenner controls) are summarized in Fig. 3C. Experiment C was performed on 10 August with rose scent. The wind was from 191° at 0.9 mile per hour. The sky was clear, the pressure was 752.7 mm-Hg, the humidity was 61 percent, and the temperature was 22°C. Experiment D was performed on 11 August with rosemary scent. The wind was from 358° at 3.4 miles per hour. The sky was clear, the pressure was 772.7 mm-Hg, the humidity was 67 percent, and the temperature was 23°C.

Experi- ment	Period (min)	For- ager visits	For- ager dances	Recruit arrivals						
				FS	22½°	35°	47½°	60°	72½°	85°
A	30	92	70	4	1	0	1	0	5	19*
	30	98	99	7	7	31*	7	0	3	12
	30	90	83	3	0	11	6	28*	3	1
B	30	60	27	3	0	0	0	1	10*	3
	30	59	49	8	3	10*	4	0	1	0
C	30	52	29	19	5	6	9	14	8	13*
	30	52	36	22	18*	14	16	22	10	7
D	30	60	39	22	11*	8	14	15	7	6
	30	52	36	22	12	18	32	27	12	18*

ment B). In all cases, recruits favored the end of the array indicated by the dancing (Fig. 4B).

In the 400-m direction experiments, the mean recruit error was 4.2° (29 m). The smaller angular error for 400 m is consistent with von Frisch's results, as well as with the observation that the scatter of direction indications between cycles of individual dances is smaller at 400 m. Why the dance scatter should be so large at short distances, but small at large distances is another question (29).

Wenner's Techniques

When Wenner and his colleagues perform their experiments, they obtain quite different results. A number of factors might be responsible for this disparity. Von Frisch suggested (7) that Wenner's training technique might be the cause. Wenner trains his bees on a concentrated sucrose solution containing the scent which will later be used during the experiment (30). Virtually all other experimenters use a dilute solution with either no scent, or a scent other than the one to be used later. With Wenner's technique the foragers dance throughout training, and recruits are constantly being captured at the forager station. When techniques with less concentrated solutions are used, however, little or no dancing and recruitment occur. With Wenner's technique, the hive accumulates stores of solution containing the experimental scent for hours or even days before the experiment begins. The other technique does not result in any accumulation of experimental odor.

To test the effects of Wenner's techniques, the 150-m direction experiments were repeated after foragers were trained on a 2M sugar solution containing the experimental scent. In the first experiment the light was first placed 85° to the left of vertical for 30 minutes, then moved to 22½° left for the same period (Table 1, experiment C). In the second experiment, the procedure was reversed (Table 1, experiment D).

In both cases recruits showed virtually no preference for the direction indicated by the dances (Fig. 3C); the mean error was 29° (77 m). This result might be due, as von

Table 2. Direction experiments at 390 to 400 meters. Asterisks indicate the station signaled by the dancing. In experiments C and D the daggers indicate the direction of the forager station. Experiments A and B are summarized in Fig. 4B. Experiment A was performed on 31 July 1974. The experimental scent was rose. The wind was from 171° at 2.9 miles per hour. The sky was clear, the barometric pressure was 765.3 mm-Hg, the relative humidity was 71 percent, and the temperature was 28°C. Experiment B was performed on 1 August with rosemary scent. The wind was from 167° at 2.3 miles per hour. The sky was clear, the pressure was 766.5 mm-Hg, the humidity was 67 percent, and the temperature was 25°C. The results of experiments C and D are shown in Fig. 4, C and D. Experiment C was performed on 2 August with sassafras scent. The wind was from 037° at 0.6 mile per hour. The sky was hazy, the pressure was 768.3 mm-Hg, the humidity was 69 percent, and the temperature was 27°C. Experiment D was performed on 4 August with ilang-ilang scent. The wind was from 006° at 1.1 miles per hour. The sky was partly cloudy, the pressure was 766.6 mm-Hg, the humidity was 76 percent, and the temperature was 29°C. Abbreviation: FS, forager station.

Experi- ment	Period (min)	For- ager visits	For- ager dances	Recruit arrivals						
				FS	45°	42°	39°	36°	33°	30°
A	40	89	69	12	16	19*	8	5	6	3
	40	83	66	9	4	4	5	11	16*	12
B	40	104	76	5	5	2	4	8	12*	13
	40	111	87	1	16	16*	10	2	4	1
C	30	75	45	33*†	8	15	31*†	10	10	6
	30	76	47	55*†	11	18	46*†	19	10	8
D	50	95	80	82†	30*	24	12	15	34	68†

Frisch suggests (7), to recruits becoming lost while the forager station is being moved rapidly away from the hive during training; or, as Free implies (31) and von Frisch had shown earlier (6, pp. 23, 33–34, 257–264), the results might be due to recruitment to a prevailing food odor in the hive. In any case, a small change in experimental technique seems sufficient to switch recruitment from a largely dance-directed form to a largely odor-directed one. Apparently honey bees are sufficiently complex to find food by more than one strategy.

Von Frisch's Techniques

From analyzing dance films, von Frisch and Jander (32) were able to calculate that a recruit would have to receive and use without error the information from five to six dance cycles in order to account for the accuracy inferred from von Frisch's famous step and fan experiments (6, pp. 84–96, 156–163). As von Frisch points out, "bees are not computers that work without making mistakes" (6, p. 108). Since recruits typically follow individual dances for about six cycles, von Frisch concluded that they must follow two or more different dances and average the information. Since recruits follow only one dance before flying out in most cases (14, 15), Lindauer proposed that bees must be able to average dances separated by many minutes (12). Another possibility, of course, is that the step and fan data do not represent precise measures of recruit accuracy.

Von Frisch did not capture recruits that approached his scent plates. Hence, a recruit could be counted any number of times (until, that is, it landed at the forager station and was killed). In order to examine the possible effects of von Frisch's techniques, my recruit stations were modified to prevent recruits from feeding on the scented food, and to allow them to escape through the top of the station. Hence, recruits could be counted more than once. The stations contained no cage of anesthetized foragers to provide bee odors. As always, recruits arriving at the ocelli-treated forager station were captured.

The recruit stations were again set out at the north end of the field, 390 m from the hive at 3° intervals (Fig. 4A). This time, however, the forager station was placed near the array as it had been in von Frisch's work. In the first experiment the forager station was placed 20 m beyond the 39° station, while the control station was placed 50 m farther out, along the same line. The experiment lasted for 60 minutes, and during this time the light was in the "up" position. Under these circumstances, no misdirection of recruits would be expected. Just as von Frisch found in his fan experiments, recruits favored the station indicated by the dancing—that is, the station immediately in front of the forager station (Fig. 4C). The mean error was not as large as in the 400-m direction experiments: 3.3° instead of 4.2° (22 m versus 29 m).

In order to examine more closely the effect of having the forager station near the array, I performed a second experiment. In this case the forager station was placed 20 m beyond the 33° station—the most easterly in the array. The control station was placed 50 m beyond the forager station. The light was moved to 15° right of vertical in order to direct recruits to the station at the other end of the array—the 45° station in Fig. 4A. The experiment lasted 50 minutes. Recruits favored *both* ends of the array—not only the station indicated by the dancing, but also the one near the forager station (Fig. 4D and Table 2, experiment D).

Taken together, these experiments suggest that von Frisch's techniques can affect the observed distribution of recruits, exaggerating the apparent accuracy. If recruits

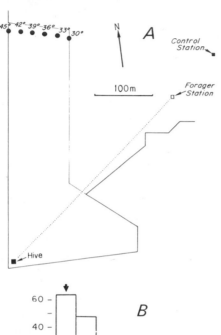

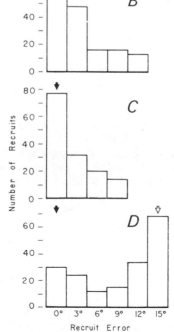

Fig. 4. Thirty-nine direction experiments at 390 to 400 meters. The array (A) consisted of recruit stations set out at 3° intervals as shown. In the misdirection experiments, the ocelli-treated forager station and control station were set out as illustrated. The data from two experiments (Table 2, experiments A and B) were combined to summarize the recruit distribution (B). The left bar—indicated by the arrow—represents the station specified by the dancing, while the bars to its right represent the averages for single stations at various distances away. In the two von Frisch controls, the forager station was moved near the array, and the control station was placed 50 m beyond it. In the first control (Table 2, experiment C), the forager station was 20 m behind the 39° station. The recruits displayed a stronger preference than before for the "correct" direction (C). In the second control, the forager station was placed 20 m behind the 30° station, but the dances indicated the 45° station. Recruits preferred both the indicated station (solid arrow in D) and the one near the forager station (open arrow).

are not as accurate as the step and fan experiments suggest, then von Frisch's conclusion that recruits must average separate dances may not be correct. If recruits were to attend more than one dance, averaging the information for a more precise estimate of distance and direction, then these recruits would average the directions of separate dances; and, since the second dance is attended only after an unsuccessful flight, recruits should take quite some time to locate the food. Both predictions were tested experimentally.

Dance Integration?

Lindauer proposed that recruits might average—"integrate" in his terminology—two or more dances to separate targets. This hypothesis gives rise to testable expectations. If one station at the end of a direction array were being signaled, recruits would receive information for only that one direction regardless of how many dances they attended. If two olfactorally identical stations at opposite ends of the array were being signaled by similar numbers of dances, and recruits attended only one dance each, then the distribution of recruits in the array would be composed of two *separate* single-station distributions —one the mirror image of the other— summed across the array. If, however, recruits arrived after attending two dances each, averaging the information from both, quite a different distribution would be expected. The chance of following two dances to one particular station would be 0.25, while the chance of attending one dance to each station would be 0.50. The overall recruit distribution in this case would be the sum of three distributions— one about each end station, and a third, with twice as many bees, about the center of the array.

This intriguing prediction was tested with the use of the array pictured in Fig. 5A. The five recruit stations were set out at 7½° intervals. Two stations of ocelli-treated foragers were established to the north. The light was placed 58° to the left

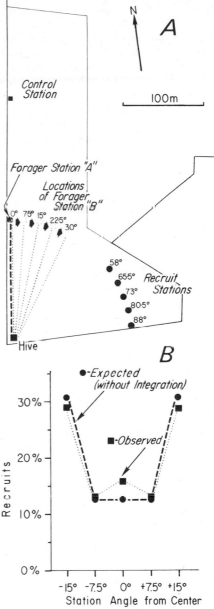

Fig. 5. Integration experiments. The array (A) consisted of recruit stations set out at 7½° intervals as shown. Two ocelli-treated forager stations were set out. One forager station (A) was always to the north, while the other (B) was moved to the various positions indicated. The data, when the dancing by both groups of foragers indicated the same end of the array, was used to predict the distribution if no integration of separate dances was taking place. This model and the results when both ends of the array were being signaled simultaneously are compared (B).

325

of vertical, thereby directing recruits to the 58° station. The distribution of recruit arrivals in this part of the experiment was taken as a measure of recruit accuracy (Table 3). After 20 minutes in the first experiment and 30 minutes in the second, one of the two ocelli-treated forager stations was moved 7½° to the right every 10 minutes until it reached a point 30° to the right of the other ocelli-treated forager station. At this stage, dances in the hive indicated both the 58° and 88° stations. This part of the experiment lasted 40 minutes. Recruits favored both ends of the array (Table 3).

If any integration of the two dance directions was taking place, more recruits would be expected at the center stations than would be predicted on the basis of normal recruit error alone. By taking the data when only one end station was being indicated (and, hence, no integration could be occurring), and adding it to its "mirror image," a model was generated for recruit distribution when both end stations are being indicated simultaneously, but *no* integration is taking place. By comparing the actual results with the model, the prediction was tested (Fig. 5). Under these conditions, little averaging of separate dances appears to have taken place.

Recruit Search Times

How can it be that in these experiments recruits did not integrate separate dances, while in the experiments of Esch and Bastian (*14*), Gould *et al.* (*11*), and Mautz (*15*), recruits were often observed to attend several different dances before being captured at the forager station? The critical difference between their experiments and mine cited above lies in the fact that they used single isolated food sources, whereas my experiments and von Frisch's step and fan experiments used wide arrays. With an array, perhaps even a recruit with relatively inaccurate information is likely to arrive near *some* station, whereas the same recruit searching far from the single forager station offered in other experiments (*14, 15*) might not find it and return to the hive

to attend another dance. If this is true, recruits in array experiments should take considerably less time to find a station than the 15 minutes required by recruits in the single-station experiments (*14, 15*).

To test this prediction, I examined the change in recruit arrival rates. When the light was moved, recruits were instantaneously presented with dances indicating a new target. The time course over which recruit arrivals at the newly indicated station increased and arrivals at the formerly indicated station decreased provides a measure of recruit search dynamics. The combined data for six experiments are presented in Fig. 6 (*33*). The simple model shown in the inset fits the data fairly well. This model supposes that a constant number of recruits were dispatched from the hive during each minute of dancing, but that individual recruits took varying amounts of time to reach the station. The average time of 6 minutes indicated by the data agrees well with the results of Mautz (*15*) and Esch and Bastian (*14*) for the duration of successful flights. Again, recruits presented with an array rather than a single station seem to arrive with neither the need nor the opportunity of averaging multiple dances (*34*).

As was mentioned above, von Frisch discovered that his recruits in the step and fan experiments performed too well to have attended only a single dance. Either von Frisch's recruits attended additional dances, or the accuracy which he measured was an artifact of his experimental technique. The second alternative is far more likely since, as demonstrated above, recruits in array experiments attend only *one* dance, and von Frisch's techniques *do* artifically enhance recruit accuracy.

Do Recruits Make Mistakes?

Since the data from the experiments reported here represent recruit accuracy after attending only a single dance and with the forager station well away from the array, and since the scatter of forager dances may be measured directly from video

Table 3. Integration experiments. Each asterisk indicates a station signaled by one group of foragers. Figure 5B summarizes the data. Experiment A was performed on 8 August 1974 with jasmine as the experimental scent. The wind was from 003° at 2.8 miles per hour. The sky was partly cloudy, the barometric pressure was 769.5 mm-Hg, the relative humidity was 76 percent, and the temperature was 24°C. Experiment B was performed on 9 August with lavender scent. The wind was from 004° at 4.6 miles per hour. The sky was partly cloudy, the pressure was 768.5 mm-Hg, the humidity was 80 percent, and the temperature was 23°C.

Experiment	Period (min)	Forager visits		Recruit arrivals				
		A	B	58°	65½°	73°	80½°	88°
A	20	13	18	19**	10	6	1	2
	10	8	10	8*	6*	3	2	0
	10	7	12	5*	5	3*	2	2
	10	10	10	6*	1	4	4*	1
	40	44	41	21*	13	12	9	25*
B	30	17	16	18**	4	2	1	0
	10	11	9	8*	3*	2	1	1
	10	14	15	6*	5	3*	0	0
	10	9	15	3*	1	4	4*	1
	40	38	37	21*	9	12	10	21*

recordings, it is possible to estimate how much error is introduced into the system by recruits using the dance information. The data from the 150-m direction experiments have been used for the following calculations since the dance scatter is larger, and therefore easier to measure, at shorter distances.

The mean scatter, averaged over six dance cycles (as we suppose recruits to be doing) was about 8.1 percent. The mean total recruit scatter was about 11.9 percent. Since recruit errors may cancel dance errors, it is appropriate to use the sum-of-the-squares method to calculate the mean total recruit error. With this technique, recruit error over and above the error in the dance was 8.8 percent. Thus, von Frisch's intuitive prediction that recruits must also contribute errors seems to be confirmed. In his step and fan experiments, the presence of the forager station in the array combined with the nature of his recruit stations probably served to "focus" the recruit distribution.

On the basis of these data, the general form of recruit distributions may be represented as

$$\frac{1}{2\pi} \exp\left[-\frac{1}{2}(x^2 + y^2)\right]$$

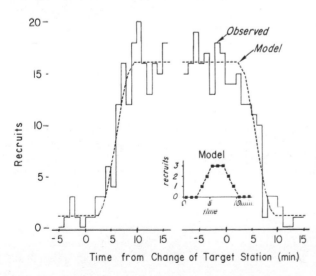

20 –

15 –

Recruits

10 –

5 –

0 –

   -5   0   5   10   15   -5   0   5   10   15

Time from Change of Target Station (min)

Observed

Model

Model

Fig. 6. Recruit search times. The number of recruits arriving at various times before and after a new station was signaled (at 0 minute) is shown at left. Before the station was indicated, a background level of recruitment can be observed. Recruitment reaches a maximum after about 10 minutes of dancing. The situation when the station is no longer signaled is illustrated at right. (The dancing to the station stops at 0 minute.) Recruitment continues for a time, then falls off to a background level. The inset depicts a simple model of recruit search-time scatter for bees sent out during each minute of dancing. The model matches the data fairly well.

where x and y are the distances from the target. The y-axis may be taken to be along the line from the hive to the target, so that $y = 1$ for 1 standard deviation (S.D.) of recruit distance scatter (or approximately $1/2 \sqrt{2}$ of the S.D. of the scatter of six-cycle averages of the distance correlations in the dance). The x-axis may be taken as the horizontal line orthogonal to the y-axis, where $x = 1$ for 1 S.D. in direction scatter of arriving recruits (or approximately $1/2 \sqrt{2}$ of the S.D. of direction scatter averaged over six cycles of the dance; minor corrections must be made since the target is only a finite distance away from the hive). Distance and direction scatter in the dance vary independently, giving rise to the prediction that recruits should in general be distributed about the target in an ellipse, whose major axis should be along the x-axis near the hive, along the y-axis at intermediate distances, and along the x-axis again at still greater distances. Although my data tend to support this prediction, the wind must be taken into account in any attempt to describe recruitment more precisely. The odors for which recruits are searching are presumably moved about and distributed in the field depending on wind direction and speed (35). Two-dimensional arrays might be useful in this regard, but for a complete description of recruit behavior the path taken by individual recruits as they get from the hive to the food must be directly determined.

Conclusions

Depending upon conditions, honey bee recruits use either the dance language *and* odor information, or odors alone. Wenner's conclusion that "one cannot have it both ways" (36)—that bees can have only one strategy for recruitment which they must use under all circumstances—is clearly incorrect. On the other hand, recruitment to odors alone might be the usual system in honey bee colonies not under stress. (Even in these experiments, a substantial number of recruits arrived at the forager station, presumably without having used the dance information.) For example, when a single, abundant, and (presumably) extensive crop is available, the odor of that source in the hive might grow strong enough to eliminate dance-language recruitment. However, even in the early stages (just after the food is first located), the dance language would be necessary to alert foragers to the new source.

By their very different training techniques, von Frisch and Wenner may have been sampling two stages of the same process: exploitation of an abundant food source. Von Frisch's experiments could be seen as examining the early phase, while Wenner's would be exploring a later phase.

The inherently "spectacular" nature of the dance language may have helped to emphasize it out of proportion to its actual place in the ecology and dynamics of foraging. Alternatively, single, abundant, extensive food sources may be more typical of current agricultural practices than of the tropical forests in which the honey bee and its language evolved. In that case, the ability to direct recruits to a distant, isolated patch of food quickly—either before it was found by another colony or another species, or before the potentially brief blooming period ended—might have been a real advantage. Only further work can establish whether the dance-language communication is common or rare under normal circumstances. Even a rare phenomenon may still be real, however, and the dance language is a real and very significant phenomenon indeed.

Some of the resistance to the idea that honey bees possess a symbolic language seems to have arisen from a conviction that "lower" animals, and insects in particular, are too small and phylogenetically remote to be capable of "complex" behavior. There is perhaps a feeling of incongruity in that the honey bee language is symbolic and abstract, and, in terms of information capacity at least, second only to human language (37). Despite expectations, however, animals continue to be more complex than had been thought, or than experimenters may have been prepared to discover (38). Especially in ethology, it is difficult to avoid the unprofitable extremes of blinding skepticism and crippling romanticism.

References and Notes

1. Aristotle, *History of Animals*, book IX, chapter 40.
2. For example, M. Maeterlinck, *The Life of the Bee* (Dodd, Mead, New York, 1901), pp. 107–113.
3. K. von Frisch, *Zool. Jahrb. Abt. Allg. Zool. Physiol. Tiere* 40, 1 (1923).
4. _____, *Oesterr. Zool. Z.* 1, 1 (1946); *Bull. Anim. Behav.* 5, 1 (1947).
5. A. M. Wenner, *Science* 155, 847 (1967); D. L. Johnson, *ibid.*, p. 844.
6. K. von Frisch, *The Dance Language and Orientation of Bees* (Harvard Univ. Press, Cambridge, Mass., 1967).
7. _____, *Science* 158, 1072 (1967).
8. Although considered conclusive by some, the study by R. Boch [*Z. Vgl. Physiol.* 40, 289 (1957)] is not cited by von Frisch. This experiment sought to take advantage of "dialects" in the distance correlations. Two races of bees with different distance dialects were mixed in a hive. When foragers of one race danced, recruits of the other race would have been expected to misinterpret the information and fly either too far or not far enough, depending on which race had the quicker dance tempo at that distance. Two scent plates were set out—one between the hive and the forager station, and the other one beyond. In all, 644 recruits arrived at the station predicted on the basis of differences in racial dialects while 667 arrived at the other station. When compared to the distribution of recruits of the foragers' own race, however, a slight tendency in the expected direction may be computed.
9. Another little-known study, again considered conclusive by some, is pointedly ignored by von Frisch and colleagues—perhaps with good reason. L. S. Gonçalves [*J. Apic. Res.* 8, 113 (1969)] sought to take advantage of A. R. Bisetzky's discovery [*Z. Vgl. Physiol.* 40, 264 (1957)] that foragers trained to walk to a food source perform dances whose tempos are appropriate for far greater distances. If the dance language were used, recruits would be expected to fly to scent plates beyond the forager station, but in the same direction. Gonçalves arranged a circle of scent plates around a hive, and forced his foragers to walk through a short tube to their station. Although most recruits flew to the forager station, the remaining bees showed some preference for the scent plates in the "correct" direction. Both Bisetzky and Gonçalves used the Italian race of honey bee. Bisetzky showed that, for distances up to 2.5 m, these foragers perform exclusively round dances—the dance form without any direction information. Although in his first seven experiments Gonçalves used walking distances of only 1 and 2 m, his recruits did as well at finding the "correct" station as the recruits in the two experiments with a distance of 3 m—a condition yielding, according to Bisetzky, about 80 to 90 percent waggle dances. Indeed, only in these latter experiments were the dance angles recorded. Since recruits seem to have done as well with or without the dance information, this experiment is unconvincing. The preference for the direction of the forager station was probably due to the fact that the wind always blew in that direction.
10. J. L. Gould, thesis, Rockefeller University (1975).
11. _____, M. Henerey, M. C. MacLeod, *Science* 169, 544 (1970).
12. M. Lindauer, *Am. Nat.* 105, 89 (1971).
13. Since foragers tend to expose their scent glands at rich food sources, both experimenters sealed these glands shut. Gould *et al.* sealed the glands of both groups of foragers, while Lindauer sealed the glands of only the dancing group. Since the odor of shellac is a potent odor cue for recruits (*6*, pp. 22–23) and was to be found on the bodies of the dancing bees, and only at the station with the more concentrated sucrose, Lindauer's stations may not have been olfactorally identical. (The same criticism may be applied to many of von Frisch's experiments.)
14. H. Esch and J. A. Bastian, *Z. Vgl. Physiol.* 68, 175 (1970).
15. D. Mautz, *ibid.* 72, 197 (1971).
16. A. M. Wenner and D. L. Johnson, *Science* 158, 1076 (1967); A. M. Wenner, P. H. Wells, D. L. Johnson, *ibid.* 164, 84 (1969); D. L. Johnson and A. M. Wenner, *J. Apic. Res.* 9, 13 (1970); P. H. Wells and A. M. Wenner, *Physiol. Zool.* 44, 191 (1971); A. M. Wenner, in *Nonverbal Communication*, L. Krames, P. Pliner, T. Alloway, Eds. (Plenum, New York, 1974).
17. For example, one could suppose that there had been no site odor in the morning, but by the afternoon the distinctive odors had appeared to guide the recruits. Trampled vegetation around the experimental stations or afternoon-blooming flowers could, in the terms of this explanation, have accounted for the results. Alternatively, one could suppose that scent gland odors explain the data. The exposure of the gland is related to the quality of the food with respect to natural sources. If natural competition was high in the morning, but low in the afternoon, scent gland odor would be present to guide recruits only in the afternoon. Since recruits seem not to have been caught, it could also be supposed that experience played a role. After first following the food odor upwind (north) in the morning without finding any food, recruits might be supposed to have searched elsewhere, downwind (south) for example, in the afternoon.
18. D. A. T. New, *J. Insect Physiol.* 6, 196 (1961); See _____ and J. K. New, *J. Exp. Biol.* 39, 271 (1962); P. H. Wells and A. M. Wenner, *Nature (Lond.)* 241, 171 (1973).
19. C. D. Michener, *The Social Behavior of the Bees* (Harvard Univ. Press, Cambridge, Mass., 1974), p. 175.
20. G. Birukow, *Z. Vgl. Physiol.* 36, 176 (1954) (beetles); V. G. Dethier, *Science* 125, 331 (1957) (flies); I. Tenckhoff-Eikmanns, *Zool. Beitr.* 4, 307 (1959) (beetles); A. D. Blest, *Behaviour* 16, 188 (1960) (moths); H. Markl, *Z. Vgl. Physiol.* 48, 552 (1964) (ants); H. Esch, I. Esch, W. E. Kerr, *Science* 149, 320 (1965) (stingless bees).
21. H. Esch, *Z. Vgl. Physiol.* 48, 534 (1964).
22. M. Lindauer and B. Schricker, *Biol. Zentralbl.* 82, 721 (1963); B. Schricker, *Z. Vgl. Physiol.* 49, 420 (1965).
23. N. E. Gary and P. C. Witherell, *Ann. Entomol. Soc. Am.* 64, 448 (1971).
24. M. Lindauer and H. Martin, *Z. Vgl. Physiol.* 60, 219 (1968).
25. In all cases, 25 μl of scent was used per liter of solution. The scents were the gift of International Flavors and Fragrances, Inc.
26. M. Renner, *Z. Vgl. Physiol.* 42, 449 (1959).
27. J. B. Free, *Behaviour* 37, 269 (1970).
28. For results of one such experiment, see J. L. Gould, *Nature (Lond.)* 252, 300 (1974). For additional distance experiments, see J. L. Gould, *J. Comp. Physiol.*, in press.
29. One possibility might be that the reduced angular scatter compensates for increasing distance; and that, taken with the nearly constant distance scatter, could serve to generate the same recruit distribution about a target regardless of its distance from the hive. (A formula for calculating the distribution is offered later.) If food sources exist in "patches," it could be to a colony's advantage to spread out its searching recruits to exploit the patches more efficiently. The area around the hive for which round dances are performed might be thought of as the first such "patch." Italian bees have both a smaller round dance zone and recruit

scatter than Carniolan bees. If this difference has (or had) an adaptive value, one might suppose that the two races evolved in habitats with different patch sizes or colony spacings, or that some compensating difference in search strategy in the field may exist. Such speculations might be best tested by looking for racial differences in any of the three species of honey bees that inhabit the tropical forests of Asia, India, and the Philippines.

30. A. M. Wenner, *Bee World* **42**, 8 (1961).
31. J. B. Free, *Nature (Lond.)* **222**, 778 (1969).
32. K. von Frisch and R. Jander, *Z. Vgl. Physiol.* **40**, 239 (1957).
33. Two minor adjustments were necessary to make the data from all six experiments comparable. Since the stations in the 400-m direction arrays were 36 seconds flying time further from the hive than the stations in the 150-m array, the times for the 400-m bees were shortened by that amount. Data were also used from two 90- to 210-m distance-array experiments. In these cases the forager station, rather than the light, was moved in order to signal a new station (*28*). Hence, the new station (that is, the new distance) was not signaled instantaneously, but only after the foragers then and subsequently feeding returned to dance. This period— 30 to 60 seconds—was measured from the video recordings, and the corresponding corrections were made to recruit arrival times.
34. Since food sources such as flowers may commonly exist as extended patches rather than as discrete point sources, array experiments may represent a particularly "natural" way to measure recruitment.
35. W. H. Bossert and E. O. Wilson, *J. Theor. Biol.* **5**, 443 (1963); J. S. Kennedy and D. Marsh, *Science* **184**, 999 (1974).
36. A. M. Wenner, *The Bee Language Controversy* (Educational Products Improvement Corp., Boulder, Colo., 1971), p. 59.
37. The repertoire size of the dance language—that is, the number of unique messages it can communicate—may be calculated and compared with that of other species. Each dance may be considered as a sentence specifying a food source's distance, direction, type, quality, and odor. For a graded system, the number of "bits" in a signal is given by [log (*R*/S.D.)/log 2] – 2.047 [J. B. S. Haldane and H. Spurway, *Insectes Soc.* **1**, 247 (1954)], where *R* is the range of signal values (0° to 360° in direction, for example) and S.D. is 1 standard deviation of recruit scatter. A "bit" is taken to be the amount of information necessary to distinguish between two equally probable alternatives. The S.D. for direction scatter in recruit arrivals varies with dis-

tance, but a conservative average is 4°. By the formula, this corresponds to 9.3 bits of information. The range of distance indications is 0 to 12 km (*6*, p. 73). An average value for the S.D. of distance accuracy is 60 m (extrapolating the relationship between dance and recruit scatters as measured in my distance experiments). This corresponds to 5.6 bits of information. The goal may be either nectar, pollen, propolis, or nesting cavities (water being taken as merely dilute nectar)—2 bits of information. The quality of the goal is given by the unquantified "vigor" of the dance. This factor may be conservatively estimated as adding 2 bits of information. The dance also contains information about the odor of the food. Von Frisch (*3*) showed that bees could distinguish all 46 floral odors he had available. On another occasion, recruits successfully located the correct plants from among 700 other flowering species (*6*, p. 48). As a conservative estimate, the number of different odors will be taken to be 100—that is, 6.5 bits. From these estimates, the lower limit to the repertoire size of the dance is 25.4 bits, or 4 × 10⁷ discrete "sentences." This is far higher than the value which can be calculated for any other known nonhuman system. The closest competition comes from the recent attempts to teach chimpanzees to use human language [for example, R. A. Gardner and B. T. Gardner, *Science* **165**, 664 (1969); D. M. Rumbaugh, T. V. Gill, E. C. von Glaserfeld, *ibid.* **182**, 731 (1973)]. If Washoe, for example, could use all of her 130-odd signs to form four-word sentences in the pattern noun-verb-modifier-noun, and if her repertoire of mostly nouns were actually to consist of 70 nouns, 30 verbs, and 30 modifiers, then 3.25 × 10⁶ sentences would be theoretically possible. (Of course, only a fraction of these would make any sense; Lana's considerable aptitude in learning experiments may allow less crude estimates in the future.) The minimum equivalent figure for the repertoire size of a 10,000-word human vocabulary is more than 10²² seven-word sentences, or 74 bits.

38. A phenomenon admirably reviewed by D. R. Griffin [in *Animal Communication*, T. A. Sebeok, Ed. (Indiana Univ. Press, Bloomington, ed. 2, 1975)].
39. I thank N. E. Gary for technical advice, M. Brines for technical assistance, R. O'Connell for obtaining the scents, M. Rosetto for designing the electronics, J. Crane for lending the video tape equipment, and P. Marler, C. G. Gould, F. Nottebohm, and especially D. R. Griffin for their criticisms and encouragement. Supported in part by the Mary Flagler Cary Charitable Trust, Millbrook, New York.

37 James E. Lloyd

Aggressive Mimicry in Photuris Fireflies: Signal Repertoires by Femmes Fatales

Reprinted from 7 February 1975, Volume 187, pp. 452-453

Abstract. *Females of* Photuris versicolor *prey on males of other species by mimicking the flash responses of the prey's own females. They adjust their responses according to the male pattern, and attract males of four species with distinctively different flashed responses. The capabilities of the firefly brain are more complex than previously suspected. The mimicry is quite effective, and females seldom answered more than ten males without catching one.*

Females of at least 12 species of *Photuris* fireflies are predators of male fireflies of the genera *Photinus, Photuris, Pyractomena,* and *Robopus.* Evidence for this comes from observations of females eating (*1*) as well as capturing (*2, 3*) the males. The females lure the males to them by mimicking the mating signals of the prey species' females. The males are then seized and devoured. Although I have observed different *Photuris versicolor* females capturing males of four species, these observations do not alone demonstrate the mimicry versatility of these females since (i) *Photuris* species are difficult to identify and the behavior of different species could be attributed to a single one (*4*) and (ii) individual females might only be capable of mimicking the signal of a single prey species, variation among the females accounting for the inferred multiple mimicry. I have been able to demonstrate by field observation and experimentation on individual *P. versicolor* females that they do indeed have signal repertoires.

Females used for experimentation were found by their flashed answers to passing males or to penlight simulations of prey species' signals, and were on the ground or low vegetation. They were left in situ for experimentation. Experiments could not follow a rigid protocol because of the variable conditions under which the females were found. Also, since the experiments were conducted in the field, the females were continually influenced by the signals of free-flying males in the area. (On seven occasions experimental subjects interrupted experiments involving simulated patterns by flashing appropriate answers to different patterns that were emitted by passing males.) A female, once located, was first presented with an artificial flash pattern simulating the signal of one of the prey species. After several responses had been elicited, the pattern of another prey species was presented. Experiments were terminated if a female failed to respond to 15 to 20 presentations.

The species involved have a common general pattern of mating behavior. Males fly about in their habitat emitting their species-specific flashing pattern. Females flash answers from perches. Timing elements of the emissions are important for recognition. Important parameters in the male patterns are flash number, rate, and duration, and in the female response, flash length, and the delay at which it occurs after the male pattern. Advertising males repeat

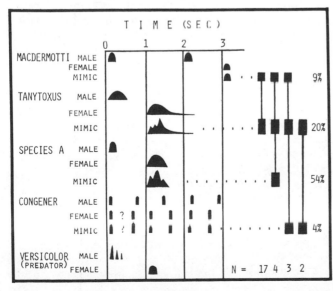

Fig. 1. Luminescent signals of fireflies. Response used by predator is shown below female answer it mimics. Vertical bars at right indicate observed individual repertoires; *N* is the number of females exhibiting the repertoire. Capture rates (percentages) are adjacent to prey species. The flash rate of the *congener* female is variable, and the specific nature of the coding is unknown (see text).

their patterns at characteristic intervals. When a male receives an answer to his signal he flies toward it and emits his pattern again. A flash-answer dialogue continues for five to ten exchanges and until the male reaches the female. Figure 1 illustrates the mating signal codes of the species involved (5), and tabulates the observed mimicry repertoires observed.

Two females were observed in natural "experiments" answering passing males of both *Photinus tanytoxus* and *Photuris congener*, and a third female answered a simulated *Photinus macdermotti* pattern and was found to be already eating a *Photuris* sp. *A* (6) male.

Eleven females answered the *macdermotti* simulation appropriately and then answered the *tanytoxus* simulation appropriately. This change required altering the length of their emissions as well as flashing after each flash rather than only after the second, as necessary for *macdermotti* (Fig. 1). For example, one female answered 11 consecutive *macdermotti* patterns with flashes 0.12 to 0.16 second in duration (at intervals of 8 to 10 seconds). On the first nine she flashed after each pulse of each pattern (7); she answered the next two correctly. She was then presented with several 0.5-second flashes (*tanytoxus*). She did not answer the first seven, but on numbers eight to ten responded with a short flash. On the 11th she produced a dim glow after the flash, and on the 12th her flash length was 0.6 second and the glow was held more than 9 seconds. She was then given the *macdermotti* pattern, which she answered properly on the first presentation, but with an intermediate flash length of 0.24 second (8) and no afterglow. Her next flash was 0.16 second in duration. In other words, she had immediately switched to the *macdermotti* response. Then, given a 0.5-second (*tanytoxus*) flash, she answered it with the appropriate long answer, a 0.6-second flash. In both cases she had made a rapid adjust-

ment on the basis of the duration of the stimulus flash. Females sometimes responded immediately to pattern changes, but occasionally as many as 15 presentations had to be made before a response to the new signal could be elicited.

Six other females were given the simulated *tanytoxus* pattern first and then after one to eight presentations responded to the *macdermotti* pattern. One would not answer the *macdermotti* pattern and one always ($N = 15$) flashed after both pulses of the pattern. Females occasionally stopped answering any pattern or flew away, and those tested undoubtedly differed with respect to age, condition of ovaries, number of successful predations, exposure to flashes of foreign males (kinds and numbers), and genetic makeup.

Apparently the mimicry is not perfect, although comparative figures cannot be given since attraction rates for conspecific interactions are unknown. One female captured the 12th *macdermotti* male she answered. Another answered 20 *congener* males, and then moved to a different perch several meters away and answered more than 20 additional males before she captured one. Another female caught the 21st *congener* male that she was observed to answer. Capture rates were higher for prey belonging to other species: on five occasions I observed the demise of *Photuris A* males; two females captured the first male answered, one caught the second, one the tenth, and one female got the 11th, although she had seized the seventh male and it had escaped. Two other females captured the fifth *tanytoxus* males that they answered.

What is the evolutionary origin of the false signals? Two independent sources are suggested. The flashed responses to *Photuris A*, *Photinus tanytoxus*, and *Photinus macdermotti* males appear to be similar in delay timing to the predator's own mating responses. False signals could have been derived originally from mating responses and subsequently modified. Responses to the flashes of *Photuris congener* males are similar to the flashes that the predaceous females, and those of many other *Photuris* species, commonly emit when they walk, land, or take flight (9). These "locomotion" flashes would need little if any modification to attract some *congener* males. (The flashes of the *congener* female, unlike those of other species, do not bear a specific relation to each flash of the male.) I am able to attract about one male in ten to the 0.08-second flashes of a free-running oscillator with a period like that of the males. I once observed a lycosid spider eating a *congener* male that continued to emit his rhythmic pattern; two additional *congener* males were attracted to the flashes of the captive, and were also seized by the spider. I offer this not as an example of a tool-using spider, for I doubt that it is repeated with regularity, but as an indication of how a physiologically inappropriate but trophically fortuitous activation of the locomotion flash mechanism by the flashes of a passing *congener* male could immediately put the female into the aggressive mimicry role. These observations indicate that the capabilities of the firefly brain are more complex than hitherto suspected.

JAMES E. LLOYD

Department of Entomology,
University of Florida, Gainesville 32611

References and Notes

1. J. E. Lloyd, *Coleopt. Bull.* **27**, 91 (1973); L. L. Buschman, *ibid.* **28**, 27 (1974).
2. J. E. Lloyd, *Science* **149**, 653 (1965).
3. E. G. Farnworth, thesis, University of Florida (1973).
4. *Photuris versicolor* is a complex of several morphologically similar species which are widely distributed in the eastern and central United States. Extensive field investigations indicate that probably only one species is present in Gainesville.
5. The mating signals of prey species are discussed in more detail in J. E. Lloyd, *Univ. Mich. Mus. Zool. Misc. Publ. No. 130* (1966), pp. 1–95; *Fla. Entomol.* **52**, 29 (1969).
6. This *Photuris* is apparently a new species. Revisional studies and a Latin binomen will be reported at a later date (J. E. Lloyd, in preparation).
7. These interposed flashes were occasionally observed during actual predation of this species on *macdermotti* (2) and could be eliminated from the responses of some females when the stimulus patterns were spaced at intervals of 8 to 10 seconds.
8. Both 0.16 and 0.24 second are within the range of *macdermotti* flash responses. The 0.24-second flash was intermediate only with respect to the responses this female emitted.
9. J. E. Lloyd, *Entomol. News* **79**, 265 (1968).
10. I thank E. G. Farnworth and R. S. Lloyd for assistance in the field; T. J. Walker for helpful discussion, comments on the manuscript, and the loan of photographic equipment; A. Owens for photographic technical assistance; and the National Science Foundation (grant GB 7407) for financial assistance. Florida Agricultural Experiment Station Journal Series No. 5447.

16 August 1974

38 Stephen T. Emlen

Celestial Rotation: Its Importance in the Development of Migratory Orientation

Reprinted from 11 December 1970, Volume 170, pp. 1198-1201

Abstract. *Three groups of indigo buntings were hand-raised in various conditions of visual isolation from celestial cues. When they had been prevented from viewing the night sky prior to the autumn migration season, birds tested under planetarium skies were unable to select the normal migration direction. By contrast, when they had been exposed as juveniles to a normal, rotating, planetarium sky, individuals displayed typical monthly directional preferences. The third group was exposed to an incorrect planetarium sky in which the stars rotated about a fictitious axis. When tested during the autumn, these birds took up the "correct" migration direction relative to the new axis of rotation. These results fail to support the hypothesis of a "genetic star map." They suggest, instead, a maturation process in which stellar cues come to be associated with a directional reference system provided by the axis of celestial rotation.*

The ontogenetic development of animal orientation abilities has received very little study. Early workers were impressed by the fact that the young of many species of birds migrate alone, setting out on a course they have never traveled before without the benefit of experienced companions. This suggested that directional tendencies must develop without any prior migratory experience and therefore must be entirely genetically predetermined (*1*).

Field studies, however, point to a dichotomy of navigation capabilities between young and adult birds. When birds of several species were captured and displaced from their normal autumnal migration routes, the adults corrected for this displacement and returned to the normal winter quarters but immatures (birds on their first autumnal migration) did not (*2*). Prior migratory experience improved orientation performance.

I arrived at a somewhat similar conclusion from studies of the migratory orientation of caged indigo buntings; the consistency and accuracy of the orientation exhibited by adults was greater than that of young, hand-raised birds. (*3*). Furthermore, young birds prevented from viewing celestial cues during their premigratory development showed weaker orientation tendencies than those exposed to the natural surroundings, including the day-night sky. I speculated that the maturation process was a complex one, which involved the coupling of stellar information with some secondary set of reference cues.

The following experiments were designed to test more precisely the ability of hand-reared birds to use celestial cues and to determine the possible importance of celestial rotation in providing an axis of reference for direction determination.

Twenty-six nestling indigo buntings between the ages of 4 and 10 days were removed from their nests and hand-raised in the laboratory, where their visual experience with celestial cues was carefully controlled. I housed the birds in 2 by 2 by 2 foot (65 by 65 by 65 cm) cages in an 8 by 8 foot (2.4 by 2.4 m) room equipped with a hung ceiling made of translucent plastic. The birds thus were prevented from ever viewing a point source of light during their development. Both fluorescent and incandescent lights were present above the artificial ceiling, and the day length was controlled by an astronomical time clock to simulate the day length outdoors.

The birds were hand-fed at frequent intervals until approximately 25 days of age (15 days postfledging), when they became self-sufficient. I then placed them in one of three experimental groups. The first, group A, never left the 8 by 8 foot living quarters until I tested their orientational tendencies during the autumn migration season. These birds were never allowed to view either the sun or the night sky.

The birds of group B also were prevented from viewing the sun. However, these individuals were taken into the Cornell research planetarium and exposed to the normal night sky during the months of August and September (*4*). The artificial sky was set to duplicate the sky outdoors and was changed appropriately to simulate the seasonal changes of hour-angle positions that occur between August and the migration season. The Spitz star projector was modified to rotate at a speed of one revolution per 24 hours, thus duplicating the normal pattern of celestial rotation. The young buntings continued to live in the room described above, but three times a week they were removed to the planetarium at 9:00 p.m.

and returned to their normal cages between 4:30 and 5:00 a.m. (EDT).

The birds of group C were also subjected to planetarium exposure. They were taken on three different nights each week and exposed for a similar length of time to an artificial sky. However, this artificial sky was abnormal in several respects.

Once again I had modified the star projector, this time by constructing a special attachment arm that allowed the celestial sphere to be rotated about any axis of my choosing. For group C, I selected the bright star Betelgeuse as the new "pole star," and the constellation Orion became the dominant pattern in the new "circumpolar" area of the sky.

This new sky setting was selected for several reasons. First, a bright star is located at the pole of the new axis. Second, a very bright constellation is located in the "circumpolar" area. This area has been determined to be of special importance in the celestial orientation process of this species (5). Third, the hour-angle position was selected carefully so that the actual northern circumpolar stars (in particular, the constellations Ursa Major and Cassiopea and the star Polaris) were present in this artificial sky. They are located just to the south of the new "celestial equator" and move progressively across the sky from east to west as the night progresses.

The logic behind the experiment is this. If celestial rotation provides a reference axis for migratory orientation, then the birds of group C might adopt this incorrect axis and orient their migratory activity in an inappropriate direction. On the other hand, if young birds possess a genetically predetermined star map as has been proposed by some authors (6), then the birds should orient "south" with reference to the normal circumpolar area of the sky. These two "south" directions should be easily distinguishable

since they range from 110 to 180 degrees apart in the planetarium settings of group C.

During their exposure to these planetarium skies, the buntings were placed in small, funnel-shaped orientation cages (7). In this way they became accustomed to the experimental apparatus that would be used later. The birds from group A were given comparable experience in the orientation cages but always in the isolation room. I did not record behavior during these sessions; thus, the degree of nocturnal activity or attentiveness to the artificial skies prior to the migration season is unknown.

Each individual bird from groups B and C received 22 nights of exposure to the appropriate planetarium sky.

The birds completed the postjuvenal molt and acquired visible subcutaneous fat deposits in late September. This development was taken as a criterion for migratory readiness and experiments were conducted throughout the month of October and into early November. I placed the birds under the same planetarium sky to which they had previously been exposed with the exception that the sky was now held stationary. By preventing direct access to rotational information, I tested whether the birds had integrated information from celestial configurations with the potential reference framework provided by the axis of rotation. The experimental design called for testing the same birds under a rotating sky if no directional preferences appeared under these conditions.

I tested as many as seven buntings simultaneously, placing their funnel cages as close to the centrally located star projector as possible in order to minimize any distortion of the artificial sky. The only change from the "exposure" situation was that the cages now had a freshly inked floor that permitted the accumulation of directional information by the footprint technique

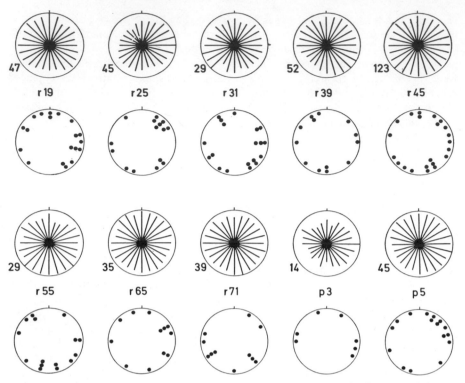

Fig. 1. Orientation of young indigo buntings prevented from viewing celestial cues during their early development. The birds were tested under a stationary, autumnal planetarium sky set for 42°N. (Top) Vector diagram summaries plotted such that the radius equals the greatest number of units of activity in any one 15-degree sector. The number that this represents is written at the lower left of each diagram. (Bottom) Distributions of mean directions for all experiments. In all figures, arrow points attached to circles indicate mean directions of orientation when the distributions shown depart significantly from random ($P \gtrless .05$).

(7). Each experiment lasted 2 hours, and the hour-angle position of the planetarium sky was set to correspond with the midpoint of the 2-hour period.

For each data distribution, I tested the null hypothesis of randomness by the "v" modification of the Rayleigh test (8), with the expected orientation being southward. Mean direction was calculated by vector analysis (9).

The results are shown in Figs. 1 through 3. Of the ten individual buntings of group A, *none* demonstrated a clear-cut directional tendency. This was true whether one analyzed the total activity of each bird or the mean directions taken during replicate tests (Fig. 1). These results argue against

the existence of a hereditary star map that the buntings can refer to for navigational information. Rather, they suggest that visual-celestial experience during early ontogeny is important for the normal maturation of stellar orientation abilities.

The results from the buntings of group B support this interpretation. Of the eight birds, seven exhibited a southerly preference in their nocturnal restlessness, the appropriate direction for their first autumnal migration flight (Fig. 2). The data from the eighth bird (r47) do not deviate from random. Although the degree of scatter in the data is large (particularly for r41 and r67) the improvement over

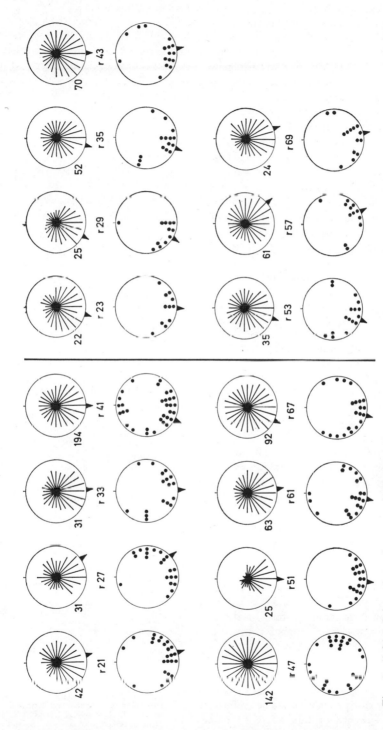

Fig. 2 (left). Orientation of young indigo buntings permitted regular viewings of a normal, rotating, planetarium sky during their early development. Fig. 3 (right). Orientation of young indigo buntings exposed to a planetarium sky that rotated about an incorrect axis during their early development. The data are plotted with the new "pole star" (Betelgeuse, of the constellation Orion, designating "north" or 0 degrees.

337

the performance of isolate birds (group A) is readily apparent.

Figure 3 shows the findings from group C when directions are plotted relative to the new axis of rotation—that is, with the position of Betelgeuse defining north. All seven birds displayed a "southerly" orientation, which indicated a realignment of directional behavior to correspond with the new axis of rotation (10). There was no tendency to move toward true stellar south. Once again, these results are inconsistent with the hypothesis of a predetermined template of star positions.

Taken together, these findings provide strong evidence that early visual experience plays an important role in the development of celestial orientation abilities in indigo buntings. I hypothesize that fledgling buntings respond to the apparent rotational motion of the night sky. The fact that stars located near the celestial axis move through much smaller arcs (have a slower linear velocity) than do those near the celestial equator allows the birds to locate a north-south directional axis. Stars and patterns of stars are of no value for direction finding until their positions are learned relative to some reference location. The axis of rotation appears to function as one such reference system. Once this coupling of stellar and rotational information has occurred, a bird can locate the rotational axis (and, hence, geographic direction) from star patterns alone. This implies that celestial motion per se should become a secondary or redundant orientational cue for adult birds. The accurate orientation of caged migrants under stationary planetarium skies supports this view (11).

This hypothesis does not explain why the young migrant orients southward on its first flight. Rotation seems merely to provide a stable reference axis. The use of this reference to select a southerly heading in preference to any other remains a topic for future investigation.

This study demonstrates the complexity involved in the maturation of one orientational system available to indigo buntings. Undoubtedly, the picture will become more complex as we learn more about additional components of avian guidance systems.

STEPHEN T. EMLEN

Section of Neurobiology and Behavior, Division of Biological Sciences, Cornell University, Ithaca, New York 14850

References and Notes

1. D. R. Griffin, *Bird Migration* (Doubleday, Toronto, 1964); K. Schmidt-Koenig, in *Advances in the Study of Behavior*, R. Hinde, D. Lehrman, E. Shaw, Eds. (Academic Press, New York, 1965); G. V. T. Matthews, *Bird Navigation* (Cambridge Univ. Press, London, ed. 2, 1968).
2. W. Rüppell, *J. Ornithol.* **92**, 106 (1944); A. C. Perdeck, *Ardea* **46**, 1 (1958); ———, *ibid.* **55**, 194 (1967); F. Bellrose, *Wilson Bull.* **70**, 20 (1958).
3. S. T. Emlen, *Living Bird* **8**, 113 (1969).
4. The Cornell research planetarium is a 30-foot (9-m) diameter, air-supported dome equipped with a Spitz A-3-P star projector.
5. S. T. Emlen, *Auk* **84**, 463 (1967).
6. F. G. Sauer, *Z. Tierpsychol.* **14**, 29 (1957); *Sci. Amer.* **208**, 44 (August 1958).
7. S. T. Emlen and J. T. Emlen, Jr., *Auk* **83**, 361 (1966). The bottoms of the cages were not inked during the "exposures" to planetarium skies during August and September.
8. D. Durand and J. A. Greenwood, *J. Geol.* **66**, 229 (1958).
9. E. Batschelet, *Amer. Inst. Biol. Sci. Monogr.* **1** (1965). Sample sizes for the Rayleigh test were determined by dividing the total number of units of activity N by a correction factor. This divisor was determined empirically and represents the interval at which activity measures become independent of one another [see appendix 5 of (3)].
10. The eighth bird in this group, r49, developed the habit of "somersaulting" in the orientation cage. Since the resulting ink smudges represented an aberrant behavior pattern, they were not included in the quantitative analysis.
11. F. G. Sauer, *Z. Tierpsychol.* **14**, 29 (1957); S. T. Emlen, *Auk* **84**, 463 (1967); and results reported in this study. The hypothesis need not imply that buntings directly perceive the slow rate of celestial motion. One can easily locate the axis of rotation by making observations over longer periods of time and comparing the degree of movement of stars located at different points in the celestial sphere.
12. Supported in part by NSF grant GB 13046 X. I thank M. Platt and C. Conley for assistance in rearing and testing the birds; I thank members of Cornell's orientation seminar group for their comments and criticisms; and I thank H. C. Howland and T. Eisner for critical readings of the manuscript.

17 August 1970; revised 8 October 1970

39

Orientation of Homing Pigeons Altered by a Change in the Direction of an Applied Magnetic Field

Charles Walcott and Robert P. Green

Reprinted from
12 April 1974, Volume 184, pp. 180-182

Abstract. *Homing pigeons were equipped with a pair of small coils around their heads. Birds with an induced field of 0.6 gauss and the south magnetic pole up, oriented toward home normally under both sun and overcast. Birds with the polarity reversed oriented toward home when the sun was visible but often flew away from home under overcast.*

Evidence has accumulated that magnetic fields may be involved in the orientation of birds. The directional preferences of ring-billed gull chicks are upset during periods when the magnetic field of the earth is disturbed (*1*). European robins show a tendency to orient in relation to a real or artificial earth field (*2*). The initial orientation of homing pigeons is often upset by small bar magnets attached to the pigeon's backs (*3*). We present evidence here suggesting that the orientation of pigeons is altered by changing the polarity of the applied magnetic field.

A flock of approximately 50 homing pigeons was kept in a small loft on the university campus at Stony Brook, New York. These birds were progressively trained along a line to the east of the loft under both sunny and overcast conditions. Experimental releases were made from three locations: Cunningham Park, 68 km, 251° (compass direction, from the loft); Hempstead, 59 km, 242°; and Spring Valley, 92 km, 287°. At the release site each pigeon was equipped with a pair of coils. One coil, 35 mm in diameter, was fitted around the pigeon's neck like a collar and the other, 23 mm in diameter, was glued to the top of the head like a hat. Each coil was made of 200 turns of No. 36 enameled wire, and the two coils were connected in series with a 1.4-volt mercury battery. This combination produced a relatively uniform magnetic field of about 0.6 gauss around the pigeon's head. A gaussmeter (Bell model 620) showed that the field between the coils varied from 1.4 gauss at either coil to 0.5 gauss in the center of the space between them. The life of the battery was 2 to 3 hours. By simply reversing the battery in its holder it was possible to reverse the flow of current through the coils and, thus, the direction of the magnetic field surrounding the bird's head. When a magnetic compass was placed between the two coils with the north seeking end of the needle pointed toward the coil glued to the top of the bird's head, we defined the coil as a Nup (north-up). A coil with the opposite polarity was a Sup. In addition to the two coils and battery, each bird carried a small radio beacon. The vanishing bearings reported here are the direction in which the radio signal was lost when the pigeon was approximately 16 km from the release point (*4*). Nups and Sups were released alternately.

Under sunny conditions the vanishing directions of the two groups, Nups and Sups, appeared to be well oriented toward the home loft (Table 1 and Fig. 1). Surprisingly, a significance test for the mean direction (5) shows that the Nups chose a direction that differed at the 1 percent level from home. A Watson test (5) shows that the two distributions of vanishing points differ ($P = .005$) although the mean vectors are only 11° apart. Despite these statistical measures, we had the impression that the orientation of Nups and Sups under sunny conditions was quite similar

When the sun was not visible through heavy overcast, the difference in the behavior of the two groups was much more pronounced. Neither group was as well oriented toward the loft as the group released under sun. As Fig. 1 shows, birds with Sup coils vanished predominantly toward home [V-test, $P = .008$ (5)], whereas birds with Nup coils were oriented away from the direction toward home. The Watson test gives the probability that the two samples were drawn from the same distribution as less than .005. Despite this difference in initial orientation, the homing performance of the two groups was similar (Table 2). Considering the relatively short life of the battery when compared to the long time it took the birds to home this is not surprising.

Pigeons that returned from these overcast releases were now taken to the same release point for a second time. Birds that had worn Sup coils before were equipped with Nup coils, and birds that had worn Nup coils were now given Sup coils. Once again when released under total overcast the Sup birds were oriented toward home. About half the Nup birds oriented toward home, and about half flew in the opposite direction. This result demonstrates that the behavior of at least half of the birds could be altered by reversing the magnetic field. It also suggests that previous experience at the release site may have reduced the

tendency of Nup birds to fly in the wrong direction.

Taken together these results indicate that varying the direction of an artificial magnetic field around a pigeon's head has an effect on its initial orientation although it has little apparent effect on its total homing performance.

Since each pigeon had identical apparatus that differed only in the direction of current flow, it is hard to see how the results could be an artifact of the experimental arrangement. Further-

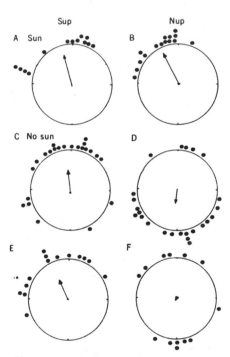

Fig. 1. Vanishing bearings of pigeons equipped with Helmholtz coils. Each dot represents the direction of a pigeon at 10 miles from the release point. Home is at the top of the circle, and the arrow is the mean vector of the distribution; its length is proportional to the degree of clumping of the vanishing bearings. In diagrams A and B, the birds were released under sun. In C and D the birds were released under total overcast for the first time at this release point; E and F are the vanishing directions of the same birds as in C and D but with opposite treatments released for the second time at this release point.

Table 1. The results of each release under the various experimental conditions. N is the number of birds tracked, B is the 10-mile vanishing bearing (degrees) relative to the direction to the loft; l is the length of the unit vector (arbitrary units), and P is the probability that the sample was drawn from a random distribution.

Release site	Date	Sup				Nup			
		N	B	l	P	N	B	l	P
	Sun								
Hempstead	17 Oct 71	6	305	.86	.007	6	323	.89	.004
Cunningham Park	27 May 72	3	16	.98	.04	4	314	.91	.03
	22 Oct 72	4	4	.98	.01	5	352	.96	.005
	Summary	13	342	.79	.00007	15	331	.88	6×10^{-7}
	Inexperienced birds under overcast								
Cunningham Park	31 Oct 71	5	358	.92	.009	4	124	.57	.27
Spring Valley	20 Nov 71	5	4	.64	.12	4	228	.89	.03
Cunningham Park	1 Nov 71	4	261	.72	.12	4	73	.17	.89
	13 Nov 72	4	9	.51	.36	5	200	.48	.32
	28 Nov 72	1	10			1	169		
	4 Dec 72	0				2	129	.98	.13
	20 Dec 72	0				1	244		
	30 Mar 73	1	44			0			
	8 May 73	3	39	.56	.39	3	262	.39	.45
	Summary	23	352	.52	.001	24	188	.36	.04
	Experienced birds under overcast								
Cunningham Park	4 Dec 72	2	334	.99	.13	1	179		
	20 Dec 72	3	270	.97	.05	1	169		
	3 Mar 73	3	346	.77	.16	4	216	.14	.92
	30 Mar 73	3	342	.65	.28	4	107	.15	.91
	4 Apr 73	3	1	.92	.07	3	216	.19	.90
	Summary	14	330	.71	.0003	13	207	.09	.91

more, since the effect of differing fields is more pronounced on overcast than on sunny days, it may be that pigeons are, in some way, using the earth's magnetic field as a compass when the sun is not visible.

It is interesting to compare these results with the orientation of European robins (*Erithacus rubecula*) described by Wiltschko (6). He reports that his birds use the inclination of the magnetic field of the earth and interpret the direction in which the magnetic field vector and gravity vector make the smallest angle, as north. The earth's magnetic field in the Long Island area has an intensity of about 0.6 gauss and an inclination of 70°. This field would combine with the field from our coils to produce a total field around

Table 2. Homing success.

Time	The number of birds which homed on:					
	Sun		Overcast			
			Inexperienced		Experienced	
	Sup	Nup	Sup	Nup	Sup	Nup
Day of release	9	13	3	1	7	7
1 day after	3	1	3	2	3	2
2 days after	0	0	6	7	0	0
3 days after	0	1	6	2	0	0
After 3 days	0	0	0	2	0	0
Never homed	1	0	5	10	4	4

the pigeon's head. The intensity and inclination of the resulting field would depend on the orientation of the pigeon's head with respect to the earth. This in turn would vary with the direction that the pigeon was flying and with the angle at which it held its head. We have estimated the pigeon's head angle from photographs of flying pigeons; the coils should produce a field oriented from about 10° to 45° above the horizon. We then plotted the net vector resulting from the combination of the earth's field and the field of the coils carried by the pigeon. The resultant obviously depends on the direction in which the bird flies, but the interesting point is that a bird with a Sup coil finds the total magnetic field vector most steeply inclined (the smallest angle between the magnetic field vector and gravity) when flying north, whereas a bird with a Nup coil finds the steepest vector when flying south. If pigeons were using the direction of the steepest inclination of the field to indicate north, then, perhaps this would explain why there was a tendency for pigeons with Nup coils to fly 180°

away from home when the sun was not visible.

CHARLES WALCOTT
ROBERT P. GREEN
Department of Cellular and Comparative Biology, State University of New York, Stony Brook 11790

References and Notes

1. W. E. Southern, *Condor* **71**, 418 (1969); *BioScience* **22**, 476 (1972).
2. W. Wiltschko and F. W. Merkel, *Verh. Dtsch. Zool. Ges.* **1965**, 362 (1966); W. Wiltschko, H. Hock, F. W. Merkel, *Z. Tierpsychol.* **29**, 409 (1971); W. Wiltschko, in *Animal Orientation and Navigation*, S. R. Galler, K. Schmidt-Koenig, G. J. Jacobs, R. E. Belleville, Eds. (Government Printing Office, Washington, D.C., 1972), p. 569.
3. W. Keeton, *Proc. Nat. Acad. Sci. U.S.A.* **68**, 102 (1971); in *Animal Orientation and Navigation*, S. R. Galler, K. Schmidt-Koenig, G. J. Jacobs, R. Belleville, Eds. (Government Printing Office, Washington, D.C., 1972), p. 579.
4. The distance at which the transmitter could no longer be detected at the release site was determined by following the pigeon in a light airplane [see M. Michener and C. Walcott, *J. Exp. Biol.* **47**, 99 (1967)].
5. E. Batschelet, *Statistical Methods for the Analysis of Problems in Animal Orientation and Certain Biological Rhythms* (American Institute of Biological Sciences, Washington, D.C., 1965); in *Animal Orientation and Navigation*, S. R. Galler, K. Schmidt-Koenig, G. J. Jacobs, R. Belleville, Eds. (Government Printing Office, Washington, D.C., 1972), p. 61; G. S. Watson, *Biometrika* **49**, 57 (1962).
6. W. Wiltschko and R. Wiltschko, *Science* **176**, 62 (1972).
7. Supported in part by NIH grant 5 RO1 NS 08708-03.

11 June 1973; revised 11 July 1973

PART 6

Learning

Now that we have studied the genetics, physiology, and development of animal behavior, with an excursion into sensory processes, communication, and orientation, we turn to what has historically been the psychologist's favorite topic: learning. The literature on animal learning is vast and selection of representative papers is particularly difficult. In reality many of those who have studied animal learning were really interested in *human* learning. They were using animals as substitutes for humans in a search for laws of learning that could apply across species.

Others, however, have approached the topic from a different point of view. It should be obvious, at this point, that learning is a biological adaptation. It is just as much a product of evolution as any other character. Therefore, learning may be studied as a process that may operate in the behavior of an animal in its natural environment. The reports that follow are primarily of this sort.

It has been said that farmers know many things about animal behavior, and some of them are even true. In actuality, it generally pays to listen to those whose livelihood depends upon animals. In an increasing number of cases this scientifically unacceptable, anecdotal lore of animal behavior has proven correct. We shall cover some examples.

The first is related to stories that certain animals can become "poison-shy." This seemed particularly difficult to believe because it violated widely accepted principles of learning. Consider what is required. An animal begins to eat food that is poisoned, but, for some reason, eating behavior is interrupted before a lethal dose has been ingested. Some time later, the animal sickens but does not die. Henceforth, it avoids the food that previously tasted good. The fact that the aversive consequences of eating the food occur some time after its ingestion violates the supposed requirement of close temporal association between stimulus and response, at least if rapid learning is to occur. Yet for an animal to become poison-shy, rapid learning is required since the animal will not survive repeated trials. Of course if such learning is simply a myth, one does not need to be concerned about explaining it. Unfortunately for the theorists (and for the food-growers), learning closely approximating that required for the avoidance of poisoned bait has been demonstrated in the laboratory.

In Reading 40 such learning is described within a broader context. In an important paper, Martin E. P. Seligman here reviews an assumption of general process learning theory, rejects it, and substitutes in its place a "continuum of preparedness which holds that organisms are prepared to associate certain events, unprepared for some, and contraprepared for others." Rats, for example, are viewed as genetically prepared to associate tastes with illness, regardless of violations of the "laws of learning," as they were previously understood.

Reading 41 is related to Reading 40 in that specific experiments to condition aversions in coyotes are described. As authors Carl Gustavson et al. note, such methods may aid in saving both prey and predator species. Learning of a different sort—through the mechanisms by which one species avoids the predation of another species—has been the topic of some previous readings. In Reading 42, L. P. Brower and J. Brower describe a form of mimicry whose "purpose" is to avoid predation. Success depends on the learning ability of the predator, in this case, toads. Most classical experimental psychologists would predict that toads would be very slow learners, yet Brower and Brower demonstrate that toads are capable of learning a visual discrimination in a very few trials. Again we see the relationship to the principles put forth in Reading 40.

Other frequently heard anecdotes concern the capacity of animals to learn from one another—a specific example is the education of the young by the mother. In Reading 43, Phyllis Chesler presents experimental evidence that kittens are indeed more apt to imitate their mother than a strange cat. Note that this particular experiment does not prove that mother cats actively educate their young (although this may well occur), but rather indicates that the young can be educated by observation and imitation of the mother.

Reading 44, by B. A. Campbell and J. Jaynes, provides experimental support for a new principle of retention. The authors describe the principle, called *reinstatement*, as "obvious and disarmingly simple," and the reader will probably agree. Nevertheless, this is the first clear statement and description of a phenomenon that is doubtless of great importance both to animals in their natural environments and to human beings.

In Reading 45, R. A. Hinde describes four different models of motivation, all of which use the concept of energy in explanations of behavioral determination. Comparing several different aspects of these models, he then questions whether an energy concept is necessary to explain behavior.

Psychological Review
1970, Vol. 77, No. 5, 406–418

40

ON THE GENERALITY OF THE LAWS OF LEARNING [1]

MARTIN E. P. SELIGMAN [2]

Cornell University

That all events are equally associable and obey common laws is a central assumption of general process learning theory. A continuum of preparedness is defined which holds that organisms are prepared to associate certain events, unprepared for some, and contraprepared for others. A review of data from the traditional learning paradigms shows that the assumption of equivalent associability is false: in classical conditioning, rats are prepared to associate tastes with illness even over very long delays of reinforcement, but are contraprepared to associate tastes with footshock. In instrumental training, pigeons acquire key pecking in the absence of a contingency between pecking and grain (prepared), while cats, on the other hand, have trouble learning to lick themselves to escape, and dogs do not yawn for food (contraprepared). In discrimination, dogs are contraprepared to learn that different locations of discriminative stimuli control go–no go responding, and to learn that different qualities control directional responding. In avoidance, responses from the natural defensive repertoire are prepared for avoiding shock, while those from the appetitive repertoire are contraprepared. Language acquisition and the functional autonomy of motives are also viewed using the preparedness continuum. Finally, it is speculated that the laws of learning themselves may vary with the preparedness of the organism for the association and that different physiological and cognitive mechanisms may covary with the dimension.

Sometimes we forget why psychologists ever trained white rats to press bars for little pellets of flour or sounded metronomes followed by meat powder for domestic dogs. After all, when in the real world do rats encounter levers which they learn to press in order to eat, and when do our pet dogs ever come across metronomes whose clicking signals meat powder? It may be useful now to remind ourselves about a basic premise which gave rise to such bizarre endeavors, and to see if we still have reason to believe this premise.

The General Process View of Learning

It was hoped that in the simple, controlled world of levers and mechanical feeders, of metronomes and salivation, something quite

general would emerge. If we took such an arbitrary behavior as pressing a lever and such an arbitrary organism as an albino rat, and set it to work pressing the lever for food, then *by virtue of* the very arbitrariness of the environment, we would find features of the rat's behavior general to real-life instrumental learning. Similarly, if we took a dog, undistracted by extraneous noises and sights, and paired a metronome's clicking with meat, what we found about the salivation of the dog might reveal characteristics of associations in general. For instance, when Pavlov found that salivation stopped occurring to a clicking that used to signal meat powder, but no longer did, he hoped that this was an instance of a *law*, "experimental extinction," which would have application beyond clicking metronomes, meat powder, and salivation. What captured the interest of the psychological world was the possibility that such laws might describe the general characteristics of the behavior acquired as the result of pairing one event with another. When Thorndike found that cats learned only gradually to pull strings to escape from puzzle boxes, the intriguing hypothesis was that animal learning in general was by trial and error. In both of these situations, the very arbitrariness and unnaturalness of the experiment was assumed to guarantee generality, since the situation

[1] The preparation of this manuscript was supported in part by National Institute of Mental Health Grant MH 16546-01 to the author. The author gratefully acknowledges the helpful comments of R. Bolles, P. Cabe, S. Emlen, J. Garcia, E. Lenneberg, R. MacLeod, H. Rachlin, D. Regan, K. Kosinski, P. Rozin, T. A. Ryan, R. Solomon, and F. Stollnitz.
[2] Requests for reprints should be sent to Martin E. P. Seligman, Department of Psychology, Morrill Hall, Cornell University, Ithaca, New York 14850.

would be uncontaminated by past experience the organism might have had or by special biological propensities he might bring to it.

The basic premise can be stated specifically: In classical conditioning, the choice of CS, US, and response is a matter of relative indifference; that is, any CS and US can be associated with approximately equal facility, and a set of general laws exist which describe the acquisition, extinction, inhibition, delay of reinforcement, spontaneous recovery, etc., for all CSs and USs. In instrumental learning, the choice of response and reinforcer is a matter of relative indifference; that is, any emitted response and any reinforcer can be associated with approximately equal facility, and a set of general laws exist which describe acquisition, extinction, discriminative control, generalization, etc., for all responses and reinforcers. I call this premise the assumption of equivalence of associability, and I suggest that it lies at the heart of general process learning theory.

This is not a straw man. Here are some quotes from three major learning theorists to document this assumption:

It is obvious that the reflex activity of any effector organ can be chosen for the purpose of investigation, since signalling stimuli can get linked up with any of the inborn reflexes [Pavlov, 1927, p. 17].

any natural phenomenon chosen at will may be converted into a conditional stimulus . . . any visual stimulus, any desired sound, any odor, and the stimulation of any part of the skin [Pavlov, 1928, p. 86].

All stimulus elements are equally likely to be sampled and the probability of a response at any time is equal to the proportion of elements in S' that are connected to it. . . . On any acquisition trial all stimulus elements sampled by the organism become connected to the response reinforced on that trial [Estes, 1959, p. 399].

The general topography of operant behavior is not important, because most if not all specific operants are conditioned. I suggest that the dynamic properties of operant behavior may be studied with a single reflex [Skinner, 1938, pp. 45–46].

A Reexamination of Equivalence of Associability

The premise of equivalence places a special premium on the investigations of arbitrarily related, as opposed to naturally occurring, events. Such events, since they are supposedly uncontaminated by past experience or by special propensities the organism brings to the situation, provide paradigms for the investigations of general laws of learning. More than 60 years of research in both the instrumental and classical conditioning traditions have yielded considerable data

suggesting that similar laws hold over a wide range of arbitrarily chosen events: the shape of generalization gradients is pretty much the same for galvanic skin responses classically conditioned to tones when shock is the US (Hovland, 1937), and for salivating to being touched at different points on the back when food is the US (Pavlov, 1927). Partial reinforcement causes greater resistance to extinction than continuous reinforcement regardless of whether rats are bar pressing for water or running down alleyways for food. Examples of analogous generality of laws could be multiplied at great length.

Inherent in the emphasis on arbitrary events, however, is a danger: *that the laws so found will not be general, but peculiar to arbitrary events.*

The Dimension of Preparedness

It is a truism that an organism brings to any experiment certain equipment and predispositions more or less appropriate to that situation. It brings specialized sensory and receptor apparatus with a long evolutionary history which has modified it into its present appropriateness or inappropriateness for the experiment. In addition to sensory-motor capacity, the organism brings associative apparatus, which likewise has a long and specialized evolutionary history. For example, when an organism is placed in a classical conditioning experiment, not only may the CS be more or less perceptible and the US more or less evocative of a response, *but also the CS and US may be more or less associable.* The organism may be more or less prepared by the evolution of its species to associate a given CS and US or a given response with an outcome. If evolution has affected the associability of specific events, then it is possible, even likely, that the very *laws* of learning might vary with the preparedness of the organism from one class of situations to another. If this is so, investigators influenced by the general process view may have discovered only a subset of the laws of learning: the laws of learning about arbitrarily concatenated events, those associations which happen in fact to be equivalent.

We can define a continuum of preparedness operationally. Confront an organism with a CS paired with US or with a response which produces an outcome. Depending on the specifics, the organism can be either prepared, unprepared, or contraprepared for learning about the events. *The relative preparedness of an organism for*

348

learning about a situation is defined by the amount of input (e.g., numbers of trials, pairings, bits of information, etc.) *which must occur before that output* (responses, acts, repertoire, etc.), *which is construed as evidence of acquisition, reliably occurs.* It does not matter how input or output are specified, as long as that specification can be used consistently for all points on the continuum. Thus, using the preparedness dimension is independent of whether one happens to be an S-R theorist, a cognitive theorist, an information processing theorist, an ethologist, or what have you. Let me illustrate how one can place an experimental situation at various points on the continuum for classical conditioning. If the organism makes the indicant response consistently from the very first presentation of the CS on, such "learning" represents a clear case of instinctive responding, the extreme of the prepared end of the dimension. If the organism makes the response consistently after only a few pairings, it is somewhat prepared. If the response emerges only after many pairings (extensive input), the organism is unprepared. If acquisition occurs only after very many pairings or does not occur at all, the organism is said to be contraprepared. The number of pairings is the measure that makes the dimension a continuum, and implicit in this dimension is the notion that "learning" and "instinct" are continuous. Typically ethologists have examined situations in the prepared side of the dimension, while general process learning theorists have largely restricted themselves to the unprepared region. The contraprepared part of the dimension has been largely uninvestigated, or at least unpublished.

The dimension of preparedness should not be confused with the notion of operant level. The frequency with which a response is made in a given situation is not necessarily related to the associability of that response with a given outcome. As will be seen later, frequent responses may not be acquired when they are reinforced as readily as infrequent responses. Indeed, some theorists (e.g., Turner & Solomon, 1962) have argued that high-probability, fast-latency responding may actually antagonize operant reinforceability.

The first empirical question with which this paper is concerned is whether sufficient evidence exists to challenge the equivalence of associability. For many years, ethologists and others (for an excellent example. see Breland & Breland, 1966) have gathered a wealth of evidence to challenge the general

process view of learning. Curiously, however, these data have had little impact on the general process camp, and while not totally ignored, they have not been theoretically incorporated. In view of differences in methodology, this is perhaps understandable. I do not expect that presenting these lines of evidence here would have any more effect than it has already had. More persuasive to the general process theorist should be the findings which have sprung up within his own tradition.. Within traditional conditioning and training paradigms, a considerable body of evidence now exists which challenges the premise. In reviewing this evidence, we shall find the dimension of preparedness to be a useful integrative device. It is not the intent of this article to review exhaustively the growing number of studies which challenge the premise. Rather, we shall look within each of the major paradigms which general process learning theorists have used and discuss one or two clear examples. The theme of these examples is that all events are not equivalent in their associability: that although the organism may have the necessary receptor and effector apparatus to deal with events, there is much variation in its ability to learn about relations between events.

CLASSICAL CONDITIONING

The investigation of classical aversive conditioning has been largely confined to the unconditioned response of pain caused by the stimulus of electric shock (cf. Campbell & Church, 1969), and the "laws" of classical conditioning are based largely on these findings along with those from salivary conditioning. Recently, Garcia and his collaborators (Garcia, Ervin, & Koelling, 1966; Garcia, Ervin, Yorke, & Koelling, 1967; Garcia & Koelling, 1966; Garcia, McGowan, Ervin, & Koelling, 1968), and Rozin and his collaborators (Rodgers & Rozin, 1966; Rozin, 1967, 1968, 1969) have used illness as an unconditioned response and reported some intriguing findings. In the paradigm experiment (Garcia & Koelling, 1966), rats received "bright-noisy, saccharin-tasting water." What this meant was that whenever the rat licked a drinking tube containing saccharine-flavored water, lights flashed and a noise source sounded. During these sessions the rats were X irradiated. X irradiation makes rats sick, but it should be noted that the illness does not set in for an hour or so following X-raying. Later the rats were tested for acquired aversions to the elements of the compound CS. The rats had acquired a strong aversion to the taste of saccharine, *but had not acquired an aversion to the "bright-noise."* The rats had "associated" the taste with their illness, but not the exteroceptive noise-light stimuli. So that it could not be argued that saccharin is such a salient event that it masked the noise and light, Garcia and Koelling ran the complementary experiment: "Bright and noisy saccharin-tasting water" was again used as a CS, but this time electric shock to the feet was the US. The rats were then tested for aversion to the elements of the CS. In this case, the bright noise became aversive, but the saccharin-tasting water did not. This showed that the bright noise was clearly perceptible; but the rats associated only the bright noise with the exteroceptive US of footshock, and not the taste of saccharin in spite of its also being paired with shock.

In the experiment, we see both ends as well as the middle of the preparedness continuum. Rats are prepared, by virtue of their evolutionary history, to associate tastes with malaise. For in spite of a several-hour delay of reinforcement, and the presence of other perceptible CSs, only the taste was associated with nausea, and light and noise were not. Further, rats are contraprepared to associate exteroceptive events with nausea and contraprepared to associate tastes with footshock. Finally, the association of footshock with light and sound is probably someplace in the unprepared region. The survival advantage of this preparedness seems obvious: organisms who are poisoned by a distinctive food and survive, do well not to eat it again. Selective advantage should accrue, moreover, to those rats whose associative apparatus could bridge a very long CS-US interval and who could ignore contiguous, as well as interpolated, exteroceptive CSs in the case of taste and nausea.

Does such prepared and contraprepared acquisition reflect the evolutionary results of selective pressure or does it result from experience? It is possible that Garcia's rats may have previously learned that tastes were uncorrelated with peripheral pain and that tastes were highly correlated with alimentary consequences. Such an argument involves an unorthodox premise: that rats' capacities for learning set and transfer are considerably broader than previously demonstrated. The difference between a position that invokes selective pressure (post hoc) and the experiential set position is testable: Would mating those rats who were most proficient at learning the taste–footshock association produce offspring more capable of such

learning than an unselected population? Conversely, would interbreeding refractory rats select out the facility with which the taste–nausea association is made?

Supporting evidence for preparedness in classical conditioning has come from other recent experiments on specific hungers and poisoning. Rodgers and Rozin (1966) and Rozin (1967, 1968) have demonstrated that at least part of the mechanism of specific hungers (other than sodium) involves conditioned aversion to the taste of the diet the rats were eating as they became sick. Deficient rats spill the old diet and will not eat it, even after they have recovered. The association of the old taste with malaise seems to be made in spite of the long delay between taste of the diet and gradual onset of illness. The place and the container in which the old diet was set, moreover, do not become aversive. The remarkable ability of wild rats who recover from being poisoned by a novel food, and thereafter avoid new tastes (Barnett, 1963; Rozin, 1968), also seems to result from classical conditioning. Note that the wild rat must be prepared to associate the taste with an illness which does not appear for several hours in only one trial; note also that it must be contraprepared to associate some contiguous CSs surrounding the illness with malaise.

Do these findings really show that rats can associate tastes and illness when an interval of many minutes or even hours intervenes or are they merely a subtle instance of contiguity? Peripheral cues coming either from long-lasting aftertastes or from regurgitation might bring the CS and US into contiguity. Rozin (1969) reported evidence against aftertaste mediation: rats received a high concentration of saccharin paired with apomorphine poisoning. Later, the rats were given a choice between the high concentration and a low concentration. The rats preferred the low concentration, even though the aftertaste that was purportedly contiguous with malaise should be more similar to the low concentration (since it had been diluted by saliva) than the high concentration.

Not only do rats acquire an aversion for the old diet, on which they got sick, but they also learn to prefer the taste of a new diet containing the needed substance. This mechanism also seems to involve prepared conditioning of taste to an internal state. Garcia et al. (1967) paired the taste of saccharin with thiamine injections given to thiamine deficient rats, and the rats acquired a preference for saccharin. So both the rejection of old foods and acceptance of new foods in specific hungers can be explained by prepared conditioning of tastes to internal state.

INSTRUMENTAL LEARNING

E. L. Thorndike, the founder of the instrumental learning tradition, was by no means oblivious to the possibility of preparedness in instrumental learning, as we shall see below. He also hinted at the importance of preparedness in one of his discussions of classical conditioning (Thorndike, 1935, p. 192–197): one of his students (Bregman, 1934) attempted to replicate the results of Watson and Rayner (1920), who found that little Albert became afraid of a white rat, rabbit, and dog which had been paired with a startling noise. Bregman was unable to show any fear conditioning when she paired more conventional CSs, such as blocks of wood and cloth curtains, with startling noise. Thorndike speculated that infants at the age of locomotion were more disposed to manifest fear to objects that wiggle and contort themselves than to motionless CSs.

Thorndike's parallel views on instrumental learning rose from his original studies of cats in puzzle boxes. As every psychologist knows, he put cats in large boxes and investigated the course of learning to pull strings to escape. What is less widely known is that he put his cats in not just one puzzle box, but in a whole series of different ones (incidentally in doing this he seems to have discovered learning set—Thorndike, 1964, pp. 48–50). In one box the cats had to pull a string to get out, in another a button had to be pushed, in another a lever had to be depressed, etc. One of his boxes—Box Z—was curious: it was merely a large box with nothing but a door that the experimenter could open. Thorndike opened the door in Box Z whenever cats licked themselves or scratched themselves. The cat is known to use both of these frequently occurring responses instrumentally: it scratches itself to turn off itches, and licks itself to remove dirt. In addition, Thorndike had established that getting out of a puzzle box was a sufficient reward for reinforcing the acts of string pulling, button pushing, and lever clawing. In spite of this, Thorndike's cats seemed to have a good deal of trouble learning to scratch themselves or lick themselves to get out of the boxes.

A reanalysis of the individual learning curves presented by Thorndike (1964) for each of the seven cats who had experience in Box Z documents the impression: of the 28 learning curves presented for these seven cats in the boxes other than Z, 22 showed faster learning than in Z, three showed approximately equal learning, and only three showed slower learning. While all of the cats eventually showed improved speeds of licking or scratching for escape, such learning was difficult and irregular. Thorndike noted another unusual property of licking and scratching:

There is in all these cases a noticeable tendency . . . to diminish the act until it becomes a mere vestige of a lick or scratch . . . the licking degenerated into a mere quick turn of the head with one or two motions up and down with tongue extended. Instead of a hearty scratch, the cat waves its paw up and down rapidly for an instant. Moreover, if sometimes you do not let the cat out after the feeble reaction, it does not at once repeat the movement, as it would do if it depressed a thumb piece, for instance, without success in getting the door open [Thorndike, 1964, p. 48].

Contemporary investigators have reported related findings. Konorski (1967, pp. 463–467) attempted to train "reflex" movements, such as anus licking, scratching, and yawning, with food reinforcement. While reporting success with scratching and anus licking, like Thorndike, he observed spontaneous simplification and arhythmia in the responses. More importantly, he reported that reinforcement of "true yawning" with food is very difficult, if not impossible. Bolles and Seelbach (1964) reported that rearing could be reinforced by noise offset, but not punished by noise onset, exploration could be modified by both, and grooming by neither. This difference could not be accounted for by difference in operant level, which is substantial for all these behaviors of the rat.

Thorndike (1964) speculated that there may be some acts which the organism is not neurally prepared to connect to some sense impressions:

If the associations in general were simply between situation and impulse to act, one would suppose that the situation would be associated with the impulse to lick or scratch as readily as with the impulse to turn a button or claw a string. Such is not the case. By comparing the curves for Z on pages 57–58 with the others, one sees that for so simple an act it takes a long time to form the association. This is not the final reason, for lack of attention, a slight increase in the time taken to open the door after the act was done, or *an absence of preparation in the nervous system for connections between these particular acts and definite sense impressions* [italics added] may very well have been the cause of the difficulty in forming the associations [p. 113].

This speculation seems reasonable: after all, in the natural history of cats, only behavior such as manipulating objects which maximized chances for escaping traps would be selected, and licking is not in the repertoire which maximizes escape. At minimum, Thorndike demonstrated that the emission of licking paired with an event which could reinforce other emitted acts was not sufficient to reinforce licking equally well. In the present terms, Thorndike had discovered a particular instrumental training situation for which cats are relatively contraprepared.

Brown and Jenkins (1968, Experiment 6) have reported findings which appear to come from the opposite end of the dimension. Pigeons were exposed to a lighted key which was paired with grain delivered in a lighted food hopper below the key. But unlike the typical key-pecking situation, the pigeons' pecking the key did not produce food. Food was contingent only on the key's being lit, not on pecking the key. In spite of this, all pigeons began pecking the key after exposure to the lighted key, followed by grain. Moreover, key pecking was maintained even though it had no effect on food. One can conclude from these "autoshaping" results that the pigeon is highly prepared for associating the pecking of a lighted key with grain.

There is another curiosity in the history of the instrumental learning literature which is usefully viewed with the preparedness dimension: the question of why a reinforcer is reinforcing. For over 20 years, disputes raged about what monolithic principle described the necessary and sufficient conditions for learning. Hull (1943) claimed that tissue-need reduction must occur for learning to take place, while Miller (1951) held that drive reduction was necessary and sufficient. Later, Sheffield, Roby, and Campbell (1954) suggested that a consummatory response was the necessary condition. More recently, it has become clear that learning can occur in the absence of any of these (e.g., Berlyne, 1960). I suggest that when CSs or responses are followed by such biologically important events as need reducers, drive reducers, or consummatory responses, learning should take place readily because natural selection has prepared organisms for such relationships. The relative preparedness of organisms for these events accounts for the saliency of such learning and hence the appeal of each of the monolithic principles. But organisms *can* learn about bar pressing paired with light onset, etc.; they are merely less pre-

pared to do so, and hence, the now abundant evidence against the earlier principles was more difficult to gather.

Thus, we find that in instrumental learning paradigms, there are situations which lie on either side of the rat's bar pressing for food on the preparedness dimension. A typical rat will ordinarily learn to bar press for food after a few dozen exposures to the bar press—food contingency. But cats, who can use scratching and licking as instrumental acts in some situations, have trouble using these acts to get out of puzzle boxes, and dogs do not learn to yawn for food even after many exposures to the contingency. On the other hand, pigeons acquire a key peck in a lighted key–grain situation, even when there is no contingency at all between key pecking and grain. These three instrumental situations represent unprepared, contraprepared, and prepared contingencies, respectively. Later we shall discuss the possibility that they obey different laws as a function of different preparedness.

DISCRIMINATION LEARNING

The next two paradigms we consider— discrimination learning and avoidance learning—combine both classical and instrumental procedures. In both of these paradigms, findings have been reported which challenge the equivalence of associability. We begin with some recent Polish work on discrimination learning in dogs. Lawicka (1964) attempted to train dogs in either a go right–go left differentiation or a go–no go differentiation. Whether such differentiation could be acquired depended on the specific discriminative stimuli used. For the left–right differentiation, if the S— and the S+ differed in location (one speaker above the dog; one speaker below), the dog readily learned which way to go in order to receive food. If, however, the stimuli came from the *same* speaker and differed only in pitch, the left–right differentiation was exceedingly difficult. Topographical differences in stimuli, as opposed to qualitative differences, seem to aid in differentiating two topographically different responses. The dog seems contraprepared, moreover, for making a left–right differentiation to two tones which do not also differ in direction. Lest one argue that the two tones coming out of the same speaker were not discriminable, Lawicka (1964; like Garcia & Koelling, 1966) did the complementary experiment: dogs were trained to go and receive food or stay with two tones coming out of the same speaker. One tone was the S+ and the other tone the S—. The dogs learned this readily. Thus, using the same tones which could not be used to establish a left–right differentiation, a go–no go differentiation was established. The author then attempted to elaborate the go–no go differentiation to the same tone differing in location of speakers. As the reader should expect by now, the dogs had trouble learning the go–no go differentiation to the difference in location of S+ and S—. Dogs, then, are contraprepared for learning about different locations controlling a go–no go differentiation although they are not contraprepared for learning that the same locations control a left–right differentiation. Dogs are contraprepared for learning that qualitative differences of tone from the same location control a left–right differentiation, but not contraprepared for using this difference to govern a go–no go differentiation. Dobrzecka and Konorski (1967, 1968) and Szwejkowska (1967) have confirmed and extended these findings.

Emlen (personal communication, 1969) reported discrimination (or at least perceptual) learning that is prepared. It is known from planetarium experiments that adult indigo buntings use the northern circumpolar constellations for migration, since blocking these from view disrupts directed migration. One might have thought that the actual constellations were represented genetically. If young birds are raised under a sky which rotates around a fictitious axis, however, they use the arbitrarily chosen circumpolar constellations for migration and ignore the natural circumpolar constellations. Thus, it appears that indigo buntings are prepared to pay attention to and learn about those configurations of stars which rotate most slowly in the heavens.

AVOIDANCE LEARNING

Data from avoidance learning studies also challenge the equivalence of associability. Rats learn reasonably readily to press bars to obtain food. Rats also learn very readily to jump (Baum, 1969) and reasonably readily to run (Miller, 1941, 1951) from a dangerous place to a safe place to avoid electric shock. From this, the premise deduces that rats should learn readily to press bars to avoid shock. But this is not so (e.g., D'Amato & Schiff, 1964). Very special procedures must be instituted to train rats to depress levers to avoid shock reliably (e.g., D'Amato & Fazzaro, 1966; Fantino, Sharp, & Cole, 1966). Similarly,

pigeons learn readily to peck lighted keys to obtain grain: too readily, probably, for this to be considered an unprepared or arbitrary response (see Brown & Jenkins, 1968). But it is very difficult to train pigeons with normal laboratory techniques to key peck to avoid shock. Hoffman and Fleshler (1959) reported that key pecking was impossible to obtain with negative reinforcement; Azrin (1959) found only temporary maintenance of key pecking in but one pigeon; and Rachlin and Hineline (1967) needed 10–15 hours of patient shaping to train key pecking to remove shock. This probably attests more to a problem specific to the response and reinforcer than to some inability of the pigeon to learn about avoidance contingencies. Ask anyone who has attempted to kill pigeons (e.g., by electrocution or throwing rocks at them), how good pigeons are at avoiding. Pigeons learn to fly away to avoid noxious events (e.g., Bedford & Anger, 1968; Emlen, 1970). In contrast, it is hard to imagine a pigeon flying *away* from something to obtain food.

Bolles has recently (1970)—and quite persuasively—argued that avoidance responses as studied in laboratory experiments are not simple, arbitrary operants. In order to produce successful avoidance, Bolles argues, the response must be chosen from among the natural, *species-specific* defensive repertoire of the organism. Thus, it must be a response for which the organism is prepared. Running away for rats and flying away for pigeons make good avoidance responses, while key pecking and bar pressing (which are probably related to the appetitive repertoire) do not.

It might be argued that these difficulties in learning avoidance are not due to contrapreparedness but to competing motor responses Thus, for example, rats have trouble pressing levers to avoid shock because shock causes them to "freeze" which is incompatible with bar pressing. A word of caution is in order about such hypotheses: I know of no theory which specifies in advance what competes with what; rather, response competition (or facilitation) is merely invoked post hoc. When, and if, a *theory* of topographical incompatibility arises it may indeed provide an *explanation* of contrapreparedness, but at the present time, it does not.

Let us review the evidence against the equivalence of associability premise: in classical conditioning, rats are prepared to associate tastes with nausea and contraprepared to associate taste with footshock. In instrumental learning, different emitted responses are differentially associable with different reinforcers: pigeons are prepared to peck lighted keys for food, since they will acquire this even in the absence of any contingency between key pecking and food. Cats are contraprepared for learning to scratch themselves to escape, and dogs for yawning for food. In discrimination learning, dogs are contraprepared to learn that different locations control a go–no go differentiation, and contraprepared for different qualities controlling a left–right response. In avoidance learning, those responses which come from the natural defensive repertoire of rats and pigeons are prepared (or at least unprepared) for avoiding shock. Those responses from the appetitive repertoire seem contraprepared for avoidance.

TWO FAILURES OF GENERAL PROCESS LEARNING THEORY: LANGUAGE AND THE FUNCTIONAL AUTONOMY OF MOTIVES

The interest of psychologists in animal learning theory is on the wane. Although the reasons are many, a prominent one is that such theories have failed to capture and bring into the laboratory phenomena which provide fertile models of complex human learning. This failure may be due in part to the equivalence premise. By concentrating on events for which organisms have been relatively unprepared, the laws and models which general process learning theories have produced may not be applicable beyond the realm of arbitrary events, arbitrarily connected. This would not be an obstacle if all of human learning consisted of learning about arbitrary events. But it does not. *Homo sapiens* has an evolutionary history and a biological makeup which has made it relatively prepared to learn some things and relatively contraprepared to learn others. If learning varies with preparedness, it should not be surprising that the laws for unprepared association between events have not explained such phenomena as the learning of language or the acquisition of motives.

Lenneberg (1967) has recently provided an analysis of language, the minimal conclusion of which is that children do not learn language the way rats learn to press a lever for food. Put more strongly, the set of laws which describe language learning are not much illuminated by the laws of the acquisition of arbitrary associations between events, as Skinner (1957) has argued. Unlike such unprepared contingencies as bar pressing for food, language does not require careful training or shaping for its acquisition. We do not need to arrange sets of linguistic contingencies carefully to get children to speak and understand English. Programmed training of speech is relatively ineffective, for under all but the most impoverished linguistic environments, human beings learn to speak and understand. Children of the deaf make as much noise and have the same sequence and age of onset for cooing as children of hearing parents. Development of language seems roughly the same across cultures which presumably differ widely in the arrangement of reinforcement contingencies, and language skill is not predicted by chronological age but by motor skill (see Lenneberg, 1967, especially pp. 125–158, for a fuller discussion).

The acquisition of language, not unlike pecking a lighted key for grain in the pigeon and the acquisition of birdsong (Petrinovich, 1970), is prepared. The operational criterion for the prepared side of the dimension is that minimal input should produce acquisition. One characteristic of language acquisition which separates it from the bar press is just this: elaborate training is not required for its production. From the point of view of this paper, it is not surprising that the traditional analyses of instrumental and classical conditioning are not adequate for an analysis of language. This is not because language is a phenomenon *sui generis*, but because the laws of instrumental and classical conditioning were developed to explain unprepared situations and not to account for learning in prepared situations. This is not to assert that the laws which govern language acquisition will necessarily be the same as those governing the Garcia phenomenon, birdsong, or the key peck, but to say that species-specific, biological analysis might be fruitfully made of these phenomena.

It is interesting to note in this context the recent success that Gardner and Gardner (1970) have had in teaching American sign language to a chimpanzee. The Gardners reasoned that earlier failures to teach spoken English to chimpanzees (Hayes & Hayes, 1952; Kellogg & Kellogg, 1933) did not result from cognitive deficiencies on the part of the subjects, but from the contraprepared nature of vocalization as a trainable response. The great manual dexterity of the chimpanzee, however, suggested sign language as a more trainable vehicle. Hayes (1968) has recently reanalyzed the data from Vicki (the Hayes' chimp) and confirmed the suggestion that chimpanzees' difficulty in using exhalation instrumentally may have caused earlier failures.

Language is not the only example of human learning that has eluded general process theory. The extraordinary persistence of acquired human motives has not been captured in ordinary laboratory situations. People, objects, and endeavors which were once unmotivating to an individual acquire and maintain strongly motivating properties. Fondness for the objects of sexual learning long after sexual desire is gone is a clear example. Acquisition of motives is not difficult to bring into the laboratory, and the extensive literature on acquired drives has often been taken as an analysis of acquired human motivation. A rat, originally unafraid of a tone, is shocked while the tone is played. Thereafter, the rat is afraid of the tone. But the analogy breaks down here; for once the tone is presented several times without shock, the tone loses its fear-inducing properties (Little & Brimer, 1968; Wagner, Siegel, & Fein, 1967). (The low resistance to extinction of the conditioned emotional response should not be confused with the high resistance to extinction of the avoidance response. This inextinguishability probably stems from the failure of the organism to stay around in the presence of the CS long enough to be exposed to the fact that shock no longer follows the CS, rather than a failure of fear of the CS to extinguish.) Yet, acquired motivators for humans retain their properties long after the primary motivation with which they were originally paired is absent. Allport (1937) raised the problem for general process theory as the "functional autonomy of motives." But in the 30 years since the problem was posed, the failure of acquired human motives to extinguish remains unanalyzed experimentally.

The notion of preparedness may be useful in analyzing persistent acquired motivation. Typically, investigations of acquired drives have paired arbitrary CSs with arbitrary primary motivators. It seems possible that if more prepared CSs were paired with primary motivators, the motivational properties of such CSs might be unusually resistant to extinction. Seligman, Ives, Ames, and Mineka (1970) conditioned drinking by pairing compound CSs with injections of hypertonic saline-procaine in rats. When the CS consisted only of an interoceptive stimuli (white box, white noise), conditioning occurred, but extinguished in a few days. When the interoceptive CS of one-hour water deprivation was added to the compound, conditioning occurred and

persisted unabated for two months. It seems possible that preparedness of mild thirst for association with rapidly induced strong thirst may account for the inextinguishability of acquired drinking.

Are humans prepared to associate a range of endeavors and objects with primary motivators, and are such associations unusually persistent after the original motivators have left the scene? Here, as for language, viewing persistent acquired motives as cases of preparedness may make human motivation—both adaptive and maladaptive—more amenable to study.

PREPAREDNESS AND THE LAWS OF LEARNING

The primary empirical question has been answered affirmatively: The premise of equivalence of associability does not hold, *even in the traditional paradigms for which it was first assumed.* But does this matter? Do the same laws which describe the learning of unprepared events hold for prepared, unprepared, and contraprepared events? Given that an organism is prepared, and therefore learns with minimal input, does such learning have different properties from those unprepared associations that the organism acquires more painstakingly? Are the same mechanisms responsible for learning in prepared, unprepared, and relatively contraprepared situations?

We can barely give a tentative answer to this question, since it has been largely uninvestigated. Only a few pieces of evidence have been gathered to suggest that once a relatively prepared or contraprepared association has been acquired, it may not display the same family of extinction curves, values for delay of reinforcement, punishment effects, etc., as the lever press for food in the rat. Consider again the Garcia and Koelling (1966) findings: the association of tastes with illness is made with very different delays of reinforcement from ordinary Pavlovian associations. Unlike salivating to sounds, the association will be acquired with delays of up to one hour and more. Detailed studies which compare directly the delay of reinforcement gradients, extinction functions, etc., for prepared versus unprepared associations are needed. It would be interesting to find that the extinction and inhibition functions for prepared associations were different than for unprepared associations. If preparation underlies the observations of functional autonomy, prepared associations might be highly resistant to extinction, punishment, and other changes in instrumental contingencies. Breland and

Breland (1966) reported that many of the "prepared" behaviors that the organisms they worked with acquired would persist even under counterproductive instrumental contingencies. To what extent would the autoshaped key pecking responses of Brown and Jenkins (1968) be weakened by extinction or punishment, as bar pressing for food is weakened? Williams and Williams (1969) reported that autoshaped key-pecking responses persist even when they actually "cost" the pigeon reinforcement.

Does contraprepared behavior, after being acquired, obey the same laws as unprepared behavior? Thorndike (1964) reported that when he finally trained licking for escape, the response no longer looked like the natural response, but was a pale, mechanical imitation of the natural response. Would the properties of the response differentiation and shaping of such behavior be like those of unprepared responses? The answer to this range of questions is presently unknown.

Preparedness has been operationally defined, and it is possible that different laws of learning may vary with the dimension. How can the dimension by anchored more firmly? Might different cognitive and physiological mechanisms covary with dimension?

Acquired aversions to tastes following illness is commonplace in humans. These Garcia phenomena are not easily modified by cognition in contrast to other classically conditioned responses in humans (e.g., Spence & Platt, 1967). The knowledge that the illness was caused by the stomach flu and not the Sauce Bearnaise does not prevent the sauce from tasting bad in the future. Garcia, Kovner, and Green (1970) reported that distinctive tastes can be used by rats as a cue for shock avoidance in a shuttlebox; but the preference for the taste in the home cage is unchanged. When the taste is paired with illness, however, the preference is reduced in the home cage. Such evidence suggests that prepared associations may not be cognitively mediated, and it is tempting to speculate that cognitive mechanisms (expectation, attention, etc.) come into play with more unprepared or contraprepared situations. If this is so, it is ironic that the "blind" connections which both Thorndike and Pavlov wanted to study lie in the prepared realm and not in the unprepared paradigms they investigated.

We might also ask if different neural structures underlie differently prepared learning. Does elaborate prewiring mediate prepared associations such as taste and nausea, while more plastic structures mediate unprepared and contraprepared associations?

We have defined the dimension of preparedness and given examples of it. To anchor the dimension we need to know the answers to three questions about what covaries with it: (a) Do different laws of learning (families of functions) hold along the dimension? (b) Do different cognitive mechanisms covary with it? (c) Do different physiological mechanisms also covary with preparedness?

PREPARATION AND THE GENERAL PROCESS VIEW OF LEARNING

If the premise of equivalence of associability is false, then we have reason to suspect that the laws of learning discovered using lever pressing and salivation may not hold for any more than other simple, unprepared associations. If the laws of learning for unprepared association do not hold for prepared or contraprepared associations, is the general process view salvageable in any form? This is an empirical question. Its answer depends on whether *differences* in learning vary systematically along the dimension of preparedness; the question reduces to whether the preparedness continuum is a nomological continuum. For example, if one finds that the families of extinction functions vary systematically with the dimension, then one might be able to formulate *general* laws of extinction. Thus, if prepared CRs extinguished very slowly, unprepared CRs extinguished gradually, and contraprepared CRs extinguished precipitously, such a systematic, continuous difference in *laws* would be a truly general law of extinction. But before such general laws can be achieved, we must first investigate what the laws of prepared and contraprepared associations actually are. If this were done, then the possibility of general laws of learning would be again alive.

REFERENCES

ALLPORT, G. The functional autonomy of motives. *American Journal of Psychology*, 1937, 50, 141–156.

AZRIN, N. J. Some notes on punishment and avoidance. *Journal of the Experimental Analysis of Behavior*, 1959, 2, 260.

BARNETT. S. *The rat: A study in behavior.* London: Methuen, 1963.

BAUM, M. Dissociation of respondent and operant processes in avoidance learning. *Journal of Comparative and Physiological Psychology*, 1969, 67, 83–88.

BEDFORD, J., & ANGER, D. Flight as an avoidance response in pigeons. Paper presented at the meeting of the Psychonomic Society, St. Louis, October 1968.

BERLYNE, D. E. *Conflict, arousal, and curiosity.* McGraw-Hill: New York, 1960.

BOLLES, R. Effects of escape training on avoidance learning. In F. R. Brush (Ed.), *Aversive conditioning and learning.* New York: Academic Press, 1970, in press.

BOLLES, R., & SEELBACH, S. Punishing and reinforcing effects of noise onset and termination for different responses. *Journal of Comparative and Physiological Psychology,* 1964, 58, 127–132.

BREGMAN, E. An attempt to modify the emotional attitude of infants by the conditioned response technique. *Journal of Genetic Psychology,* 1934, 45, 169–198.

BRELAND, K., & BRELAND, M. *Animal behavior.* New York: Macmillan, 1966.

BROWN, P., & JENKINS, H. Autoshaping of the pigeon's key-peck. *Journal of the Experimental Analysis of Behavior,* 1968, 11, 1–8.

CAMPBELL, E. A., & CHURCH, R. M *Punishment and aversive behavior.* New York: Appleton-Century-Crofts, 1969.

D'AMATO, M. R., & FAZZARO, J. Discriminated lever-press avoidance learning as a function of type and intensity of shock. *Journal of Comparative and Physiological Psychology,* 1966, 61, 313–315.

D'AMATO, M. R., & SCHIFF, J. Long-term discriminated avoidance performance in the rat. *Journal of Comparative and Physiological Psychology,* 1964, 57, 123–126.

DOBRZECKA, C., & KONORSKI, J. Qualitative versus directional cues in differential conditioning. I. Left leg-right leg differentiation to cues of a mixed character. *Acta Biologiae Experimentale,* 1967, 27, 163–168.

DOBRZECKA, C., & KONORSKI, J. Qualitative versus directional cues in differential conditioning. *Acta Biologiae Experimentale,* 1968, 28, 61–69.

EMLEN, S. The influence of magnetic information on the orientation of the indigo bunting. *Animal Behavior,* 1970, in press.

ESTES, W. K. The statistical approach to learning theory. In S. Koch (Ed.), *Psychology: A study of a science.* Vol. 2. New York: McGraw-Hill, 1959.

FANTINO, E., SHARP, D., & COLE, M. Factors facilitating lever press avoidance. *Journal of Comparative and Physiological Psychology,* 1966, 63, 214–217.

GARCIA, J., ERVIN, F., & KOELLING, R. Learning with prolonged delay of reinforcement. *Psychonomic Science,* 1966, 5, 121–122.

GARCIA, J., ERVIN, F., YORKE, C., & KOELLING, R. Conditioning with delayed vitamin injections. *Science,* 1967, 155, 716–718.

GARCIA, J., KOVNER, R., & GREEN, K. F. Cue properties versus palatability of flavors in avoidance learning. *Psychonomic Science,* 1970, in press.

GARCIA, J., & KOELLING, R. Relation of cue to consequence in avoidance learning. *Psychonomic Science,* 1966, 4, 123 124.

GARCIA, J., McGOWAN, B., ERVIN, F., & KOELLING, R. Cues: Their relative effectiveness as a function of the reinforcer. *Science,* 1968, 160, 794–795.

GARDNER, B., & GARDNER, A. Two-way communication with an infant chimpanzee. In A. Schrier & F. Stollnitz (Eds.), *Behavior of nonhuman primates.* Vol. 3. New York: Academic Press, 1970, in press.

HAYES, K. J. Spoken and gestural language learning in chimpanzees. Paper presented at the meeting of the Psychonomic Society, St. Louis, October 1968.

HAYES, K. J., & HAYES, C. Imitation in a home-raised chimpanzee. *Journal of Comparative and Physiological Psychology,* 1952, 45, 450–459.

HOFFMAN, H. S., & FLESHLER, M. Aversive control with the pigeon. *Journal of the Experimental Analysis of Behavior,* 1959, 2, 213–218.

HOVLAND, C. The generalization of conditioned responses. I. The sensory generalization of conditioned responses with varying frequencies of tone. *Journal of Genetic Psychology,* 1937, 17, 279–291.

HULL, C. L. *Principles of behavior.* New York: Appleton-Century-Crofts, 1943.

KELLOGG, W. N., & KELLOGG, L. A. *The ape and the child.* New York: McGraw-Hill, 1933.

KONORSKI, J. *Integrative activity of the brain.* Chicago: University of Chicago Press, 1967.

LAWICKA, W. The role of stimuli modality in successive discrimination and differentiation learning. *Bulletin of the Polish Academy of Sciences,* 1964, 12, 35–38.

LENNEBERG, E. *The biological foundations of language.* New York: Wiley, 1967.

LITTLE, J., & BRIMER, C. Shock density and conditioned suppression. Paper presented at the meeting of the Eastern Psychological Association, Washington, D. C., April 1968.

MILLER, N. E. An experimental investigation of acquired drives. *Psychological Bulletin,* 1941, 38, 534–535.

MILLER, N. E. Learnable drives and rewards. In S. S. Stevens (Ed.), *Handbook of experimental psychology.* New York: Wiley, 1951.

PAVLOV, I. P. *Conditioned reflexes.* New York: Dover, 1927.

PAVLOV, I. P. *Lectures on conditioned reflexes.* New York: International Publishers, 1928.

PETRINOVICH, L. Psychobiological mechanisms in language development. In G. Newton & A. R. Riesen (Eds.), *Advances in psychobiology.* New York: Wiley, 1970, in press.

RACHLIN, H. C., & HINELINE, P. N. Training and maintenance of key pecking in the pigeon by negative reinforcement. *Science,* 1967, 157, 954–955.

RODGERS, W., & ROZIN, P. Novel food preferences in thiamine-deficient rats. *Journal of Comparative and Physiological Psychology,* 1966, 61, 1–4.

ROZIN, P. Specific aversions as a component in specific hungers. *Journal of Comparative and Physiological Psychology,* 1967, 63, 421–428.

ROZIN, P. Specific aversions and neophobia resulting from vitamin deficiency or poisoning in half wild and domestic rats. *Journal of Comparative and Physiological Psychology,* 1968, 66, 82–88.

ROZIN, P. Central or peripheral mediation of learning with long CS-US intervals in the feeding system. *Journal of Comparative and Physiological Psychology,* 1969, 67, 421–429.

SELIGMAN, M. E. P., IVES, C. E., AMES, H., & MINEKA, S. Conditioned drinking and its failure to extinguish: Avoidance, preparedness, or functional autonomy? *Journal of Comparative and Physiological Psychology*, 1970, 71, 411–419.

SHEFFIELD, F. D., ROBY, T. B., & CAMPBELL, B. A. Drive reduction versus consummatory behavior as determinants of reinforcement. *Journal of Comparative and Physiological Psychology*, 1954, 47, 349–354.

SKINNER, B. F. *The behavior of organisms.* New York: Appleton-Century-Crofts, 1938.

SKINNER, B. F. *Verbal behavior.* New York: Appleton-Century-Crofts, 1957.

SPENCE, K. W., & PLATT, J. R. Effects of partial reinforcement on acquisition and extinction of the conditioned eye blink in a masking situation. *Journal of Experimental Psychology*, 1967, 74, 259–263.

SZWEJKOWSKA, G. Qualitative versus directional cues in differential conditioning. II. Go–no go differentiation to cues of a mixed character. *Acta Biologiae Experimentale*, 1967, 27, 169–175.

THORNDIKE, E. L. *Animal intelligence.* New York: Hafner, 1964. (Originally published: New York: Macmillan, 1911.)

THORNDIKE, E. L. *The psychology of wants, interests, and attitudes.* New York: Appleton-Century, 1935.

TURNER, L., & SOLOMON, R. L. Human traumatic avoidance learning: Theory and experiments on the operant-respondent distinction and failures to learn. *Psychological Monographs*, 1962, 76(40, Whole No. 559).

WAGNER, A., SIEGEL, L., & FEIN, G. Extinction of conditioned fear as a function of the percentage of reinforcement. *Journal of Comparative and Physiological Psychology*, 1967, 63, 160–164.

WATSON, J. B., & RAYNER, R. Conditioned emotional reactions. *Journal of Experimental Psychology*, 1920, 3, 1–14.

WILLIAMS, D. R., & WILLIAMS, H. Auto-maintenance in the pigeon: Sustained pecking despite contingent non-reinforcement. *Journal of the Experimental Analysis of Behavior*, 1969, 12, 511–520.

(Received November 5, 1969)

41

Coyote Predation Control by Aversive Conditioning

Carl R. Gustavson, John Garcia, Walter G. Hankins and Kenneth W. Rusiniak

Reprinted from
3 May 1974, Volume 184, pp. 581-583

Abstract. *Conditioned aversions were induced in coyotes by producing lithium chloride illness in them following a meal, and the effects upon eating and attack behavior were observed. One trial with a given meat and lithium is sufficient to establish a strong aversion which inhibits eating the flesh of that prey. One or two trials with a given flesh (lamb or rabbit) specifically suppresses the attack upon the averted prey but leaves the coyote free to attack the alternative prey. A method of saving both prey and predator is discussed.*

Predation of lambs by feral coyotes, in the public lands of the western United States, has led to a sharp controversy between naturalists who wish to see this carnivore survive in its natural habitat and stockmen who wish to reduce sheep losses to predatory coyotes. The principal method of controlling predation has been simply to kill the coyotes, employing bounty hunters, traps, and lethal poisons. These methods do not distinguish between the sheep killers and other coyotes (1). We are devising behavioral methods which spare both prey and predator.

If an animal eats a poisoned meal and survives, it will develop an aversion for the flavor of that meal. Such conditioned aversions have been studied most extensively in the rat, which is an omnivore specialized in seeking and testing new sources of food, including many plants which could be toxic (2). Several general questions arise: (i) Can such aversions be as readily established in a feral carnivore which preys principally on animals? (ii) Will gustatory aversions inhibit attack behavior directed at olfactory, auditory, and visual aspects of living prey? (iii) Can the inhibitory effect be limited to a specific prey (for example, lambs), leaving other options (for example, rabbits) open to the predator?

Our subjects were seven adult coyotes, approximately 2 to 4 years of age, maintained in individual dog runs. Five coyotes were removed from their feral den at approximately 3 weeks of age. Three (male "Brujo" and females "Luna" and "Coty") had hunted rabbits in the desert, while two (male "Feisty" and female "Dizzy") had no hunting experience. All were relatively domesticated and readily attacked both lambs and rabbits. The male "Sam" and female "Mary" were trapped as adults and were shy feral animals, each with an amputated forepaw, a result of the trapping. After several months of habituation to captivity Sam attacked and ate rabbits, but refused to attack lambs even when they were placed in his pen for hours. Mary would not attack either animal.

Our general procedure was to maintain the animals on one "safe food" and then present a "poison food" laced with lithium chloride or followed by an intraperitoneal injection of lithium chloride. In the hamburger experiment, four coyotes (Brujo, Coty, Luna, and Mary) were maintained on dog food and given fresh hamburger on two occasions. On the first hamburger trial the meat contained 6.0 g of lithium chloride in 31 capsules (No. 4 gelatin coated with a cellulose acetate compound). All the animals became ill and vomited after that meal (3). Three days later the animals were presented with unpoisoned hamburger and refused to eat the meat. After the hamburger test, they ate a full ration of dog food, indicating the aversion was specific to hamburger (see Fig. 1). During the attack experiment, each coyote was placed in an individual pen approximately 4 by 5 by 4 feet

high (1 foot = 0.3 m). The individual pens were located on the perimeter of a large grassy enclosure approximately 30 by 40 feet, where the animals were exercised daily. Latencies to attack, to eat, and to vomit after lithium treatment were timed by stop watch and photographed with still or motion picture cameras. During this period the coyotes were given a safe meal of rabbit every 48 hours and lamb was substituted on test days. Attacks on lambs and rabbits were usually fatal.

On day 1, a rabbit was released into the large enclosure in view of the coyote. The coyote pen door was opened and timing started for the attack latency. On day 3 this test was repeated with a lamb. On day 5, each coyote was given a bait packet of minced lamb flesh wrapped in a fresh woolly lamb hide. On day 7, they were tested with a rabbit which they attacked immediately and then consumed. On day 9, this test was repeated with a live lamb. Brujo and Coty immediately attacked and killed their lambs but the latency to begin feeding increased markedly (Brujo, from 270 seconds to 1200 seconds; Coty, from 600 seconds to 1680 seconds), and the rate of feeding was decreased. In contrast, Luna did not attack; instead she actively avoided the lamb. After 15 minutes the lamb was removed and a rabbit

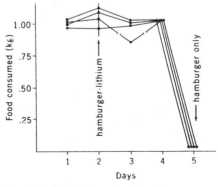

Fig. 1. Individual graphs, for four coyotes, of food consumed: dog food (days 1, 3, and 4), poisoned hamburger (day 2), and unpoisoned hamburger (day 5).

was presented. Luna immediately attacked the rabbit and subsequently ate it. Luna's attack data are summarized on line 1 of Table 1.

The two coyotes whose attacks were not suppressed vomited soon after eating the bait packet containing lithium (Brujo, 42 minutes; Coty, 30 minutes). Luna vomited later (80 minutes) and thus may have absorbed more lithium chloride. Therefore, on day 13 we gave Brujo and Coty a second lamb meat–lithium treatment · followed by an intraperitoneal injection of 2.5 g of lithium chloride in 100 ml of water (4). On day 15, the two coyotes were tested with a rabbit which they attacked immediately and then consumed. On day 17, they were presented with a lamb which they did not attack (Table 1, lines 2 and 3).

We then tested two more animals (Dizzy and Feisty) which had no prior experience with lithium illness. The procedures were essentially the same, except that dog food was the safe food provided on the intervening days and a rabbit carcass perfused with 5 g of lithium chloride in 50 ml of water was the poison food. The poison meal was followed by the immediate injection described above. The results paralleled the earlier studies. One treatment did not suppress attacks upon the rabbit but did suppress feeding during the 15-minute period. When presented with a dead rabbit during the second treatment, they mouthed it but did not actually eat it. They were then given a second injection. Subsequently they refused to attack the rabbit but immediately attacked the lamb (Table 1, lines 4 and 5). The feral male (Sam) would not attack a lamb, so he was tested with rabbits only on the same schedule. One meal of the toxic rabbit was sufficient to suppress his attack upon live rabbits (Table 1, line 6).

The behavior of the coyotes in the posttreatment tests and the extinction tests in which the averted animals were

presented at weekly intervals revealed some interesting patterns. After becoming sick on hamburger, they buried their vomitus. When presented with hamburger, Luna sniffed and licked it and then turned away. Brujo buried his hamburger. Coty urinated on her hamburger and repeated this behavior with the lamb she killed after the first flesh-lithium treatment.

After a single lamb flesh–lithium treatment, Luna loped up to the lamb, sniffed it, and circled back, hiding under a cage and vomiting within 1 minute. She then retreated into her pen. When the lamb followed her into her pen, she growled and snapped at the lamb, then gagged and retched, breaking off the attack with no injury to the lamb. Luna died several days later. Since she showed good activity and appetite following her last treatment, and her autopsy was negative, the cause of death was uncertain.

Coty made a lunge at the lamb after the second treatment but stopped abruptly, turned her back to the lamb, and spent the rest of the test period eating grass. Brujo also ate grass on his last three extinction trials. This behavior was observed only during posttreatment tests with lambs, thus it may be an example of "displacement behavior" often observed when motivation or drive is high but consummatory behavior is blocked. A week later, Coty attacked a lamb but interrupted her attack without killing. Brujo was tested for 8 weeks without a single attack.

The group conditioned with rabbit and lithium was also variable in extinction. Feisty attacked in 140 seconds after 4 weeks, Dizzy in 400 seconds after 2 weeks, and Sam in 120 seconds after 1 week. In the identical pattern, each coyote carefully first chewed off and consumed the ears, then ceased eating for about 30 minutes. Next, he ate the head and waited another half hour before returning to consume the

Table 1. The time taken by coyotes to attack a live rabbit and a live lamb before and after experimental treatment, which consisted of pairing lamb or rabbit flesh with illness-producing lithium chloride. No att., no attack.

Coyote	Before treatment		Treatment: LiCl paired with	After treatment	
	Rabbit	Lamb		Rabbit	Lamb
Luna	4 sec	1 sec	Lamb flesh (once)	1 sec	No att.
Brujo	1 sec	1 sec	Lamb flesh (twice)	1 sec	No att.
Coty	1 sec	1 sec	Lamb flesh (twice)	2 sec	No att.
Dizzy	1 sec	1 sec	Rabbit flesh (twice)	No att.	2 sec
Feisty	61 sec	6 sec	Rabbit flesh (twice)	No att.	2 sec
Sam	231 sec	No att.	Rabbit flesh (once)	No att.	No test

rest of the body. Before treatment, the coyotes attacked the back of the neck and began consuming the neck, head, and finally the ears without long periods between bouts of eating before commencing on the body. The animals appear to be reacquiring a taste for rabbit meat, as eating is now followed by nutritious effects without toxic effects.

We propose a two-phase conditioning process in mammals on the basis of previous work (5). In phase one, the flavor of food becomes aversive after one illness, after which the sights and sounds of the prey may still elicit attack but the aversive flavor inhibits feeding. Phase two occurs when the auditory, visual, and olfactory cues from the prey become associated with the now aversive flavor, thus subsequent attacks are inhibited and perhaps a second treatment is unnecessary. In some cases emesis may forge an association between vomited gustatory cues and odors of the vomitus which is sufficient for the second phase of conditioning.

For suppressing sheep predation, we could scatter baits that smell like sheep, taste like sheep, and contain a non-lethal emetic toxin. We could also perfuse carcasses of lambs and sheep with lithium. The predator could then be expected to subject himself to repeated trials until the flavor and the spoor of sheep becomes aversive. Subsequently, when foraging it would turn away before it sights sheep. Thus in

open range, the aversive effect may be much more durable than our extinction data indicate. Our coyotes had no other food and few other activity options in the small enclosure. The lambs persisted in following the coyotes, and occasionally a rabbit literally leaped into a coyote's jaws.

In addition, the feeding habits of the mother coyote averted to sheep might be transmitted to her pups, via flavor which her diet imparts to her milk, and by their early experience with the prey she brings to the den. Similar mechanisms have been demonstrated in the rat (6). This method should be effective against other predators, such as large cat species and eagles. Studies (7) indicate that birds form aversions to the visual as well as gustatory aspects of food, so the infused lamb may be the method of choice for eagles. Finally, since it is known that flavors are enhanced when beneficial effects follow ingestion (8), this method could also be used to change the food preferences of some species that are endangered because their naturally preferred food is diminishing owing to ecological change.

CARL R. GUSTAVSON
Department of Psychology, University of Utah, Salt Lake City 84112
JOHN GARCIA
WALTER G. HANKINS
KENNETH W. RUSINIAK
Departments of Psychology and Psychiatry, University of California, Los Angeles 90024

References and Notes

1. F. Wagner, "Coyotes and sheep," 44th Honor Lecture Faculty Association, Utah State University, Logan, January 1972.
2. For general reviews, see J. Garcia *et al.* and P. Rozin and J. W. Kalat, in *Biological Boundaries of Learning*, M. E. P. Seligman and J. Hager, Eds. (Appleton-Century-Crofts, New York, 1973).
3. One coyote (Mary) gulped the entire hamburger with 6.0 g of lithium chloride; the others gingerly separated out some of the capsules; however, they all ingested 3.0 g or more. The dose absorbed cannot be specified because of vomiting. Other tests indicate that for intraperitoneal injections of 0.12M LiCl, 100 ml induces vomiting and 250 ml produces a strong aversion in coyotes weighing 9 to 13 kg. The cellulose-covered capsules were designed to pass through the stomach into the intestine, thus avoiding ejection by vomiting.
4. To establish maximal learning in one or two trials and thus minimize attack testing, we employed the combination of (i) LiCl (6.0 g) treated food in case the animal vomited and reingested the vomitus and (ii) intraperitoneal injection to ensure an effective absorbed dose. Since 0.12M LiCl is similar in flavor to physiological saline, lithium treatment does not radically alter the flavor of the prey.
5. J. Garcia, J. C. Clarke, W. G. Hankins, in *Perspectives in Ethology*, P. P. G. Bateson and P. H. Klopfer, Eds. (Plenum, New York, 1973).
6. B. G. Galef and M. M. Clark, *J. Comp. Physiol. Psychol.* **78**, 220 (1972).
7. L. P. Brower, *Sci. Am.* **220**, 22 (February 1969); H. C. Wilcoxon, W. B. Dragoin, P. A. Kral, *Science* **171**, 826 (1971).
8. J. Garcia, R. F. Ervin, C. H. Yorke, R. A. Koelling, *Science* **122**, 716 (1967); K. W. Green and J. Garcia, *ibid.* **173**, 749 (1971); S. F. Meier, *Psychonomic Sci.* **17**, 309 (1969).
9. Supported in part by PHS grant 1RO1 NS 11041-01. We thank Ecodynamics for supplying and maintaining coyotes, and Buzz Moss for the use of his coyotes and his photography. Address correspondence to J.G., Department of Psychiatry, Neuropsychiatric Institute, University of California, Los Angeles 90024.

4 January 1974

LINCOLN P. BROWER and
JANE VAN ZANDT BROWER

42 Investigations into Mimicry

Just one hundred years ago, the great English naturalist H. W. Bates published a classical paper interpreting some of the observations he had made on mimicry in tropical American butterflies, which was based on eleven years of research in the Amazon Valley He presented his report to the Linnaean Society of London, and within a short time, mimicry became a key support of the theory of evolution by natural selection, which Charles Darwin and Alfred Russel Wallace had outlined to the society three years earlier, in 1858.

Bates had returned to England from South America with vast collections consisting of individuals of nearly 15,000 species, no less than 8,000 of which were then new to science. Contemplating the facts he had gathered, Bates saw that his observations contained unique evidence in support of Darwin's and Wallace's new theory. His own hypothesis concerned mimetic analogies, which were, in the naturalist's own words, "resemblances in external appearance, shapes, and colours between members of widely distinct families. . . ."

His idea of mimicry was based particularly on the tropical butterflies of the family Heliconiidae. In South America he had noticed that among the myriad insects these common butterflies with brilliant coloration and slow, conspicuous flight were never eaten by the abundant birds and lizards of the jungle. Surprisingly, other butterflies of a second, quite unrelated family appeared superficially identical to the Heliconiids. To explain this curious relationship, Bates proposed that the common Heliconiids were distasteful to insect-eacting vertebrate predators. Upon trying one, Bates postulated, a bird would find it unpleasant and would reject it. The bird would then remember this bad experience, and when encountering the butterfly's bright color pattern again would refuse the insect on sight. Now, if a palatable species looked like a common, distasteful one, the predator would mistake the palatable butterfly for the unpalatable insect. Thus, a palatable butterfly could, by deception, escape being eaten. Such a deceptor is called the mimic, and the distasteful species it resembles is termed the model. Bates also discovered that the mimic is much rarer than the model. This is reasonable because, in learning what is good to eat, birds would most often attack the common model, and so learn quickly that its color pattern was associated with an unpleasant taste. Mimicry would thus be effective.

The question of whether or not vertebrate predators, and especially birds, really do eat butterflies became a subject of controversy among naturalists, because neither ornithologists nor entomologists had recorded many observations of birds attacking these insects in nature. Two camps developed. One stated flatly that birds never eat butterflies, that any similarities in appearance of unrelated butterfly species are due to chance, and that mimicry is a figment of man's imagination. The proponents of mimicry countered this argument by gathering extensive observations, particularly in the Tropics, where some kinds of birds feed heavily on butterflies during certain seasons of the year. These men also found that model butterflies they had

caught in their nets often bore V-shaped notches in their wings or V-shaped scars where the powdery scales normally covering the wings had been removed. It was soon discovered that birds sometimes chased and snapped at other individuals belonging to the model butterflies' species, but then let them go. And when these butterflies were caught, it became clear immediately that the torn or rubbed areas on the wings were caused by the birds' beaks. The beak marks, besides showing that birds do attack butterflies, also provided indirect evidence that the models were unpalatable: they were tried but then rejected. Moreover, recent experiments by H.B.D. Kettlewell, in England, have shown conclusively that birds prey heavily on species of moths that had rarely been seen to be eaten by birds. Similarly, in the past two years we have observed a flock of red-winged blackbirds, *Agelaius phoeniceus* (Linné), catching and devouring many large swallowtail butterflies in a swamp in south central Florida. These findings that birds do eat Lepidoptera in nature point to the danger of drawing false conclusions based on insufficient observations.

Shortly after the turn of the century, some observers attempted to perform experiments with caged predators, including birds, lizards, and monkeys, to which model butterflies were offered. Almost invariably, the brightly colored models proved to be unpalatable. Without doubt, the greatest co-ordinator of these observational studies of mimicry was Sir Edward Poulton, for many years Hope Professor at Oxford University in England. It was Poulton who, while lecturing in 1909 to the Entomological Society of America, drew attention to three relatively simple instances of mimicry among North American butterflies, one of which will be described below, and pointed out the opportunities for research they offered. For in spite of a vast literature about observational studies, actual experimental evidence for the existence of mimicry was strikingly inadequate, as

A. E. Emerson, the famous American ecologist, emphasized as recently as 1949.

As though the lack of conclusive experimental evidence for mimicry were not enough of a problem, the very origin of mimicry puzzled scientists at first. While Bates' hypothesis offered an explanation of the incredible resemblances between a number of insects, considerable time passed before the mechanisms of the early evolution of mimicry began to be understood. As mentioned previously, a mimic appears superficially to resemble its model, but in fact the two are not closely related. When a mimic butterfly is examined in detail, the careful observer can see that in characteristics such as the number and the arrangement of wing veins, the structure of the legs, or the sexual organs, it is like other members of its own family and not like the model butterfly at all. How, then, does it happen that in color pattern alone the mimic is altered so as to differ radically from its relatives and resemble the model? This can be explained by reference to a well-known example of mimicry in the North American butterflies. The familiar monarch and viceroy frequent summer gardens throughout the eastern United States, and related pairs of butterflies occur in Florida and in the Southwest. The monarch, *Danaus plexippus* (Linné), is the model, and it is known from experimental studies to be distasteful to some birds, such as the Florida scrub jay, *Aphelocoma coerulescens coerulescens* (Bosc). The viceroy, *Limenitis archippus archippus* (Cramer), is the palatable mimic. Both model and mimic are orange in ground color with black and white markings, and look remarkably alike. However, the viceroy's non-mimicking relatives are all basically blue-black in color. They include several species of *Limenitis* from the Northeast, the Rocky Mountain states, and the Far West. Because of the prevalence of blue-black coloration among the viceroy's relatives, it is considered the ancestral shade of the viceroy. The problem, then, is to account for a change in the viceroy from its probable original dark color to its present orange, mimetic coloration, which resembles the monarch's.

The study of heredity has disclosed that spontaneous, heritable changes called mutations occur in the germ plasm of all living things. These mutations may result in visible differences in characteristics, such as the color pattern of a butterfly's wing. Knowing this, one can start with the presumably dark viceroys, and reconstruct the manner in which their orange coloration has theoretically evolved. Suppose among the dark ancestral viceroys a few slightly orange individuals arose through mutation. These, along with the commoner, blue-black ones, would both make up the viceroy population. Now, if an experienced predator that had learned to avoid orange-colored monarchs were to come upon these viceroys, it would be more likely to eat the blue-black form of the viceroy than it would the pale orange one, because even a hint of orange would remind the predator of the distasteful monarch. Thus, more of the new, orange variant viceroys would survive to produce offspring than would the blue-black form, even though the sum total of individuals of both forms would be the same in each new generation. If we extend this situation by imagining not just one mutation for pale orange, but rather a series of small color and pattern changes occurring at intervals during a very long period of time, it can be seen that through the agency of discriminating predators, the color variant of the viceroy most like the unpalatable model would tend always to prevail. In this way, apparently, the present, orange viceroy, so like the monarch, has gradually evolved.

Butterflies belonging to the insect order Lepidoptera were the subject of Bates' classical theory. There are numerous other instances of striking resemblances between different kinds of insects. For example, many bees (Hymenoptera), noxious because of their stings, are models for harmless flies (Diptera) that look, act, and even sound very much like bees.

While conducting our research program in south central Florida at the Archbold Biological Station, we became interested in two instances of mimicry involving bees and their fly mimics. The first concerns the bumblebee, *Bombus americanorum* (Fabricius), which is mimicked by a robberfly, *Mallophora bomboides* (Wiedemann). The two insects are seen together quite frequently in fields where blooming plants of the pea family are found. Like the bee, the robberfly has a black and light color pattern, a plump, fuzzy body, and hairy legs. On the third pair of legs it even has two patches of light hairs that simulate the pollen baskets of the bumblebee. The second model-mimic pair that we studied is the honeybee, *Apis mellifera* Linné, and the dronefly, *Eristalis vinetorum* (Fabricius). The dronefly has a narrow, black marking along the middle of its back that creates the impression of the honeybee's "wasp-waist." The dronefly also has beelike yellow and black rings girdling its abdomen. The buzzing of the dronefly, as well as its habit of feeding along with honeybees at certain composite flowers, make it a very convincing mimic of the honeybee to the human observer.

We wanted to know if this similarity is also confusing to insect-eating predators. Are the bee models really noxious? Is it the sting that makes them so? And what will happen when a predator that has encountered a bee is then given a harmless fly mimic? In order to test the effectiveness of mimicry of the bumblebee and of the honeybee, we carried out laboratory experiments in which caged toads, *Bufo terrestris* (Bonnaterre), were used as predators. Known as the southern toad, this animal was a particularly good subject because it is a common insect eater in the southeastern United States and is abundant in the vicinity of the Archbold Biological Station, where we caught them for our experiments. The toads were taken to the laboratory and were confined singly in cubic cages twelve inches on a side. The bottom and back of each cage was made of plywood, the two sides and front of gray plastic screening, and the top was a removable piece of glass that allowed access to the inside. Cardboard partitions separated the cages so that the toads could not see one another, thus precluding the possibility that the behavior of one might influence another visually. Each cage was equipped with a three-quarter-inch-deep water dish.

In preparation for each day's test, which was conducted in the evening when the toads were naturally active, bumblebees and robberflies or honeybees and droneflies were collected in fields near the Archbold Station. They were stored in a cold room until needed, and then were anesthetized lightly with carbon dioxide so they could be handled easily and prepared for presentation to the toads. In addition to the models and mimics, we also needed edible insects that we knew were acceptable to toads. By presenting these insects to the toads, we could make certain that the animals were hungry enough to eat palatable insect food, even if they should reject a model or mimic. For the bumblebee-robberfly experiments we used large dragonflies, *Pachydiplax longipennis* (Burmeister), as edible insects, and for the honeybee-dronefly tests we gave the toads beetle larvae, called mealworms, *Tenebrio molitor* Linné, which have approximately the same bulk as the models and mimics.

After trying many different methods of presenting the insects to the toads, we finally settled upon a technique that proved very satisfactory. Toads will eat food only if it moves. To standardize the motion of the insects, each was strung with a fine needle and 50-gauge, gray cotton thread so that it could be suspended and moved in front of a toad, to be seized or rejected. The animal had thirty seconds to eat each food item lowered into its cage. If the food was not eaten, the thread was pulled up, withdrawing the insect. During an experiment, model, mimic, and edible insects were presented in such an order that the toad could not learn to anticipate

365

what was coming next. A sample random sequence for two successive days in the bumblebee experiment was as follows: edible, model; mimic, edible/mimic, edible; model, edible.

The experiments on the bumblebee-robberfly complex were conducted with six toads. Of these, three were experimental animals and three were control animals. The experimental toads were given ten live bumblebees and ten dragonflies, singly, at the rate of four insects a day. At first, each readily seized a bumblebee, but in so doing was severely stung on the tongue and roof of the mouth. The toad reacted by making violent movements with its tongue, by blinking, by listing toward the side of the injury, by puffing up the body, and by ducking the head, which produced a generally flattened appearance. After one or two such experiences, the three experimental toads learned that bumblebees were noxious and they would not strike at the others that were offered. They consistently ate the dragonflies, however, showing that they could distinguish between noxious and edible insects. The crucial part of the experiment was the substitution of robberfly mimics for bumblebees. Would the toads eat the flies, or would they confuse them with the bumblebees and reject them? Of the thirty robberflies that were presented to the three experimental toads, only one robberfly (3 per cent) was eaten.

Meanwhile, the three control toads were playing a key role in the experiments. They were used to determine the actual edibility of the robberfly mimic. These toads were never exposed to the sting of a bumblebee. They were offered only robberflies and dragonflies. What were their reactions to the mimic? Two of the three ate the robberfly readily every time it was presented. The third toad was bumped in the face by the first robberfly it was given. This experience seemingly affected the toad, since it subsequently rejected all mimics, although, like the other two controls, it ate all the dragonflies. To summarize, then, of the fifty-one robberflies presented to the control toads, thirty-four (67 per cent) were eaten.

One further aspect of mimicry remained to be tested. Was the sting really the source of the bumblebee's noxiousness? It is quite easy—with the aid of a dissecting microscope and fine watchmaker's forceps—to remove a bee's stinging mechanism. First the bee is anesthetized with carbon dioxide. Then it is placed on its back on the stage of the microscope. The abdomen is pressed carefully with the forefinger, which causes the sharp point of the stinger to protrude. This is grasped with the forceps, and the whole stinging apparatus and the sac containing the liquid poison is pulled out. Thirty-six bumblebees were prepared in this way and were then presented to the three control toads. Twenty-four of the harmless bumblebees were eaten by the two toads that had eaten robberflies. The third toad, which had been bumped by the robberfly, rejected the twelve operated bumblebees presented to it. These results told us two things: first, that the sting was the source of noxiousness in bumblebees, and second, that a toad that refused robberflies would also reject bumblebees. This indicated that toads fail to distinguish between bumblebee models and robberfly mimics.

If we now look at the results of this experiment as a whole, we can reach several conclusions that answer the initial questions we set out to test. The experimental toads showed us that live bumblebees were highly noxious to them and that toads, after being stung, could learn to reject bumblebees on sight. When the robberfly mimic was offered instead of a bee, the toads also refused the mimic. However, two control toads that never experienced bumblebees ate robberflies readily. Therefore, we concluded that the rejection of the robberfly by the experimental toads could be attributed to mimicry. These toads apparently learned that the bumblebee's color pattern was associated with the noxious sting and confused the robberfly's coloration with that of the bumblebee. Mimicry was thus shown to be effective.

A second experiment was performed with honeybee models and their dronefly mimics. A large number of toads served as caged predators. We wanted to be able to compare the reactions of toads that ate mimics freely at the beginning of the experiment with the same toads' reactions to mimics after experiencing the model. In order to do this, it was necessary for each toad to pass a qualification test before being allowed to participate in the experiment; each had to begin by eating mimics as well as the edible mealworms. Of the sixty-seven toads that were brought in from the wild and tested, forty-four qualified. The other twenty-three did not, because they failed to eat mealworms or droneflies, or both. Fourteen of this group ate their mealworms but refused the first dronefly, and some also exhibited a reaction of rejection to the droneflies by ducking and puffing up. Without further experiments it is not possible to say conclusively why the toads rejected the initial mimics. However, their behavior suggested strongly that these animals were already agents in mimicry: that the toads had had experience with honeybees before being caught, and when they were confronted with mimics in the laboratory, they confused the two. This is possibly a valid conclusion, because all fourteen of the toads ate the mealworms, which shows that they were discriminating between the food items offered, and were not simply rejecting everything.

In addition to the qualification test to select only those toads that would initially eat the mimics, a second precaution was taken to insure that only visual mimicry was involved in the investigation. All the droneflies were killed by deep-freezing before they were presented to the toads. In this way, we eliminated the possibility that auditory mimicry, caused by the similarity of the buzz of the dronefly mimic and the honeybee model, might influence

TABLE 1

Two typical toads' reactions in honeybee-dronefly tests.

SUCCESSIVE INSECTS PRESENTED	EXPERIMENTAL TOAD (LIVE HONEYBEES WITH STINGING APPARATUS INTACT)	CONTROL TOAD (DEAD HONEYBEES WITH STINGING APPARATUS REMOVED)
First Day		
Mealworm	Eaten	Eaten
Dronefly	Eaten	Eaten
Dronefly	Eaten	Eaten
Mealworm	Eaten	Eaten
Honeybee	Eaten (Stung)	Eaten
Mealworm	Eaten	Eaten
Mealworm	Eaten	Eaten
Honeybee	Rejected	Eaten
Second Day		
Honeybee	Eaten (Stung)	Eaten
Mealworm	Eaten	Eaten
Mealworm	Eaten	Eaten
Honeybee	Rejected	Eaten
Honeybee	Rejected	Eaten
Mealworm	Eaten	Eaten
Dronefly	(Rejected)	(Eaten)
Mealworm	Eaten	Eaten

the toads' reactions. To begin the experiment, the forty-four qualified toads were divided into two groups: half were designated as experimental subjects, and the others as controls. The experimental animals were given a series of five live honeybees and five mealworms. The control toads were given five dead honeybees from which the stinging apparatus had been removed, and five mealworms. These bees had also been killed by deep-freezing to eliminate their buzzing, which in itself might have caused the controls to reject them. The order in which the insects were presented to all the toads in both groups and the reactions of a typical control toad and a typical experimental toad are shown in Table 1, upper left.

The experiments showed that the live

367

honeybees did indeed sting the experimental toads, although apparently not as severely as the larger bumblebees had stung the toads in the other tests. A few of the toads ate the live honeybees without evident discomfort, but most, after receiving two or three stings, rejected the honeybees on sight. The control toads, for the most part, readily ate their frozen bees from which the stinging mechanism had been removed. This shows that the sting of the honeybee, perhaps reinforced by its buzzing, accounts for its rejection as food by the experimental toads. Both groups of toads continued to eat mealworms throughout the experiment. This indicates that the experimental toads that rejected the bees actually had learned to tell the difference between the two. In the final test, both the controls and the experimentals were given a last dronefly mimic, followed by a final mealworm. All forty-four toads ate the mealworm, which indicates that they accepted insect food readily to the end of the experiment. The vital question is: did more control toads than experimental toads eat the last dronefly mimic? The results are summarized in Table 2, which shows that there was a striking difference between the reactions of the two groups of toads. Nearly all (86 per cent) of the control toads ate the final dronefly, whereas less than half (41 per cent) of the experimental toads ate their last dronefly. We can conclude that the experience of the experimental toads with live honeybees greatly reduced the likelihood that they would eat the dronefly mimics. To a large degree the noxious model did protect the mimic from being eaten. The experiment, therefore, offered strong evidence in favor of Bates' theory. The final, essential support, so long overlooked by naturalists, depended again upon the control toads. By eating the droneflies at the beginning and end of the experiment, they proved that the dronefly was palatable, and was thus a true, visual, Batesian mimic.

It is interesting to compare the results

TABLE 2

Numbers of experimental toads (fed intact, live honeybees) and control toads (fed dead honeybees, with stinging apparatus removed) eating and rejecting a final dronefly mimic.

	EXPERIMENTAL TOADS	CONTROL TOADS	TOTAL
Ate Final Mimic	9 (41%)	19 (86%)	28
Rejected Final Mimic	13 (59%)	3 (14%)	16
Totals	22 (100%)	22 (100%)	44

of the honeybee-dronefly experiments with those of the bumblebee-robberfly tests. Were the two models equally noxious? Were the mimics in each instance protected to the same extent from being eaten by the toads? The results showed that the three experimental toads attacked a total of nine out of fifty-one bumblebees (18 per cent), whereas in the experiments with honeybees, the twenty-two experimental toads seized seventy-seven out of one hundred and ten (70 per cent). This indicated that the bumblebee is more noxious to toads than is the honeybee, because fewer trials were needed to teach the toads to reject the former on sight. This is undoubtedly attributable, in part, to the bumblebees' being larger and possessing more toxic substance than do honeybees. In addition, bumblebees can inject the poisonous fluid into an attacker repeatedly, while honeybees can sting only once. Was this difference also reflected in the percentage of mimics eaten by the experimental toads in the two experiments? The results indicated that it was: only two out of a total of thirty robberflies (7 per cent) were attacked; however, nine out of twenty-two droneflies (41 per cent) were seized. These results fit one part of the mimicry theory very nicely. Students of natural selection have reasoned that the more noxious the model, the more protected the mimic would be from predation. The bumblebee and honeybee appear to exemplify, for the first time, a situation in which two levels of noxiousness do confer differing degrees of immunity to attack by toads.

These experiments are only the beginning of a vast amount of basic biological research necessary to understand more fully the evolution of mimicry. One particularly interesting but as yet unsolved problem concerns the possibility of an additional reason for the mimicry of the bumblebee by the robberfly. These flies are commonly known as bee killers because they prey on Hymenoptera by preference, although if bees are scarce, the flies will eat large beetles, bugs, and grasshoppers. In the course of our field research, we kept records of the kinds of prey that robberflies were seen to attack, and found that the bumblebee was the favorite food of *Mallophora bomboides*, the robberfly that resembles the bumblebee. The manner in which the robberfly seized its prey was precise and swift. It would perch on a stalk in a vertical position about one to three feet from the ground. As a bumblebee rose from feeding at a nearby blossom, the fly rapidly flew to the bee from above and behind it and grasped the dorsum with its long hairy legs. Then the robberfly immediately drew the bumblebee toward its body, inserted its mouthparts into the bee's thorax, and injected a substance that paralyzed the bee almost instantaneously. The fly then returned to a stalk, often the same one from which it had begun the attack, and, in a vertical position once more, proceeded to digest the prey externally by pumping its digestive juices into the bee. This caused the soft tissues of the bee to liquefy, and the fly sucked in the resulting fluid. The feeding process took approximately five to ten minutes to

complete. Afterward, the empty exoskeleton of the bumblebee might be left adhering to the plant stem.

We have suggested that the mimicry of the bumblebee by the robberfly may facilitate the mimic's exploitation of its model as food. This means that bumblebees would tend to defend themselves more successfully against those forms of the robberfly that least resemble it. On the other hand, those robberflies that closely imitate the bumblebees would tend not to be noticed until it was too late for a bee to defend itself or to escape. This would favor the survival of robberflies that look like bumblebees, and thus would bring about the evolution of mimicry. Experiments to test this idea have not yet been conducted. A possible procedure would be to confine the insects in a room-size screen cage. Thus, one could observe large numbers of a robberfly species that closely mimics the bumblebees actually attacking them. Then, robberflies of a species that does *not* resemble bumblebees could be introduced into a similar cage with bumblebees. The relative success of both robberfly species in attacking bumblebees could then be compared. Batesian mimicry, which was described in the toad experiments, and the suggested aggressive mimicry, both favor the enhancement of the resemblance between the mimic and the model. There is no reason why the two selective forces could not work together and be cumulative in their effect.

Further experiments on the degree of protection afforded the mimic by a very noxious model in comparison to one less so, the duration of memory of the predators, and their discriminatory ability are a few of the important points to be studied in the laboratory and in the natural environments of the animals. Complicated and fascinating groups of tropical insects, and even Bates' classical butterflies, still remain to be investigated.

43

Maternal Influence in Learning by Observation in Kittens

Abstract. *Kittens who observed their mothers perform a stimulus-controlled response (lever pressing to a visual stimulus for food) acquired and discriminated that response sooner than kittens who observed a strange female cat's performance. Kittens exposed to a trial and error condition never acquired the response. Initial differences in attentiveness to demonstrator performances disappeared by the second day. "Altruism" (food sharing) and other forms of social behavior were exhibited by both mother and stranger demonstrators.*

In several animal species, including man, mothers care for their young for a long time after birth. During this time, the young develop sensory and motor functions and acquire skills which are necessary for survival. The mother's role in teaching her young a specific skill, such as acquisition of food, has often been observed (1) but has not been experimentally demonstrated. Several investigators have suggested that infant mammals may learn from their mothers (2), and from their elders (3), primarily by observation. We have previously shown that learning by observation in adult cats is a more efficient method of learning than conventional shaping procedures (4). In this study, we undertook to determine whether the speed and efficiency of observation learning is improved by the use of a mother cat as demonstrator.

The subjects were 18 kittens, all between 9 and 10 weeks old when observation began. Each kitten lived with its mother and littermates in a home or homelike laboratory environment, or both, from birth until the end of the experiment. Group I consisted of six kittens who observed their mother's performance (M kittens); group II consisted of six kittens who observed the same strange female's performance (S kittens); group III consisted of six kittens exposed to a trial and error condition (TE kittens). The members of a given litter were randomly distributed to at least two of these three groups,

and where possible, to all three groups. All littermates began testing on the same day. Five female demonstrator cats (three mothers and two strangers) were used. Their task performances were equivalent and practically without error throughout the experiment.

The task was a lever press performed within 20 seconds after onset of a flickering light (4 cycle/sec). The lever was made of plexiglass and extended 12.5 cm beyond the front panel of a standard operant conditioning cage. A plexiglass partition divided the cage evenly into a demonstrator and observer compartment. A dipper that delivered a blended mixture of milk and meat was located 3.75 cm away from the lever in the demonstrator compartment.

After being familiarized with the cage, a kitten that had been deprived of food for 24 hours was placed in the demonstrator compartment alone and given one "free" food reward. The demonstrator cat (mother or strange female) was then introduced and performed ten stimulus-controlled lever presses. Although both M and S kittens had physical access to the food during these ten observation trials, they generally did not eat at this time. In fact, the occasional one, or at most two, rewards eaten by an M or S kitten during these ten trials, does not seem to constitute a determinant in their motivation or attentiveness. After these observation trials, the kitten was removed to the adjacent observer compartment for the opportunity to ob-

370

serve 30 more lever presses. The number of times the kitten oriented toward (paid attention to) the demonstrator cat was recorded for the 40 observation trials.

The demonstrator cat was then removed, and the kitten was placed back in the demonstrator compartment. Using a blind procedure, an assistant presented ten randomly spaced trials of the visual stimulus. This overall procedure was repeated daily until the kitten had pressed the lever in eight of the ten trials. When this occurred, it was given 20 additional presentations. When the kitten achieved 90 percent criterion for these 30 trials, acquisition was considered to have taken place and it was removed. Thirty trials were then presented daily, without further observation, until stimulus discrimination was achieved. Discrimination was decided to have taken place when the kitten made five or fewer interstimulus presses each day for three consecutive days. No kitten remained just below criterion in acquiring the response. Every kitten stabilized at or above the criterion level. All kittens were tested for 30 days or until they had discriminated the response.

The TE kittens were subjected to the same procedure except that no demonstrator cat was present. A TE kitten received one "free" food reward in the demonstrator compartment and ten presentations of the stimulus, after which it was placed in the observer compartment for 30 trials. During this time in the observer compartment, the stimulus was presented at random intervals and was terminated with the sound of the food dipper, as if a demonstrator cat were performing. The kitten was then placed back in the demonstrator compartment and presented with ten trials.

This procedure was abbreviated any time a kitten started to press the lever spontaneously during the first ten observation trials when it had access to the lever. The demonstrator cat (if any) was removed, the 30 additional observation trials were bypassed, and the kitten was tested alone. Three M kittens and three S kittens achieved criterion performance in this way.

The M kittens acquired the lever-pressing response faster (median of 4.5 days) than did S kittens (median of 18.0 days) (Fig. 1). One M kitten performed the response at criterion on the first day after observing 29 demonstrator performances. A second M kitten spontaneously performed the response at criterion on the second day, after having observed 16 demonstrator performances on the first day. Two S kittens never acquired the response. No TE kitten ever acquired the response. Once lever pressing was achieved, M kittens brought it under stimulus control within a median of 3.5 days as compared to 14.0 days for S kittens. The M kittens never fell below acquisition criterion once it was reached; two of the four S kittens did so briefly before they discriminated the response.

Kittens acquire and discriminate a lever-pressing response more rapidly and efficiently by observing their mothers than by observing a strange female or by a trial and error procedure. Such rapid learning on the part of M kittens, occurring with relatively little prior reinforcement or practice, suggests that some unique representational process is operative during their observation period. However, it is likely that a representational process also exists in S kittens. Despite the variable rate with which M and S kittens acquired the response, if and when the response appeared, it was accompanied and defined by specific and identical behavior in all kittens: (i) Both M and S kittens made their initial lever presses at criterion with a directness, sureness, and minimum latency indicative of informationally motivated behavior. For example, the average latency of the first

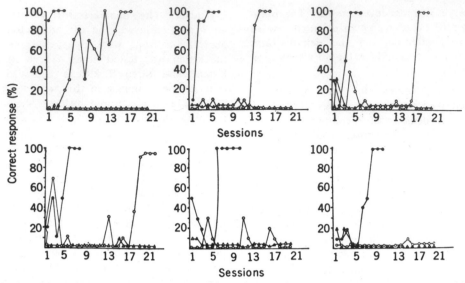

Fig. 1. Acquisition of an approach response (lever press) by observation learning in 18 kittens. Solid circles, kittens who observed their mothers (M kittens); open circles, kittens who observed strangers (S kittens); triangles, trial and error (TE) kittens.

lever press made on the first day of response acquisition was 3.5 seconds for both M and S kittens. (ii) Both M and S kittens were similarly attentive—in terms of body orientation and eye movements —to those demonstrator performances that directly preceded their own response acquisition. (iii) With one exception, both M and S kittens had a characteristically sharp response-acquisition curve (Fig. 1). All observing kittens acquiring the response moved from lever pressing at or below 50 percent to lever pressing at 90 percent or criterion as a step function. Thus, whereas the speed, efficiency, and success of response acquisition and discrimination were influenced by whether the kitten observed his mother or a strange female, when the response appeared it was invariably accompanied by the above behavior.

The mother may function as a more effective demonstrator for several reasons. These include her having nursed the kittens, having provided contact proximity, having some kind of maternal "teaching instinct" (5), providing

a still lactating and therefore stimulating or arousing presence during the observation period (6), and providing a familiar and therefore rewarding or relaxing presence during the observation period. All or any combination of the above might constitute a social or affective bond that enhances learning by observation.

Perhaps response acquisition depends on the existence of or, in the case of S kittens, on the eventual formation of an affective or social bond with the demonstrator. In fact, both M and S kittens displayed what are considered friendly relations (7) with the demonstrator cat. Both mothers and strangers were generally nonaggressive toward the kittens, licked them, and exhibited "altruistic" behavior by pressing the lever and either sharing or allowing the kitten to eat the entire reward. Also, whereas M kittens observed a mean of 16 demonstrator performances on the first day, as compared with a mean of 7 for the S kittens, this initial difference in attentiveness disappeared by the second day, when M kittens observed a mean of 18 demonstrator performances and S kittens a mean of 16. This suggests that any distraction caused by the strange demonstrator's presence was quickly reduced or eliminated.

In conclusion, these data show that a mother cat may function as an important vehicle for information transmission, via observation. Perhaps the suggested primacy of learning by observation in the adult cat (8) and in other mammals (9), as opposed to trial and error learning or operant conditioning, stems from the particular social and biological responses developed in the infant by a period of mother-dependence (10).

PHYLLIS CHESLER

Brain Research Laboratory, New York Medical College, New York and Department of Psychology, Richmond College, City University of New York, Staten Island 10301

References and Notes

1. R. G. Burton, The Book of the Tiger (Hutchinson, London, 1933); R. F. Ewer, Nature 222, 698 (1969); R. F. Ewer, Z. Tierpsychol. 20, 570 (1963); K. R. C. Hall, Brit. J. Psychol. 54 (3), 201 (1963); K. Imanishi, Psychologia 1, 47 (1957); P. Leyhausen, Z. Tierpsychol. Beineft, 2, (1956); H. L. Rheingold, Ed., The Maternal Behavior of Mammals (Wiley, New York, 1963); G. B. Schaller, The Deer and the Tiger (Univ. of Chicago Press, 1967); T. C. Schneirla, J. S. Rosenblatt, E. Tobach, in The Maternal Behavior of Mammals, H. Rheingold, Ed. (Wiley, New York, 1963); E. F. V. Wells, Lions, Wild and Friendly (Viking Press, New York, 1934); C. Wilson and E. Weston, The Cats of Wildcat Hill (Duell, Sloan and Pearce, New York, 1947).
2. Z. Y. Kuo, J. Comp. Psychol. 11, 1 (1930); ibid. 25, 1 (1938); see also K. R. C. Hall (1) and K. Imanishi (1).
3. E. J. Corner, Proc. Roy. Inst. 36, 1 (1955); H. W. Nissen and M. P. Crawford, J. Comp. Physiol. Psychol. 22, 283 (1936).
4. E. R. John, P. Chesler, I. Victor, F. Bartlett, Science 159, 1489 (1968).
5. R. F. Ewer, Nature 222, 698 (1969).
6. One of the mothers responded to nursing attempts by her kittens by walking away or by cuffing them. She continued to press the lever, indifferent to the kitten's meowing.
7. H. Winslow, J. Comp. Psychol. 37, 297 (1944).
8. H. A. Adler, J. Genetic Psychol. 86, 159 (1955); M. J. Herbert and C. M. Harsh. J. Comp. Psychol. 37, 81 (1944); see also E. R. John et al. (4).
9. A. L. Bandura, in Nebraska Symposium on Motivation, M. R. Jones, Ed. (Univ. of Nebraska Press, Lincoln, 1962); ———, in Advances in Experimental Social Psychology, S. Berkowitz, Ed. (Academic Press, New York, 1965), vol. 2; J. A. Corson, Psychon. Sci. 7, 197 (1967); M. P. Crawford and K. W. Spence, J. Comp. Psychol. 27, 133 (1939); K. K. Hayes and C. Hayes, J. Comp. Physiol. Psychol. 45, 450 (1952); W. N. Kellogg and L. A. Kellogg, The Ape and the Child (Hafner, New York, 1967 reprint of the 1933 edition); D. Mainardi and A. Pasquali, Soc. Ital. Sci. Natur. Milan. 107, 2 (1968); C. J. Warden and T. A. Jackson, J. Genet. Psychol. 46, 103 (1935); C. J. Warden, H. A. Fjeld, A. M. Koch, ibid. 56, 311 (1940).
10. K. R. C. Hall, Brit. J. Psychol. 54 (3), 201 (1963).
11. I thank E. R. John, A. Rabe, N. Chesler, N. Jody, F. Burgio, and P. Walker for their assistance. Supported by PHS grant MH 08579 and by the New School for Social Research.

30 June 1969; revised 10 September 1969

44

THEORETICAL NOTE

REINSTATEMENT [1]

BYRON A. CAMPBELL AND JULIAN JAYNES

Princeton University

Reinstatement is defined as periodic partial repetition of an experience such that it maintains the effects of that experience through time. This principle is demonstrated in a developmental study on the effects of early fear in rats, and is then discussed in relation to clinical and developmental theory.

[1] This research was supported in part by Public Health Service Grant M-1562 from the National Institutes of Mental Health and by National Science Foundation Grant GB 2814.

In most of the phyla from arthropods to man early experience exerts a multiplicity of effects on adult behavior (Beach & Jaynes, 1954; Scott, 1962). Sometimes such effects are the simple persistence in adult behavior of habits formed early in life. In other instances it may be that early experience influences later behavior by structuring the individual's perceptual or response capacities. And in still others, there is a critical period of development during which some aspect of behavior, on which later behaviors depend, is learned and molded for life.

In this paper we suggest yet another mechanism. Although obvious and disarmingly simple, it yet seems to the authors of such neglected importance as to warrant this note and the coining of a term for it. By *reinstatement* we denote a small amount of partial practice or repetition of an experience over the developmental period which is enough to maintain an early learned response at a high level, but is not enough to produce any effect in animals which have not had the early experience. The following experiment is meant as a demonstration of this phenomenon in a commonly studied instance of learning.

Method

The subjects were 30 albino rats of the Wistar strain born and raised in the Princeton colony. They were divided into three groups of 10 each, with an equal number of males and females in each. The apparatus used was one commonly used in fear experiments (Campbell & Campbell, 1961). It consisted of two compartments separated by a door, a black one with a grid floor, and a white compartment with a solid metal floor. Shock could be administered to the grid of the black compartment. To two of the three groups an early fear-arousing experience was given in the black compartment. This consisted of placing the rat just after weaning, when approximately 25 days old, on the grid side of the apparatus with the door fixed so that the rat could not escape, then giving the rat 15 2-second 170-volt shocks on a 20-second variable interval schedule, taking aproximately 5 minutes, then removing the animal and placing him on the nonshock side for 5 minutes, and then repeating this entire procedure once. Thus each animal received a total of 30 shocks. At the end of this period the rat was removed and placed in a home cage. A third control group was run through this procedure without any shock being administered to the grid. During the next month a total of three shocks—the reinstatements—were given to one of the early experience groups and to the control group. These shocks were administered 7, 14, and 21 days after the original training session. The procedure was to administer, at some random number of seconds up to a minute after the animal was placed on the grid side of the apparatus, a single 2-second shock of the same intensity as before. The rat was then placed in the white compartment for an identical period of time and then returned to its home cage. On alternate weeks the animal was placed first on the nonshock side of the cage and then on the shock side, with half of the animals being placed on the shock side for the first reinstatement procedure and half on the safe side. Otherwise this procedure was precisely the same as the training procedure except that only 1 instead of 30 shocks was administered. The second pretrained group was given the same procedure except that no shock was administered. One week after the third reinstatement procedure, when the animals were 53 days of age, they were all tested for the effects of their early experience. This was done by placing them individually in the black compartment (where all of them had been shocked at one time or another) with the door removed so that the animal could run freely into the white compartment. The time spent in the white compartment over the ensuing hour was then recorded.

Results and Discussion

The results were unequivocal. As seen in Figure 1, the group that had received the early fearful experience followed by three 2-second shocks administered at weekly intervals, spent an increasing percentage of its time in the white compartment during the 1-hour test period, thus showing the effects of the early fearful experience with the black compartment. In contrast, the group that had had a similar early experience just after weaning, but no reinstatement of it in the intervening month, failed to show any significant fear of the black compartment, spending on the average all but about 10 minutes of the hour on that side. Similarly the group which had not had any early traumatic experience, but had received the three brief shocks over the month, failed to acquire any significant

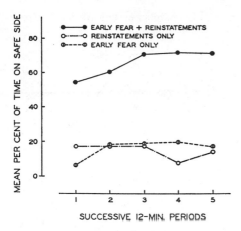

MEAN PER CENT OF TIME ON SAFE SIDE

● EARLY FEAR + REINSTATEMENTS
○ REINSTATEMENTS ONLY
◐ EARLY FEAR ONLY

SUCCESSIVE 12-MIN. PERIODS

FIG. 1. The effect of reinstatement of early fear on later behavior.

fear of the black compartment. The difference between the first group and the other two groups is, as it appears on the graph, highly reliable statistically ($p < .01$, Mann-Whitney U test).

There is nothing dramatically surprising about this finding. It is indeed what anyone thinking carefully about learning and practice would expect, namely, that there is some small amount of practice over certain time intervals which could maintain a previously learned response and yet not be enough to train naïve animals to perform that response. The possibility that this mechanism of reinstatement has wide and important applicability in the ontogeny of behavior in many vertebrate species seems beyond question.

In theoretical analyses of human growth and development traumatic events in infancy and childhood have long occupied a central, if controversial, role. In Freud's early analyses, traumatic events in childhood were considered a major cause of adult behavior disorders. With time, this view was gradually modified such that White writing in 1956 summed up current opinion by stating:

Undoubtedly it is true that some adult neuroses have their origin in violently frightening events. . . . The theory has long since been abandoned, however, that all neuroses, or even a majority of neuroses, take their start from traumatic events [1956, p. 238].

The early trauma theory has inconsistencies with certain facts of memory and learning as well. First, on a mere phenomenological level, we know that memory becomes more and more dim the further back into our childhood we try to remember. Second, in rats, the earlier in life that a fearful experience is given the animal, the more likely it is to be forgotten in adulthood (Campbell & Campbell, 1962). Third, in chickens, the earlier in the critical period that the chick is imprinted, the more likely it is to be forgotten when the animal reaches the juvenile stages (Jaynes, 1957). This evidence seems to indicate that the organism is constantly forgetting, time or neurological maturation or perhaps other processes constantly changing the mnemonic traces of events and feelings. And all the evidence suggests that the earlier the experience has occurred, the more profound and the faster the forgetting.

In this context reinstatement is proposed as a major mechanism by which the effects of early experiences can be perpetuated and incorporated into adult personality. Following an early experience, either pleasant or unpleasant, three developments may occur. First, the experience may be gradually forgotten as described above. Second, it may be remembered and persist indefinitely if it is occasionally reinstated. The language-based cultures of human societies are particularly rich in methods of such reinstatement, including ones so simple as occasionally reminding a child of a previous event or feeling. Even the child may occasionally reinstate the experience himself under the prompting of his ethical value system. A third possibility is the active repression of the experience, and we suggest here that the repression itself —as well as the experience—may undergo either forgetting or maintenance by reinstatement in exactly the same way. Again, the language-based cultures of man contain many reinstatement-of-repression mechanisms such as parental conversational taboos, etc., which determine what repressions are maintained into adult life. In a general sense, we propose that any learned response, whether acquired in infancy or adulthood,

375

conscious or unconscious, instrumental or autonomic, joyful or traumatic, can be maintained at a high level by an occasional reinstatement.

Moreover, reinstatement as a principle has considerable adaptive significance, particularly in the learning of fear. Young organisms, at least after a short initial period of apparent fearlessness in some species, become highly vulnerable to the acquisition of fears. These fears have, of course, great survival value in keeping the young organism away from danger. But if they all persisted and could not be forgotten, they would imprison the animal in his own prior experience, making adult adaptive behavior impossible. It is thus essential to adult activity that most early experiences be forgotten, and that only those experiences which are periodically reinstated by a particular habitat or culture be retained.

REFERENCES

BEACH, F. A., & JAYNES, J. Effects of early experience upon the behavior of animals. *Psychological Bulletin,* 1954, 51, 239–263.

CAMPBELL, B. A., & CAMPBELL, E. H. Retention and extinction of learned fear in infant and adult rats. *Journal of Comparative and Physiological Psychology,* 1962, **55,** 1–8.

JAYNES, J. Imprinting: The interaction of learned and innate behavior: II. The critical period. *Journal of Comparative and Physiological Psychology,* 1957, 50, 6–7.

SCOTT, J. P. Critical periods in behavioral development. *Science,* 1962, **138,** 949–958.

WHITE, R. W. *The Abnormal Personality.* (2nd ed.) New York: The Ronald Press, 1956.

(Early publication received March 21, 1966)

R. A. HINDE

45 Energy Models of Motivation

Introduction

The problem of motivation is central to the understanding of behaviour. Why, in the absence of learning and fatigue, does the response to a constant stimulus change from time to time? To what is the apparent spontaneity of behaviour due? This paper is concerned with one type of model which has been developed to help answer such questions—namely that in which changes in the organism's activity are ascribed to changes in the quantity or distribution of an entity comparable to physical, chemical or electrical energy.

Such models have been developed by theoreticians with widely differing backgrounds, interests and aims, and the frameworks of ideas built round them diverge in many respects; but in each case the energy treatment of motivation is a central theme (cf. Carthy, 1951; Kennedy, 1954). They have had a great influence on psychological thought, and although they are unlikely to continue to be useful, it is instructive to examine their nature, their achievements and their limitations.

The Models

The four models or theories to be discussed here are those of Freud, McDougall, Lorenz and Tinbergen. They are only four of many in which energy concepts are used, but in them the energy analogy is made explicit in terms of a mechanical model, instead of being merely implied by a 'drive' variable which is supposed to energize behaviour. The models were designed to account for many features of behaviour in addition to the phenomena of motivation, and here it will be necessary to extract only those aspects relevant to the present theme.

In the psycho-analytic model (Freud, 1932, 1940) the id is pictured as a chaos of instinctive energies which are supposed to originate from some source of stimulation within the body. Their control is in the hands of the ego, which permits, postpones or denies their satisfaction. In this the ego may be dominated by the super-ego. The energy with which Freud was particularly concerned—the sexual energy or libido—is supposed not to require immediate discharge. It can be postponed, repressed, sublimated, and so on. The source of this energy lies in different erogenous zones as the individual develops, being successively oral, anal and phallic, and it is in relation to these changes that the individual develops his responses to the external world. The instinctual energy is supposed to undergo various vicissitudes, discussions of

which often imply that it can be stored, or that it can flow like a fluid. It may become attached to objects represented by mental structures or processes (libidinal cathexes) and later withdrawn from them in a manner that Freud (1940) likened to protoplasmic pseudopodia: it has also been compared with an electric charge. Thus some of the characteristics of the energy depend on its quantitative distribution.

McDougall (1913) envisaged energy liberated on the afferent side of the nervous system, and held back by 'sluice gates.' If the stimuli necessary to open the gates are not forthcoming, the energy 'bubbles over' among the motor mechanisms to produce appetitive behaviour. On receipt of appropriate stimuli, one of the gates opens, and the afferent channels of this instinct become the principal outlet for all available free energy. Later (1923) he used a rather more complex analogy in which each instinct was pictured as a chamber in which gas is constantly liberated. The gas can escape via pipes leading to the executive organs when the appropriate lock(s) is opened. The gas is supposed to drive the motor mechanisms, just as an electric motor is driven by electrical energy.

The models of Lorenz and Tinbergen have much in common with McDougall's. Lorenz's 'reaction specific energy' was earlier (1937) thought of as a gas constantly being pumped into a container, and later (e.g. 1950) as a liquid in a reservoir. In the latter case it is supposed that the reservoir can discharge through a spring-loaded valve at the bottom. The valve is opened in part by the hydrostatic pressure in the reservoir, and in part by a weight on a scale pan which represents the external stimulus. As the reservoir discharges, the hydrostatic pressure on the valve decreases, and thus a greater weight is necessary to open the valve again.

Tinbergen (1951) pictured a heirarchy of nervous centres, each of which has the properties of a Lorenzian reservoir. Each centre can be loaded with 'motivational impulses' from a superordinated centre and/or other sources. Until the appropriate stimulus is given the outflow is blocked and the animal can show only appetitive behaviour: when the block is removed the impulses can flow into the subordinate centre or be discharged in action.

It is important to emphasize again that the theories of these authors have little in common except for the energy model of motivation—they were devised for different purposes, and the more recent authors have been at pains to emphasize their differences from the earlier ones. For instance, for McDougall the most important feature of instinct was the 'conative-affective core', while for Lorenz it was the stereotyped motor pattern. Furthermore, the models differ greatly in the precision with which they are defined. The Freudian model is a loose one: its flexibility is perhaps necessary in view of the great range of behavioural and mental phenomena it comprehends, but makes it very difficult to test. The other models are more tightly defined, but differ, as we shall see, in their supposed relations to the nervous system.

In spite of such differences, all these models share the idea of a substance, capable of energizing behaviour, held back in a container and subsequently released in action.[*] In the discussion which follows, I shall be little concerned with the other details of the models, or with the ways in which the theories based on them differ. Furthermore, I shall disregard the niceties of terminology, lumping instinctual energy, psychophysical energy, action specific energy and motivational impulses together as, for present purposes, basically similar concepts.

Reality Status of the Models

Until recently, students of the more complex types of behaviour could get little

[*] It will be clear that in some respects the postulated entity has the properties of a material substance, rather than energy. However it is on the 'energy' properties of flowing and 'doing work' that the models primarily depend.

help from physiology, and had to fashion their concepts without reference to the properties of the nervous system. Many, indeed, advocated this course from preference, either on grounds of expediency, suggesting that knowledge of the nervous system was still too primitive and might be misleading, or on principle, claiming that physiology and behaviour were distinct levels of discourse. At present the models and theories used in attempts to understand, explain or predict behaviour range from those whose nature is such that physiological data are irrelevant (Skinner, 1938) to those which consist of a forthright attempt to describe psychological data in physiological terms (Hebb, 1947, 1955). The former type may be applicable over a wide range of phenomena, but only at a limited range of analytical levels: the latter may point the way to analysis at lower levels, but their expectancy of life depends on their compatibility with the phenomena found there.

The originators of all the models discussed here regard them .s having some relation to structures in the nervous system, but vary in the emphasis which they lay on this. Tinbergen, although freely emphasizing the hypothetical status of his model, clearly regards his 'centres' as neural structures, and his 'motivational impulses' as related to nerve impulses. He speaks of his hierarchical scheme as a 'graphic picture of the nervous mechanisms involved'. McDougall likewise regards the relationship between model and nervous system as a close one, for he localizes the 'sluice gates' in the optic thalamus. Lorenz, on the other hand, usually treated his model in an 'as if' fashion—he did not suggest that we should look for reservoirs in the body. He did, however, bring forward physiological evidence in its support —quoting, for instance, Sherrington's work (1906) on spinal contrast, and von Holst's (1936) work on endogenous rhythms in fishes; and he sometimes uses such terms as 'central nervous impulse flow' as a synonym for 'reaction specific energy'. His use

378

of physiological evidence was, however, *post hoc*—the model was based on behavioural data and the physiological evidence came later.

Freud's model developed from physiology, in particular from a sensory-excitation-motor-discharge picture of nervous function, and its basic postulates are almost a direct translation of such ideas into psychological terms—excitation into mental energy, the discharges of excitation into pleasure, and so on (Peters, 1958). However, Freudian theory developed far beyond these primitive notions, and then bore little or no relation to physiology, even though the instincts were supposed to have an ultimately physiological source.

Thus two of these models (Tinbergen and McDougall) had explicitly physiological implications; that of Lorenz was usually used in an 'as if' fashion; and Freud's, although it had physiological roots, became divorced from any supposed structures or functions in the nervous system. However, as we shall see, all have been influenced by the covert introduction of existence postulates concerning the explanatory concepts used.

The Relation of Behavioural Energy to Physical Energy

In these theories the concept of energy, earlier acquired by the physical sciences from everyday observation of behaviour, is reclaimed for use in its original context. In accounting for the organism's changing responsiveness, the theorist is concerned with its capacity for doing work and an energy concept seems an obvious choice. The use of such a concept, however, brings with it the temptation of ascribing to the behavioural energy the various properties of physical energy. Thus it may be said to flow from point to point, or to exist in more than one form (bound or free, in Freudian theory). It is of fundamental importance to the theorist to recognize that such properties are additional postulates in the behaviour theory: because behavioural energy is postulated to account for the activity of organisms, it *need* share no other properties with the energy postulated to account for the movement of matter. The distinction is particularly important in that students of behaviour, while using a concept of behavioural energy to explain the behaviour of the whole animal, may simultaneously be concerned in establishing bridgeheads with physiologists, who use energy in a manner closely similar to the physicists.

Freud ascribed many of the properties of physical energy to mental energy, which was stored, flowed, discharged, and so on, but he did recognize the importance of distinguishing between them. Thus he wrote (1940) 'We have no data which enable us to come nearer to a knowledge of it [mental energy] by analogy with other forms of energy'. The use of the phrase 'other forms' is revealing, and finds an echo in the work of the psychoanalyst Colby (1955), who discussed this question in detail and elaborated an even more complex energy model. Colby regards mental energy as a postulated form of energy in addition to mechanical, thermal, electrical and chemical, and states that it *'does not disobey'* the principles formulated for other forms of energy, though conversion into these other forms is not 'yet' possible. It appears that for him psychic energy would be expected to obey the Laws of Thermodynamics but for the fact that organisms are open systems. Colby, however, is clearly ambivalent on this issue, for elsewhere he emphasizes that psychic energy is not mechanical, thermal, chemical or electric, and writes 'Perhaps we have no right to speak of energy at all'. Another analyst (Kubie, 1947), with a highly sophisticated attitude to energy concepts, draws a sharp distinction between psychic and physical energies, although he does invest them with similar properties:

'It is therefore scientifically necessary to keep clearly in mind the fact that the psychodynamics dealt with in psychoanalysis refer to something which is loosely analogous to, but still very far from, the exacter field of thermodynamics. These psychodynamics deal with an effort to estimate *(a)* the sources of energy, *(b)* the kinds of quantities of energy, *(c)* the transformations of energy, one into another, and *(d)* the distributions of energy. But the "energy" referred to here means not what is intended by the physicist, but *simply apparent intensities of feelings and impulses, or in psychoanalytic terms "the libido".'*

McDougall was less reserved than Freud on this issue, and clearly regarded his 'psycho-physical energy' as a form of physical energy. 'We are naturally inclined to suppose that it is a case of conversion of potential energy, stored in the tissues in chemical form, into the free or active form, kinetic or electric or what not; and probably this view is correct.' He further suggested that there is a positive correlation between the flow of energy and the 'felt strength' of the impulse.

With Lorenz and Tinbergen this question was concealed. Lorenz's reaction specific energy, pictured as a liquid, was clearly only distantly related to physical energy. Since the model was primarily an 'as if' one, the question of conversion to physical energy did not arise. Tinbergen's motivational impulses, although supposedly related to physical energy exchanges in the nervous system, were not regarded as physical energy sums themselves. In spite of this, a number of properties of physical energy came to be ascribed to them—thus they could be 'discharged in action,' 'stored,' 'released,' 'flow,' 'spark over' and so on (Hinde, 1956). This has undoubtedly influenced the course of research (see below).

Thus although, in the four schemes at present under consideration, the relation between behavioural energy and physical energy was not particularly close, some properties of the latter insinuated themselves almost unnoticed into the behavioural theories. Physical and behavioural energy have in fact often been confused in theories of motivation: for instance, Brown

(1953) uses as evidence for an energizing function of drives the 'marked disproportionality between the energy content of a stimulus and the energy expended in response'.

The transposition of properties from physical to behavioural energy could be helpful, suggesting new questions which open further avenues of research or coordinating previously unrelated facts. However, their presence is a danger, and likely to lead to sterile endeavour, if their nature is not recognized, and if they are introduced not as stated postulates but as known properties of physical energy which are therefore without further thought ascribed to a hypothetical behavioural energy. Some examples of confusions which have arisen in this way are discussed below.

Number of Forms of Energy Postulated

An important issue in these models is the number of forms of energy postulated. The behaviour of an organism is diverse: is a different form of energy to be postulated for each type of behaviour, or is there only one form producing behaviour differing according to the structure within which it acts?

Freud, in an early model, postulated two basic types of energy—sexual energy (libido) and energy pertaining to the self-preservative instinct.† Later, sexual and self-preservative were grouped together as 'Life' instincts, in contrast to the 'Death'

† If the difference between the energies of the sexual and self-preservative instincts is one of kind, then the dichotomy is distasteful to biologists, for there is no reason for supposing that different examples of sexual (in a broad sense) behaviour on the one hand, and self-preservative behaviour on the other, classified together on functional grounds, have anything causally in common. Lloyd Morgan (1912) made a similar dichotomy in his top level instincts of self-preservation and race maintenance, which also involves a confusion of functional and causal categories.

380

instincts, which become manifest when directed outwards as the instinct of destruction.‡ Within each major group of instincts are recognized component instincts differing in their source, aim, object and in their quantitative distribution: but it is not always clear whether these differences are thought to lie in the nature of the energy, or in the structure within which the energy acts. Sometimes Freud stated that the instincts differ primarily because of the differing quantities of excitation accompanying them (Freud, 1915), but later he often implied differences of quality in the energy itself (Freud, 1924). Colby (1955), who, as we have seen, associated the concept of behavioural energy closely with physical energy, emphasized that all energy must be neutral, its 'aim', etc. being acquired only when it acts through a structure.

McDougall is ambivalent on the number of sources of energy, stating the possibilities on the one hand that the instincts each have their own energy, and on the other that they all draw on a common supply: he inclines towards the second view.

Lorenz is concerned primarily with limited sequences of behaviour, and not with a synthesizing model of the behaviour of the whole organism. It was therefore sufficient for him to talk about action specific energy. From his earlier writings it was not clear whether the specificity of this energy was supposed to be due to its nature or to the structure within which it acts, but later, because of the occurrence of displacement activities (see below), he concluded that the specificity was due to the structure. Furthermore, Lorenz did not suppose that the reservoirs for functionally related activities were fed from a common source: he (1937) emphatically

‡ There remains, by implication, a third type of energy—Ego—which may oppose both sexual and destructive instincts. Freud thought of the Ego as deriving energy from the Id, but there are divergent views on this. Indeed there are numerous variants on this theme—some psychoanalysts, for example, postulate a neutral energy.

opposed McDougall's view of superordinated instincts which employ motor mechanisms as means to an end, regarding such instincts merely as functional categories. Rather Lorenz emphasized the individuality of each type of response, and ascribes to the external situation the integration of discrete responses into functional units.

Tinbergen's scheme differs from that of Lorenz in this matter, for the motivational impulses were supposed to descend the hierarchical system of nervous centres: each such system constituted an 'instinct'. The impulses were thus general at least to all the activities of one hierarchical system, and, since he regarded (1952) 'sparking over' from one system to another as possible, perhaps to all. Since Tinbergen suggests that each activity is supplied by motivational impulses both from the superordinated centre and from its own specific source, his scheme combines features from those of McDougall and Lorenz.

The importance of this question for energy theorists is emphasized by Thorpe (1956). Most behaviour is directive in the sense that variable means are used to a constant end. If all behaviour depends on one type of energy, then the directiveness must be a consequence of the structure in which the energy acts: since the motor patterns used to achieve a certain goal may be diverse, this seems to demand a fantastic complexity of structure. Thorpe therefore prefers to think of there being an element of directiveness in the drive itself—and thus prefers Lorenz's model, in which the energy is specific to the action, to Tinbergen's scheme, in which the motivational impulses flow down a hierarchical scheme containing a limited number of channels.

Utility and Level of Applicability of Energy Models

These energy models of motivation were developed with a minimum of reference to physiological data—they were intended for the understanding or prediction of behaviour from behavioural data.

It is therefore on this level that, in the first instance, they must be assessed. It has been said that it does not help to ascribe feeding behaviour to a feeding drive, or to feeding specific energy, any more than it helps to postulate a locomotive force to explain the progress of a railway engine. This type of criticism is based on a misunderstanding. Although the mere postulation of a locomotive force may be of little use, there is a level—that of classical dynamics—in which the language of forces and so on helps a great deal in predicting the behaviour of railway engines. We can say, for instance, what will happen if the engine meets a stationary truck on the line (see also McDougall, 1923).

In a similar way, energy models have been surprisingly successful. The Freudian energy model not only accounts for the more general properties of motivated behaviour, such as its apparent spontaneity and persistence, but also for the manner in which instincts can change their aim (displacement in the psychoanalytical, not the ethological, sense) and the way in which component instincts can replace each other. Similarly, Lorenz's reservoir analogy and Tinbergen's hierarchy comprehend the relation between the threshold of stimulation necessary to elicit a response and the time since it was last released; the occurrence of appetitive behaviour, responses to normally inadequate stimuli, and ultimately vacuum activities (i.e. responses in the absence of the appropriate stimuli) if the releasing stimuli are withhheld; the variations in intensity at which instinctive activities appear; the initial 'warming up' phase and the after-response shown by many responses, and many other aspects of changing responsiveness. Further, differences between the characteristics of response patterns can be related in an 'as if' fashion to differences in the dimensions of the reservoir. These models are thus of value in illustrating diverse properties of behaviour in a simple manner, and can be used for explanation and exposition. In addition, the analytical study of behaviour forms

only a first stage in its understanding—the products of the analysis must be re-synthesized so that the relations between them can be understood: for this such models can be an invaluable aid.

However, we have seen that some of the models purport to go further than this —they are not just 'as if' models of the mechanisms underlying the behaviour, but representations of those mechanisms themselves. They must thus be assessed also by their compatibility with our knowledge of the nervous system. Indeed even the 'as if' models must ultimately be assessed in this way, for only if the model is close to the original will the question it poses be relevant, and only then will it continue to be of service as analysis of the original proceeds. To take an example from physics, the ray theory of light, used originally for explanations of shadow-casting, etc., suggested questions (e.g. 'What is it that travels?') which paved the way for the corpuscular and wave theories. Although the latter is essential in some contexts (such as the explanation of diffraction), the ray theory retains its usefulness, for a treatment of shadow-casting in terms of the wave theory would be unnecessarily clumsy. Further, the two theories remain compatible with, and translatable into, each other (Toulmin, 1953). For similar reasons, it is important to assess these energy models of motivation not only at the behavioural level, but also in terms of their compatibility with lower ones. Although a model must not resemble the original too closely, or it will lose just those properties of simplicity and manipulability which makes it useful, it must approximate to it, or the questions it suggests will be irrelevant.

Difficulties and Dangers

In the following paragraphs we shall consider some of the difficulties and dangers inherent in the use of an energy model of motivation. These arise in part from misunderstandings of the nature of the model, and in part from incompatibilities between the properties of the model and those of the original.

(1) CONFUSION BETWEEN BEHAVIOURAL ENERGY AND PHYSICAL ENERGY

We have already seen that behavioural energy, postulated to account for changes in activity, need share no properties with physical energy. Not only is there no necessary reason why it should be treated as an entity with any of the properties of physical energy, but the question of its convertibility into physical energy is a dangerous red herring. The way in which the properties of the model may be confused with those of the original have been discussed for Freudian theory by Meehl & McCorquodale (1948). Concepts like libido or super-ego may be introduced initially as intervening variables without material properties, but such properties have a way of creeping into discussion without being made explicit. Thus Meehl & McCorquodale point out that libido may be introduced as a term for the 'set of sexual needs' or 'basic strivings,' but subsequently puzzling phenomena are explained in terms of properties of libido, such that it flows, is dammed up, converted, regresses to earlier channels, and so on. Such properties are introduced surreptitiously as occasion demands, and involve a transition from admissible intervening variables, which carry no existence postulates, to hypothetical constructs which require the existence of decidedly improbable entities and processes.

Such difficulties are especially likely to occur when a model which purports to be close to the original, like that of Tinbergen, develops out of an 'as if' model, like that of Lorenz. This case has been discussed elsewhere (Hinde, 1956). To quote but one example, ethologists have called behaviour patterns which appear out of their functional context 'displacement activities'. These activities usually appear when there is reason to think that one or more types of motivation are strong, but unable to find expression in action: instead, the animal shows a displacement activity, which seems to be irrelevant. Thus when a chaffinch has conflicting tendencies to approach and avoid a food dish, it may show preening behaviour. Such irrelevant activities were explained on the energy model by supposing that the thwarted energy 'sparked over' into the displacement activity—sparking over being a property of (electrical) energy which was imputed to the behavioural energy. This idea hindered an analytical study of the causal factors underlying displacement behaviour. Thus it has recently become apparent that many displacement activities are not so causally irrelevant as they appear to be, for those factors which elicit the behaviour in its normal function context are also present when it appears as a displacement activity. For example, some displacement activities appear to be due to autonomic activity aroused as a consequence of fear-provoking stimuli or other aspects of the situation. The displacement activity may consist of the automatic response itself (e.g. feather postures in birds) or of a somatic response to stimuli consequent upon autonomic activity (Andrew, 1956; Morris, 1956). In other cases the displacement behaviour consists of a response to factors continuously present, which was previously inhibited through the greater priority of the incompatible behaviour patterns which are in conflict (van Iersel & Bol, 1958; Rowell, 1959). Of course it remains possible that the intensity of the apparently irrelevant behaviour is influenced by factors not specific to it, including those associated with the conflicting tendencies (see also Hinde, 1959).

Similarly in psychoanalytical theory we find not only that within one category of instincts (e.g. sexual) the constitutent instincts can change their aim, but also that 'they can replace one another—the energy of one instinct passing over to another' (Freud, 1940). Explanations of this type may be useful at a descriptive level, but are misleading as analysis proceeds.

383

In all energy models, the energy is supposed to build up and subsequently to be released in action. McDougall, Lorenz and Tinbergen, all of whom were influenced by Wallace Craig, compare the releasing stimulus to a key which opens a lock. This apparent dichotomy between releasing and motivating effects is a property of the model, and may not be relevant to the mechanisms underlying behaviour. Although many factors appear to have both a motivating and a releasing effect on the responses they affect—they appear both to cause an increase in responsiveness, and to elicit the response—this does not necessarily imply that two distinct processes are at work. For example, if a given input increased the probability of a certain pattern of neural firing, it might appear in behaviour both that the responsiveness was increased and that the behaviour was elicited.

This sort of difficulty is the more likely to arise, the more precisely the model is portrayed. Thus McDougall, who did not work out his model in such detail as Lorenz and Tinbergen, implied that motivation and release were in fact one process when he wrote 'The evoking of the instinctive action, the opening of the door of the instinct on perception of its specific object, increases the urgency of the appetite'.

(III) IMPLICATIONS ABOUT THE
CESSATION OF ACTIVITY

In all these theories, the cessation of activity is ascribed to the discharge of energy—the behavioural energy flows away as a consequence of performance. Influenced by the analogy with physical energy, Freud held that the main function of the nervous system is to reduce excitation to the lowest possible level. McDougall, Lorenz and Tinbergen imply a similar view, and the two latter emphasize that it is the performance of more or less stereotyped motor patterns which involve the discharge of the energy.§

This view of the cessation of activities comes naturally from models in which the properties of physical energy are imputed to behavioural energy. It is, however, also supported by another type of argument, also involving a *non sequitur*. Much behaviour is related to an increase in stimulation. Therefore, it might be argued, all activity is due to an increase in stimulation, and cessation of activity is related to a decrease. On an energy model, stimulation may increase the energy, and thus decrease in activity is related to a decrease in energy.

Such a view is incompatible with the data now available on two grounds. First, cessation of activity may be due to the animal encountering a 'goal' stimulus situation, and not to the performance of an activity. If this goal stimulus situation is encountered abnormally early, the behaviour which normally leads to it may not appear at all. McDougall recognized this, and indeed defined his instincts in terms of the goals which brought about a cessation of activity. This, however, made it necessary for him to be rather inconsistent about his energy model. While the energy was supposed to drive the motor mechanisms, it was apparently not consumed in action, but could flow back to other reservoirs or to the general source. The more precisely described Lorenz/Tinbergen models, on the

§ In doing so they did not imply that behavioural energy is converted into physical energy—thus Tinbergen (1952) suggests that even sleep is an activity, in which, presumably, behavioural energy is discharged.

The view that a fall in responsiveness is normally due to the performance of a stereotyped activity is not a necessary consequence of the use of energy models—Freud and McDougall did not suggest that energy could be discharged only in this way—but their use makes such errors more likely. Lorenz and Tinbergen were apparently also influenced by over-generalizing from the observation that performance of some stereotyped activities leads to a fall in responsiveness, to the conclusion that all falls in responsiveness are due to such activities.

other hand, do not allow for reduction in activity by consummatory stimuli: reduction in responsiveness occurs only through the discharge of energy in action. These models are misleading because they are too simple—energy flow is supposed to control not only what happens between stimulus and response, but also the drop in responsiveness when the response is given. In practice, these may be due to quite different aspects of the mechanisms underlying behaviour: for instance the energy model leaves no room for inhibition (Kennedy, 1954). Further, even if the cessation of activity is in some sense due to the performance, many different processes may be involved: the mechanism is not a unitary one, as the energy model implies (see below).

Secondly, if activity is due to the accumulation of energy and cessation to its discharge, the organisms should come to rest when the energy level is minimal. In fact, much behaviour serves the function of bringing the animal into conditions of increased stimulation. This has been shown dramatically with humans subjected to acute sensory deprivation—the experimental conditions are intolerable in spite of the considerable financial reward offered (Bexton, Heron and Scott, 1954). Energy theories are in difficulty over accounting for such 'reactions to deficit' (Lashley, 1938).

(IV) UNITARY NATURE OF EXPLANATION

In these energy models, each type of behaviour is related to the flow of energy. Increase in strength of the behaviour is due to an increased flow of energy, decrease to a diminished flow. The strength of behaviour is thus related to a single mechanism. It is, however, apparent that changes in responsiveness to a constant stimulus may be due to many different processes in the nervous system and in the body as a whole —for instance, the changes consequent upon performance may affect one response or many, may or may not be specific to the stimulus, and may have recovery periods varying from seconds to months. Energy

models, by lumping together diverse processes which affect the strength of behaviour, can lead to an over-simplification of the mechanisms underlying it, and distract attention from the complexities of the behaviour itself. Similarly, energy models are in difficulty with the almost cyclic short-term waxing and waning of such activities as the response of chaffinches to owls, the song of many birds, and so on.

Kubie (1947) has emphasized this point with reference to the psychoanalytic model. Changes in behaviour are referred to quantitative changes in energy distribution, but in fact so many variables are involved (repression, displacement, substitution, etc.) that it is not justifiable to make easy guesses about what varied to produce a given state. Similar difficulties in relation to other models have been discussed by Hinde (1959).

Precht (1952) has elaborated the Lorenzian model to allow for some complication of this sort. Analysing the changes in strength of the hunting behaviour of spiders, he distinguishes between 'drive' which depends on deprivation, and 'excitatory level', which is a function of non-release of the eating pattern. The distinction is an important one, but it may be doubted whether the elaborate hydraulic system which he produced is really an aid to further analysis.

Tinbergen's model translated the Lorenzian reservoir into nervous 'centres'. Changes in response strength are ascribed to the loading of these centres. Now for many types of behaviour it is indeed possible to identify *loci* in the diencephalon whose ablation leads to the disappearance of the behaviour, whose stimulation leads to its elicitation, and where hormones or solutions produce appropriate effects on behaviour. There is, however, no evidence that 'energy' is accumulated in such centres, nor that response strength depends solely on their state. Indeed the strength of any response depends on many structures, neural and non-neural, and there is no character by character correspondence between such postulated centres and any structure in the brain.

Although the greatest attraction of these energy models is their simplicity—a relatively simple mechanical model accounting for diverse properties of behaviour —there is a danger in this, for one property of the model may correspond to more than one character of the original. This difficulty has in fact arisen in many behaviour systems irrespective of whether they use an energy model of motivation. Thus a single drive variable is sometimes used not only with reference to changes in responsiveness to a constant stimulus, but also to spontaneity, temporal persistence of the effects of the stimuli, after-responses (i.e. the persistence of activities after the stimulus is removed), the temporal grouping of functionally related activities, and so on. As discussed elsewhere (Hinde, 1959), there is no *a priori* reason why these diverse characters of behaviour should depend on a single feature of the underlying mechanism: an over-simple model may hinder analysis.

(v) INDEPENDENCE OF ACTIVITIES

Another difficulty which arises from the use of energy models, though by no means peculiar to them, is due to the emphasis laid on the independence of different activities. Lorenz & Tinbergen (1938) write 'If ever we may say that only part of an organism is involved in a reaction, it may confidently be said of instinctive action'. Activities are interpreted as due to energies acting in specific structures, and not as responses of the organisms as a whole. Both types of attitude carry disadvantages, but an over-emphasis on the independence of activities leads to a neglect of, for instance, sensory, metabolic or temperamental factors which affect many activities.

Is an Energy Concept Necessary?

We have seen that these energy models will account for diverse properties of behaviour, but that they meet with serious difficulties when the behaviour is analysed more closely. They have also been strangely sterile in leading to bridgeheads with physiology. These shortcomings of energy models have been emphasized by a number of other writers (e.g. Kubie, 1947; Deutsch, 1953; Bowlby, 1958). Energy concepts are useful in descriptions of changes in behaviour, but are they necessary? Colby states that 'a dynamic psychology must conceive of psychic activities as the product of forces, and forces involve energy sums. It is thus quite necessary that metapsychology have some sort of energy theory'. Is this really so?

Kubie (1947) has pointed out that psychological phenomena are the product of an interplay of diverse factors. A rearrangement of these factors can alter the pattern of behaviour without any change in hypothetical stores of energy. Such a view is in harmony with the known facts about the functioning of the nervous system. The central nervous system is not normally inert, having to be prodded into activity by specific stimuli external to it. Rather it is in a state of continuous activity—a state supported primarily by the non-specific effects of stimuli acting through the brainstem reticular system. Factors such as stimuli and hormones which affect specific patterns of behaviour are to be thought of as controlling this activity, of increasing the probability of one pattern rather than another. Changes in strength or threshold can thus be thought of as changes in the probability of one pattern of activity rather than another, and not as changes in the level of energy in a specific neural mechanism. This involves some return to a 'telephone exchange' theory of behaviour, but with emphasis on the non-specific input necessary to keep the switch mechanism active, and with switches which are not all-or-none, but determine the probability of one pattern rather than another. Furthermore, switching does not depend solely on external stimuli—i.e. we are not concerned with a purely reflexological model. This is not the place to pursue this view further: it suffices to say that it seems possible and preferable to formulate behaviour theories in which concepts of energy, and of drives which energize behaviour, have no role.

Summary

1. Phenomena of motivation have often been explained in terms of an energy model.

2. The energy models used by Freud, McDougall, Lorenz and Tinbergen are outlined briefly.

3. The extent to which these models are considered by their authors to correspond with structures in the nervous system is discussed.

4. The relation between physical energy and the postulated behavioural energies are examined.

5. The number of forms of energy postulated by each author is discussed.

6. These models have had considerable success in discussions of the behaviour of the whole animal.

7. They have, however, certain grave disadvantages. In particular, these arise from a confusion between the properties of physical and behavioural energy, and from attempts to explain multiple processes in terms of simple unitary mechanisms.

8. It seems doubtful whether an energy concept is in fact necessary at all.

Acknowledgments

I am grateful to Drs. J. W. L. Beament, John Bowlby, Charles Kaufman and W. H. Thorpe for their comments on the manuscript.

REFERENCES

ANDREW, R. J., 1956 Some remarks on conflict situations, with special reference to *Emberiza* spp. *Brit. J. Anim. Behav.* 4, 41–45.

BEXTON, W. H., W. HERON, and T. H. SCOTT, 1954 Effects of decreased variation in the sensory environment. *Canad. J. Psychol.* 8, 70–76.

BOWLBY, J., 1958 The nature of the child's tie to his mother. *Internat. J. Psychoanalysis,* 39.

BROWN, J. S., 1953 Problems presented by the concept of acquired drives. In: *Current Theory and Research in Motivation,* a symposium. Univ. of Nebraska.

CARTHY, J. D., 1951 Instinct. *New Biology,* 10, 95–105.

COLBY, K. M., 1955 *Energy and Structure in Psychoanalysis.* New York: Ronald Press Co.

DEUTSCH, J. A., 1953 A new type of behaviour theory. *Brit. J. Psychol.* 44, 304–317.

FREUD, S., 1915 Instincts and their vicissitudes. *Collected Papers,* Vol. IV.

FREUD, S., 1923 *The Ego and the Id.* London: Hogarth Press, 1947.

FREUD, S., 1924 The economic problem in masochism. *Collected Papers,* Vol. II, XXII.

FREUD, S., 1932 *New Introductory Lectures on Psychoanalysis.* London: Hogarth Press, 1946.

FREUD, S., 1940 *An Outline of Psychoanalysis.* New York. London: Hogarth Press, 1949.

HEBB, D. O., 1947 *The Organization of Behaviour.* New York: Wiley.

HEBB, D. O., 1955 Drives and the C.N.S. (Conceptual Nervous System). *Psych. Rev.* 62, 243–254.

HINDE, R. A., 1956 Ethological models and the concept of drive. *Brit. J. Philos. Sci.* 6, 321–331.

HINDE, R. A., 1959 Unitary drives. *Anim. Behav.* 7, 130–141.

HINDE, R. A., 1960 (In press.)

HOLST, E. VON, 1936 Versuche zur Theorie der relativen Koordination. *Pflüg. Arch. ges. Physiol.* 237, 93–121.

IERSEL, J. J. A. VAN, and A. C. BOL, 1958 Preening of two tern species. A study on displacement. *Behaviour* 13, 1–89.

KENNEDY, J. S., 1954 Is modern ethology objective? *Brit. J. Anim. Behav.* 2, 12–19.

KUBIE, L. S., 1947 The fallacious use of quantitative concepts in dynamic psychology. *Psychoanalytic Quart.* 16, 507–518.

LASHLEY, K. S., 1938 Experimental analysis of instinctive behaviour. *Psychol. Rev.* 45, 445–471.

LLOYD MORGAN, C., 1910 *Instinct and Experience.* London: Methuen.

LORENZ, K., 1937 Über die Bildung des Instinktbegriffes. *Naturwiss* 25, 289–300, 307–318, 324–331.

LORENZ, K., 1950 The comparative method in studying innate behaviour patterns. *Sym. Soc. Exp. Biol.* IV, 221–268.

LORENZ, K., and N. TINBERGEN, 1938 Taxis und Instinkthandlung in der Eirollbewegung der Graugans. *Z. Tierpsychol.* 2, 1–29. Translated in: C. H. SCHILLER, 1957. *Instinctive Behaviour*. London: Methuen.

McDOUGALL, W., 1913 The sources and direction of psychophysical energy. *Amer. J. Insanity*. Not consulted. Quoted in McDOUGALL, 1923.

McDOUGALL, W., 1923 *An Outline of Psychology*. London: Methuen.

MEEHL, P. E., and K. McCORQUODALE, 1948 On a distinction between hypothetical constructs and intervening variables. *Psych. Rev.* 55, 95–107.

MORRIS, D., 1956 The feather postures of birds and the problem of the origin of social signals. *Behaviour* 9, 75–113.

PETERS, R. S., 1958 *The Concept of Motivation*. London: Routledge and Kegan Paul.

PRECHT, H., 1952 Über das angeborene Verhalten von Tieren. Versuche an Springspinnen. *Z. Tierpsychol.* 9, 207–230.

ROWELL, C. H. F., 1959 The occurrence of grooming in the behaviour of Chaffinches in approach-avoidance conflict situations, and its bearing on the concept of "displacement activity." *Ph.D. Thesis*. Cambridge.

SHERRINGTON, C. S., 1906 *Integrative Action of the Nervous System*. New York: Scribner.

SKINNER, B. F., 1938 *The Behavior of Organisms*. New York: Appleton Century.

THORPE, W. H., 1956 *Learning and Instinct in Animals*. London: Methuen.

TINBERGEN, N., 1951 *The Study of Instinct*. Oxford.

TINBERGEN, N., 1952 Derived activities: their causation, biological significance, origin and emancipation during evolution. *Quart. Rev. Biol.* 27, 1–32.

TOULMIN, S. E., 1953 *The Philosophy of Science*. London: Hutchinson.

PART 7

Evolution and Function of Behavior, and Social Behavior

In Reading 3, Tinbergen's four basic questions regarding animal behavior were reviewed. These questions concern the causation, development, evolution, and function (survival value) of behavior. Causation and development have been the topics of most of the previous readings, although problems in the evolution and function of behavior have also been covered. But now we turn to readings that emphasize questions of evolution and function, as well as the related topic of social behavior.

Much of the animal behavior that we observe results from an individual's interaction with members of its own species, social behavior. This behavior exists in several diverse groups of animals because it has biological adaptiveness. Thus, questions concerning the evolutionary past and present function of the behavior are particularly appropriate in this area. For this reason, ethologists have been interested in social behavior, and much of their research has been concerned with it. Many of the readings included in Part 7 are examples of such research.

The section begins with a most interesting discussion of the evolution of tool-use by feeding animals. As author John Alcock notes, such comparative studies may contribute to an understanding of the evolution of our own tool-using and tool-making species.

The next two readings concern the function of particular behavior patterns. In Reading 47, Niko Tinbergen, who has successfully championed research on survival value, describes a series of experiments concerned with the black-headed gull's removal of empty eggshells from its nest. The aspects of the shell resulting in its removal are determined by the use of models, a frequent technique in ethological research. More work with gulls is reported in Reading 48, again using models to study the survival value of a behavior pattern. In this case, the curious attraction of gull prey toward their predators is examined.

The term sociobiology has recently become popular, primarily because of E. O. Wilson's book by that name. Reading 49 is a good example of sociobiological research, much of which is descriptive and analytical, rather than experimental. Authors J. F. Eisenberg et al. describe selected data on primate social systems and present modifications of existing concepts about the adaptive value of these systems. They find that one-male social systems are much more common than was previously supposed. Comparing these uni-male systems with age-graded-male and multi-male social structures, the authors then relate the social structures to the ecology of the various species.

Older than the concept of sociobiology is that of territoriality, an important aspect of social behavior in the lives of many animals. The functional advantages of holding a territory have been the subject of considerable speculation. Territoriality may act as an antipredator device; it may aid in the protection of the young; it may reduce the likelihood of the spread of disease. The spacing out of individuals by the establishment of territories may insure enough food, nest material, or space to conduct reproductive displays, or it may increase a territory-holding male's chances of successful mating. In Reading 50, Stephen Emlen describes territorial behavior in a species of bullfrog in which a male who holds a territory close to other territory-holding males has an increased chance of mating.

Not all animals establish individual territories, and for those who don't, dominance hierarchies may develop within a group occupying a common space. As with territoriality, these hierarchies may have different functions in different species. Reading 51 describes a most unusual function of dominance, that of controlling the biological sex of the members of the group.

One of the most important recent developments in the study of animal behavior has been the great increase in the number of field studies such as those described in Reading 52. Field studies are particularly important for threatened species. In Reading 52, Valerius Geist summarizes some of his observations on the North American mountain sheep. The author points out that the social behavior of this species is a major reason why their numbers are not increasing. Historically, the study of such social behavior in animals assumed great importance after Darwin pointed out the continuity of animal life, including the human being. While the structures of several different animal societies have been used as models to explain human society, as it is or as it is supposed to be, it quickly became apparent, as observations accumulated, that a great variety of animal societies exist. Even among nonhuman primates there are major differences from species to species. Moreover, anthropological studies have shown large differences among various human cultures. Yet, are these differences only variations on a theme? Are there underlying similarities such as territoriality, dominance relationships, and, consequently, aggressive responses? Several scientists think so and have written books, designed for the general public, in which the findings of research on animal behavior have been extrapolated to man. In our final reading, Niko Tinbergen discusses two of these books in a paper appropriately titled "On War and Peace in Animals and Man."

46 THE EVOLUTION OF THE USE OF TOOLS BY FEEDING ANIMALS

JOHN ALCOCK

Department of Psychology, University of Washington, Seattle, Washington 98195

Received October 8, 1971

Reprinted from EVOLUTION, Vol. 26, No. 3, November 14, 1972 pp. 464–473

The purpose of this paper is to examine cases of tool-using behavior by animals feeding under natural conditions and to discuss the origin, transmission, and subsequent evolution of this behavior. Hall's important paper (1963) on this subject dealt largely with weapon using by animal species. Van Lawick-Goodall's broad ranging review (1970) is devoted entirely to birds and mammals with heavy emphasis on the chimpanzee. I wish to focus attention on feeding tools and to examine these cases from an evolutionary perspective.

In addition, I plan to explore the significance of tool-using behavior by lower animals for an understanding of the evolution of this trait by humans. Hall's central purpose was to debunk the idea that all examples of tool-using are in some way homologous. He was especially successful in demonstrating the illogic of the belief that because humans are intelligent, adaptable tool-users therefore other species which happen to use tools must also possess special intelligence and adaptability. However, Hall, perhaps in reaction to the many ludicrously anthropomorphic misconceptions surrounding the topic, argues that non-human tool-using behavior is virtually irrelevant for an understanding of human evolution. I hope to make a modest case for the opposite position.

A DEFINITION

Students of tool-using have had a great deal of difficulty deciding just what should be included in this category. Hall's paper (1963) is based on a definition which embraces such things as the use of other animals to obtain a goal. Givens (1963) objects, properly I think, that this definition

embraces almost all social behavior. Gruber (1969) feels that the use of natural objects "as if they were tools" does not constitute true tool-using which he would restrict entirely to humans because only this species skillfully and habitually alters objects for use as tools. This seems unnecessarily anthropocentric. Just when object manipulation by primates in agonistic displays becomes tool-using has been hotly debated by Washburn and Jay (1967) and Kortlandt (1967). Finally such diverse behavior as string-pulling by birds (Thorpe, 1963), nest building by primates and birds (Lancaster, 1968), and web building and net making by spiders and caddis fly larvae as well as case construction by caddis fly larvae (Wheeler, 1930) have been offered as potential candidates for the tool-using category. Van Lawick-Goodall (1970) notes that string-pulling and the like are really only skillful manipulation of objects and materials and not tool-using.

Recognizing then that any definition of tool-using behavior is unlikely to satisfy everyone, I prefer one similar to that of Millikan and Bowman (1965) and van Lawick-Goodall (1970). Tool-using involves the manipulation of an inanimate object, not internally manufactured, with the effect of improving the animal's efficiency in altering the position or form of some separate object. This definition would exclude all aspects of social behavior and such things as the construction of spider webs.

The list of tool using species which follows is meant to be exhaustive although I recognize the possibility that I have overlooked some rare cases of tool-using behavior.

ANIMAL SPECIES WHICH USE TOOLS WHILE FEEDING UNDER NATURAL CONDITIONS

Myrmeleon spp. (ant-lions, Neuroptera). *Lampromyia* spp., *Vermilio* spp. (worm-lions, Diptera), Wheeler (1930).

Sand particles are showered on prey walking by or at the top of a small sand pit constructed by a larva which resides at the bottom of the pit. The sand is propelled by a head throwing movement and has the effect of knocking prey to the bottom of the pit.

Toxotes jaculatrix (and other members of the genus?, archer fish), Luling (1958, 1963).

Water droplets are shot at terrestrial prey from the mouth of the fish knocking insects and spiders off vegetation onto the surface of the water where they can be easily captured.

Cactospiza pallida (woodpecker finch), Lack (1947), Eibl-Eibesfeldt (1961).
Cactospiza heliobates (mangrove finch), Curio and Kramer (1964).
Certhidea olivacea (warbler finch), Hundley (1963).
Geospiza sp.? (ground finch), Hundley (1963).

Twigs, cactus spines, and petioles are used to dislodge prey hidden in crevices and cavities of dead and decaying trees. The trait is performed frequently by *C. pallida*, occasionally by *C. heliobates*, and very infrequently by the other two species.

Hamirostra melanosterna (black-breasted buzzard), Chisholm (1954).

There are several reports of this hawk carrying rocks and dirt lumps into the air and dropping them onto eggs, especially of the ground nesting emu. The birds feed on opened eggs.

Neophron percnopterus (Egyptian vulture), van Lawick-Goodall and van Lawick (1966).

Some vultures have been seen and photographed taking rocks in their beaks and hurling them onto ostrich eggs. They then feed upon the broken eggs.

Sitta pusilla (brown-headed nuthatch), Morse (1968).

This bird has been observed on several occasions to use a piece of pine bark to dislodge other pieces of bark from trees and to feed upon insects uncovered by this action.

Enhydra lutris (sea otter), Kenyon (1959), Hall and Schaller (1964).

Stones and shells are collected from the ocean floor, carried to the surface, placed on the animal's chest while it floats on its back, and used as an anvil to crack open mussels and other hard shelled molluscs.

Macaca fascicularis (crab-eating macaque), Chiang (1967).

These animals sometimes use leaves to wipe the surface of potential food items removing sticky or noxious substances.

Saimiri sciureus (squirrel monkey), Kortlandt and Kooij (1963).

This species was once seen using a stick to dislodge ants from some fruit.

Pongo pygmaeus (orang-utan), Kortlandt and Kooij (1963).
Gorilla gorilla (gorilla), Kortlandt and Kooij (1963).

These species have been seen once or twice using a branch to rake in fruit they could not easily reach.

Cebus sp.? (cebus monkey), Thorington (1968), in Lancaster.
Cercocebus sp.? (mangabey), Kortlandt and Kooij (1963).
Colobus sp.? (colobus monkey), Kortlandt and Kooij (1963).
Papio? sp.? (baboon), Kortlandt and Kooij (1963).
Pan troglodytes (chimpanzees), Suzuki (1966), van Lawick-Goodall (1968, 1970), Jones and Pi (1969).

Individuals of all these species have been observed using pieces of wood, twigs, or branches as probes usually at ant or termite nests. However, with the exception of the chimpanzee, reports of this behavior are very brief and involve only one or two animals. In the case of the chimpanzee, tool-using is not exceptional. Animals in several widely dispersed populations are known to use fairly large sticks to probe ant, termite and bee nests. Those insects or the honey that clings to the tools are eaten. In addition, the chimps studied by van Lawick-Goodall also use fine grass stems and twigs which they insert into the tunnels of the nests of some termite species. In fact, they may on occasion actually modify a grass stem or twig constructing a more efficient tool for collecting termites.

Van Lawick-Goodall (1970) describes these activities in more detail and provides a few other possible cases of tool-using by feeding wild animals.

DISCUSSION

Preadaptations for Tool-Using

Those animals which use tools seem preadapted for this behavior. For example, many primates have grasping hands, behavior patterns which involve object manipulation, a high level of exploratory and learning abilities all of which may be seen in various ways as a foundation for tool-using behavior (Hall, 1963; Washburn, 1963; Goodall, 1964; van Lawick-Goodall, 1968; Jay, 1968). The sea otter's habit of feeding on its back provides it with a flat surface for rock pounding (Hall and Schaller, 1964). Egyptian vultures perform egg-throwing movements with eggs of suitable sizes, movements which are very similar to those employed in rock throwing (van Lawick-Goodall and van Lawick, 1966; Brown and Urban, 1969). Darwin's finches collect and manipulate small twigs in constructing a nest (Lack, 1947). Finally ant-lions and worm-lions perform head tossing movements to throw out sand in the formation and maintenance of their pit traps (Wheeler, 1930). These examples should

be sufficient to illustrate a variety of possible preadaptations possessed by tool-using species. However, many other animals have similar adaptations or other traits which could conceivably be utilized in the manipulation of objects as tools. The question is, how can one plausibly account for the origin of tool-using behavior in the very few species which actually practice it?

The Origins of Tool-Using

The evolution of tool-using behavior can be divided into three stages: (1) the initial appearance of the trait in a population of animals, (2) the initial transmission of the trait through the population, and (3) the subsequent evolution of the behavior over time.

Here I shall argue that most examples of the use of tools *could* have had their origin in the novel use of a pre-existing behavior pattern. These behavior patterns may have been performed in a way that more or less accidentally involved the use of objects as tools in special situations. What follows is acknowledged to be highly speculative.

Ant-lions and worm-lions.—The larvae of these species remove sand from their prey traps in sand dunes and other areas with a stereotyped head tossing movement. They also periodically remove sand which has blown into their traps (Wheeler, 1930). All that would be required for the origin of sand throwing at prey is an individual with a low threshold for the performance of the action. A passing prey or a struggling one which dislodged some sand grains might trigger sand throwing, be struck by some particles, and be captured by the pioneer tool-users in ant-lion and worm-lion populations.

Archer fish.—This fish produces a jet of water droplets simply by rapidly closing its gill covers, an action which *mechanically* forces the water in its gill chambers through a narrow channel between the tongue and the roof of the mouth. To capture prey on vegetation overhanging the water, the fish maneuvers into position very close to the water's surface, assumes an almost perpendicular pose, and then contracts its muscles shutting its gill covers forcefully. Adult fish can accurately strike objects three or four feet above the water (Luling, 1963).

It is difficult to imagine what the initial form of such a highly specialized trait might be. However, Luling notes that the fish leap out of the water on occasion in pursuit of insects on vegetation close to the surface. Leaping must be preceded by closing the gill covers. If an individual fish had an innately high threshold for leaping, it might orient toward an insect, pull its gill covers shut, but fail to jump after the insect. Nevertheless, the preparatory movements would produce a stream of water *if* the tongue and roof of the mouth were preadapted to form a channel of some sort. (Luling describes an incident in which a fish repeatedly shot water at a sunken mealworm. Aerial water shooting might be a secondary adaptation based on structural characters which had evolved for underwater spitting—a behavior which stirred up underwater debris uncovering hidden food.)

It may be argued that, even if this scenario is correct, the initial water spitter would have produced such a feeble jet that it would have been ineffective in striking and knocking an insect off a leaf. However, if the prey were only a few inches from the surface then the original weak spitting fish might occasionally have been successful while avoiding the energetic expense of jumping out of the water. The fact that juvenile archer fish also water shoot even though they can fire only a few inches (Luling, 1963) suggests that streams of water of this length may sometimes be effective. Otherwise selection should have acted to favor individuals which did not water shoot *until* they were large enough to produce a longer stream (because such individuals would have conserved the time and energy costs of ineffective water shooting).

The woodpecker finch and other Darwin's finches.—Elsewhere I have suggested

(Alcock, 1970) that tool-using by the woodpecker finch stems from conflict behavior. Finches unable to secure insect larvae and adults which they could see but not reach in crevices in wood may have performed displacement nest material gathering in the breeding season. Reattracted to to the crevice while still carrying a collected twig, a bird might jab beak and, inadvertently, twig into the space accidently dislodging the insect. A bird might in this way learn to associate twig use with food reinforcement and habitually use tools in this situation.

It is possible that the other finches which have been observed using twig tools acquired the trait independently in a fashion similar to the woodpecker finch. However, it may be that tool-using by these species stems from observation of the woodpecker finch. (Interspecific observational learning would be unusual but given the great similarity between the different species of finches it is at least possible.) A discussion of the role of observation in the transmission of a trait follows later in the paper.

Egyptian vulture.—The vulture's ability to open ostrich eggs may well also have its origin in conflict behavior. A bird attracted to a clutch of eggs but unable to pick one up to throw might redirect its egg throwing movements to a rock lying nearby. These movements, which successfully open those eggs the birds can pick up, consist of lifting the object and then swinging the head down while releasing it. A bird engaged in this behavior with a rock substitute but oriented toward the highly attractive ostrich eggs could accidentally strike an egg and open it. In this way rock-throwing could be reinforced leading to learned performance of the behavior (van Lawick-Goodall, 1970; Alcock, 1970).

Black-breasted buzzard.—Very similarly the hawk's habit of dropping rocks and lumps of dirt onto emu eggs which it is unable to carry off because of their size may have its origin in redirected egg carrying behavior. The bird is known to transport smaller eggs to its nest (Chisholm, 1954).

A bird which attempted to lift an emu egg but failed might fly up with substitute objects, such as rocks. If these were dropped as the bird was reattracted to the real eggs an egg might be accidentally broken. With food reinforcement the hawk could be conditioned to repeat its behavior.

Brown-headed nuthatch.—Morse (1968) has suggested that these birds learn to use bark flakes as tools when they accidentally chip off pieces of bark while hammering open pine seeds placed in bark crevices. Small insects are sometimes exposed when this occurs thus reinforcing the use of objects as wedges and levers. Morse does not explain how the switch from pine seeds to bark tools might take place but possibly this happens when a piece of bark is only partially displaced by a pine seed wedge. If the bark were pulled off and no prey found underneath the bird might carry the bark bit to the next hammering site instead of selecting a new pine seed.

Sea otter.—The use of rocks as tools by sea otters may well have occurred initially in situations when an animal was thwarted in attempts to open a hard shelled mollusc. Such an animal might strike one mollusc against another or against a rock attached to some shellfish retrieved from the bottom. Fisher (1939) and Kenyon (1959) have observed Californian and Alaskan sea otters using shells as tools. If an action originally taken in a conflict situation should happen to open a prey a sea otter might readily learn to repeat its behavior.

Primates.—Inasmuch as the primates using tools in feeding situations have been seen doing so only once or twice (with the exception of the chimpanzee) it does not seem useful to speculate on the origin of each case. It might be worth noting however that the reported cases all deal with *potential* conflict situations—attractive food covered with unpleasant substances or ants, fruits which were out of reach, or ants and termites which might be observed in tunnels or known from past experience to be in a nest where they cannot be secured with

the animal's fingers. Hungry animals confronted with these situations in which they were thwarted by the nature of the food or its location may have sometimes engaged in conflict behavior involving leaf gathering or twig or branch manipulation. These activities could be reinforcing in some cases and so learned. For example, a chimpanzee frustrated by a hidden or inaccessible termite could have performed displacement nest material gathering, food collecting or even perhaps displacement threat display with branches. If these objects happened to be thrust into termite tunnels the chimp might be rewarded by the capture of some insects and be conditioned to use tools.

However, it may be more likely, as van Lawick-Goodall (1968) suggests for the chimpanzee, that the use of sticks, twigs and leaf stem probes may be the outcome of essentially random exploratory manipulations of these objects at ant and termite nests.

In summary, it is clear that one need not invoke insight behavior by an animal to provide a *possible* explanation for the origin of tool-using in the cases discussed. Indeed, it seems probable that most examples originated as a result of fortuitous accidents involving the performance of a pre-existing behavior pattern in a slightly altered way in a novel situation. In particular, as Hall (1963) pointed out largely on the basis of tool-using by primates in aggressive situations, conflict behavior may have played a key role in the origin of the use of tools. Animals thwarted while attempting to feed on food items they could see but not capture or consume may have performed displacement or redirection activities with naturally occurring objects, activities which proved to be rewarding. It should be noted however that in some cases conflict behavior is an unlikely candidate for the foundation of tool-using behavior (e.g., brown headed nuthatch, ant-lions). No single explanation can account for the origin of all examples of tool-using by feeding animals.

The Initial Spread of Tool-Using Behavior in a Population

Although it is possible that the increase in the frequency of tool-using in some species required a long period of evolution, in other species it may have spread rapidly. First, a favorable set of environmental conditions may have resulted in the independent discovery of tools by a large number of individuals over a short period of time. Second, observation of an experienced conspecific may have facilitated the spread of the trait in a variety of ways (van Lawick-Goodall, 1970). Theoretically this could occur in much the same way that other novel behavior patterns, such as milk bottle opening by European titmice and other species (Hinde and Fisher, 1952) and potato washing by macaques (Kawamura, 1959), have occurred. Just as milk bottle opening and potato washing are limited to some populations of titmice and macaques, so tool-using is often restricted to one or a few populations or is distinctively different from population to population within a tool-using species (e.g., Egyptian vultures, nuthatches, sea otters, chimpanzees).

Again an observer animal need not display insight in order to benefit from observation of a companion (Thorpe, 1963; Alcock, 1969a, b). It is often enough that an animal merely attend to the activities of another and in so doing have its attention directed to some object or part of the environment that it would otherwise have ignored. Then through trial and error the observer can learn for itself what is significant about the object or part of its environment.

The fact that most species which learn to use feeding tools are social animals and/or have a period of close parent-offspring relationship would facilitate learning from observation. All tool-using birds and primates feed in groups. The performance of a tool-using activity may release this behavior in an observer animal in much the same way that satiated chickens are stimulated to peck at food by watching hungry chickens do so. Socially facilitated behavior could provide the basis for learned tool-use. For example, a naive Egyptian vulture, stimulated to throw rocks by a companion might learn to associate the behavior with food even if it was actually the companion's rock which opened an ostrich egg (Alcock, 1970).

Even more important perhaps, young animals which have been subject to selection pressure to learn a variety of things directly or indirectly from their parents may be especially likely to acquire tool-using behavior as a result of watching others. The sea otter and the chimpanzee have particularly close and prolonged ties between adults and their progeny. But even young birds are often fed for weeks after fledging and may accompany their parents in loose groups for still longer periods of time. Thus young tool-using birds, otters, and primates have ample opportunity to observe experienced tool-users and to manipulate objects their parents have been using. Van Lawick-Goodall (1968, p. 209) provides examples of chimpanzees doing exactly this.

In summary, tool-using behavior, once fortuitously discovered, may have spread rapidly through some populations of birds and mammals as a result of socially facilitated learning.

The Subsequent Evolution of Tool-Using Behavior

The selective advantage enjoyed by the original tool-users in any population may be great enough to set in motion a chain of evolutionary events which can affect the frequency of the trait in a population and a complex of factors underlying the behavior. However, it should be stressed that for many species tool-using is a rare event which probably is simply the by-product of selection for certain general abilities which underlie many different behavior patterns. One questions whether the use of feeding tools can provide even a minor focal point for selection in the case of the nuthatch, Darwin's finches other than the woodpecker finch, the buzzard, and most primates. For other species tools may be effectively used more often permitting a significant expansion of exploitable food resources.

Sand-throwing by ant- and worm-lions seems highly adaptive. For these species, in which the behavior may have originated simply because some individuals possessed unique thresholds for a particular behavior pattern, selection appears to have shifted the threshold for sand-tossing downward for entire populations.

Luling (1963) argues that water shooting by the archer fish can be of no significance because the fish spends much of its life at sea. Nevertheless the ability to capture prey with reduced energy expenditure (assuming that leaping from the water is energetically expensive) is surely advantageous even if only practiced for that portion of the animal's life spent in mangrove swamps. Thus in the archer fish selection may have first effected a shift in the threshold of a behavior pattern for an entire population. Individuals which tended to close the gill covers but not leap from the water would become the norm. Given the present very stereotyped nature of the behavior it would appear that there is one best way to water shoot most accurately and effectively. Selection has evidently acted to perfect and stereotype the relationship between what has become a fixed action pattern and a set of releasing stimuli. Of course correlated with these behavioral changes has been the remarkable specialization of mouth morphology to enhance the spitting ability of the fish (Luling, 1958).

In contrast to the fish and insects, learning may be a much more important component of tool-using by some birds and mammals. When learning plays a role in the development of the trait, selection can operate in several ways to increase the probability that animals will learn quickly to use tools skillfully.

(1) *Selection for specific play and exploratory behavior patterns.*—Since it has long been realized that facility in the use of sticks as tools by captive apes depends on their having had opportunities to manipulate and play with sticks (Kohler, 1927), selection may favor the development of specific kinds of manipulatory play and exploratory behavior (Jay, 1968). The sea otter is known to be an exceptionally playful animal (Kenyon, 1959) as is the chimpanzee. It would be intriguing to observe fledgling and juvenile woodpecker finches in the wild. Eibl-Eibesfeldt (1967) has observed captive finches playing with tools when satiated.

(2) *Selection for specific conflict behavior in conflict situations.*—When conflict behavior of a certain sort is the key to tool-using it is reasonable to assume that selection will favor those individuals which perform that kind of behavior most readily in the appropriate situations. It may be that selection of this sort is occurring now in some populations of Egyptian vultures. Woodpecker finches may have been under selection to gather twigs when thwarted by visible but unreachable prey. Millikan and Bowman (1967) note that twig manipulation occurs as a displacement activity very frequently suggesting that this behavior pattern has a low threshold in conflict situations. It is conceivable in the case of the finch that selection has acted to "ritualize" conflict behavior in tool-using situations by standardizing the behavior pattern, the releasing situation, and by divorcing the behavior from its original motivational state.

(3) *Selection for attentiveness to parents and/or companions.*—If acquisition of tool-using is aided by or dependent upon observation of other tool-users, selection should favor those individuals which are especially attentive to the activities of companions or parents. Moreover, there may be an additional advantage for those animals with a tendency to perform actions similar to those they observe. An isolated woodpecker finch failed to learn to manipulate twig

tools, even when eventually permitted to watch others when older (Millikan and Bowman, 1967). If this case is truly representative of the species, not only is tool-using linked to observation of others but there is a critical period for learning to take place. As previously noted, sea otters stay with their mothers for many months and van Lawick-Goodall (1968) has seen young chimps watching adults carefully and duplicating their behavior.

(4) *Selection for ease of conditioning in tool-using situations.*—Finally, when a learned behavior pattern is of real importance to an animal there is substantial evidence that selection can lead to the evolution of animals with a highly specific ability to learn *that* behavior pattern easily (Tinbergen, 1951; Lorenz, 1970). It seems probable that woodpecker finches and sea otters in particular, and also chimpanzees and Egyptian vultures (from tool-using populations) can learn to use their tools much more readily than other species, even closely related ones. However, Millikan and Bowman (1967) argue that the woodpecker finches' use of tools stems from purely ecological factors (an abundance of wood-dwelling larvae and the absence of woodpeckers on the Galapagos) and not genetic ones. They cite as evidence the fact that the finches proved no more adept at learning to pull a string to retrieve a prey tied at the end than such birds as titmice which do not use tools naturally. This argument ignores the specificity of selection which has surely been directed at favoring finches able to learn to manipulate sticks and thorns skillfully with their beaks. Finches capable of coordinating movements of their beaks and feet in string pulling exercises probably have not enjoyed any particular selective advantage.

The Significance of Tool-Using Behavior by "Lower" Animals For an Understanding of Human Evolution

Thinking about human tool-using appears to have passed through a number of stages. It was first thought that only humans used tools and that the pioneer tool-users must have been individuals of exceptional intellect. Subsequently it was realized that other animals used tools. This was then taken to indicate that these species were extraordinarily intelligent ones, forerunners in the animal kingdom of human intelligence. It is now recognized that simply because some animals use tools does not mean that these species have anything essential in common with humans. Tool-using has obviously evolved independently many times with many different underlying mechanisms (Hall, 1963). Is then the fact that other animals use tools basically irrelevant for an understanding of human evolution? I think not, for the study of other animals suggests that:

(1) The origin of tool-using in the human evolutionary line need not have involved a high level of conceptual ability on the part of the pioneer individuals. In all likelihood protohumans began using tools accidentally when certain conflict and/or exploratory behavior patterns involving object manipulation happened to be reinforcing.

(2) The initial transfer of the trait from individual to individual likewise could have occurred without insightful behavior on the part of the observer. This might have happened simply because protohumans tended to watch what others did and to investigate the properties of objects used by their companions or parents. It may well be that the selective advantage of tool-using by our ancestors stemmed in large measure from the fact that not only the innovator but also his offspring and relatives were likely to acquire the trait and the improved efficiency in food gathering it permitted.

(3) Once tool-using originated much of its peculiar importance to protohumans lay in the fact that it provided access to a substantial food resource, mammalian prey, previously unavailable for the most part to members of the species. The use of feeding tools by chimpanzees is significant to the extent that it indicates a potential for the behavior in the human evolutionary line.

400

But the animals which are most revealing from a comparative standpoint are not our primate relatives but unrelated species subject to ecological pressures broadly similar to those operating on our human ancestors. It is surely significant that the major tool-using species, sea otters, woodpecker finches, and humans, have all invaded niches that are not at all characteristic of their phylogenetic groups. The sea otter is the only otter to have entered the marine environment. It did so rather recently and thus faced competition from many other established marine mammals. The woodpecker finch is in all likelihood one of the most recently evolved Darwin's finch and the only one which specializes in the un-finch-like habit of removing larvae from decaying wood. Competition from the many other existing species of the Geospizinae has almost certainly been critical in its evolution (Lack, 1947). Humans and their immediate ancestors are the only primates to feed on other mammals to a significant degree. This niche was invaded recently at a time when there were many other accomplished savannah-dwelling predators to compete with protohumans for mammalian prey. In each case tool-using compensates for the finch, otter, and protohumans' lack of biological equipment for the exploitation of wood-dwelling larvae, hard shelled molluscs, and moderately large mammals respectively. Thus the use of tools may have been a key adaptation permitting coexistence between these species and potential competitors already present in their environments. When tool-using behavior achieves this importance it can create selection pressures which have the manifold effects outlined in the previous section.

In the case of humans, the structure of the hand, the neurophysiology of the brain, the kinds of play, exploratory, and conflict behaviors, the attentiveness to and apparent imitation of others, and specific learning abilities have surely been affected by selective forces associated with tool-using (Washburn, 1959; Campbell, 1966).

Thus although the simple fact that other animals use tools may be relatively insignificant, comparative studies of tool-using behavior can illuminate aspects of the evolutionary process which apply equally to woodpecker finches and human beings.

SUMMARY

This paper reviews the known cases of the use of tools by animals feeding under natural conditions and discusses the possible origins, transmission, and subsequent evolution of this behavior. It seems probable that the pioneer tool-users of various species were simply using pre-existing behavior patterns in a way which accidentally involved the use of an object as a tool. The transmission of the trait in some populations of birds and mammals may have taken place fairly rapidly if inexperienced animals could benefit from observation of tool-using parents or companions. In the relatively few cases where the use of tools makes a substantial contribution to the food economy of a species selection for facility in tool-using may be evident. This may have taken the form of shifts in thresholds for selected behavior patterns in the insect and fish tool-users or it may have led to special learning abilities in the bird and mammal tool-users. Comparative studies of this behavioral trait may contribute to an understanding of human evolution. Sea otters, woodpecker finches, and humans may be especially proficient in the use of feeding tools because all have invaded unusual niches for their phylogenetic groups and have faced potentially severe competition from species well established in their environments. Under these circumstances selection may operate at a variety of levels to improve the tool-using abilities of populations of animals.

ACKNOWLEDGMENTS

I thank Dr. Robert Lockard for having read the paper and for his suggestions. This research was supported by NIH Mental Health Small Grant MH19777-01 and NSF grant GB 28714X.

401

LITERATURE CITED

ALCOCK, J. 1969a. Observational learning in three species of birds. Ibis 111:308–321.

——. 1969b. Observational learning by fork-tailed flycatchers (*Muscivora tyrannus*). Anim. Behav. 17:652–657.

——. 1970. The origin of tool-using by Egyptian vultures *Neophron percnopterus*. Ibis 112:542.

BROWN, L. H., AND E. K. URBAN. 1969. The breeding biology of the great white pelican *Pelecanus onocrotalus roseus* at Lake Shala, Ethiopia. Ibis 111:199–237.

CAMPBELL, B. 1966. Human Evolution. Aldine, Chicago.

CHIANG, M. 1967. Use of tools by wild macaques in Singapore. Nature 214:1258.

CHISHOLM, A. H. 1954. The use by birds of "tools" or "instruments." Ibis 96:380–383.

CURIO, E., AND P. KRAMER. 1964. Vom Mangrovefinken (*Cactospiza heliobates* Snodgrass und Heller). Ergebnisse der Deutchschen Galapagos-Expedition 1962/63 I. Zeits. Tierpsych. 21:223–234.

EIBL-EIBESFELDT, I. 1961. Uber den Werkzeuggebrauch des Spechtfinken. Zeits. Tierpsych. 18:343–346.

——. 1967. Concepts of ethology and their significance in the study of human behavior. *In* Early Behavior: Comparative and Developmental Approaches. (ed.) H. W. Stevenson. Wiley, New York.

EIBL-EIBESFELDT, I., AND H. SEILMANN. 1962. Beobachtungen am Spechtfinken *Cactospiza pallida*. J. Ornithol. 103:92–101.

FISHER, E. M. 1939. Habits of the southern sea otter. J. Mammal. 20:21–36.

GIVENS, R. D. 1963. Comment on "Tool-using performances as indicators of behavioral adaptability." Curr. Anthropol. 4:489.

GOODALL, J. 1964. Tool-using and aimed throwing in chimpanzees. Nature 201:1264–1266.

GRUBER, A. 1969. A functional definition of primate tool making. Man 4:573–579.

HALL, K. R. L. 1963. Tool-using performances as indicators of behavioral adaptability. Curr. Anthropol. 4:479–494.

HALL, K. R. L., AND G. B. SCHALLER. 1964. Tool-using behavior of the Californian sea otter. J. Mammal. 45:287–298.

HINDE, R. A., AND J. FISHER. 1952. Further observations on the opening of milk bottles by birds. Brit. Birds 44:393–396.

HUNDLEY, M. H. 1963. Notes on methods of feeding and the use of tools in the Geospizinae. Auk 80:372–373.

JAY, P. 1968. Primate field studies and human evolution. *In* Primates, Studies in Adaptation and Variability. (ed.) P. Jay. Holt, Rinehart, and Winston, New York.

JONES, C., AND J. S. PI. 1969. Sticks used by chimpanzees in Rio Muni, West Africa. Nature 223:100.

KAWAMURA, S. 1959. The process of subculture propagation among Japanese macaques. J. Primatol. 2:43–60.

KENYON, K. W. 1959. The sea otter. Ann. Rep. Smith. Inst. 1958:399–407.

KOHLER, W. 1927. The Mentality of Apes. Kegan Paul, Trench, Trubner and Co., London.

KORTLANDT, A. 1967. Reply to "More on tool-using among primates." Curr. Anthropol. 8:253.

KORTLANDT, A., AND M. KOOIJ. 1963. Protohominid behavior in primates. *In* The Primates. Symp. Zool. Soc. Lond. 10:61–88.

LACK, D. 1947. Darwin's Finches. Cambridge University Press, Cambridge.

LORENZ, K. 1970. Innate bases of behavior. *In* On the Biology of Learning. (ed.) K. H. Pribham. Harcourt, Brace, and World, New York.

LULING, K. H. 1958. Morphologische-anatomische und histologische Untersuchungen am Auge des Schutzenfisches *Toxotes jaculatrix* nebst Bemerkungen zum Spuckgehaben. Z. Morph. Oekol. Tiere 47:529–610.

——. 1963. The archer fish. Sci. Am. 209:100–109.

MILLIKAN, G. C., AND R. I. BOWMAN. 1967. Observations on Galapagos tool-using finches in captivity. Living Bird 6:23–42.

MORSE, D. H. 1968. Use of tools by brown-headed nuthatches. Wilson Bull. 80:220–224.

SUZUKI, A. 1966. On the insect-eating habits among wild chimpanzees living in the savanna woodlands of western Tanzania. Primates 7:481–487.

THORPE, W. H. 1963. Learning and Instinct in Animals. Methuen and Co., London.

VAN LAWICK-GOODALL, J. 1968. The behaviour of free-living chimpanzees in the Gombe Stream Reserve. Anim. Behav. Monog. 1:165–311.

——. 1970. Tool-using in primates and other vertebrates. *In* Advances in the Study of Behavior (vol. 3). D. S. Lehrman, R. A. Hinde, and E. Shaw (eds.). Academic Press, New York.

VAN LAWICK-GOODALL, J., AND H. VAN LAWICK. 1966. Use of tools by Egyptian vultures. Nature 12:1468–1469.

WASHBURN, S. L. 1959. Speculations on the inter-relations of tools and biological evolution. Human Biol. 31:21–31.

——. 1963. Comment on "Tool-using performances as indicators of behavioral adaptability." Curr. Anthropol. 4:492.

WASHBURN, S. L., AND P. C. JAY. 1967. More on tool-using among primates. Curr. Anthropol. 8:253.

WHEELER, W. M. 1930. Demons in the Dust. W. W. Norton and Co., New York.

NIKO TINBERGEN

47   The Shell Menace

When a chick is about to hatch, it cracks the shell near the obtuse end. Then, by rhythmic stretching movements (for which chicks have a muscle that degenerates after hatching) it lifts off a small "lid" and half-rolls, half-crawls out of the shell.

In some species, such as gallinaceous birds and ducks, the parents rarely if ever pay any attention to the empty egg shells; they lead the young away as soon as they are dry, leaving nest and shells behind. Most other birds dispose of the shells in one way or another. Hawks are reported to eat them, as a general rule; grebes take them in their bills and "drown" them some distance from the nest. Most birds fly away with each shell and drop it a good distance from the nest, although not at any particular place. Black-headed Gulls (the Old World cousins of Bonaparte's Gull) do this regularly, and while there is variability in the time, few if any fail to remove a shell within two hours after hatching; sometimes they do it within a matter of minutes.

One could hardly imagine a more trivial response, for it takes no more than, at most, thirty seconds of a bird's time each year—between three and ten seconds for each of its three eggs. Yet, as biologists, we have gradually become convinced that very few such regular occurrences are really insignificant. When my friends and I began to look more closely at egg shell removal, we began to suspect that it must in some way be a very useful response. The argument was indirect. We had noticed that several predators, such as Carrion Crows, Herring Gulls, and even neighboring Black-headed Gulls, were often quick to seize an egg or a newly hatched chick, in spite of the fact that their khaki coloring and dark blotching make them difficult to detect. It is not for nothing, therefore, that broods are rarely left unguarded. Parent gulls take turns at the nest and the "on duty" bird seldom leaves before having been relieved by its partner. When a gull flies away from the nest to dispose of the egg shell, and so leaves the brood unguarded for something between three and ten seconds, even this short absence can be heavily penalized when a crow dashes down for a "grab and fly" robbery. So when, despite this threat, all members of the species take the risk, we must suspect that there are advantages that outweigh the disadvantage, or the habit would have been eliminated by natural selection. Therefore, we speculated, there must be something that penalizes the broods that have "untidy" parents.

As to the nature of the penalty, there were several possibilities. An empty egg shell might slip over an unhatched egg, and so trap the chick inside. This has actually been observed in birds that lay eggs with strong shells. The sharp edge of the broken shell might injure the chick. This, too, has been reported—by poultry breeders, for instance. Another possibility was that three shells left in the nest might interfere with the parents' efficiency in brooding the chicks. After all, a gull has only three brood spots—one for each egg or chick. Neither of these three possibilities seemed probable in our species, which has extremely thin shells that are easily crushed. Nor did we think much of the possibility that the moist, organic material that is sometimes left in the shell could be a breeding ground for pathogenic bacteria; the shells usually dry quickly. We were rather inclined to think of a fifth possibility: the shell, by being white inside and thus, at least to us, very conspicuous, might attract the attention of predators—such as crows and Herring Gulls, which hunt by sight—by helping to reveal the otherwise camouflaged brood. The gulls will remove not only egg shells but many other objects from their nests (such as mussel shells, bits of paper, leaves, and even bottle tops), which seemed to support this theory. Even more suggestive was the fact, reported by my co-workers Dr. and Mrs. Cullen, that while many species of gulls and terns remove their egg shells, the Sandwich Tern and the Kittiwake lack the response—and these two species are exactly those members of the group that have no camouflaged broods. The eggs of Sandwich Terns, although blotched, are very conspicuous, and the down of a Kittiwake chick is almost uniformly silvery-white.

When, in 1959, some of our guest workers were keen to tackle this problem with me, we decided to study it systematically in the large gulleries near Ravenglass, which is situated on a sandy peninsula on the Irish Sea coast of Cumberland, England. This was the beginning of a three-year study, in which I was joined at one time or another by Dr. G. J. Broekhuysen of Capetown, Miss C. Feekes of Utrecht, J. C. W. Houghton of Leeds, H. Kruuk of Utrecht, Miss M. Paillette of Paris, Dr. E. Szulc of Warsaw, and Dr. R. Stamm of Basel.

Two questions were posed. First, does or does not an egg shell, if left in or near the nest, expose the brood to increased predation? Second, how does a gull "recognize" an empty egg shell and how does it distinguish that shell from intact eggs, chicks, and nest material, none of which we ever observed a gull to carry?

We worked on the assumption that the egg shell would make the brood more conspicuous. Although this is so, at least to human eyes, in no species with allegedly camouflaged broods had it ever been put to the test with respect to the species' natural predators.

We laid out, singly and scattered over an open dune valley, equal numbers of natural-colored gulls' eggs and gulls' eggs that we had painted white. From an observation blind erected on a dune top we observed which predators would take them, and which of the two types of egg they would take most often. The area was regularly patrolled by several Herring Gulls and by Carrion Crows, which found and ate a number of the eggs. Each test was broken off when roughly half the eggs had been discovered. After a few weeks the results became clear: although both types of eggs were taken, both the gulls and the crows found more white than natural-colored eggs (Figure 1). Not surprisingly, the crows, members of a tribe known for intelligence and keen sight, were better at finding eggs than were gulls. Also, the crows were extremely helpful to us, for unlike the gulls they did not stop looking for eggs when they were satiated, but continued to take and bury them, returning to such caches weeks after.

Later we repeated this experiment with painted hens' eggs. Half of them were uniformly khaki—a color that matched the base color of the gulls' eggs; the other half had irregular dotting added on the base color. Again, the uniformly colored eggs were taken in larger numbers than were the dotted eggs. The eggs, therefore, derive some protection from their coloration. However, since many of the natural eggs were found, the camouflage cannot be

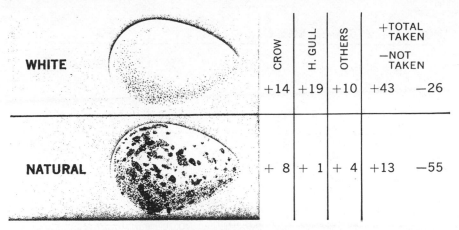

	CROW	H. GULL	OTHERS	+TOTAL TAKEN −NOT TAKEN	
WHITE	+14	+19	+10	+43	−26
NATURAL	+ 8	+ 1	+ 4	+13	−55

FIG. 1. White eggs proved to be more subject to predation than did those naturally camouflaged. Plus and minus signs indicate numbers of each taken or left.

called very effective. Without the efficient manner in which Black-headed Gulls attack marauding crows, few gull eggs would survive.

Our next step was to test whether the presence of an empty egg shell next to a natural-colored gull's egg would endanger such an egg. Here we used a trick that I feel I must justify. We had seen that even a natural-colored egg could be found with relative ease. In the natural situation (that is, in the gulleries themselves) crows that entered the colony spent much of their time dodging the violent attacks of the gulls, while in our test area they were free of such attacks and could, as a result, look down continuously. We further realized that the prey, which in the natural situation would be betrayed by the egg shells, could be either eggs or chicks. Chicks are much better camouflaged than are eggs; moreover, they crouch in cover when the parents sound the alarm. So we decided to make things more difficult for the crows. All the eggs we laid out in the dune valley were covered with a few straws of marram grass, which improved the camouflage most strikingly (in fact, some plovers use this trick themselves). Half the eggs were single; empty egg shells were placed at some 10 cm. distance from each of the other half. We weighted the situation a little

against the expected results by covering the eggs that had an egg shell next to them slightly better than we did the single eggs. Thus, the single eggs were a little easier to find. In spite of this, our predators—again Carrion Crows and Herring Gulls—found 65 per cent of the eggs accompanied by the extra shell, and only 22 per cent of the single eggs.

Because some gulls move the shell no more, perhaps, than a couple of feet from the nest, we experimented with varying the distance between the egg and the shell. The result was clear-cut: the farther the shell from the egg, the less danger to the egg (Figure 2). In these tests we saw the crows alight near a shell, walk round for awhile, and then leave without having found the egg.

Could we now conclude the hypothesis had been proved? In our opinion, not quite. We had shown that the presence of a shell endangered the egg, but as we had conducted our experiments outside the gullery, there had been no gulls to guard the eggs. In the gullery, however, removal of the egg shell, apart from making the brood less conspicuous, also involves abandoning the brood for a brief time, thus exposing it to predation. We were not entitled to judge the over-all effect of egg shell removal, because in our tests neither category

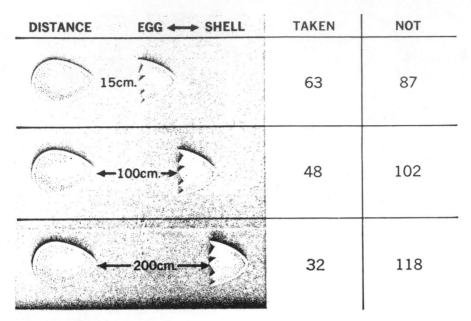

DISTANCE EGG ⟷ SHELL	TAKEN	NOT
15cm.	63	87
←100cm.→	48	102
←200cm→	32	118

FIG. 2. Predation was much heavier on those eggs that were close to a broken shell. Chart shows varying distances between the two, and the numbers taken or not.

of eggs was ever guarded by a gull, and the effects of continuous guarding were therefore not comparable with those of interrupted guarding. However, we have good reasons to suppose that the advantage of removing the egg shell outweighs the disadvantage. For one thing, gulls either sit on the nest or attack when a crow or a Herring Gull enters the gullery; for another, if they did not remove the shells at all, these would be attractants for days, whereas the danger caused by removal lasts only a few seconds. Whether or not any of the other advantages mentioned above will later prove to exist as well, our experiments have convincingly shown that, at least in the Black-headed Gull, egg shell removal can be considered a behavior component of camouflage, and part and parcel of the species' defense against predators.

We next wanted to find out the stimuli whereby the gulls recognize the egg shell. Although they remove a great variety of miscellaneous objects in addition to shells this does not mean that they respond with equal promptness to all objects. We presented the gulls with a variety of dummy shells, and compared their responses to them. First we determined the length of the season during which the gulls would remove egg shells we provided. At various times in the breeding season, from weeks before the eggs were laid until well after the chicks had hatched, we gave hundreds of gulls an egg shell each on the rim of their nests and checked, after a standard period of six hours, how many of them had removed the shell. We found that the first responses, few and incomplete, appeared three weeks before the first egg was laid. After a gradual buildup, a high level of removal was reached while the eggs were being laid; this level was maintained throughout the incubation period (about twenty-four days), and it did not drop off until after the chicks had hatched. The level was so constant and so high that we knew we could test throughout the breeding season.

We could now begin to compare the different dummies. We made simple dummies that could easily be transported—rectangular strips of metal, 2 x 5 cm., and bent at right angles in the middle. Some of these we painted a light khaki; others dark khaki (equal to the color of eggs viewed from such a distance that the dots blurred); some white, some black, others bright red, yellow, or blue and others green. We marked a large number of nests and divided them into as many groups as we had models, with equal numbers of nests in each group. In one trial we gave each nest one dummy on its rim, with a different color for each group. After an hour we revisited all the nests and noted how many of each type of dummy had been removed. After having given the gulls a few hours rest, we repeated the test, continuing until each nest had received each color once. The different groups were presented with the dummies in a different order. After such a test we could directly compare the responses of all gulls to each model.

The results were rather surprising. The gulls did remove some of all the dummies, but some colors were removed much more consistently than others. It was clear that conspicuousness was no criterion. On the one hand, red, blue, and black were not carried away particularly often. On the other hand, both white and the khakis had a very high score, although white was extremely conspicuous and khaki was about the most cryptic color possible. Since we knew from other tests that the gulls had a good color sense, the only possible conclusion was that objects that had the same colors as real egg shells were most frequently removed, not those that were most conspicuous. The various green models (some of which were uniform, while others were dotted) had a very low score. This was particularly true of a shade of green that was very similar to that of young vegetation. The dotting made no difference, nor did contrast within one dummy (such as that offered by one that was khaki outside and white inside) raise the score.

These responses to color were obviously adaptive: strong responses to the natural main colors of the egg shell; a moderate response to objects of various bright colors that might occur in the natural situation (for instance, yellow, pink, and blue shells

of snails and mussels are frequently kicked into the nest); and a low response to green. This latter reaction, we think, is adaptive, because the gulls showed little preference for particular shapes, and were quite willing to remove flat paper disks. Had they responded readily to green, they would probably also strip leaves from surrounding vegetation and so remove useful cover, in addition to leaving the nest unguarded far too often.

We tested various other properties of objects in much the same way. In some series we varied the shape of the models, in others the size, in others the distance between the egg shell and the nest. We found that the shape response was best to the real egg shell, second to halved ping-pong balls, third to the cylindrical rings, fourth to the "angles" we had used for our color tests, and fifth to the flat cardboard or metal strips.

Egg shells placed at various distances from the nest gave interesting results. As standard egg shells we used either real gulls' shells (gathered in masses at the end of each season and kept until the next year) or broken hens' eggs painted khaki.

The gulls' removal response fell off sharply with increased distance, just as did the predatory response of the crows. In this way we gradually acquired a rather good idea of the types of stimuli that to the gulls characterized an egg shell to be removed.

We now turned our attention to two other aspects of behavior. As we had seen, the gulls carried away, with varying degrees of "enthusiasm," many different types of objects. In fact, one could say that any alien object tended to be removed—that is, any object not resembling an egg, a chick, or nest material. But we had not yet found out enough to be able to say what characteristics caused the birds to distinguish egg shells from each of these other things.

We knew that they would remove flat cardboard dummies, but nest material (mainly dry marram grass) is also practically flat and yet is not carried away. We decided to offer the gulls flat strips of different proportions. Four types of strips were made, all with a surface area of nine square centimeters, but ranging from squares of 3 x 3 cm. to rectangles of 18 x ½ cm. We found that those of 9 x 1 cm. were carried off most often, and that both the longer and squatter strips were carried less often. Why strips of 9 x 1 were carried more than strips of 4.5 x 2 we do not know, although we have some theories. The very long and thin strips of 18 x ½ cm. elicited interesting behavior. In only 66 out of 280 cases were they removed. Of those not removed, however, 46 per cent were built into the nest. We also made some direct observations from our blinds, and found that the matter was still more complex, for some of the gulls tried to eat the long strips. It is easy to be wise after the event, and we realized that the shape was not dissimilar to that of earthworms—the staple diet of this gull colony.

In order to discover how the gulls distinguished between shells and intact eggs we proceeded rather differently. When an egg is presented on a nest's rim, the gull usually retrieves it. The movement is characteristic: the bird stretches its neck, brings the bill down behind the egg, and rolls it into the nest, balancing it against the narrow underside of the bill. This response is not always prompt or complete; it can also misfire if the gull fails to balance the egg properly and loses it before it is rolled in. By direct observation one could notice even incipient responses that could not have been concluded from the position of the egg afterward. Similarly, an egg shell is not always completely removed, but incipient movements can be recognized: instead of bending the head over and beyond the object, the gull may nibble at the broken edge, or may pick it up only to drop it immediately. We wanted to observe all these responses, and therefore our next series of experiments involved watching individual birds from a blind.

In these tests the procedure is frankly anthropomorphic, as indeed it is in all the preceding tests. That is we first analyze the differences *we* see between an egg and an egg shell, and then try out the gulls' responses to each of these one by one. Now, an egg shell differs from an egg by weight, in having a broken contour when seen from the side, by being hollow, and by having a thin, ragged edge. Since the gull's response can be recognized by us before the egg is touched, the first response must be visual, and weight cannot influence the response at this stage. It was a further lucky circumstance for us that an empty, that is, blown, egg was retrieved as well as a normal one, showing that weight, too, had little, if any, effect on retrieving.

To test whether the broken contour had an effect on the gull, we compared the birds' responses to a true egg with those to an egg shell filled to the rim with plaster. This plaster-filled model did not offer a hollow space or a thin edge, since the plaster adhered exactly to the rim. Both models were always rolled in; no bird ever even nibbled at the rim of the plaster-filled egg. This meant that the broken contour was, to the gull, not a shell characteristic. When we filled egg shells with cotton wool — which fills the shell, but does not fit exactly to the rim, and so leaves a thin edge visible without, however, leaving a hollow space of any significance—the response was quite different: these models were nibbled at and removed. This meant that the hollowness of the shell had very little effect. It looked as if the thin edge was the principal stimulus. To test this we offered blown eggs as controls, and for comparison similarly intact blown eggs on which was glued about one square centimeter of egg shell, which stood out at right angles from the egg's surface. This model had all the characteristics of a whole egg, but in addition offered a thin edge. No hollow was visible. In two out of every three tests this model was taken by the "flange" and removed; in the other tests it was rolled in. Often a bird alternated between the two responses, showing bits of each in turn. It seemed, therefore, that the thin edge was the main character by which egg shell was distinguished from egg.

This, however, raised a new problem. If the birds respond mainly to the thin edge, and not to hollowness, how does it happen that a newly hatched, wet chick, which has not yet fully left the shell, is not removed with the shell? While we do not yet know what characteristics of the chick might play a part, we found that the chick's weight has a profound effect. We gave the gulls egg shells in which pieces of lead that weighed as much as a chick were placed near the pointed end. Such eggs were always nibbled at, but as soon as they were lifted, the behavior simply ended. The bird might nibble again and again, but we never saw gulls remove such shells. This could not be because the bird saw the little piece of lead inside, for even a shell filled to the rim with cotton wool was carried off. The weight alone, therefore, is sufficient to stop a shell from being removed as long as the chick is not completely free. Probably the thin edge of such leaded shells was the reason none was ever rolled into the nest, as was the plaster-filled shell.

Once our interest in this behavior had been aroused, we began observations of egg shell removal patterns in other birds. It struck us that Ringed Plovers and Oyster-catchers—which also nest on the peninsula and lay well-camouflaged eggs—removed their egg shells much more promptly than did the Black-headed Gulls. Of ten gulls we observed closely during the whole period of hatching, when there was no outside disturbance, two removed the shell one minute after hatching and one fifteen minutes after. The rest took from one to over three hours. Again we argued: if there is such a premium on removing the shell, why don't the birds do it more promptly, as the waders do? We believe we have found the solution. The Black-headed Gull is a colonial species. It is also a predator, and we found that some individual birds made it a practice to rob their neighbors whenever they could. They are not much attracted to fresh eggs and are, in fact, very inexpert at outing their contents. But pipped eggs and newly hatched chicks are often taken and swallowed whole. However, as soon as chicks are dry and fluffy, the robber gulls seem to lose interest in them. We think that

this predation by "rogue" gulls is the environmental pressure that has prevented the parent gulls from removing the shells while the chick is vulnerable. Oystercatchers and Ringed Plovers, being solitary breeders, are not so delayed in shell removal because they do not have predatory neighbors.

Thus we gradually built up a picture of this seemingly insignificant activity. It certainly has survival value, and the analysis of the gulls' behavior revealed how beautifully adapted its control is to the needs. How do gulls acquire this efficiency? Is it "innate" response, or is it learned behavior? We believe that (as is usual with problems of development) the question is too simple.

We have found that on the one hand a gull in its first breeding season removes an egg shell, even if it is presented on the nest's rim before the first egg is laid. An egg presented to such an inexperienced bird is rolled in. It is true that we have so far tested only three such birds, but they responded as promptly and as completely as did more experienced birds.

On the other hand, we also have evidence of learning. We placed three plaster eggs in each of a group of sixty nests. The eggs were painted pitch-black and were introduced into the nests before the birds had laid eggs of their own. Some of the birds deserted; others settled on the dummy eggs. Of the latter, some reacted by laying no eggs at all (a well-known effect of sitting on such dummies); others accepted the black eggs but added some eggs of their own. We removed the real eggs within half a day after they were laid, and were left with a group of fifty-six nests with black eggs. Similarly, we placed green eggs in a group of fifty-three nests. A control group of sixty pairs were left with their own eggs. The birds were allowed to incubate for, on the average, seventeen days. We then tested the responses of all birds to ring-shaped dummies (one of our earlier types) of the three colors involved: black, green, and khaki. We found that the birds that had sat on black eggs removed

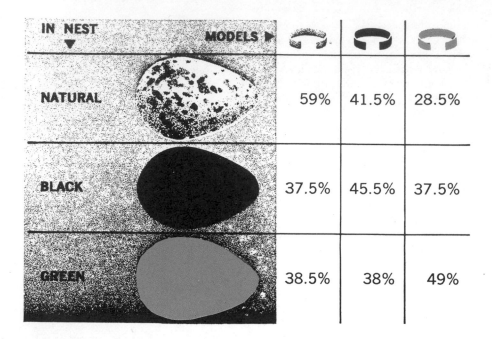

IN NEST ▼	MODELS ▶		
NATURAL	59%	41.5%	28.5%
BLACK	37.5%	45.5%	37.5%
GREEN	38.5%	38%	49%

FIG. 3. Percentage of egg shell dummies of three colors removed by gulls that had incubated eggs of the same colors, at the left, shows the effect of experience.

more black rings than either green or even khaki ones; the "green egg birds" removed more green than either of the other colors; and the controls carried more khaki than either green or black (Fig. 3). By changing the contents of the nest at the moment of the tests, we made sure that the birds had not been matching the color of the eggs in the nest with that of the ring—these birds had really learned the color of their eggs during incubation. They showed the same acquired preference in their egg-rolling response, and had transferred this experience to the egg shell removal response.

Even though we cannot claim to have done more than skim the surface of the problem, it is already clear that the control of egg shell removal is extremely complex, and beautifully adapted to the requirements. And yet it is only a minor part of the total set of devices by which these gulls defend themselves against predators. For one thing, there are other aspects to camouflage, such as the wonderfully adapted

color glands that, in some unknown way, manage to lay down the pattern of blotching on the eggs; the protective color patterns of the chicks and their ability to crouch in response to the alarm calls of the parents and even to hide under cover; the tendency of the adults to leave their nests at the slightest disturbance and thus, unlike camouflaged birds such as curlews and nightjars, to rely on the camouflage of the brood. But the protection of the brood does not depend on camouflage alone; I have already mentioned the massed attacks gulls make on crows.

Through our work on egg shell removal we have now become interested in the antipredator system of the Black-headed Gull in all its aspects, and we begin to get indications that such things as synchronized laying, the characteristic pattern of nest-spacing, the seasonal and daily rhythms in the selection of the habitat, and several other traits of the gulls are related to defense against predators. And yet, losses through predation are heavy; thousands of young gulls fall victims to crows, Herring Gulls, hedgehogs, and foxes, to mention only the worst of the predators we have seen at work. Our observations on egg shell removal are but the first step in the unraveling of the highly complex relationships that apparently exist between these gulls and all their predators.

48

REPRINT No. 6 from *Animal Behaviour*, **24**, 1, February, 1976

Anim. Behav., 1976, **24**, 146–153

THE BIOLOGICAL FUNCTION OF GULLS' ATTRACTION TOWARDS PREDATORS

BY HANS KRUUK

Department of Zoology, University of Oxford*

Abstract. Many species of birds and mammals are attracted towards some of their natural predators, often without overt aggression. The biological function of this predator attraction was studied in a colony of herring gulls and lesser black-backed gulls, with the aid of predator models. More birds flocked above the predator model when it was presented together with a dead gull; also the birds did not land as close to a predator with a dead gull as they did to a predator alone. Having seen a predator with a dead gull had a clear effect on the distance at which birds later avoided that particular predator on the ground when it was presented to them alone. It is suggested that predator-attraction enables animals to collect information about a potential enemy; in this information the experience of conspecifics with the predator is taken into account; subsequent behaviour of the animals depends partly on this information. The increase in the avoidance distance with respect to a predator with a dead bird may be an adaptation to carnivores' habit of surplus killing.

When a stoat (*Mustela erminea*) ventures into a gull's breeding colony a large number of birds may gather above it, making the animal very conspicuous. This is a well known phenomenon, and it has been described for the black-headed gull (*Larus ridibundus*) in response to a fox (*Vulpes vulpes*), stoat and hedgehog (*Erinaceus europaeus*); gulls were only attracted towards this last animal if it was on or near a nest of a gull, and being attacked by the owner (Kruuk 1964). Birds appear to approach the potential predator from a great distance, and except those whose nests are immediately around the predator, show very little or no tendency to attack; on the contrary, from their postures and calls an increased readiness to flee is often indicated. Birds may hover or circle over the predator for some time, perhaps even land near by, then return to their territories; while they are near the predator, they appear to keep it continuously in sight. This non-aggressive attraction to a possible predator occurs in many birds and mammals; it is conspicuous, for instance, in ungulates on the African savannahs, who exhibit it towards the large carnivores (Kruuk & Turner 1967; Walther 1969; Kruuk 1972a; Schaller 1972). In this last situation carnivores have occasionally been observed to kill ungulates so attracted (Walther 1969; Kruuk 1972a), and sometimes gulls leave their nest to approach a predator, indicating that this behaviour is not without disadvantages. This makes it all the more tempting to speculate on

*Present address: Institute of Terrestrial Ecology, Banchory, Kincardineshire, Scotland.

its survival value, and several possibilities suggest themselves. For instance, an important function of the attraction might be merely to keep the predator in sight; otherwise it might sneak up and approach unseen. It seems unlikely, however, that this would require the very close approach which one observes; also such a function would not be really compatible with observations of birds briefly coming in to 'have a look'. Also, in many situations, predator-avoidance would be a better strategy than merely keeping the enemy in sight. A second possible biological function of predator-attraction has been discussed by Smythe (1970), who speculates that it, and various alarm signals, might entice the predator to engage in pursuit, thus exposing himself, whilst the prey shows its invulnerability. This explanation could, perhaps, be relevant for ungulates, but not in the gull colony. Thirdly, prey animals might be gaining information which might lead to increased success in escaping from future attempts at predation; this possibility has been investigated in the present study.

It has been shown by Kramer & von St Paul (1951) that learning may play a role in the predator-avoidance response of bullfinches (*Pyrrhula pyrrhula*); the behaviour of individual birds changed after being shown by a predator model. It might be that in social animals such learning would play an even more important part; birds such as gulls which nest in large colonies could clearly benefit greatly if they were able to gain information from interactions between their conspecifics and their predators.

411

There is evidence that social learning effects play an important role in various Corvids (Lorenz 1931; Kramer 1941).

Gulls react to predator models in approximately the same way as to live predators (Kruuk 1964), and this has led me to investigate whether the reactions of these birds to predator models change after they have seen such models with a 'captured' other gull. In the present paper I will describe experiments carried out in a mixed colony of herring gulls and lesser black-backed gulls in which predator models are presented to the birds in various situations to test the hypothesis that gulls learn from other birds' experiences with a predator.

Methods

Observations were carried out in the breeding season of 1974 in a large colony of herring gulls (*Larus argentatus*) and lesser black-backed gulls (*Larus fuscus*) on Walney Island, Cumbria, England; the area, and the gulls' breeding season have been described earlier by Brown (1967b), Tinbergen & Falkus (1970) and MacRoberts & MacRoberts (1972). The two species of gull nest in approximately equal numbers amongst each other, practically completely mixed (although they have somewhat different site preferences). We estimated approximately 47 000 pairs, taking the two species together, at a mean density of 0·034 pairs per square metre (count in May 1974, organized by Dr M. Norton-Griffiths). In general, the vegetation is very short and there are large open areas where models can easily be presented.

The behaviour of herring gulls and lesser black-backed gulls is in many respects almost identical (Brown 1967a), and I could not detect any differences in their responses to predators, except in such aspects as the quality of their calls. Everywhere in the colony and in all the model presentations both species of gulls were involved; thus, for the purpose of this paper I have considered them together and have treated them as one species.

The experiments described here consist of presentations of predator models to the gulls in the colony. The models used were a stuffed hedgehog, stoat and fox and have been described earlier (Kruuk 1964). Dead gulls (*L. fuscus* or *L. argentatus*) were presented alone or with the predator model. The presentations were made at various places anywhere well inside the colony by the observer who was either on foot or in a car; after a model was put down the observer walked or drove to 50 or 100 m away, watching the reactions of the gulls from a point where his presence obviously did not affect the birds near the model (judging from the behaviour of birds in between). The dead bird was presented on its back, and if presented together with a predator model it was immediately next to the head of the latter. The birds usually settled well within $\frac{1}{2}$ min of the experimenter leaving; the first observation in each series was made 1 min after the model was put down, and most tests lasted 5 min, with observations at 1-min intervals. Care

was taken that the models were not seen by the gulls until they were in position in the colony; models were always covered by a sack until put down on the ground. Each test was done with 'naïve' birds, i.e. in a part of the colony where no models had been presented before, except when this is explicitly stated.

Most of the information collected on the gulls' behaviour after presentation of the model was on the number of gulls in the flock of birds flying above the model, and on the distance at which birds settled again on the ground after the experimenter had left ('flock size' and 'alighting distance'). In observations of small flocks (less than twenty birds) gulls were counted individually; in large flocks numbers were estimated by extrapolating from a small, countable part of the flock. Checking this method by counting gull flocks from photographs taken simultaneously showed no systematic bias and an error within 10 per cent. The alighting distance was estimated, often with sticks or landscape features at measured intervals, between the second or third nearest bird and the model (usually all nearest birds were settling equally far from the model, but sometimes one gull or a pair settled well within the vacated area and such birds were discarded by using the nearest but one gull).

All observations were made in April and May during the laying and incubation periods of the gulls; there were no detectable differences in the birds' behaviour to the models over this period (reactions of gulls to a stoat model, April compared with May, flock size $U = 173·5$, $z = 0·72$, $P > 0·23$; alighting distance, $U = 206$, $z = 0·43$, $P > 0·33$; so all data over the whole period have been lumped. Here, as well as for most other statistical evaluation non-parametric tests were applied, mostly the Mann–Whitney U-test (Siegel 1956).

Results

The supposition that the flock of birds above a predator in the colony consists of only those birds which nest in the vacated area immediately around the predator is not borne out. Many birds from territories considerable distances away were observed to circle above the model a number of times and then leave again. Hence it is not surprising that the correlation between the size of the area vacated by the gulls around a stoat model, and the number of birds overhead, is not a very good one (means over 5 min observations, $n = 40$, Spearman rank correlation co-efficient $r_s = 0·45$, $P < 0·01$) although the birds' density in the colony is relatively uniform. The independent variation of flock size and alighting distance can be demonstrated in many series of observations at individual presentations (Table I). Thus, a large part of the flock above a carnivore often consists of long-distance participants, as well as of birds nesting near by.

From casual observations of the circling flock it is clear that there is a continuous turnover of participating individuals; birds are joining

Table I. Two Examples of the Variation in Numbers of Gulls in the Flock Above a Stuffed Stoat, Largely Independent of the Size of the Area Vacated by the Birds

| | Minutes after first exposing model | | | | | | | |
	1	2	3	4	5	6	7	8
(a) No. of gulls in flock	0	150	55	16	100	11	0	0
Alighting distance (m)	3	5	4	5	5	3	3	3
(b) No. of gulls in flock	0	13	0	22	0	21	0	4
Alighting distance (m)	3	3	3	8	8	8	8	8

and leaving. This is in contrast to the situation on the ground where individual birds alight and stay at a certain distance for long periods. The sudden increases in the number of birds overhead as exemplified in Table I, sometimes clearly follow on sudden changes in the behaviour of gulls nesting nearby; for instance, a gull suddenly flying up, uttering 'kak' calls (see below) may immediately attract a crowd of gulls from further away which then circles overhead. Such a reaction to the behaviour of nearby nesting gulls was shown clearly in the black-headed gull to conspecifics attacking a hedgehog (Kruuk 1964).

Herring and lesser black-backed gulls showed very little aggression to any of the models (in contrast to the black-headed gulls, Kruuk 1964), nor did they show this to live stoats or hedgehogs in the colony. Calls from birds in the flock overhead consisted of 'kiauws' and occasional 'kaks' both associated with a tendency to flee (Tinbergen 1953).

Experiment 1

The purpose of this experiment was to investigate whether a predator model was treated by the gulls differently if it was presented together with a dead gull. The tests were carried out with a stuffed hedgehog, stoat and fox, using naïve birds for every 5 min presentation; some of the birds' reactions are represented in Figs 1, 2 and Table II. The results show that:

(a) Each of the three predator models presented alone and the dead gull alone caused the gulls to flock above it, and to vacate an area around it.

(b) There were significant differences in the number of birds in the flock above each of the three models (without the dead gull) and in the alighting distances; the hedgehog had the smallest effect, the fox the largest and the stoat was intermediate.

(c) Presenting a dead gull next to the stuffed predator caused a significant increase in flock size above the models and in the alighting distances.

(d) The mean response to a predator model with a dead gull (over 5-min tests) was much greater than the sum of the effects of that model alone and of a dead gull alone.

Experiment 2

This experiment attempts to answer the question whether the birds' reactions to the predator model changed after they had seen this model with a dead conspecific. Only the stuffed stoat was used. This was presented to a group of naïve birds for 5 min; 30 min later it was presented with a dead gull for 5 min and at the same spot; then after another 30-min interval it was presented again alone for another 5 min. As a control the stoat model was presented alone three times for 5 min, with ½-hr intervals, to another group of naïve birds. Thus, each series of presentations lasted 75 min; the two types of presentation were each repeated ten times with different groups of experimental and control birds. Figure 3 shows that the flock size above

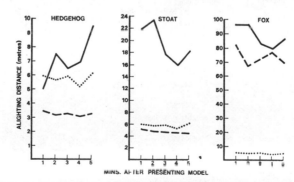

Fig. 1. Experiment 1. Mean alighting distances of gulls from different predator models, from a dead gull, and from models together with a dead gull, in 5-min tests. For number of presentations, and statistical evaluation see Table II. – – – – = response to predator model only; · · · · · = response to dead gull only; —— = response to predator model with dead gull.

413

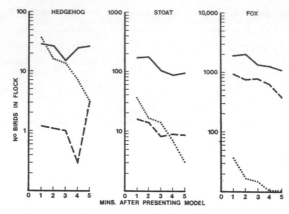

Fig. 2. Experiment 1. Mean flock sizes of gulls (log-scale) above different predator models and a dead gull. Legends as in Fig. 1.

the model decreased rapidly with successive presentations of the stoat alone (the control experiments); in fact virtually no birds were attracted to the stoat after the third minute of the second presentation (difference in flock size between first and third presentations: Mann–Whitney $U = 5.5$, $P < 0.001$). But the birds that had been exposed to the model with the dead bird still showed a flocking tendency even during the third presentation; the mean flock size above the model after exposure to model plus dead bird is 7.5 compared with

12.1 before (this difference is not significant; Mann–Whitney $U = 45.5$, $P > 0.05$). A Mann–Whitney test carried out on the ratios

$$\frac{\text{mean flock size in third presentation}}{\text{mean flock size in first presentation}}$$

shows that this ratio was significantly larger when the second presentation had included a dead gull ($U = 11$, $P < 0.01$). This effect was even more striking for the alighting distances

Table II. Mean Flock Sizes and Alighting Distances in Responses of Herring and Lesser Black-backed Gulls to Predator Models, and to a Dead Gull

Model	Flock size ($\pm$ 1SE)	Alighting distance (m $\pm$ 1SE)	No. of presentations
Hedgehog	1.4 $\pm$ 0.7	2.4 $\pm$ 0.2	22
Hedgehog with dead gull	24.6 $\pm$ 8.2	7.0 $\pm$ 1.3	11
Stoat	11.4 $\pm$ 2.0	4.8 $\pm$ 0.6	40
Stoat with dead gull	121.8 $\pm$ 25.7	24.8 $\pm$ 3.7	18
Fox	698.3 $\pm$ 195.4	74.3 $\pm$ 5.1	6
Fox with dead gull	1533.3 $\pm$ 88.8	88.7 $\pm$ 4.3	6
Dead gull alone	15.7 $\pm$ 4.4	5.8 $\pm$ 1.0	17

Mann–Whitney tests:

Hedgehog versus stoat	Flock size:	$U = 727.5$	$z = 4.23$	$P < 0.0001$
	Alighting distance:	$U = 676$	$z = 3.47$	$P < 0.001$
Stoat versus fox	Flock size:	$U = 0$	$z = 3.91$	$P < 0.0001$
	Alighting distance:	$U = 0$	$z = 3.91$	$P < 0.0001$
Hedgehog versus hedgehog with dead gull	Flock size:	$U = 34.5$	$z = 4.47$	$P < 0.0001$
	Alighting distance:	$U = 34.5$	$z = 4.47$	$P < 0.0001$
Stoat versus stoat with dead gull	Flock size:	$U = 37.5$	$z = 5.42$	$P < 0.0001$
	Alighting distance:	$U = 11.5$	$z = 5.86$	$P < 0.0001$
Fox versus fox with dead gull	Flock size:	$U = 1.5$		$P < 0.004$
	Alighting distance:	$U = 7$		$P < 0.05$

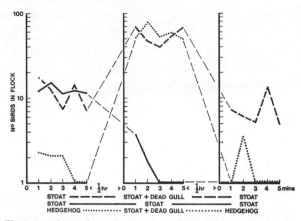

Fig. 3. Experiments 2 and 3. Mean flock size of gulls (log-scale) over predator models before, during and after presentation with a dead gull, and a control. For each point $n = 10$.

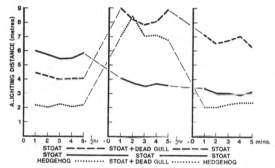

Fig. 4. Experiments 2 and 3. Mean alighting distance of gulls from predator models. Legends as for Fig. 3.

(Fig. 4); these were considerably larger in response to the model after exposure to model plus dead gull (mean 6·7 m compared with 4·2 m; $U = 20·5, P < 0·025$). If the second presentation was a stuffed stoat only (the control experiment) the responses to the third presentation were smaller than the responses to the first presentation, but the difference was not significant (mean distance 3·1 m compared with 5·8 m; $U = 32·5, P > 0·05$). Here too, the ratios

$$\frac{\text{alighting distance in the third presentation}}{\text{alighting distance in the first presentation}}$$

were significantly different between the experimental and control series ($U = 7, P < 0·001$).

In one series of presentations a group of birds which had been tested with the stuffed stoat, the stoat plus dead gull, then again the stoat alone, were presented with the stoat alone 24 hr later. The result suggested strongly that the effect of seeing the predator with the dead gull persists over at least 1 day, but more observations are needed to verify this.

Experiment 3

The results of experiment 2 do not show to what extent the change in the response of the gulls, after being exposed to a predator model with a dead gull, is a change in the response to the predator model itself; it might be that the gulls react differently to the place and experimental situation after seeing the model and the dead bird, or after seeing the dead bird alone. If this were the case one would expect the change in the birds' responses to the stoat model, after seeing it with a dead gull, to occur also in the gulls' responses to another kind of predator if presented in the same place. To test this possibility I presented naïve groups of birds for 5 min with a stuffed hedgehog, gave those same birds 30 min later a 5-min exposure to a stuffed stoat with a dead gull and after another 30 min presented them again with a stuffed hedgehog. This experiment was carried out almost simultaneously with experiment 2, and the results are also presented in Figs 3 and 4, using the results of experiment 2 as a control. The gulls' reactions to the hedgehog did not change as a consequence of seeing a stoat with a dead gull in

415

that same place (mean alighting distance before and after was 2·2 m, $U = 49$, $P > 0·05$; mean flock size before and after was 1·5 and 1·2 m, $U = 43·5$. $P > 0·05$). Thus, the results of experiment 2 should be interpreted as showing a change in the gulls' reactions towards the specific predator involved, not towards the presentation site or the experimental situation.

Discussion

The observations reported here show how herring and lesser black-backed gulls on the ground in the colony keep a certain distance from a potential predator, whilst in the air they congregate above it. A model of a stoat has a greater effect than a model of a hedgehog, and the effect of a fox model is larger still; these differences correspond with the ecological significance of these predators to the gulls, similar to what was found in black-headed gulls (Kruuk 1964). The flock above the predator consists not just of birds which have their territories immediately around the place where the predator is, but also of gulls which have flown in from a long distance; numbers in the flock are to a considerable degree independent of the size of the area around the model vacated by the gulls. The tendency to congregate above the predator is considerably higher towards a predator model accompanied by a dead gull than towards a model alone. The same can be said for the alighting distance around a predator model; gulls keep much further away from a predator with a dead gull than from either the dead gull or the predator singly. But after being exposed to a predator with a dead gull the birds' attraction (as measured by flock size) towards the predator alone changes little, whereas their alighting (= fleeing) distance increases considerably.

From these observations we may conclude that gulls do obtain information about a predator's behaviour when it turns up in the colony, and that this information is important in determining later reactions towards that particular predator species. The observations do not show whether the predator-attraction itself is important in this learning process; it might be that birds which fly away immediately, rather than stay around and circle over the enemy, show the same kind of learning process. This is not likely, however, but the possibility should be tested. Considerably more information is needed, too, about the behaviour of individual gulls, about the stimulus-situation for this predator-attraction, etc.

Whatever the mechanism of the learning process, it is clear that gulls are able to make use of their conspecifics' bad experiences with a potential enemy, a trait which could have very considerable adaptive significance. In order to judge to what extent there is survival value in the increased fleeing distance towards a predator with a dead gull we would have to make predictions about the likely foraging behaviour of a predator not associated with a dead gull, and of a predator which has recently killed one.

The stoat's usual diet does not include birds of the size of large gulls (Southern 1964; Day 1968; Müller 1970; Ewer 1973), although it is able to kill them and does so (personal observations; see also Hewson & Healing 1971). Individual stoats may specialize on relatively large prey (Müller 1970), a phenomenon found in several other species of carnivore (Ewer 1973); also, in stoats as in other carnivores, 'surplus killing' is known to occur (Hewson & Healing 1971). Thus, having killed one victim does not or hardly decreases the likelihood of killing another, on the contrary, it may actually show up the individual stoat as a gull killer rather than for example a mouse eater. Surplus killing is well-documented from foxes in a colony of black-headed gulls (Kruuk 1964, 1972b) and amongst roosting herring gulls (Goethe 1956); from the pattern in which the kills occurred in the black-headed gull breeding season it was clear that they were highly 'clumped' in time; in other words, once a fox had killed a gull he was more, rather than less, likely to kill another soon after. Observations since then, in the same black-headed gull colony, indicate that some individual foxes are more likely to kill large numbers of gulls than others; this, too, is evidence that there is a greater likelihood of a fox attempting to kill more gulls once it has been seen to kill one, compared with a fox which has not killed a gull. Hedgehogs are not known to kill birds of herring gull size, although they do kill almost fully grown, black-headed gulls' chicks (Kruuk 1964). Proper surplus killing does not seem to occur in this species; however, it may eat as little as one preening gland from a victim, then capture another.

I suggest, therefore, that an individual predator which has killed a gull is more likely to kill another than a potential predator which has not done so. It follows that the increased alighting (= fleeing) distance of gulls to a predator with a dead gull could be an adaptation to protect birds against a typical carnivore behaviour pattern, surplus-killing. The main survival value of the attraction towards predators, then, might be that it brings birds into a position where they collect information about a potential source of danger. This information influences further anti-predator reactions immediately and for some time after; was partly based on the experience of conspecifics.

Acknowledgments

This study was financed from a grant of the Natural Environment Research Council to Professor N. Tinbergen, F.R.S. I am grateful to the South Walney Naturalists' Trust for permission and facilities to work in the Walney gulleries; and to Dr Richard Dawkins for critical comments on the manuscript.

REFERENCES

Brown, R. G. B. (1967a). Species isolation between herring gull *Larus argentatus* and lesser black-backed gull *Larus fuscus*. *Ibis*, **109**, 310–317.
Brown, R. G. B. (1967b). Breeding success and population growth in a colony of herring and lesser black-backed gulls *Larus argentatus* and *L. fuscus*. *Ibis*, **109**, 502–515.

Day, M. G. (1968). Food habits of British stoats (*Mustela erminea*) and weasels (*Mustela nivalis*). *J. Zool., Lond.*, **155**, 485–497.

Ewer, R. F. (1973). *The Carnivores*. London: Weidenfeld and Nicholson.

Goethe, F. (1956). Fuchs, *Vulpes vulpes*, reibt Schlafgesellschaft von etwa sechzig jugendlichen Silbermöwen (*Larus argentatus*) auf. *Säuget. kdl. Mitt.*, **4**, 58–60.

Hewson, R. & Healing, T. D. (1971). The stoat *Mustela erminea* and its prey. *J. Zool. Lond.*, **164**, 239–244.

Kramer, G. (1941). Beobachtungen über das Verhalten der Aaskrähe (*Corvus corone*) zu Freund und Feind. *J. Ornithol.*, **89**, 105–131.

Kramer, G. & St Paul, U. von (1951). Über angeborenes und erworbenes Feinderkennen beim Gimpel (*Pyrrhula pyrrhula* L.). *Behaviour*, **3**, 243–255.

Kruuk, H. (1964). Predators and anti-predator behaviour of the black-headed gull (*Larus ridibundus* L.). *Behaviour Suppl.*, **11**, 1–129.

Kruuk, H. (1972a). *The Spotted Hyena*. Chicago: University of Chicago Press.

Kruuk, H. (1972b). Surplus killing by carnivores, *J. Zool., Lond.*, **166**, 233–244.

Kruuk, H. & Turner, M. (1967). Comparative notes on predation by lion, leopard, cheetah and wild dog in the Serengeti area, East Africa. *Mammalia*, **31**, 1–27.

Lorenz, K. (1931). Beiträge zur Ethologie sozialer Corviden. *J. Ornithol.*, **79**, 67–122.

MacRoberts, M. H. & MacRoberts, B. R. (1972). Social stimulation of reproduction in herring and lesser black-headed gulls. *Ibis*, **114**, 495–506.

Müller, H. (1970). Beiträge zur Biologie des Hermelins. *Säuget. kdl. Mitt.*, **18**, 293–380.

Schaller, G. B. (1972). *The Serengeti Lion*. Chicago: University of Chicago Press.

Siegel, S. (1956). *Nonparametric Statistics for the Behavioral Sciences*. New York: McGraw-Hill.

Smythe, N. (1970). On the existence of 'pursuit invitation' signals in mammals. *Am. Nat.*, **104**, 491–494.

Southern, H. N. (1964). *The Handbook of British Mammals*. Oxford: Blackwell.

Tinbergen, N. (1953). *The Herring Gull's World*. London: Collins.

Tinbergen, N. & Falkus, H. (1970). *Signals for Survival*. Oxford: Clarendon Press.

Walther, F. R. (1969). Flight behaviour and avoidance of predators in Thomson's gazelle (*Gazella thomsonii* Guenther 1884). *Behaviour*, **34**, 184–221.

(*Received* 7 *February* 1975; *revised* 15 *March* 1975; *MS. number*: 1403)

49

The Relation between Ecology and Social Structure in Primates

J. F. Eisenberg, N. A. Muckenhirn, R. Rudran

Reprinted from 26 May 1972, Volume 176, pp. 863-874

Copyright© 1972 by the American Association for the Advancement of Science

Selected data on primate social systems will be discussed and reinterpreted in this article, with a view toward modifying the existing concepts about the adaptive nature of these social systems (*1–3*) In light of recent data, we limit the concept of the multi-male troop (*2*) and call attention to the reproductive group as an organic unit that shows stages of growth and decline which may vary under different environmental circumstances. To this end, we introduce a new category of

social structure for the primates—the age-graded-male troop. Before proceeding with the definitions and discussions, we must put the data into a historical framework.

The Existence of "Species Typical" Social Organization

The cornerstones of Carpenter's theories of primate social structure (*4*) can be summarized as follows: (i) primate troops tend to have more or less exclusive home ranges (*5*); (ii) the average size of a troop tends to be typical for a given species (the term "apoblastosis" defines the process whereby primate troops divide to restore a spe-

Dr. Eisenberg is resident minist, National Zoological Park, Smithsonian Institution, Washington, D.C. 20009; Miss Muckenhirn is a research associate, National Zoological Park, Smithsonian Institution; and Mr. Rudran is a graduate student, department of zoology, University of Maryland.

417

cies-specific balance in numbers); and (iii) the composition of the troop, with respect to the proportions of sex and age classes, tends to be relatively invariant, regardless of troop size. The ratio of males to females was termed the "socionomic sex ratio." This ratio tended to be species-specific, and, for most species, there were more adult females than adult males in a given troop. Two corollaries followed: first, there was a strong polygynous trend in most primate species; second, extra adult males were excluded from the troop by some process, to dwell either in a peripheral subgroup of their own or as solitary individuals.

In line with Carpenter's generalizations and some early theories of social evolution, most field workers believed that the behavioral attributes of a species were relatively constant and reflected species-specific adaptations. It was assumed that social structures resulting from the defined patterns of interaction of a given species would manifest themselves as rather predictable entities (1, 2). This assumption still has heuristic value, but the task is much more complex, since, for most primate species, social structure varies with habitat. This has been amply demonstrated for the olive baboon, *Papio anubis* (6, 7), the gray langur, *Presbytis entellus* (8–10), and the vervet monkey, *Cercopithecus aethiops* (11, 12). On the other hand, some species appear to have a standardized form of social organization—for example, the lar gibbon, *Hylobates lar* (13, 14), and the hamadryas baboon, *Papio hamadryas* (15, 16).

The causes of intraspecific variation in social structures are related, in part, to differences in habitat (especially factors of food availability and predation), as well as to differences intrinsic to the troop itself (3); however, once the range in the variation of troop structure is described, a "modal" social organization for a given species can often be

discerned, thus facilitating comparisons with other species. It would appear that one generalization is possible: those species that exhibit a wide range of adaptation to differing habitats often show an equally wide range in social structure [for example, *Papio anubis* (6) and *Presbytis entellus* (9)]. On the other hand, species that show a uniform adaptation to specific kinds of habitats often show a corresponding uniformity in their grouping tendencies (1). In fact, when a group of allopatric species shares the same relatively narrow range of adaptation, then this group begins to exhibit a predictable "adaptive syndrome" with respect to feeding, antipredator behavior, spacing mechanisms, and social structure. Examples from two separate "syndromes" are (i) the arboreal, leaf-eating monkeys of Africa and Asia [*Presbytis cristatus* (17), *Presbytis johni* (18), *Presbytis senex* (19), and *Colobus guereza* (20)] and (ii) the slow-moving, insectivorous lorisoids of Africa and Asia [*Arctocebus, Perodicticus, Nycticebus,* and *Loris* (21–23)].

Grades of Social Structure—

A Reassessment

It has not been uncommon to find primate societies classified into grades based on supposed increases in social complexity. The implicit suggestion was that higher grades were achieved through evolutionary stages, but it was well recognized that, within and among each major primate taxon, considerable parallel evolution had occurred (1, 2). Even within the morphologically conservative Prosimii, social organizations equal in complexity to those of the Cebidae and Cercopithecidae were formed (24, 25).

The so-called solitary-living species are characterized by a minimum amount of direct social interaction with conspecifics of either sex in the same age

Table 1. Range of social organization and feeding ecology for selected primate species (see text for descriptions).

Solitary species	Parental family	Minimal adult ♂ tolerance* (uni-male troop)†	Intermediate ♂ tolerance‡ (age-graded-male troop)†	Highest ♂ tolerance§ (multi-male troop)†
A. Insectivore-frugivore	A. Frugivore-insectivore	A. Arboreal folivore	A. Arboreal folivore	A. Arboreal frugivore
Lemuridae	Callithricidae (Hapalidae)	Cebidae	Colobinae	Indriidae
Microcebus murinus	*Saguinus oedipus*	*Alouatta palliata*	*Presbytis cristatus*	*Propithecus verreauxi*
Cheirogaleus major	*Cebuella pygmaeus*	Colobinae	*Presbytis entellus*	Lemuridae
Daubentoniidae	*Callithrix jacchus*	*Colobus guereza*	Cebidae	*Lemur fulvus*
Daubentonia madagascarensis	Cebidae	*Presbytis senex*	*Alouatta palliata*	B. Semiterrestrial frugivore-omnivore
Lorisidae	*Callicebus moloch*	*Presbytis johni*	B. Arboreal frugivore	Cercopithecidae
Loris tardigradus	*Aotus trivirgatus*	*Presbytis entellus*	Cebidae	*Cercopithecus aethiops*
Perodicticus potto	B. Folivore-frugivore	B. Aboreal frugivore	*Ateles geoffroyi*	*Macaca fuscata*
B. Folivore	Indriidae	Cebidae	*Saimiri sciureus*	*Macaca mulatta*
Lemuridae	*Indri indri*	*Cebus capucinus*	Cercopithecidae	*Macaca radiata*
Lepilemur mustelinus	Hylobatidae	C. Semiterrestrial frugivore	*Miopithecus talapoin*	*Papio cynocephalus*
	Hylobates lar	Cercopithecidae	C. Semiterrestrial frugivore-omnivore	*Papio ursinus*
	Symphalangus syndactylus	*Cercopithecus mitis*	Cercopithecidae	*Papio anubis*
		Cercopithecus campbelli	*Cercopithecus aethiops*	*Macaca sinica*
		Cercocebus albigena	*Cercocebus torquatus*	Pongidae
		Cercopithecidae	*Macaca sinica*	*Pan satyrus*
		Erythrocebus patas	D. Terrestrial folivore-frugivore	
		Theropithecus gelada	Pongidae	
		Mandrillus leucocephalus	*Gorilla gorilla*	
		Papio hamadryas		

* Troop with one adult male and strong intolerance to maturing males.

† "Troop" refers to the basic social grouping of adult females and their dependent or semidependent offspring.

‡ Troop typically showing age-graded-male series.

§ Troop with several mature, adult males and age-graded series of males.

419

class. Typically, the "mother family" (that is, an adult female and her dependent offspring) forms the only cohesive social unit that indulges in daily, intimate interaction. Nevertheless, solitary species, whether primates, carnivores, or rodents, have a social life (*26*), and indirect communication is maintained among adults that have neighboring or overlapping home ranges. The communication channels of solitary species are characterized by olfactory and auditory modalities which maintain spacing except at mating times (*1, 22*). The terms "solitary," "asocial," and "dispersed" have been objected to (*27*) because they obscure the fact that a given pair of adults and their subadult descendents can share a home range completely or partially, even though they do not indulge either in communal nesting or regular physical contact. In addition, when there are overlapping home ranges, the same adult pair can reproduce in subsequent years; thus, although individuals are dispersed, a family structure and relatively closed breeding unit (*28*) are maintained. For example, in *Microcebus murinus* and *Galago demidovii* (*29*), the home ranges of adult females may overlap considerably. A reproductive male's home range includes the home ranges of from one to six females and their juvenile offspring, while extra males live on the periphery of the dominant male's home range, either as solitary individuals or as a noncohesive bachelor group. The spatial distribution of the adults (*29*) implies a polygynous breeding system (see Fig. 1). Other nocturnal prosimians appear to exhibit a similar spacing system—for example, *Cheirogaleus major, Daubentonia madagascarensis, Loris tardigradus, Perodicticus potto,* and *Lepilemur mustelinus* (*22–24, 30*) (see Table 1).

The parental family structure is characterized in its extreme form by a bonded pair of adults and their immature descendents. This bonded state occurs rarely within the order Primates [assuming the term "bond" includes only social relationships between specific individuals based on the performance of mutually reinforcing activities, in addition to mating behaviors (*31*)]. It follows then that grooming, huddling, and other nonsexual behavior engaged in on a daily basis by two individuals of the opposite sex defines the pair bond.

Marmosets of the genera *Saguinus, Cebuella,* and *Callithrix* are examples of the bonded parental family (*32, 33*). These marmosets exhibit a unique form of parental care, in that the male participates to an extent unparalleled by the other families of Primates (*1, 32*). The male marmoset typically transports the young from the time they are born, transferring them to the female only for nursing (*33*).

The gibbons, *Hylobates* and *Symphalangus* (*13, 14, 34*), as well as some species of the family Cebidae [notably the Titi monkey, *Callicebus moloch* (*35*)], show grouping tendencies similar to those of the marmosets in that a single adult pair and their dependent offspring occupy a given home range, but the male's participation in the care of the young is limited. For example, *Callicebus* and *Aotus* males (*36*) transport the young to a certain extent and thus appear to be somewhat closer to the marmoset pattern (*33*), while gibbons and *Indri* males participate little in the rearing of infants. Thus, at least two subvariants of the parental spacing system must be designated: (i) mutual participation by the male and female in the rearing of offspring and (ii) limited or no direct participation by the male in the rearing of dependent young (see Table 1).

Parental groups have probably evolved by at least two different pathways (Fig. 1). The parental group, a more cohesive form of primate social structure. can easily be derived, in a phylogenetic sense, from the more dispersed systems of some of the prosimians, for example, *Galago demidovii* or *Microcebus murinus* (29).

If only adult female primates were to form affiliations, an extended mother family would exist, given that the exclusion of excess adult males takes place through the aggressive action of the parental adult male. If the parental male were closely bonded to the extended group of mothers and daughters, the result would be a typical uni-male group (Fig. 1). Extra-group males would exist on the periphery of the reproductive units. Recent research on terrestrial African primates (the Patas monkey, *Erythrocebus patas*; *Papio hamadryas*; and the Gelada baboon, *Theropithecus gelada*) has indicated the existence of troops composed of several females, their dependent offspring, and one sexually mature, adult male

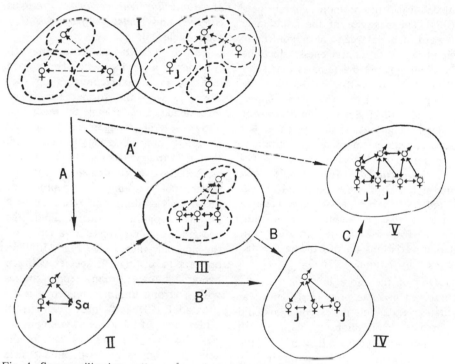

Fig. 1. Space utilization patterns for selected mammals with hypothetical evolutionary pathways. (I) Solitary pattern: adult males and females have separate centers of activity and encounter infrequently. The extended ranges of the males overlap with the home ranges of the females. Only polygynous patterns are presented; *Microcebus murinus* (29). (II) Family group: a bonded pair and their subadult offspring travel as a homogeneous unit in an exclusive home range; *Saguinus oedipus* (32). (III) Uni-male group (extended mother family): an adult male is in periodic contact with a cohesive group of adult females and their progeny; *Ateles geoffroyi* (46). (IV) Uni-male group: an adult male is in relatively constant contact with a cohesive group of adult females and their progeny; *Papio hamadryas* (16). (V) Multi-male group: a cohesive group of several adult males and females with their progeny; *Papio anubis* (49). Hypothetical phylogenetic steps in the formation of mobile, cohesive groupings are indicated by arrows. Routes A, B', and C or A', B, and C are the most probable steps in the formation of cohesive, multi-male groups (1). The arrows formed from dashed lines are less probable evolutionary pathways (1). For simplicity, the symbols J (juveniles) and Sa (subadult) have been included only occasionally to indicate the presence of immature animals.

(16, 37, 38). In such species, troops can also contain a few maturing males as a normal transitional stage in the life cycle of the troop; however, the young males generally leave the parental troops as subadults and join bachelor bands.

The uni-male troop is more complex in its structure than is the parental group because there is an increased representation of sex and age classes in the troop. Adult females must be tolerant of one another and have affiliation mechanisms to promote cohesiveness (39). The behavior of the adult male toward females can also contribute significantly to cohesiveness, since he may "discipline" the females that stray from his group, thereby keeping the group intact [for example, *Papio hamadryas* (16, 40)]. Such herding behavior by the adult male is not typical of all species with a uni-male configuration; it has not been reported for *Cercopithecus campbelli* (41), for example. New uni-male troops can be formed in *Papio hamadryas* when a solitary younger male is able to "capture" a subadult female from a structured breeding unit. Other variations on this theme are discussed by Kummer (16).

One of the more intriguing aspects observed in some primate societies was that several adult males could and did associate continuously with adult females and young. Because of the contrast with other mammalian taxa that have complex social structures in which males are not permanent members of the group, the "multi-male group" came to be thought of as an advanced and almost unique characteristic of higher primates (1, 2, 4). While some species do have a multi-male troop, it is obvious that the concept has been applied too broadly (42). An intermediate form of social organization, between the uni-male and the multi-male structures, should be recognized. This may be termed the age-graded-male troop. Although several males of varying ages coexist in such troops, there are proportionately fewer males in these troops than there are in true multi-male troops (whose sex ratio may approach 1 : 1) (6). The linear male dominance order is based on the age of the males, with no definable subunit of several males in the oldest age bracket. The lack or absence of fully adult males of equivalent age is the characteristic that defines an age-graded-male troop [for example, troops of *Mandrillus leucocephalus*, *Presbytis entellus*, and *Ateles geoffroyi* (43–47) (see Table 1)].

The age-graded-male troop may be considered a phylogenetic step toward the true multi-male configuration (see Fig. 1), with the former having an intermediate level of male tolerance that allows several young males to mature longer within the troop of their birth than do young males in uni-male troops. Nevertheless, a fundamental tendency toward polygyny and the possibility of the troop's splitting and returning to a uni-male condition remain. Thus, the age-graded-male troop is a variation on the uni-male theme (see Fig. 2). What appears to distinguish species with an age-graded-male troop from species with a strong uni-male tendency is (i) the adult "leader" male exhibits a wider range of tolerance of young males near his own age and (ii) part of the tolerance shown by the dominant male appears to derive from the fact that these species generally have larger troops. In addition, the larger troops have larger home ranges and therefore more possibilities for dispersing into subgroups while foraging. This very tendency toward fractionation can generate new troops by apoblastosis (4).

Members of the genus *Papio*, particularly the species *ursinus*, *anubis*, and *cynocephalus*, are adapted to savannas and forage a great deal on the ground, although they retire to rocky places or

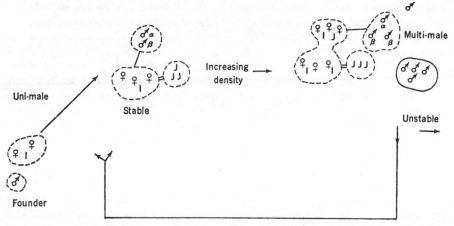

Fig. 2. Hypothetical diagram of troop growth for arboreal primates. The assumption is that a uni-male tendency is the most typical configuration at moderate population densities. Given a founder situation at the left, consider an adult male attached to a cohesive unit of two adult females and their young. The troop grows by recruitment, yielding a subgroup of juveniles and a beta male that is subdominant to the alpha male or father. The two older males form a subgroup in their own right. At greater densities, younger males may form a peripheral subgroup of their own that has no direct contact with the basic subgroup of mothers and their young. The subgroup of adult males may now be augmented slightly to three, with the founding father still dominant. Although the troop now appears to be multi-male, it would be more correct to consider it an age-graded-male group. Splitting of the new unstable troop can lead to the original uni-male configuration (19, 47, 66).

trees for sleeping at night (48, 49). These species exhibit classic multi-male groups [as do, with some limitations, the semiterrestrial macaques, including Macaca mulatta, M. fuscata, and M. speciosa (50–52)]; however, to classify other primate species that have more than one male in the troop as multi-male and then to rank these species within the same grade of social structure as these terrestrial macaques and baboons is to oversimplify matters.

The multi-male troop is characterized by an oligarchy of adult males that are roughly equivalent in age (49). These males show affiliation behaviors, and, although they may be ranked in a dominance order, the ranking is not pronounced within the oligarchic subgroup. Cooperation exists among the superdominant males (39), and they actively oppose and dominate younger males in the subadult age class (49). Species exhibiting a multi-male config-

uration do, of course, have within them males that may be arranged in an age-graded series (53, 54). Furthermore, there may be solitary males that live independently of the troops and that join and leave troops in a semiregular pattern (55, 56).

The multi-male troop was first defined for Papio anubis, primarily in the pioneer studies of Washburn and De Vore (48, 49). Baboons were portrayed as living in cohesive troops of 30 to 50 animals; solitary males were not described originally (48), but were subsequently noted for Papio cynocephalus (56). The multi-male troop appeared to serve as an antipredator device, since the presence of many adult and subadult males permitted collective attack should the troop be menaced by a terrestrial predator.

Although the sex ratio at birth is almost equal in Papio and Macaca, the number of reproducing females for each adult male varies from one to three. Any difference in the adult sex ratio is presumed to result from both differential mortality and the differential maturation rates of the sexes. The adult males are organized in a dominance hierarchy, and the alpha male does most of the breeding during the peak of a female's estrous period (49, 54).

Recent studies by Rowell (6) and Altmann and Altmann (56) indicate that in both forest and savanna habitats baboons of the genus *Papio* show neither such strong dimorphism in social roles nor such a disparity in the male-female sex ratio as was originally described (57). In light of more recent evidence, we will probably have to alter our concept of the baboon social life. Nevertheless, the existence of several adult males of an equivalent age serves to crystallize and define the concept of the multi-male troop.

In previous reviews of primate social organizations, the uni-male groupings characteristic of *Erythrocebus patas* (37), *Papio hamadryas*, and *Theropithecus gelada* (16, 38), all three of which are adapted to arid climates, were considered specialized offshoots of the multi-male grouping that supposedly characterized most advanced primates (2, 3). This tendency to characterize most primate species by a multi-male social structure and to set aside those species exhibiting uni-male groups as cases of adaptation to extreme environments was motivated, in part, by a desire to emphasize the uniqueness of primate societies when contrasted with the social groupings of other mammals.

The preoccupation with the concept of the multi-male organization was unfortunate for at least two reasons: (i) male-male tolerance mechanisms were studied to the virtual exclusion of the equally complex spacing and affiliation mechanisms exhibited by females [as an example of such exclusive attention see (58)], and (ii) differences in the roles of males among the various species exhibiting so-called multi-male systems were masked by lumping a broad range of species-typical social organizations under the term "multi-male groups." In support of this last statement, it should be pointed out that the multi-male troops of the lemurid primates, such as *Lemur catta* and *Propithecus verreauxi* (25, 59), are organized around a female matriarchy that differs markedly in discrete social control mechanisms from the multi-male troops of the savanna baboons, *Papio anubis*, *P. ursinus*, and *P. cynocephalus* (56, 57).

The preoccupation with the "unique," the desire to relate primate behavior to human behavior, and the lack of sound ecological studies are hindrances at this stage of our understanding. What is needed more than ever is a clear appraisal of primate species as mammals exploiting an ecosystem and subject to ecological pressures similar to those acting on other mammalian species (3).

We propose that not only terrestrial primate species adapted to arid climates but also most primate species adapted to forests are characterized by either uni-male troops or age-graded-male troops. The true multi-male system is a less frequently evolved specialization, and the term "multi-male" should be restricted to those species having large troops that include several functionally reproductive adult males, as well as nonreproductive males of different ages.

Regulation of Troop Size

One of the key problems in describing a multi-male troop involves the definition of an adult male. In censusing primate troops, most workers use the category "adult male" for males that are sexually mature; however, it is well known that the age of sustained spermatogenesis does not necessarily correlate with the age of "social" maturity. In most medium-sized primates, a male that has become sexually mature at 3 to 5 years of age may not be sociologically mature or physically dominant until 8 or 9 years of age—yet all of these males are lumped together as adults. A similar social maturation sequence has been described in elephants, for which the problems of age-class definition are comparable to those encountered in primates (60, 61).

Long-term studies of Macaca fuscata and M. mulatta show how roles are established and maintained and how males succeed to leadership (54, 62–64). The rank of young males in later life is, in part, dependent on the status of their mothers (63). Fractionation of a large troop to form two new troops may be accompanied by the deposition of an old leader, the assumption of leadership by a solitary male, or the succession to leadership of second- or third-ranking males upon the removal or loss of an old leader (62, 64). Without recounting the extensive research on this, we wish to make the point that the processes of splitting do not necessarily involve extensive mortality in the infant and juvenile age classes. Furthermore, the new troops thus generated are not essentially uni-male in structure.

Figure 2 shows the hypothetical growth of a primate troop from a founder situation to eventual instability and breakup because of internal recruitment and crowding that resulted from close neighbors. More data are needed on the natural genesis of troops, since we believe that there is a strong tendency toward a polygynous, uni-male reproductive unit for most species of forest-dwelling primates in the New World and Old World (65).

The case of a Mona monkey, Cercopithecus campbelli lowei (66), troop parallels the diagram in Fig. 2. From 1964 to 1968, the troop increased in size because of births to the founder adult male and four adult females. In the years 1968 to 1970, the number of births was curtailed and four young adult males from 3 to 4 years of age emigrated. During this transition period, the tendency toward a uni-male structure was masked and the troop might have been called multi-male.

Ateles geoffroyi and A. belzebuth maintain an age-graded-male troop (46, 47, 67). In contrast to Cercopithecus campbelli lowei, adult male spider monkeys are not in continuous association with adult females and young. Instead, an Ateles "group" is composed of units that contain one or more females and their dependent young and that forage independently in a common home range. The adult male will accompany the female units when females are in estrus and when there exist special positive, dyadic relationships between individuals. The age structure of the males, based on birth intervals, lends itself to the formation of an alpha-beta-gamma dominance hierarchy. Among subadult males, mutual support may be shown during offensive and defensive behavior toward potential predators and intruders, but with increasing age some young males become peripheral.

The range of social organization encountered in Ateles geoffroyi in either space or time is given in Fig. 2. At one extreme is the founder situation, or so-called uni-male group. The social organization of this species is generally several females and a semidetached male group that is, in essence, an age-graded group, since there is a male hierarchy based on age and most of the mating during a peak of a given female's estrus is probably accomplished by the older, dominant male. With denser populations, there are larger groups of females and more peripheral males, and there may be fighting among males when associated with female groups (68).

The genesis of two new troops from a single large troop in those species exhibiting the age-graded-male system can occur without a take-over of leadership or fighting among males. For example, division of a large *Presbytis entellus* troop consisting of approximately 30 individuals would begin by a departure from the sleeping tree and a fractionation into two foraging subgroups, each subgroup being under the leadership of a different male. The largest subgroup of females would follow the dominant male, the second subgroup following a subdominant male. At the conclusion of the day's foraging, the animals would return to a common sleeping place. The females themselves would exert a powerful influence on the ultimate composition of the subgroups, depending upon which male they followed. The second stage in troop development would consist of separate foraging patterns and occasional utilization of different sleeping trees. Eventually, the subgroups would be foraging and sleeping independently; thus, two new troops would be created. Examples are further amplified for *Presbytis entellus* by Muckenhirn (*44*).

Long-term population studies on Barro Colorado Island permit some generalizations concerning the grouping tendencies of *Alouatta palliata*, the howler monkey (*69, 70*). Chivers (*71*) has summarized the data for the island population and for one troop (the laboratory group) in particular. Although this species is often cited as exhibiting a multi-male structure, peripheral or extra-group males exist and uni-male troops occur frequently. It would appear that, at low population densities, this species approximates a uni-male structure (*72*), while at higher densities a temporary age-graded-male structure appears. We must reemphasize that, when there exists an age-graded series of related maturing males with one older, dominant male, then the multi-male structure is more apparent than real. In fact, in large troops of *Alouatta palliata* with several mature,

adult males, there appears to be curtailed reproduction (*73*). This, then, leads us to consider some of the ecological factors that lead to the division of troops and what forms of social pathology may result in high-density populations.

Social Pathology and Density

For many primate species, the conditions under which splitting and male "take-overs" can occur are, in part, related to the density of, and degree of disturbance in, the population (*19*). Alternatively, fractionation of large groups can occur under ecological conditions (such as dispersed resources) that favor the maintenance of small groups and that may even approximate the founder situation (*12*).

Presbytis senex of Ceylon shows some rather instructive trends in population growth and composition when high-density populations are compared with low-density populations. As Table 2 indicates, *Presbytis senex* tends to live in uni-male reproductive units (*19, 74*) and is found in a wide range of habitat types, from a lowland dry zone (Polonnaruwa) to a highland wet zone (Horton Plains). Extra-troop males are organized into groups having an average size of 7.5 individuals. In high-density populations, the uni-male reproductive groups are subject to harassment from the peripheral bachelor groups, which occasionally results in infant mortality and leadership takeovers. Infant mortality is reflected, in part, by the different percentages of subadults and juveniles in low- and high-density populations (see Table 2). In Sugiyama and Mohnot's studies in India, *Presbytis entellus* infants were actually attacked and killed by invading adult males (*10, 45*). In the Ceylon langurs, Rudran did not witness such events directly, but infants and juveniles were found to be injured or missing

426

Table 2. Comparison of the populations of *Presbytis senex* found at Polonnaruwa and Horton Plains (*19*). (Population densities are minimum estimates that may show higher levels in more restricted sample areas.)

Presbytis senex	Polon-naruwa	Horton Plains
Number of troops studied	33.0	27.0
Total population studied	278.0	229.0
Population density (per square kilometer)	215	92.6
Percent of adults in population	63.8	51.3
Percent of subadults and juveniles in population	14.1	30.7
Percent of adult males in adult population	28.2	36.7
Percent of adult females in adult population	71.8	63.3
Ratios of adult males to adult females in total population	1 : 2.5	1 : 1.7
Average number of animals in uni-male troops	8.4	8.9
Average number of animals in predominantly male troops	7.5	7.5
Ratio of adult males to adult females in uni-male troops	1 : 4.1	1 : 3.3
Number of infant deaths*	19.0	2.0

* Based on one complete reproductive period.

after a male replacement had occurred. This type of male replacement appears to occur under conditions of high population density (*9*) or in marginal habitats (*45*); in either case, the altered age structure that results from male take-overs curtails population growth to some extent.

The suggestion that the uni-male structure is a response to crowding stress has been made before (*12*). The tendency for captive primate groups to assume a uni-male or "despot" male configuration was often assumed to be a pathological response; however, we believe that, although this uni-male condition may manifest itself at crowding densities, it is erroneous to think of the uni-male structure as being pathological.

Some Correlations between
Social Structure and Ecology

The history of primate evolution has been subject to many reviews (*75*). Beginning with an insectivore-like form exhibiting certain arboreal adaptations with rather enlarged eyes, the primates underwent an extensive radiation throughout the Paleocene, giving rise to two main branches. One major branch differentiated into the present-day galagos, lorises, and lemurs, while the other differentiated into the Old World monkeys, New World monkeys, tarsiers, pongids, and hominids. The lorises and galagos still persist in Africa and South Asia as nocturnal, forest-adapted forms. These may be considered the most morphologically conservative primates, representing most nearly Paleocene forms.

The island of Madagascar served as a reservoir for the lemuroid primates, and this isolated radiation resulted in the occupancy of feeding niches that replicate, in part, continental niches (*76*). Deriving from the second major radiation, the neotropical monkeys (Hapalidae and Cebidae) began their adaptations in the Oligocene in isolation from the Old World monkey radiation. Thus, the Madagascan, neotropical, and Palaeotropical radiations can be compared to elucidate convergences.

Judging from the habits of the living primates and the structure of fossil forms, the early primates were nocturnal and arboreal and subsisted on an

427

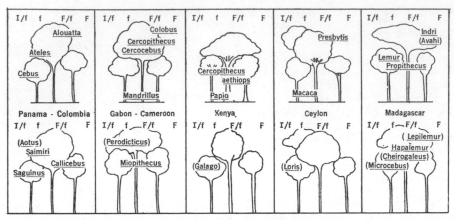

Fig. 3. Ecological equivalence for selected primate species. Horizontal scaling: *I/f*, insectivore-frugivore; *f*, frugivore; *F/f*, folivore-frugivore; *F*, folivore. Since the feeding categories are not absolute, only relative trophic preferences are indicated. The midpoint of the name should lie at the modal feeding classification. Vertical scaling: relative feeding height is illustrated by the position of the name. Lowest position: semiterrestrial; middle position: arboreal, second growth; highest position: upper canopy feeder. Names in parentheses indicate a rhythm of nocturnal activity. The upper series are all medium to large primates (adult weight, greater than 5 kilograms); the lower series are all small primates generally (adult weight, less than 2 kilograms).

omnivorous diet, including fruits, small invertebrates, and perhaps small vertebrates. The early radiation took place in a tropical forest habitat, and, in general, present-day primates remain tropical in their distribution, with the greatest diversity of species occurring in the rain forests of West Africa, Indo-Malaysia, South America, and Madagascar. The tropical rain forest, then, is the habitat in which the most complex problems of primate evolution are to be found. Those primate species that have extended their ranges into seasonally arid areas characterized by thorn scrub or savanna are often confined to areas surrounding riverine forests.

The forests were retained by primates as their primary environment. The acquisition of terrestrial habits is recent in primate history and occurred in some species of the now extinct giant lemurs, Arceolemurinae (*77*), from Madagascar, as well as in the cercopithecine genera *Papio, Mandrillus, Erythrocebus, Theropithecus, Macaca*, and *Cercocebus* in the Old World tropics. Terrestrial adaptation within the Pongi-

dae are exemplified by *Pan* and *Gorilla*. Only in South America has a truly terrestrial form not evolved.

Figure 3 compares four geographic areas with respect to the kinds of niches occupied by their respective primate genera. We have classified the genera by activity cycle, height of feeding, diet, and relative size. Many species weighing less than 2 kilograms are nocturnal and show a pronounced tendency to feed on high-energy food resources such as insects and fruit. All small primates are arboreal, and, with the exception of the neotropical forms and the West African *Miopithecus*, are derived from the morphologically conservative lorisoid or lemuroid stocks. The larger primate species are almost all diurnal, with the exception of the Madagascan genus *Avahi*. Only among the larger primate species do we find several genera exhibiting semiterrestrial adaptations; these genera include *Mandrillus, Papio, Pan, Gorilla*, and *Macaca*.

The large diurnal primates (the Lemuridae, Indriidae, Cebidae, and Cercopithecidae) display parallel evolu-

428

tion with respect to their feeding strategies; there have been trends away from dependency on energy-rich invertebrates and fruit toward the more readily available cellulose found in leaves. Gut modifications associated with the change in diet can include a larger caecum, as in *Alouatta, Indri, Avahi* (78), and *Lepilemur* (79), or a chambered stomach and bacterial symbionts, as in *Colobus* and *Presbytis* (80). The latter modifications are convergent with the physiological and morphological adaptations evolved by terrestrial ungulates. Those primate species that can utilize cellulose in leaves are referred to as folivores and are distinguished from frugivores, which cannot utilize cellulose (81). Of course, these basic categories do intergrade, since folivores may supplement their leaf diet with fruit. Conversely, frugivorous species may utilize leaves for the simple sugar in their parenchyma cells (81).

Not surprisingly, the arboreal folivores are the most numerous of the

larger forest mammals, sometimes accounting for 30 to 40 percent of the arboreal mammalian biomass (see Fig. 4, Tables 3 and 4). Only in the Neotropics does the primate biomass rank second to that of another arboreal mammalian order—the edentate grazer of the treetops, represented by the three-toed sloth *Bradypus* (82) (see Fig. 4).

Social Structure of Arboreal Primates

The smaller, nocturnal, insectivorous prosimians of Asia, Africa, and Madagascar seem to exhibit the same form of social organization. They are solitary or organized into dispersed family groups (22–24). In part, their solitary habits may result from their being nocturnal, since the coordination of groups would be difficult. The solitary state may also correlate with their insectivorous habits, since foraging patterns may demand a solitary technique (22).

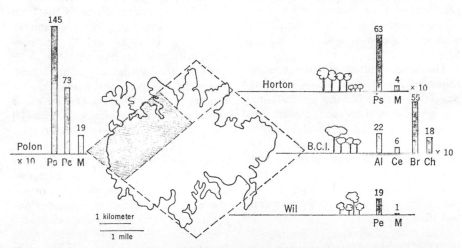

Fig. 4. Primate biomass comparisons for Ceylon and Panama. To compare the sizes of survey areas, the Ceylon study areas are superimposed on an outline map of Barro Colorado Island (B.C.I.). The dashed perimeter outlines the extent of the Wilpattu (Wil) survey area; for comparison, the Horton Plains (Horton) and Polonnaruwa (Polon) study areas are superimposed, and the latter is indicated by shading. To aid scaling on a single diagram, the biomasses for Polonnaruwa, Horton, and B.C.I. are displayed at 1/10 the actual values in kilograms per square kilometer. (Ps, *Presbytis senex*; Pe, *Presbytis entellus*; M, *Macaca sinica*; Al, *Alouatta palliata*; Ce, *Cebus capucinus*; Br, *Bradypus infuscatus*; Ch, *Choloepus hoffmani*.)

Species	Locality	Average home range (km²)	Group size (average)	Feeding class
Alouatta villosa (71)	B.C.I., Panama	< 0.08	14.6	Folivore-frugivore
Ateles geoffroyi (46, 87)	B.C.I., Panama	0.60	12.0	Frugivore
Cebus capucinus (91)	B.C.I., Panama	0.85	15.0	Frugivore
Presbytis senex (19, 74)	Ceylon, Polonnaruwa	0.06	8.4	Folivore
Presbytis entellus (88, 96)	Ceylon, Polonnaruwa	0.14	~ 20.0	Folivore
Macaca sinica (53)	Ceylon, Polonnaruwa	0.15	24.1	Frugivore
Presbytis entellus (44)	Ceylon, Wilpattu	> 1.00	~ 20.0	Folivore
Macaca sinica (53)	Ceylon, Wilpattu	> 1.00	20.0	Frugivore

Table 3. Some comparisons of home range and group sizes. [Average home range and group size will vary widely from one study area to another. We present only species groups from the same areas for the best relative comparison. (B.C.I. indicates Barro Colorado Island.)]

Lepilemur is a folivore, however, and retains the solitary pattern, which suggests that the rhythm of nocturnal activity restricts group formation and co-ordination. The neotropical night monkey *Aotus* is derived from a diurnal form, and nocturnality is a secondary adaptation. *Aotus* exhibits a parental family structure and is thus partly an exception to the rule established for the prosimians. This may indicate the retention of a ·phylogenetically old parental system that is still compatible with the rhythm of nocturnal activity (*36*).

The small diurnal primates concentrated in the New World (*Callicebus, Saguinus, Callithrix,* and *Saimiri*) are insectivore-frugivores and exhibit two grouping tendencies. *Callicebus, Saguinus,* and *Callithrix* live in parental families, while *Saimiri* lives in age-graded-male troops. The parental structure of *Callicebus* and *Saguinus* may be the retention of a phylogenetically conservative trait.

The only small diurnal primate of the Old World is *Miopithecus talapoin*. It shows strong convergences in social behavior and ecology with *Saimiri sciureus* of the Neotropics (*83, 84*). Both species live in large troops (20 to 100 individuals) that divide into subgroups based on age and reproductive classes (*83–85*). Although these species are superficially multi-male, adult males show little affiliation with each other or with females except during the breeding season, and young males may form a peripheral group. Furthermore, the form of the social structure is influenced by the breeding season, which leads to dominance among males and the formation of temporary uni-male breeding units (*86*); hence, the classic multi-male structure is not matched. Table 1 indicates our interpretation of the form of social structure for the smaller species of primates and the correlations with their feeding habits.

Given the two broadly defined feeding niches, folivores and frugivores, the following generalizations can be made about the larger arboreal primates: (i) given comparable habitats, the frugivores have larger home ranges and move more widely during their daily activities than do folivores of an equivalent size class [for example, compare *Ateles* with *Alouatta* (*87*), and *Presbytis senex* with *Macaca sinica* (*88*)] (see Table 3); (ii) both trophic types tend toward either a uni-male structure or an age-graded-male system (see Table 1), with folivores especially tending toward a uni-male organization; and (iii) many folivores [*Colobus, Alouatta,* and *Presbytis* (*19, 20, 71*)], but only some of the frugivores [*Symphalangus* and *Hylobates* (*14, 34*)], employ troop

Species	Biomass (kg/km²)	Estimated arboreal biomass* (%)
Barro Colorado Island		
Alouatta villosa	220.0	22
Cebus capucinus	60.0	6
Bradypus infuscata	550.0	54
Choloepus hoffmani	180.0	18.
Ghana		
Colobus (three species)	55.08	79
Cercopithecus (two species)	5.3	8
Cercocebus (one species)	2.6	4
Polonnaruwa, Ceylon		
Presbytis senex	1450.0	61
Presbytis entellus	730.0	31
Macaca sinica	190.0	8
Wilpattu, Ceylon		
Presbytis entellus	19.0	< 95
Macaca sinica	< 1.0	> 5
Horton Plains, Ceylon		
Presbytis senex	630.0	94
Macaca sinica	< 40.0	6

* Percentage is based on the known mammalian biomass totals but not the total arboreal mammalian biomass, which may be one-sixth again as high.

Table 4. Comparisons of arboreal biomasses. [Data for Ghana from Collins (108); data for Barro Colorado Island are preliminary estimates and represent the minimum; data for Ceylon are minimum estimates that may show higher levels in restricted sample areas.]

or individual vocalizations in maintaining spacing between adjacent troops.

Comparisons among the African, Asian, and South American forest-dwelling species show the marked tendencies toward a uni-male reproductive unit (see Table 1). In an extensive survey, Struhsaker lists almost all species of West African rain forest frugivores of the genus *Cercopithecus* as typically exhibiting a uni-male reproductive unit (65). *Cercopithecus mitis*, a frugivore studied in Uganda (89), and *Colobus guereza*, a folivore (20), show similar modes. Examples of uni-male reproductive units found among the strongly arboreal folivores of Asia include *Presbytis cristatus*, *P. johni*, and *P. senex*

(17–19). Although some troops of *P. cristautus* may, under certain circumstances, contain more than one male (90), this reflects an age-graded-male system. In Central America, the folivore *Alouatta* may exhibit a uni-male system under conditions of low population density, but at high densities several males may be included in an age-graded-male troop. The frugivorous, white-throated capuchin, *Cebus capucinus*, appears to exhibit a uni-male reproductive system (91), while *Ateles geoffroyi* tends toward an age-graded-male system (92).

The arboreal folivores may be characterized by their vocalizations, which usually take the form of dawn choruses. These announcement calls may also be produced at various times during the day, since they appear to occur prior to progressions and are also triggered by exogenous factors. The role of these calls is definitely related to intraspecific spacing. This function has been demonstrated for *Colobus* (20) and *Alouatta* (71) and is undoubtedly being served by the calling of *Presbytis* (19). The frugivorous gibbons and the siamang exhibit comparable chorusing behavior (14, 93). No investigator has specifically attempted to correlate possible seasonal increases in the frequency of chorusing behavior with breeding peaks for any of the chorusing species. McClure, however, offers observations on *Symphalangus syndactylus* and *Hylobates lar* in Malaysia that suggest such a correlation may exist (94).

Social Structure of Semiterrestrial Primates

In general, semiterrestrial species tend to live in variable habitats and are frugivorous or varied feeders. *Presbytis entellus* and *Gorilla gorilla* are the only folivorous primates that do forage extensively on the ground. Semiterrestrial species also tend to live in larger groups, compared to arboreal forms of a similar body size, and tend to form age-graded-male troops (see Tables 1 and 3).

All but one species of *Presbytis* and of *Cercopithecus* fit into the two forest niches (arboreal folivore and frugivore) described above. The exceptions are *Presbytis entellus*, which is terrestrially adapted and occupies riverine forests, seasonally dry forests, and seasonally arid scrub (*8*), and *Cercopithecus aethiops*, which occupies gallery forests and savanna areas (*11*). Both species show a variable troop structure, with *Cercopithecus aethiops* tending to exhibit an age-graded-male to multimale configuration (*11, 12*), and *Presbytis entellus* showing an age-graded-male organization in Ceylon and in some parts of India. *Presbytis entellus*, however, may also show a strictly unimale pattern when crowded into remnant patches of roadside forest in India (*8, 10, 44, 45, 95, 96*).

The tendency toward a larger group size and an increased number of males in a group is well illustrated in the forest-adapted mangabeys (*Cercocebus*) when a species that forages on the ground (*C. torquatus*) is compared to a conspecific of similar size that is primarily arboreal (*C. albigena*). The more terrestrial *C. torquatus* forms larger troops with more than one adult male per troop. The sympatric species *C. albigena* forms smaller troops and characteristically has a uni-male group structure (*97*); however, Chalmers (*98*) presents data that suggest a multi-male group for *C. albigena* in Uganda.

The tendency toward larger group sizes than those of strictly arboreal forms and an accompanying age-graded-male structure may also hold for semiterrestrial macaques and *Mandrillus*. *Macaca sinica* troops, for example, have a distinct age gradation in the so-called adult male class, as well as extragroup males (*53*). Recent research by Gartlan (*43*) suggests that the drill

(*Mandrillus leucophaeus*), which forages on the ground in dense forests, has retained the uni-male to age-graded-male structure.

Papio anubis, P. cynocephalus, P. ursinus, Macaca mulatta, and *M. fuscata* forage primarily on the ground (*7, 50, 56*). They exhibit a multi-male grouping pattern, which seems to be effective as an antipredator mechanism (*7, 49*) (see Table 1).

Terrestrial species of the genera *Theropithecus* and *Erythrocebus*, as well as *Papio hamadryas*, present other problems. *Theropithecus gelada* and *Papio hamadryas* are adapted to extremely arid environments and break up into foraging parties that are uni-male in composition; hence, the selective advantage of group attacks on terrestrial predators is lost. Crook and Gartlan (*2, 3*) have suggested that this pattern promotes feeding efficiency by reducing the competition for food that would occur if a large number of males accompanied the females and young; hence, these species have approximated a uni-male grouping at the expense of group offensive behavior toward terrestrial predators. Such foraging units reassemble at selected sleeping sites in the evening.

The mode of antipredator behavior is extremely important to a full understanding of the selective advantage of either a multi-male or uni-male group. The terrestrial *Erythrocebus patas*, for example, lives sympatrically with *Papio* but has retained the uni-male grouping tendency, with the male keeping watch while the females forage. *Erythrocebus* relies mainly on speed, distraction displays by the adult male, and dispersed hiding to avoid predation; thus, its antipredator behavior is not and never was built around a mobbing response (*37*). Since *Erythrocebus* is derived from a *Cercopithecus*-like form, it may have retained the uni-male structure in its new habitat and have undergone selection for greater speed instead of evolving a multi-male structure, as is the case in some species of the genus *Papio* (*65*).

The chimpanzee (*Pan troglodytes*) occupies a range of habitats, from rain forest to forest-grassland. Troops seem to have a loose grouping pattern (*99*), although cohesive troop behavior may be shown during long marches in savanna areas (*100*). These frugivores break up into foraging units in which strong cohesion exists mainly in the immediate mother family, although patterns of long-term affiliation are displayed when the independently foraging subunits come together (*99*, *101*).

Gorilla gorilla may be considered a semiterrestrial folivore that has a cohesive group structure. The males appear to be age-graded, with one old, silver-back male as the leader (*102*).

Roles, Group Functions, and Social Structure—Some Selective Advantages

Crook (*3*) has evaluated the interrelations between ecology and social structure in primates. We offer here some comments on critical issues that deserve further research. The fundamental questions are (i) Why do some primate species exhibit a multi-male troop composition? and (ii) Why do adult female primates find it advantageous to form extended mother families? In an evolutionary sense, the number of males in a given troop will depend on what advantage the males are to the reproducing females (*103*).

It would appear, in comparing the various species of higher primates, that there is a strong trend toward polygyny. Certainly a given male increases his individual fitness by distributing his genes among the greatest possible number of females. However, many advantages can accrue to a dominant male through the presence and actions of subordinates (*16, 39*). A given male's dependence on other males for support

in enhancing his survival value will determine, in part, the short-term advantage of having many males in a troop. Nevertheless, if fewer males can do the same task better than many males, then all things will tend to favor a polygynous mating system. Even in a multi-male troop, such as has been reported for various species of *Papio* and *Macaca*, it may well be that the descendents of the alpha male would be more predominant within the breeding unit than would those of subordinate males, since subordinate males engage in less sexual activity than do alpha males (*49, 54*).

In most species of primates, the role of the male involves little parental care. There are, however, exceptions within the families Cebidae and Hapalidae, as well as in some species of the genus *Macaca*. The functions of the adult male, either alone or in company with other males, seem to be (i) to maintain spacing with respect to neighboring troops of the same species, (ii) to reduce competition within the group by driving out younger males, and (iii) to enforce some degree of protection against predators. These behaviors may involve vigilance, mobbing, pursuit-invitation displays, or outright attack.

A fourth aspect of the adult male's activities, which has received only sporadic attention, is his role of providing leadership. By initiating and maintaining movement in a certain direction, the male is influential in promoting cohesion of the troop and serves as a focus for the troop's movements (*44, 102*). The dominant adult male is the one most likely to initiate a following response from the females, hence his leadership role is dependent upon the active participation of females and attendant juveniles in his movements (*44*).

In contrast to the roles of adult males, the roles of adult females are dominated by infant care. During the early phases of the infant's development, the mother is responsible for protecting it; when the infant enters the juvenile stages, the mother serves as a focus for its socialization. Indeed, the juvenile's status is, in part, predetermined by its mother's status within the female hierarchy (63).

To some extent, each female old enough to reproduce is in competition with males and other females of the same age class for food, sleeping space, and so on. Nevertheless, certain advantages (for example, increased feeding efficiency and antipredator behavior) may be inherent in a group of several females traveling together and maintaining a close liaison. The extent to which individual females benefit from group rearing of their progeny and the extent to which their own chances of survival are increased by associating with other females will determine the size to which the troop can profitably increase.

Ultimately, the size of the troop is, in part, a compromise between competition among members for resources that are in short supply and the advantage of having many members in locating resources that are scattered and available for restricted periods (96). The advantage of feeding in a group is often overlooked by field workers, although feeding calls, which promote aggregation in fruit trees, are a well-known phenomenon for some species. We speculate that, although primary folivores such as Colobus guereza and Presbytis senex eat considerable quantities of fruits, their feeding strategy is not predicated on a daily need of finding ripening fruit trees within their home range. Small, cohesive, uni-male social units are permitted within this strategy. However, in frugivores such as Ateles geoffroyi and Pan satyrus (46, 101), the best feeding strategy involves breaking up the troop into small, independently foraging units that spread out to locate fruit trees within their home range and then "announce"

the location of feeding spots. In Saimiri, foraging for small, dispersed canopy insects leads to the formation of subgroups that forage at different rates (84). The interaction between the distribution of primary resources (food and water) in a specific habitat and the population density in that habitat can profoundly affect social structure (12).

In rain forests, sympatric species often form mixed feeding groups that move together and show very little overt competition. The phenomenon has been described for Central America and for Gabon, West Africa (104). Such species associations should be distinguished from instances in which an occasional male of one species may associate with a troop of another species [for example, a Saimiri male in a Callicebus group (35) and two Macaca sinica males in a Presbytis entellus group (44, 95)]. The exact significance of either the casual or the frequent mixed-species group has not been ascertained, although in both types predator avoidance or feeding efficiency, or both, may be increased for all members. The whole problem requires attention (105).

We have referred to antipredator behavior throughout our consideration of selective advantages for the various social systems. Yet no single aspect of primate field studies has less supportive data than the generalizations concerning the survival value of the various presumed antipredator mechanisms. The primate troop can and does exhibit antipredator behavior. Emphasis on different patterns (including vigilance, alarm calls, distraction displays, mobbing, and attack) may vary between species and between age classes and sex classes within a species (44).

Only a few individuals in a troop need spend a great percentage of their time in vigilance for all members of the group to be benefited; typically subadult and adult males perform this

434

role. Visual scanning from high positions may serve to locate intruding conspecific males as well as predators (37, 49), and troop members may be warned of the presence of a predator by alarm calls. The manner in which monkeys respond to a potential predator depends partly upon their own size and mobility and partly upon the relative size and position of the predator.

The jumping and vocalizations of adult males are common alarm behaviors, and are described for such diverse primate species as terrestrial *Erythrocebus patas* (37) and arboreal *Presbytis senex* (19, 74). It is generally considered that such behavior distracts the predator while the females encumbered by young scatter and hide (44). Protective males typically position themselves between the group and a terrestrial predator.

Mobbing, another alarm behavior, requires the presence of a group. A predator may be harried through vocalizations, group defecation and urination (46, 69), and, in large species, branches that may fall from the weight of leaping adult males (*Presbytis*) or be purposely broken off and dropped (*Alouatta, Cebus*) (46, 69, 91).

If a predator should surprise a troop at short range, individual flight responses may scatter troop members in various directions, thus confusing a predator. This confers some selective advantage on group life, even though no altruistic tendencies can be detected.

Unlike the smaller primates, larger species attack outright, and some (for example, gorilla and chimpanzee) are more than a match for even the largest predators. Baboons exhibit a pronounced sexual dimorphism, and, since the baboons of the genus *Papio* typically forage in large multi-male groups, the larger males can form an effective attack unit and displace a predator. The responses of baboons and chimpanzees to leopards have been studied (106) and have been found to include the use of actual weapons (sticks and rocks) (107).

General Conclusions

When the major radiations in the Old World, New World, and Madagascar are compared with each other, no doubt some of the differences in social structure seen in those species adapted to similar ecological niches will be found to result from phylogenetic differences; that is, the social structures of the ancestral forms have been carried forward in the adaptive radiation of these species. For example, the consistent tendency in lemuroid primates such as *Propithecus* and *Lemur* to show multi-male groups with more males than females, a dominance of females over males, and the segregation of troops into all-male and all-female subgroups does not exist in any known continental species (25, 59). The South American radiation has produced a trend toward male participation in parental care and the formation of pair bonds between given males and females.

Although we can generalize about the selective advantages of primate social structures, we must remember that the history of the population under study, its particular adaptation to local environmental conditions, and the idiosyncratic nature of its dyadic relations (which have been ontogenetically established within the particular group) can result in a great deal of variability in social structure, even within the same species when it occurs in widely differing habitats. Hence, in making generalizations about social structure, we must remember that we are talking about behavioral modes or behavioral medians.

Summary

It has been the custom for ethologists to divide mammalian societies into grades. Each ascending level of complexity denotes an increase in the complexity of interaction patterns among the members of the group. The multi-male group traditionally represented a high level of social organization, as well as the higher primate norm, but it was defined from early studies on terrestrial primates. What we have tried to show is that the uni-male system occurs in a wide variety of primate species in both the cercopithecoid and ceboid radiations. Furthermore, we have attempted to illustrate that multi-male systems are more apparent than real and that many should be considered age-graded-male systems.

The three proposed classes (uni-male, age-graded-male, and multi-male) of social structure (above the level of the parental family) are gradations and represent an increased complexity based on an increased tolerance among adults at the maximum "sociological" age level. There are only a few species for which the data are sufficient to place them in a class. The multi-male system is apparently a specialized form of social grouping that represents a particular adaptation to terrestrial foraging by intermediate-sized primates. It is readily derived from an age-graded-male system. The multi-male system does not differ profoundly from an age-graded-male system, but the former does allow for increased affiliation and cooperation among adult males.

The uni-male system or the age-graded-male system is favored in arboreal species, both frugivores and folivores. The structure of a species' social organization is more predictable for diurnal, leaf-eating forms than it is for the frugivores. We can correlate an arboreal, diurnal, leaf-eating niche with a species having a social structure that tends toward a uni-male system with a small home range and the employment of chorusing behavior to effect spacing. Such species tend to be sedentary and are typified by *Alouatta,*

Colobus, and *Presbytis.* It would appear, then, that similar predation pressures (semiarboreal felids probably being most important) and similar foraging problems have forced similar behavioral solutions upon these species.

One should be wary of generalizations concerning the form of social structure for any given species, since social structure may vary with habitat. Similar variations have been noted in response to problems of density and habitat disturbance. Parallel trends can be noted in both the ceboid and cercopithecoid lines of evolution. More nearly accurate correlations will be possible only when we have more data concerning feeding efficiency and anti-predator mechanisms for a wide variety of species, each studied within a range of habitats.

References and Notes

1. J. F. Eisenberg, *Handb. Zool.* **8** (No. 39), 1 (1966).
2. J. H. Crook and J. S. Gartlan, *Nature* **210**, 1200 (1966).
3. J. H. Crook, in *Social Behaviour in Birds and Mammals*, J. H. Crook, Ed. (Academic Press, London, 1970), pp. 103–168.
4. C. R. Carpenter, *Trans. N.Y. Acad. Sci.* **4**, 248 (1942); *Hum. Biol,* **26**, 269 (1954).
5. In Carpenter's original formulation, he suggested that territorial behavior toward conspecific troops was shown. We now know that such exclusive patterns can be effective without overt fighting, and the generalization can be modified to state that all primate troops appear to have exclusive rights to certain areas of their home range at certain critical times of the year, even though they may share parts of their home range with neighboring troops.
6. T. E. Rowell, *J. Zool. London* **149**, 344 (1966).
7. K. R. L. Hall and I. DeVore, in *Primate Behavior*, I. DeVore, Ed. (Holt, Rinehart & Winston, New York, 1965), pp. 53–110.
8. P. Jay, in *ibid.*, pp. 197–249; S. Ripley, in *Social Communication among Primates*, S. A. Altmann, Ed. (Univ. of Chicago Press, Chicago, 1967), pp. 237–253.
9. K. Yoshiba, in *Primates*, P. Jay, Ed. (Holt, Rinehart & Winston, New York, 1968), pp. 217–242.
10. Y. Sugiyama, in *Social Communication among Primates*, S. A. Altmann, Ed. (Univ. of Chicago Press, Chicago, 1967), pp. 221–236.
11. T. T. Struhsaker, *Ecology* **48**, 891 (1967); *Univ. Calif. Publ. Zool.* **82**, 1 (1967).
12. J. S. Gartlan and C. K. Brain, in *Primates*, P. Jay, Ed. (Holt, Rinehart & Winston, New York, 1968), pp. 253–292.
13. C. R. Carpenter, *Comp. Psychol. Monogr.* **16**, 1 (1940).
14. T. Ellefson, in *Primates*, P. Jay, Ed. (Holt, Rinehart & Winston, New York, 1968), pp. 180–200.

15. H. Kummer, *Z. Psychol.*, No. 33 (1957).
16. ———, *Social Organization of Hamadryas Baboons: A Field Study* (Univ. of Chicago Press, Chicago, 1968).
17. I. Bernstein, *Behaviour* **32**, 2 (1968).
18. F. E. Poirier, in *Primate Behavior*, L. A. Rosenblum, Ed. (Academic Press, New York, 1970), vol. 1, pp. 251–383.
19. R. Rudran, thesis, University of Ceylon, Colombo (1970).
20. P. Marler, *Science* **163**, 93 (1969).
21. G. H. Manley, *Symp. Zool. Soc. London* **15**, 493 (1967).
22. P. Charles-Dominique, *Biol. Gabonica* **7**, 121 (1971); *ibid.* **2**, 347 (1966).
23. J-J. Petter and C. M. Hladik, *Mammalia* **34**, 394 (1970).
24. J-J. Petter, *Mem. Mus. Nat. Hist. Natur. Paris Ser. A Zool.* **27**, 1 (1962).
25. A. Jolly, *Lemur Behavior: A Madagascar Field Study* (Univ. of Chicago Press, Chicago, 1966).
26. P. Leyhausen, *Symp. Zool. Soc. London* **14**, 249 (1965).
27. P. K. Anderson, *ibid.* **26**, 299 (1970).
28. The objection to the term "solitary" is valid if and only if the term is taken literally. If due consideration is given to the existence of indirect communication and the possibility of the maintenance of reproductive continuity among related members of such dispersed population units (*1*), then we can maintain the older term; if not, then we may well substitute the more cumbersome term, "dispersed, noncohesive family group."
29. P. Charles-Dominique, *Fortschr. Verhaltensforschung* **9**, 7 (1972); R. D. Martin, *ibid.*, p. 43.
30. A. Petter-Rousseaux, *Mammalia* **26**, 1 (1962); P. Charles-Dominique and C. M. Hladik, *Terre Vie*, part 1 (1971), p. 3.
31. H. Fischer, *Z. Tierpsychol.* **22**, 247 (1965).
32. N. A. Muckenhirn, thesis, University of Maryland (1967).
33. G. Epple, *Folia Primatol.* **7**, 37 (1967); M. Moynihan, *Smithson. Contrib. Zool.* **28**, 1 (1970).
34. M. Kawabe, *Primates* **11**, 285 (1970).
35. W. A. Mason, in *Primates*, P. Jay, Ed. (Holt, Rinehart & Winston, New York, 1968), pp. 200–216.
36. M. Moynihan, *Smithson. Misc. Collect.* **146**, 1 (1964).
37. K. R. L. Hall, *J. Zool. London* **148**, 15 (1965).
38. J. H. Crook, in *Play, Exploration and Territory in Mammals*, P. A. Jewell and C. Loizos, Eds. (Zoological Society of London, London, 1966), pp. 237–258.
39. ———, in *Man and Beast: Comparative Social Behavior*, J. F. Eisenberg and W. Dillon, Eds. (Smithsonian Institution, Washington, D.C., 1971), pp. 235–260.
40. H. Kummer, W. Goetz, W. Angst, in *Old World Monkeys*, J. R. Napier and P. H. Napier, Eds. (Academic Press, New York, 1970), pp. 351–364.
41. The uni-male system may be maintained ontogenetically from a founding pair through the combined processes of internal recruitment by birth and antagonism exercised by the founding adult male toward his sons. The maintenance of such uni-male groupings is paralleled in the Carnivora [D. G. Kleiman and J. F. Eisenberg, "Comparisons of canid and felid social systems from an evolutionary perspective" (paper presented at the First International Symposium on World Felidae, Laguna Hills, Calif., 1971)].
42. In recognition of this fact, Crook has dis-tinguished two grades of "multi-male" social structures (*2*, *3*), but he has not elaborated on his reasons for doing so or on the criteria for separation.
43. J. S. Gartlan, in *Old World Monkeys*, J. R. Napier and P. Napier, Eds. (Academic Press, New York, 1970), pp. 445–480.
44. N. Muckenhirn, thesis, University of Maryland (1972).
45. S. M. Mohnot, *Mammalia* **35**, 175 (1971).
46. J. F. Eisenberg and R. E. Kuchn, *Smithson. Misc. Collect.* **151**, 1 (1966).
47. L. Klein and D. Klein, *Int. Zoo Yearb.* **11**, 175 (1971); J. F. Eisenberg, unpublished data.
48. I. DeVore and S. L. Washburn, in *African Ecology and Human Evolution*, F. C. Howell and F. Bourlière, Eds. (Aldine, Chicago, 1963), pp. 335–367.
49. I. DeVore and K. R. L. Hall, in *Primate Behavior*, I. DeVore, Ed. (Holt, Rinehart & Winston, New York, 1965), pp. 20–52.
50. C. H. Southwick, M. A. Beg, M. R. Siddiqi, *ibid.*, pp. 111–159.
51. K. Imanishi, *Curr. Anthropol.* **1**, 393 (1960).
52. M. Bertrand, *Bibl. Primatol.* **11**, 1 (1969).
53. W. Dittus, unpublished data (1968–1971).
54. J. H. Kaufmann, *Ecology* **46**, 500 (1965).
55. T. Nishida, *Primates* **7**, 141 (1966).
56. S. Altmann and J. Altmann, *Baboon Ecology: African Field Research* (Univ. of Chicago Press, Chicago, 1970).
57. K. R. L. Hall, *Proc. Zool. Soc. London* **139**, 283 (1962); I. DeVore, in *Classification and Human Evolution*, S. L. Washburn, Ed. (Viking Fund Publications in Anthropology, No. 37, Wenner-Gren Foundation, New York, 1963), pp. 301–319.
58. L. Tiger, *Men in Groups* (Random House, New York, 1968).
59. N. Bolwig, *Mem. Inst. Rech. Sci. Madagascar Ser. A Biol. Anim.* **14**, 205 (1960).
60. H. Hendrik, in *Dikdik und Elefanten* (Piper, München, 1970) pp. 70–77; G. M. McKay, thesis, University of Maryland (1971).
61. J. F. Eisenberg, G. M. McKay, M. R. Jainudeen, *Behaviour* **38**, 193 (1971).
62. C. B. Koford, in *Primate Behavior*, C. Southwick, Ed. (Van Nostrand, New York, 1963), pp. 136–152.
63. D. S. Sade, *J. Phys. Anthropol.* **23**, 1 (1965).
64. M. Kawai, *Primates* **2**, 181 (1960); Y. Furuya, *ibid.*, p. 149.
65. T. T. Struhsaker, *Folia Primatol.* **11**, 80 (1969).
66. F. Bourlière, C. Hunkeler, M. Bertrand, in *Old World Monkeys*, J. R. Napier and P. H. Napier, Eds. (Academic Press, New York, 1970), pp. 297–350.
67. L. L. Klein, "Ecological correlates of social grouping in Colombian spider monkeys," paper presented at the Animal Behavior Society Meeting, Logan, Utah, 1971.
68. C. R. Carpenter, *J. Mammal.* **16**, 171 (1935).
69. ———, *Comp. Psychol. Monogr.* **10**, 1 (1934).
70. ———, in *Primate Behavior*, I. DeVore, Ed. (Holt, Rinehart & Winston, New York, 1965), pp. 250–291.
71. D. J. Chivers, *Folia Primatol.* **10**, 48 (1969).
72. N. A. Collias and C. H. Southwick, *Proc. Amer. Phil. Soc.* **96**, 143 (1952).
73. J. B. Calhoun, in *Physiological Mammalogy*, W. V. Mayer and R. G. Van Gelder, Eds. (Academic Press, New York, 1963), vol. 1, p. 91.
74. G. Manley, in preparation.
75. W. E. LeGros-Clark, *The Antecedents of*

Man (Quadrangle, Chicago, 1960).
76. J. F. Eisenberg and E. Gould, *Smithson. Contrib. Zool.* **27**, 1 (1970).
77. A. Walker, in *Pleistocene Extinctions*, P. S. Martin and H. E. Wright, Eds. (Yale Univ. Press, New Haven, Conn., 1967), p. 425.
78. C. M. Hladik, *Mammalia* **31**, 120 (1967).
79. ———, P. Charles-Dominique, P. Valdebouze, J. Delort Laval, J. Flanzy, *C. R. Hebd. Seances Acad. Sci. Paris* **272**, 3191 (1971).
80. T. Bauchop and R. W. Martucci, *Science* **161**, 698 (1968).
81. C. M. Hladik and A. Hladik, *Biol. Gabonica* **3**, 43 (1967); *Terre Vie*, part 1 (1969), p. 27.
82. G. G. Montgomery and M. E. Sunquist, in preparation.
83. A. Gautier-Hion, *Folia Primatol.* **12**, 116 (1970).
84. R. W. Thorington, Jr., in *The Squirrel Monkey*, L. Rosenblum and R. W. Cooper, Eds. (Academic Press, New York, 1968), pp. 69–87.
85. F. V. DuMond, in *ibid.*, pp. 88–146.
86. J. D. Baldwin, *Folia Primatol.* **9**, 281 (1968).
87. A. Richard, *ibid.* **12**, 241 (1970).
88. C. M. Hladik and A. Hladik, *Terre Vie*, in press.
89. F. P. G. Aldrich-Blake, in *Social Behaviour in Birds and Mammals*, J. H. Crook, Ed. (Academic Press, London, 1970), pp. 79–102.
90. Z. Y. Furuya, *Primates* **3**, 41 (1961–62).
91. J. R. Oppenheimer, thesis, University of Illinois (1968).
92. The frugivorous species do not show such uniform trends in social structure. As with the folivores, there is a tendency toward a uni-male situation, but at least two kinds of group organization can be discerned: (i) the cohesive uni-male band, which is generally quite small, 20 members (for example, *Cebus capucinus* and *Cercopithecus mitis*), or (ii) the less cohesive, extended group, in which mother families and age-graded-male units forage independently (for example, *Ateles geoffroyi*), reassemble only from time to time at preferred resting loci, and so on. Although all of these species have long-distance calls, the calls are generally not given in a chorusing fashion and their relation to intraspecific spacing is not well understood.
93. H. E. McClure, *Primates* **5**, 39 (1964). It should be noted that, although the gibbons are classically considered frugivores (*13*), they do feed on leaves at certain seasons of the year (*14*). It may be that further research will lead to their being classified as folivore-frugivores. If such is the case, their spacing system is more easily comprehensible. See also D. J. Chivers, *Malayan Nature J.* **24**, 78 (1971).
94. In *Symphalangus*, the peak number of morning calling sessions heard in June was several times greater than the peak number heard between September and January, the months of minimal calling. The only observation of a newborn was recorded in February. While such evidence is obviously not sufficient to establish a birth peak for this species, it suggests that some breeding did occur in June. Likewise, 2.5 times more calls of *Hylobates* were heard in June than in November and February, the months of minimal calling.
95. J. F. Eisenberg and M. Lockhart, *Smithson. Contrib. Zool.* **101**, 1 (1972).
96. S. Ripley, in *Old World Monkeys*, J. Napier and P. Napier, Eds. (Academic Press, New York, 1970), pp. 481–512.
97. C. Jones and J. Sabater Pi, *Folia Primatol.* **9**, 99 (1968).
98. N. R. Chalmers, *ibid.* **8**, 247 (1968).
99. J. Goodall van Lawick, *Anim. Behav. Monogr.* **1**, 165 (1968).
100. K. Izawa, *Primates* **11**, 1 (1970).
101. V. Reynolds and F. Reynolds, in *Primate Behavior*, I. DeVore, Ed. (Holt, Rinehart & Winston, New York, 1965), pp. 368–424.
102. G. B. Schaller, *The Mountain Gorilla* (Univ. of Chicago Press, Chicago, 1963).
103. For a theoretical discussion, see J. E. Downhower and K. Armitage, *Amer. Natur.* **105**, 355 (1971).
104. R. W. Thorington, Jr., in *Progress in Primatology*, D. Starck, R. Schneider, H.-J. Kuhn, Eds. (International Primatological Society, Frankfurt, 1967), pp. 180–184; T. T. Struhsaker, in *Old World Monkeys*, J. Napier and P. Napier, Eds. (Academic Press, New York, 1970), pp. 365–444; J. P. Gautier and A. Gautier-Hion, *Terre Vie*, part 2 (1969), p. 164.
105. Interspecific antagonisms can occur, however, even between those species that form mixed feeding associations, since the propensity to show interspecific aggression apparently varies with respect to the history between the two populations under study and the distribution of resources in the habitat under consideration. In a broad sense, the density of primate populations, either arboreal or semiterrestrial, may be severely limited by the presence of competitors. It should not be assumed, however, that the most severe competition typically comes from other species of primates. Struhsaker (*11*) has reported that the major food competitor of *Cercopithecus aethiops* in Kenya is the elephant. This competition results from the fact that, during the dry season, the elephants push over the major species of trees that provide food and refuge for the monkeys.
106. A. Kortlandt, in *Progress in Primatology*, D. Starck, R. Schneider, H.-J. Kuhn, Eds. (International Primatological Society, Frankfurt, 1967), pp. 208–224; ——— and M. Koij, *Symp. Zool. Soc. London* **10**, 61 (1963).
107. Although the members of the group that attack the leopard may act individually, the effect of several adult males charging can be quite intimidating to a potential predator and thereby be a distinct selective advantage for a multi-male social group. Such antipredator maneuvers are generally in the province of males, and those species that tend to show responses of this kind generally exhibit a great dimorphism, with the males being much larger than the females; however, dimorphism is less pronounced in the chimpanzee than in the baboon.
108. W. B. Collins, as quoted by F. Bourlière, in *African Ecology and Human Evolution*, F. C. Howell and F. Bourlière, Eds. (Aldine, Chicago, 1963), pp. 43–54.
109. This article is an expanded version of a talk presented by J.F.E. at the Animal Behavior Society Meeting, Logan, Utah, 1971, and is an outgrowth of fieldwork that has been in progress since 1964. Our primary effort was in Ceylon from 1968 to 1970 (*95*), but studies were also conducted in Central America and Madagascar (*32, 46, 76*). In 1967, J.F.E. and Suzanne Ripley

formed a research group that studied the Ceylon primates from the standpoints of ecology and behavior. Some eight investigators participated in the project from 1968 to 1971. The adaptive patterns that have emerged will be or are being published by individual investigators (*19, 23, 44, 53, 88, 96*), and this article does not attempt to synthesize all interpretations. Research was supported, in part, by National Science Foundation grant GB-3545, awarded to J.F.E.; National Institute of Mental Health grant RolMH15673-01; research grant 686 from the National Geographic Society; and Smithsonian Institution Foreign Currency Program grant SFC-7004, awarded to J.F.E. and Suzanne Ripley. The authors wish to thank D. G. Kleiman for a critical reading of the manuscript.

50 Territoriality in the bullfrog, *Rana catesbeiana*

Stephen T. Emlen

During the breeding season, adult male bullfrogs, *Rana catesbeiana*, establish territories from which conspecific males are aggressively excluded. Stereotyped postures, approaches, and physical encounters all function in the defense of such areas. It is proposed that the highly polygamous social system present in this species creates an intense intermale competition for females, and that possession of a territory directly influences an individual male's chances of successful mating.

Introduction

THE tendency for an animal to restrict its activities to a specified area and to defend this area against other members of its species is known as territoriality. Although such behavior is extremely widespread among almost all groups of vertebrates, there are few reports of its occurrence within the class Amphibia. Unequivocal evidence for the existence of such aggressive behavior in urodeles is lacking. Among anurans, individuals of several species have been shown to inhabit specific home ranges to which they often return upon short-distance displacement (Bogert, 1947; Martof, 1953b; Jameson, 1957; Kikuchi, 1958; Ferguson, 1963; Dole, 1965). In addition, field and laboratory observations of apparent agonistic behavior have been described for several species of tropical dendrobatids (Test, 1954; Sexton, 1960; Duellman, 1966) and pipids (Rabb and Rabb, 1963a, b), as well as for *Leptodactylus insularum* (Sexton, 1962) and *Hyla faber* (Lutz, 1960). But among temperate zone species, only the green frog (*Rana clamitans*) is known to exhibit what Martof (1953a) has described as a "primitive type of territory."

During the breeding season male bullfrogs (*Rana catesbeiana*) commonly space themselves at regular intervals along the shores of millponds, small lakes, and streams, and individuals frequently remain at specific locations for periods of two to three weeks (Noble, 1931:404–405; Wright and Wright, 1949:445). The purpose of this report is to demonstrate that the areas surrounding such male bullfrogs do in fact constitute territories that are actively defended from other males by stereotyped, aggressive behaviors.

Materials and Methods

During the summers of 1965 and 1966 I studied a population of approximately 200 bullfrogs inhabiting a five-acre pond on the University of Michigan's Edwin S. George Reserve near Ann Arbor. The frogs overwintered in an adjacent marsh area and entered the pond through a narrow inlet stream during May and early June. Individual bullfrogs were trapped as they entered the pond, sexed, measured, and toe-clipped for identification. Males were additionally marked by placing colored and/or numbered nylon elasticized bands around their waists. These permitted rapid individual recognition from a distance, without the necessity of capture and consequent disturbance.

Observations were made during the evenings from late May, when the frogs first called from the pond, until late July, by which time the majority of the females had spawned. Behavioral interactions were recorded with the aid of binoculars and a Bolex H16 movie camera from both the shore line and a small rowboat. Nightly censuses were also conducted and the positions of marked males were accurately mapped.

OBSERVATIONS AND DISCUSSION

The frogs commonly spent the daylight hours in cracks or crevices in the overhanging bank of the pond or in the shallow water near shore. Calling was infrequent (except on two overcast, rainy days), and social interactions were rarely observed.

At dusk, the majority of the males left the bank and took up regular calling stations, where they remained throughout the evening, and to which they returned on consecutive nights. During the early part of the summer these stations characteristically were located in a band within 50 ft from shore, at a water depth of two–five ft. But by mid-June, when immigration had increased the male population to its peak of approximately 100 individuals, large choruses of from 14 to 37 males were formed. These aggregations extended into deeper water, farther from shore.

Males occupying stations adopted a characteristic "high" posture in which the lungs were greatly inflated with air. This caused the animals to float very high in the water with the entire head, most of the abdomen, and the dorsal portions of the hind legs resting above the water's surface. The head was raised slightly, exposing the brilliant yellow gular area.

Males exhibiting this posture called at frequent intervals and aggressively excluded other males from the areas surrounding them. The size of these defended areas varied from individual to individual, and appeared to be influenced by the availability of suitable shore-line. Measurements taken in June 1966, show the average distance between a frog and its closest male neighbor within a chorus (both exhibiting the "high" posture) to be 17.8 ft (standard deviation = 5.9 ft; n = 94). This implies an average minimum territory radius of approximately nine ft.

If an inflated neighboring male approached closer than this distance, a stereotyped challenge ensued. The resident gave a sudden, short, staccato "hiccup" vocalization, turned abruptly, faced the intruder, and swam a few feet toward him. This "hiccup" challenge was observed several hundred times and almost invariably accompanied intermale aggressive encounters. (The few exceptions occurred in the absence of any stimulus discernible to me.)

The behavior that followed this initial challenge depended upon the response of the intruding male. If he turned away and returned to his previous station, no further interaction occurred. However, if he called, continued to advance, or both, the resident continued his approach via a series of discrete challenges. Each challenge consisted of the "hiccup" note and an advance of a few ft toward the intruder; each was followed by a motionless pause, often lasting for one min or more, during which one or both males usually called. As the distance between frogs decreased, the swimming approach was replaced by a slower crawling movement through the water until the frogs came to rest only inches apart.

It should be emphasized that this sequence could be broken at any time, should either male turn away from his challenger and leave the immediate vicinity. In fact, of 79 encounters observed over a five-night period in June 1966, 58 (73.5%) terminated in this manner without physical contact.

However, if neither individual withdrew, a confrontation occurred when the distance between the males decreased to three or four inches. In this situation, one frog occasionally jumped directly at or upon the other, but much more commonly the two individuals simultaneously pushed against one another and became locked, throat to throat and venter to venter, with the fore-arms of each tightly grasped around the inflated pectoral region of the other. Locked in this manner, the frogs engaged in a violent struggle, kicking with the hind legs until one individual was forced over onto its back, at which time contact was broken. The force exerted during this struggle pushed the two males vertically out of the water until only the posterior third of the abdomen and the hind legs remained submerged.

Such struggles generally lasted only 15–20 sec, but bouts continuing for four and

440

five min were occasionally observed. Following an encounter, the frog which successfully out-pushed its opponent remained in the "high" posture and called. By contrast, the other individual stayed submerged for several sec and, upon surfacing, adopted an entirely different, "low" posture in which the lungs were deflated and only the upper portion of the head was visible above the water. If he surfaced within one or two ft of his opponent, he was usually rechallenged and responded by rapidly exiting from the vicinity. Far more commonly, however, the "low" individual swam some distance under water, surfacing several ft away. In these instances, although the deflated frog often was still within the defended area, further aggressive behavior was rare, suggesting that deflation characterizes a submissive posture which inhibits attack.

The behavior of "non-established" males, individuals which either did not possess territories or which inhabited areas far from active choruses, lends additional support to this hypothesis. The population under study included many such individuals and each evening several would move into the choruses. These males consistently maintained the "low," deflated posture when approaching and entering a chorus, and in this manner they moved among the established males without evoking any aggressive response. Yet, when such a male inflated himself and called, he was challenged and attacked immediately. Occasionally the invaders succeeded in supplanting a resident, but more frequently they were forced from the chorus, or at least to its periphery.

There is also evidence suggesting that the "high" posture functions directly as a threat. The inflated body form is undoubtedly a necessary prerequisite for effective vocalization since air is forced from the lungs through the larynx into the vocal sacs during calling. But this posture was maintained throughout the night, even during long, noncalling periods. Preliminary experiments measuring the intensity of males' aggressive responses to a model bullfrog suggest that the inflated posture (which maximizes the apparent size of the caller) is an effective stimulus eliciting aggressive challenges. The model used in these tests was a painted, latex cast of a bullfrog mounted over a styrofoam base. The "high" and "low" postures could be simulated by attaching various weights to the base, thereby altering the buoyancy of the model. When placed inside an established territory in the "high" position, the model evoked prompt challenges from six of eight resident males tested. The remaining two individuals deflated, assuming the "low" posture, and no aggressive behavior was noted. By contrast, no behavioral responses occurred when the same model was presented floating low in the water ($n = 12$). Playbacks of tape recordings of both the "hiccup" vocalization and the normal male call (neither of which are given by males in the "low" posture) also consistently evoked aggressive responses.

The adaptive significance of bullfrog territoriality is best understood in conjunction with the breeding biology of the species. Unlike many North American anurans, in which members of a given population attain peak reproductive condition synchronously and mating occupies only a brief period of days or weeks, male bullfrogs remain reproductively active throughout much of the summer. Each female, however, is sexually receptive for only a short time, and the long breeding season results from the great variation in the dates at which different females attain this state (Raney, 1940; Emlen, unpubl. obs.). In the Michigan population studied, gravid females were encountered from mid-June until late July.

The behavior of the females varied in accordance with their reproductive condition. Prior to ovulation, they were quite secretive, remaining hidden along the shore or in a shallow marsh adjacent to the breeding pond. When ripe, however, females immigrated into the pond via the small inlet stream, left the cover of the shoreline, and approached the male choruses. These approaches evidently were guided at least in part by the sound of the males' calling, since females could be attracted to a tape recorder by playbacks of chorusing. However, direct observation revealed that females were *not* strongly attracted to males calling from isolated territories; in all observed instances ($n = 16$) females passed by such males and continued on to the large choruses.

Once a chorus was reached, a female adopted an extreme "low" posture and moved from territory to territory without eliciting any noticeable response from the resident males. After many hours of remaining "low" in first one territory and then another, the female appeared to select a mate, swimming up to him and seemingly making physical contact. Only then would the male respond, by immediately seizing her in amplexus.

It is apparent from this information that an individual male bullfrog maximizes his chances of contributing genetically to the next generation by 1) congregating with other males in a large chorus, and 2) remaining in a sexually active state throughout the season when females become ripe.

Although it is assumed that all mature females in a population breed, it was only rarely that more than one or two individuals entered a chorus during the course of an evening. Consequently, the ratio of responsive males to receptive females at any one time is enormous—on the order of 40–50 to 1 in the population studied. As a result of this, male competition for females must be extremely intense.

This social system is quite similar to the "leks" present in several species of polygamous birds and mammals in which males establish territories in a communal display ground or arena (see Armstrong, 1964:431–433). Females come to these display areas and, after moving among the males, select a mate and copulate. In males of these species, as with male bullfrogs, there must be a strong selective premium placed upon rapid attraction of the opposite sex.

I suggest that the possession of a territory is one important factor increasing a male bullfrog's chance of successful mating. Perhaps this advantage is accrued by the creation of a buffer zone around the holder which allows females to approach and select a mate without interference from numerous, nearby, competing males. Such buffer zones might also yield a measure of stability to a chorus, thereby resulting in an overall decrease in energy expended in aggressive interactions. In addition, some aspect of a territory, such as its size or location, might directly enhance the attractiveness of its holder. It is even possible that the agonistic interactions associated with territoriality help maintain the prolonged state of sexual responsiveness present among males.

At present these suggestions are highly speculative. Future experiments examining the factors operative in a female bullfrog's selection of a mate will be required to provide an empirical test of this hypothesis—that the possession of a territory helps to maximize the reproductive potential of the individual holder.

ACKNOWLEDGMENTS

I thank C. F. Walker for his helpful suggestions and H. W. Ambrose, T. Eisner, and W. T. Keeton for commenting upon the manuscript. This study was supported by a National Science Foundation Graduate Fellowship.

LITERATURE CITED

ARMSTRONG, E. A. 1964. Lek display. In: A new dictionary of birds. A. L. Thomson, ed. McGraw-Hill, New York.

BOGERT, C. M. 1947. A field study of homing in the Carolina toad. Am. Mus. Novit. No. 1355, 24 pp.

DOLE, J. W. 1965. Summer movement of adult leopard frogs, Rana pipiens Schreber, in northern Michigan. Ecology 46:236–255.

DUELLMAN, W. E. 1966. Aggressive behavior in dendrobatid frogs. Herpetologica 22:217–221.

FERGUSON, D. E. 1963. Orientation in three species of anuran amphibians. Ergebn. Biol. 26:128–134.

JAMESON, D. L. 1957. Population structure and homing responses in the Pacific treefrog. Copeia 1957(3):221–228.

KIKUCHI, T. T. 1958. On the residentiality of a green frog, Rana nigromaculata Hallowell. Jap. J. Ecol. 8:20–26.

LUTZ, B. 1960. Fighting and an incipient notion of territory in male tree frogs. Copeia 1960(1):61–63.

MARTOF, B. S. 1953a. Territoriality in the green frog, Rana clamitans. Ecology 34:165–174.

———. 1953b. Home range and movement of the green frog, Rana clamitans. Ibid. 34: 529–543.

NOBLE, G. K. 1931. The biology of the amphibia. McGraw-Hill, New York.

RABB, G. B. AND M. S. RABB. 1963a. On the behavior and breeding biology of the African pipid frog Hymenochirus boettgeri. Zeits. Tierpsychol. 20:215–241.

——— AND ———. 1963b. Additional observations on breeding behavior of the Surinam toad, Pipa pipa. Copeia 1963(4):636–642.

RANEY, E. C. 1940. Summer movements of the bullfrog, Rana catesbeiana Shaw, as determined by the jaw-tag method. Am. Midl. Nat. 23: 733–745.

SEXTON, O. 1960. Some aspects of the behavior and of the territory of a dendrobatid frog, Prostherapis trinitatus. Ecology 41:107–115.

———. 1962. Apparent territorialism in Leptodactylus insularum Barbour. Herpetologica 18:212–214.

TEST, F. H. 1954. Social aggressiveness in an amphibian. Science 120:140–141.

WRIGHT, A. H. AND A. A. WRIGHT. 1949. Handbook of frogs and toads of the United States and Canada. Cornell Univ. Press, Ithaca, New York.

DEPARTMENT OF ZOOLOGY AND MUSEUM OF ZOOLOGY, UNIVERSITY OF MICHIGAN, ANN ARBOR 48104. *Present address*: SECTION OF NEUROBIOLOGY AND BEHAVIOR, DIVISION OF BIOLOGICAL SCIENCES, CORNELL UNIVERSITY, ITHACA, NEW YORK 14850.

51

Social Control of Sex Reversal in a Coral-Reef Fish

Abstract. *Males of* Labroides dimidiatus *control the process of sex reversal within social groups. Each group consists of a male with a harem of females, among which larger individuals dominate smaller ones. The male in each harem suppresses the tendency of the females to change sex by actively dominating them. Death of the male releases this suppression and the dominant female of the harem changes sex immediately. Possible genetic advantages of the system are considered.*

Reprinted from SCIENCE,
15 September 1972, volume 177, pages 1007-1009

Sex reversal is widespread in a number of tropical fishes included in the families Labridae, Scaridae, and Serranidae (*1*, *2*). In this report I describe the pattern of protogynous sex reversal in the labrid fish *Labroides dimidiatus*. The species is a member of a small but widespread genus, the species of which are termed "cleaner fish" because they remove ectoparasites from the skin of other fishes (*3*). Choat (*2*) established that the species is protogynous, with far more females than males, and that probably all the males were secondarily derived from females.

The basic social unit is a male with a harem of usually three to six mature females and several immature individuals living within the male's territory. At Heron Island, Great Barrier Reef, detailed field records were kept on 11 groups for up to 25 months; 48 sex reversals were recorded in these and another eight groups. Individual adults were recognized by unchanging variations in their color patterns.

All individuals exhibit territoriality, but its expression varies with age and sex. The largest, oldest individual is the male, which dominates all the females in the group. Larger, older females of the group dominate smaller ones, which usually results in a linear dominance hierarchy. Thus territoriality is only fully expressed in males and is directed mainly toward other males. Usually there is one dominant female in each group, but sometimes two equal-sized females are codominant and can successfully defend their territories against each other. The dominant female lives in the center of the male's territory, with the other females scattered around. The male is socially very active. It makes frequent excursions throughout its territory both to the feeding areas of the females and to

443

points on the territory border where the male is likely to meet neighboring males. During these excursions the male feeds in the females' areas and actively initiates aggressive encounters with them and other individuals. Females, on the other hand, are more sedentary and passive. When a male meets a female of the same group, the male frequently performs a distinctive aggressive display toward the female. This display has not been seen in encounters between males and only very rarely in encounters between females, when it was given by dominant females.

Some males and large females have maintained nearly the same territories and feeding sites for almost 2 years. Small adults and large juveniles are more mobile. Deaths of individuals high in the hierarchy result in more marked changes in the distribution of other high-status individuals than do deaths of low-status individuals. With the death of a high-status female, the vacated area may be incorporated into the territory of an individual of equal status or taken over by an individual immediately below the deceased in status, the lower status female deserts its own territory in the process. This shift can result in the immediate redistribution of three or four high-status females.

Sex reversal frequently occurs as a part of the reorganization of the group following the death of the male. The success of an initiated reversal depends upon both inter- and intragroup social pressures. Intergroup social pressures take the form of territory invasion and takeover attempts by neighboring males, and if these pressures are successfully resisted by the dominant female it changes sex. Groups with codominant females sometimes divide when both dominants change into males. In all, 26 cases of single dominant females reversing sex were observed (five naturally occurring and 21 experimentally induced by removing

the male), and four cases of reversal of pairs of codominants were also seen (all induced by removing the male).

Observations of five dominant females after the removal of their males have shown that the first behavioral signs of sex reversal appear rapidly and that the behavioral changeover can be completed within a few days. For approximately half an hour after the death of the male the dominant female continues to behave aggressively as a normal female. This simple female aggression then wanes to more neutral reactions to nearby subordinate females. Approximately 1½ to 2 hours after male death, maleness appears in the form of the special male aggressive display that the new "male" starts performing to the females of its group. The assumption of the male aggressive role can be virtually completed within several hours, when the "male" starts visiting its females and territory borders. The switchover to male courtship and spawning behavior takes somewhat longer but can be partly accomplished within 1 day and completed within 2 to 4 days. Other individuals also respond within a couple of hours to the altered social situation created by male death; low-status females take over vacant female territories, and neighboring males invade and attempt to take over the territory and harem.

The death of a male does not necessarily lead to a sex reversal within the group. In 11 cases intergroup pressures were apparently too great, and neighboring males invaded the territory, taking it and the female group over. In four of these cases the dominant female had started to behave like a male before the invasion but reverted to female behavior after the completion of the takeover and remained as a fully functional female. In all these successful takeovers the invading male was considerably larger than the incumbent dominant female and was able to dominate relatively easily. In the one interrupted sex reversal observed, the dominant female was under the control of a slightly larger invading male for about 2 weeks after the death of the original male. During this period the female behaviorally went through the series $♀ → ♂ → ♀ → ♂$, the final change being a successful one.

Five observed sex reversals were not associated with male death. Before reversal all five individuals were medium-large females of high status; four were subdominants and one was a co-dominant. The area of each individual was away from the main areas of social activity of the dominant male, which visited the female relatively infrequently. Consequently each female had much less social contact with the male than did females of comparable status within the same group.

Histological examinations were made of the gonads of 29 females and 35 males. The ovaries of 28 of the 29 females contained small spermatogenic crypts located close to sites of early oogenesis in the ovarian lamellae, and in 15 of these 28 females some crypts contained sperm or spermatids. Free sperm were not detected, and the spermatogenic crypts appeared to be completely enclosed. The gonads of 28 males of known "age" (age from the start of reversal) have been examined. From the small series of "young" males examined, it appears that sperm can be released 14 to 18 days after the start of reversal.

These data demonstrate social control of sex reversal in this species, with males regulating the production of males. Probably all females are capable of changing sex, and most (possibly all) have testicular elements within perfectly functional ovaries. The tendency of any female to change may be actively suppressed by more dominant individuals in the hierarchy. Non-dominant females have aggression by both sexes directed at them; the dominant female is the object of only

male-type aggression and is dominated by only a single individual. Death of the male means that the female of highest status becomes totally dominant with the group, and the tendency to change sex is no longer suppressed. The rapidity with which a new male behaviorally assumes its role is a reflection of the presence of male elements in all females and the necessity for a new male to consolidate its position quickly in the face of constant intra- and intergroup pressures. Subdominant females are also potential males and must be inhibited if the group structure is to be maintained. Neighboring males must be excluded if a harem is to be maintained. Males direct their aggression differentially toward females of different status in their hierarchies. The male is more aggressive toward those females most likely to change sex and threaten his position—that is, larger females, especially the dominant one (Table 1). The aggression directed at these higher status females is also more characteristically male. Incomplete control of high-status females such as those most peripheral in the male's territory, can result in sex reversal. High-status females probably

Females	Male aggressive acts		
	SA	HD	LD
Dominant	54	94	371
Large subdominant	36	8	48
Small subdominant	20	1	22

Table 1. Aggressive acts by males against three hierarchical classes of females [three types of aggressive acts—simple attacks (SA), high-intensity male aggressive display (HD), and low-intensity male aggressive display (LD)]. The data represent eight males and their harems. Hierarchy subdominants have been arbitrarily divided into two classes. The distribution of aggressive acts did not differ significantly between the eight social groups [goodness-of-fit test, G (5)]. The data from all groups were therefore pooled. The same method demonstrated significant heterogeneity among the three classes of females (G = 69.085, d.f. = 4, $P < .01$).

suppress females lower in the hierarchy; the latter evidently need less male control.

Field experiments with seven isolated females (the other members of naturally isolated groups were removed) indicate that the presence of a harem is not necessary for sex reversal to be accomplished, although the process may be slower. Experiments with six males similarly isolated without harems demonstrated the continuance of sperm production for up to 26 days in functional males.

Many Labridae and Scaridae are protogynous hermaphrodites with female-biased sex ratios (2), and in some species social control of sex reversal may operate in a similar manner to that in *Labroides dimidiatus*. In other species, especially schooling forms, a well-defined social structure based on individual relationships might not be possible, and sex reversal may be controlled more by endogenous factors. Male control of the production of males has been demonstrated in the laboratory in a protogynous serranid fish, *Anthias squamipinnis* (4).

At the present, discussion of the biological significance of protogyny remains speculative. Arguments have been put forward to explain protogyny and the biased sex ratio in terms of population growth, with the predominance of females increasing fecundity (4). Choat (2) advanced the idea that the biased sex ratio, maintained by protogyny, could be considered as an inbreeding mechanism because it reduced the number of genotypes available for recombination, and this would permit adaptation to specific local conditions.

My observations on the pattern of sex reversal in *L. dimidiatus* support this idea of genetic advantages for the system. The genotypes of the males are those which are maximally recombined, because each male spawns regularly with the females of its group and each female spawns, in the main, with the dominant male only. The male genotype is the genotype best adapted to local conditions because the male is derived from the oldest female of the group. Individuals enter the group and gradually move up within it, with only the best adapted females eventually being able to reverse sex. Thus the social organization is a framework within which the selective process works. The social group is a self-perpetuating system which ensures the maintenance of the biased sex ratio by controlling sex reversal. Social control of sex reversal both maximizes the genetic advantages of the process and imparts considerable flexibility to it. Males are produced ònly when they are needed, and this method overcomes the possible precariousness of a strongly biased sex ratio maintained by endogenously controlled sex reversal.

D. R. ROBERTSON

Zoology Department, University of Queensland, St. Lucia, Brisbane, Q. 4067, Queensland, Australia

References and Notes

1. J. W. Atz, in *Intersexuality in Vertebrates including Man*, C. N. Armstrong and A. J. Marshall, Eds. (Academic Press, London, 1964), pp. 145–232; R. Reinboth, *Zool. Jahrb. Abt. Allg. Zool. Physiol. Tiere* **69**, 405 (1962).
2. J. H. Choat, thesis, University of Queensland (1969), part 1; R. Reinboth, *Z. Naturforsch. B* **23**, 852 (1968).
3. J. E. Randall, *Pac. Sci.* **12**, 327 (1958).
4. L. Fishelson, *Nature* **227**, 90 (1970).
5. R. R. Sokal and F. J. Rohlf, *Biometry: The Principles and Practice of Statistics in Biological Research* (Freeman, San Francisco, 1969).
6. I thank R. Bradbury, A. Cameron, J. Choat, J. Connell, D. Dow, and J. Kikkawa for criticizing a draft of the manuscript; the Great Barrier Reef Committee for the use of Heron Island Research Station facilities; and the University of Queensland for financially supporting this research.

1 June 1972

52 A Consequence Of Togetherness

The inability of the North American mountain sheep

to disperse into new habitat stems from

the inheritance of social traditions

by Valerius Geist

During the last century, large populations of bighorn sheep grazed widely from central British Columbia to northern Mexico, and from the coastal mountains of California to the Black Hills of the Dakotas. But with the opening of the West, the bighorns (*Ovis canadensis*) dwindled in number and finally disappeared from most of their former range. Protective measures enacted at the turn of the century failed to reverse their fortunes, and ever since they have existed only in small, widely scattered bands. Some of these relic populations have occasionally increased in size, but they have never recolonized the large amount of former living space available to them. Although the ranges of Dall's and Stone's sheep (*O. dalli* and *O. d. stonei*) are remote and less vulnerable to human invasion, a similar predicament exists for these animals. In short, the North American mountain sheep appear to be incapable of dispersal.

In contrast to the sheep, moose (*Alces alces*) and white-tailed deer (*Odocoileus virginianus*) have made spectacular acquisitions of new territory and have succeeded in increasing their numbers throughout most of their North American range. Why this differential success in acquiring new living space between the sheep and these ungulates?

We know that the former ranges are still acceptable to sheep, for reintroduced populations have generally done well. From twenty-nine reintroductions made on this continent, whose success can be evaluated today, twenty-two have proven successful. And the failure of others may have been due to improper release methods rather than to unsatisfactory range conditions. However, reintroduced populations behave much like the natural relic populations. They remain small in number and generally fail to spread far from the release sites, although an occasional animal may be seen a great distance from his companions.

Nor are sheep physically limited from dispersing. They are most capable migrants and in their travels often swim lakes or rivers and cross over rock rubble and cliffs. In late winter they often travel from mountain to mountain on frozen snow crusts without breaking through. They can go anywhere deer or moose can, and many places they cannot.

Apparently, sheep dislike entering timber, and it is thought that wooded valleys act as barriers to dispersion. Bighorns, for instance, which had become so tame that they followed me at heel as I conducted my field research, rarely went more than one hundred paces into the timber before they left me and returned to the open slopes. Similarly, migratory routes tend to take the shortest distance between mountains and to skirt forest; yet, during regular, seasonal migrations of natural populations, sheep may cross miles of dense timber.

The North American sheep's failure to disperse away from established ranges becomes the more unusual when we remember that, on a worldwide basis, sheep have distributed over a great area. They are found from southwestern Europe, through Asia Minor, the Himalayas, central Asia, eastern Siberia, Alaska, western Canada, the western United States, and Mexico. No other living bovid equals, let alone surpasses, such a distribution. Since sheep have spread through so vast a range, why are they failing in North America today?

Between 1961 and 1966, field studies of several natural populations of North American sheep suggested an answer to their inability to extend their ranges. During this period, my observations indicated that sheep maintain their areas of distribution by passing on home ranges from generation to generation as a living tradition. Each generation of sheep inherits the home ranges of its elders; that is, they acquire the same habits of living in certain areas at specific seasons, and of using the same migratory routes. Exploration apparently plays a most insignificant part in establishing the home range of an individual. Moose and deer, however,

extend their areas of distribution via individual exploration. Neither traditions nor a means of passing them are discernible in the life history of these two species.

If home range traditions are to exist, then there must be a continuous association between donors and receivers, a condition satisfied by sheep. Moreover, sheep society provides for leadership by the elders, reinforces acquired habits, and, by minimizing disruptions in social life, provides no impetus for independent exploration. A comparison between the separation of mother and young in moose and sheep is instructive in this regard.

While the cow moose forms a close relationship with her calf, the female sheep and her lamb live in a rather loose association. But in an abrupt change of behavior, when the female moose is about to give birth to another calf she turns on her yearling and drives it off. The life of the young animal is changed suddenly and violently. Within a few days it changes from a social to a solitary existence, for no cow moose will tolerate the yearling close by and the bulls are little better. The calf is now on his own, and begins the wanderings that eventually lead to the establishment of a new home range.

In contrast, the young sheep experiences no such upheaval. About one week after birth, the lamb begins to associate with other lambs in a band. After it is weaned, the bond with its mother commences to weaken, and the young sheep will follow other animals, preferably an old, barren female, which may even run and play with the youngster. Eventually, the lamb loses all preference for its mother over other sheep.

The important feature of this separation is that the offspring is not chased away and thus does not lose contact with other sheep. Of 283 bighorn yearlings observed between May 15 and June 30, in 1964 and 1965, 76 per cent followed some adult ewe, 15.5 per cent followed an adult ram, 5 per cent followed a subadult sheep (two years old), 3.2 per cent followed other yearlings. Only 1 per cent were seen alone.

Observation of tagged or marked sheep has shown that females, but not males, acquire the home ranges of the female band they are born in. Only rarely do young females leave their band and accept the traditions of another group. When after a prolonged absence I returned to my Stone's sheep study area, I found known female sheep in the expected localities. They were feeding on the same slopes, resting in the same favorite beds, and entering the same caves they had frequented three years previous. In fact they behaved so much in the familiar manner that I felt I had never been away from them.

Young males gradually break with the movement pattern of their female band, by joining male bands that have other, equally stable home ranges. In his third summer, when the young ram begins to search out male bands, he shows a weak, but discernible, wandering phase. In July and August, up to one third of all such little rams may be seen alone, or in the company of a companion of equal age. After his third summer, the ram is again a dedicated follower of older sheep—and remains such till, with increasing age, he gradually becomes more independent and is then followed in turn.

Generally the leader of a ram band is an animal at least eight years old, with long, massive horns. By following these older sheep, the younger rams get to know the wintering areas, the salt licks, the rutting grounds, the summer ranges, and the migratory routes, which have been passed on through the generations. In general, by following the choices of their leaders, younger males acquire those habits that allowed their leaders to grow old: habits that are constant and predictable. Thus, it is evident that sheep society is virtually designed for the passage of habits.

Two-year-old rams can and do stray into areas uninhabited by sheep, yet they will not remain unless accompanied by other animals. If we observe such a single, wandering ram we will note that he often climbs on elevated sites and scans the country below. He calls loudly—which all lone sheep do when traveling. These young rams are apparently searching for other sheep. If they do hit upon unoccupied good habitat, they are not likely to remain, for a habitat without companions is simply not a suitable place for a sheep to live. Furthermore, young rams cannot lead other animals into a newly discovered range, for they have not yet become leaders of a band. Thus, the range extension performed by dispersing juvenile moose will not occur in mountain sheep.

Unlike moose or deer, mountain sheep seem able to disperse only as groups, if they disperse at all. We have no direct data on this, save one introduced population that moved, as a whole, twenty-five to thirty miles to a different locality, but field observations revealed a possible means by which this might have been accomplished. In spring, groups of young rams occasionally go on long excursions into the valleys. Apparently, the "group spirit" of such a band is a strong enough factor to permit the sheep to overcome their reluctance to enter even small patches of timber. Thus, a band may leave an inhabited mountain to visit a nearby, unpopulated area, if intervening timber is not particularly dense. Once a visit has been made, the rams are almost certain to return, since they experienced the new habitat in the presence of companions. However, these group excursions can only occur if the sheep population is large enough to provide large bands of young rams that will discover and retain the new range.

How is it that sheep have developed one system of range establishment and retention, while moose have adopted quite different methods? This question appears particularly apt when we consider that sheep are faring relatively poorly compared to moose. The answer seems to stem from the fact that the habitats of these two species have shaped the evolution of different social systems, which in turn have determined the means by which these animals acquire living space.

Throughout much of its range in America, the moose is found in communities of deciduous shrubs and trees, which follow in the wake of forest fires. Such moose habitat flourishes some years after a fire and allows moose to build up large populations in response to the rich food supply. As coniferous climax forests slowly displace the deciduous species on which the animals feed, the populations decline and vanish. Mature timber, hot summers, and lightning storms assure moose new living space, albeit a rather temporary one. Today, man-made forest fires and logging operations play an important role in the creation of new habitat, although man's intrusion is a relatively recent influence that probably had little effect upon the animal's social evolution.

We find moose not only in the short-lived "burn habitat" created by forest fires but also in small areas with permanent shrub communities, such as on rich alluvial soils along watercourses—particularly in deltas—and on south-facing mountain slopes above timber line, as in northern British Columbia. These limited localities assure a small amount of permanent living space to moose, where they can survive in the absence of habitat created by forest fires.

Under these conditions, the practice of juvenile dispersion serves an important function in maintaining and extending the moose population. Young animals wander out on their own, chance upon burn habitat, establish new ranges, and become the ancestors of large populations that will remain in an area until the habitat disappears. Since the vast majority of moose habitat is short-lived, traditions of range retention would be useless to them. Such traditions would return the animals to increasingly poorer habitat and finally to no habitat at all. Moreover, sizable herds of moose— needed to transmit such traditions— could not be maintained on small patches of permanent habitat. Moose are large animals, and where a few can survive, a dozen would eat themselves out of house and home. Thus, if moose were to develop a social tolerance for one another as sheep have, the resulting "group living" might cause their extinction.

Sheep habitat, whether in California's Death Valley or beyond the Arctic Circle in Alaska's Brooks Range, is always characterized by an open, mountainous landscape and stable plant communities. Most sheep are found on grasslands, which regenerate themselves and do not vanish within a few sheep generations as moose burn habitat does. Today sheep habitat is usually found in small, often widely dispersed patches. These may be separated by brush flats, glaciers, lakes, or wide belts of timber in the valleys. Natural populations of sheep link such patches of habitat via migratory routes inherited from their ancestors.

The distribution of sheep tells us that today's habitat is only the remnant of much larger grasslands that existed shortly after the glaciers withdrew. In later millennia, forests spread along the valleys and ascended mountain slopes, thereby dividing and squeezing mountain grasslands into ever smaller patches. Much sheep habitat vanished, but the remainder was retained by the sheep's home range and migratory traditions—the animals continued to exploit all available patches of habitat. Under these conditions, there is no selection in favor of dispersing juveniles since all the habitat they would normally encounter is already occupied. By leaving adults and moving off through miles of unsuitable territory, juveniles would only increase their chances of getting killed. In relic populations that make long migrations between patches of habitat, selection will favor those animals that remain in the company of others. Any behavior that disrupts this association between young and old sheep will be selected against. Traditions of habitat retention are therefore desirable for sheep, whereas juvenile dispersal is not.

Upon the sheep's evolved habits of range retention, however, must be superimposed the recent intrusion of man. After climatic and vegetal changes had forced sheep to link patches of suitable habitat by means of migratory routes, the coming of man destroyed many of these links— as well as the sheep. Thus, in much of North America, sheep now live in scattered populations, despite the presence of nearby habitat. But this set of conditions is a man made artifact not previously encountered by the species and to which it lacks a meaningful response. When sheep are exterminated or alienated from certain mountains, the migratory traditions die out, for no surviving adult will return to where it is harassed and no young sheep will be led to it. Once an area has been "forgotten" by a population, sheep are then incapable of returning, even if the area is close at hand. Furthermore, reintroductions appear to me as only a partial solution. Once sheep are stocked on an unfamiliar mountain, the animals' lack of familiarity with the new habitat tends to make them cling to the immediate vicinity. Their exploration is limited to open territory connected to the original area of introduction. They rarely cross timbered valleys and other barriers that may separate them from nearby, potential range. While in natural populations, the males move off to their own areas; in the introduced populations, they share the area (which may be a single mountain) with the ewes. This limits the available food supply via competition and prevents significant growth in numbers. The ideal solution would necessitate a knowledge of access routes to nearby habitat, but this knowledge does not exist. Thus, dispersal is unlikely. The problem is further complicated if men inhabit nearby areas, for this will generally frighten sheep and restrict their activity to the undisturbed parts of a range.

Thus it seems that the evolution of the mountain sheep's means of maintaining its population has not prepared the species for coexistence with man. The contraction of suitable habitat following the glacial retreat has brought about the traditions of range establishment and migration, which have recently been disrupted by human interference. If this interference persists, sheep face an uncertain future. But if a means is found to re-establish migratory routes between patches of available habitat, I believe the animals will overcome their earlier setbacks.

EDITOR'S NOTE: After forty-three months spent in the field researching the mountain sheep. Valerius Geist found that these animals would eat from his hand, follow him about, even play "hide and seek" with him. This behavior conflicts with the rather stereotyped notion that sheep are wild and unapproachable. However, it suggests an answer to the problem mentioned in his article—

the sheep's inability to spread into new habitat from their established home ranges. In a recent letter, Dr. Geist stated: "A little patience, tact and delicacy can turn these shy creatures into pets, or even into a nuisance. Although I have not yet tried to lead a band of sheep into an area it does not regularly visit. I am sure it can be done. I believe they can be tamed and made to use salt licks that are progressively moved into new territory, thereby establishing migratory routes which would be followed by succeeding generations."

53   On War and Peace in Animals and Man

An ethologist's approach to the biology of aggression.

N. Tinbergen

In 1935 Alexis Carrel published a best seller, *Man—The Unknown* (*1*). Today, more than 30 years later, we biologists have once more the duty to remind our fellowmen that in many respects we are still, to ourselves, unknown. It is true that we now understand a great deal of the way our bodies

Dr. Tinbergen is professor of animal behavior, Department of Zoology, University of Oxford, Oxford, England. This article is the text of his inaugural address, 27 February 1968.

function. With this understanding came control: medicine.

The ignorance of ourselves which needs to be stressed today is ignorance about our behavior—lack of understanding of the causes and effects of the function of our brains. A scientific understanding of our behavior, leading to its control, may well be the most urgent task that faces mankind today. It is the effects of our behavior that begin to endanger the very survival of our species and, worse, of all life on earth. By our technological achievements we have attained a mastery of our environment that is without precedent in the history of life. But these achievements are rapidly getting out of hand. The consequences of our "rape of the earth" are now assuming critical proportions. With shortsighted recklessness we deplete the limited natural resources, including even the oxygen and nitrogen of our atmosphere (2). And Rachel Carson's warning (3) is now being followed by those of scientists, who give us an even gloomier picture of the general pollution of air, soil, and water. This pollution is seriously threatening our health and our food supply. Refusal to curb our reproductive behavior has led to the population explosion. And, as if all this were not enough, we are waging war on each other—men are fighting and killing men on a massive scale. It is because the effects of these behavior patterns, and of attitudes that determine our behavior, have now acquired such truly lethal potentialities that I have chosen man's ignorance about his own behavior as the subject of this paper.

I am an ethologist, a zoologist studying animal behavior. What gives a student of animal behavior the temerity to speak about problems of human behavior? Of course the history of medicine provides the answer. We all know that medical research uses animals on a large scale. This makes sense because animals, particularly vertebrates, are, in spite of all differences, so similar to us; they are our blood relations, however distant.

But this use of zoological research for a better understanding of ourselves is, to most people, acceptable only when we have to do with those bodily functions that we look upon as parts of our physiological machinery—the functions, for instance, of our kidneys, our liver, our hormone-producing glands. The majority of people bridle as soon as it is even suggested that studies of animal behavior could be useful for an understanding, let alone for the control, of our own behavior. They do not want to have their own behavior subjected to scientific scrutiny; they certainly resent being compared with animals, and these rejecting attitudes are both deep-rooted and of complex origin.

But now we are witnessing a turn in this tide of human thought. On the one hand the resistances are weakening, and on the other, a positive awareness is growing of the potentialities of a biology of behavior. This has become quite clear from the great interest aroused by several recent books that are trying, by comparative studies of animals and man, to trace what we could call "the animal roots of human behavior." As examples I select Konrad Lorenz's book *On Aggression* (4) and *The Naked Ape* by Desmond Morris (5). Both books were best sellers from the start. We ethologists are naturally delighted by this sign of rapid growth of interest in our science (even though the growing pains are at times a little hard to endure). But at the same time we are apprehensive, or at least I am.

We are delighted because, from the enormous sales of these and other such books, it is evident that the mental block against self-scrutiny is weakening —that there are masses of people who, so to speak, want to be shaken up.

But I am apprehensive because these books, each admirable in its own way, are being misread. Very few readers give the authors the benefit of the doubt. Far too many either accept uncritically all that the authors say, or (equally uncritically) reject it all. I believe that this is because both Lorenz and Morris emphasize our knowledge rather than our ignorance (and, in addition, present as knowledge a set of statements which are after all no more than likely guesses). In themselves brilliant, these books could stiffen, at a new level, the attitude of certainty, while what we need is a sense of doubt and wonder, and an urge to investigate, to inquire.

Potential Usefulness of Ethological Studies

Now, in a way, I am going to be just as assertative as Lorenz and Morris, but what I am going to stress is how much we do not know. I shall argue that we shall have to make a major research effort. I am of course fully aware of the fact that much research is already being devoted to problems of human, and even of animal, behavior. I know, for instance, that anthropologists, psychologists, psychiatrists, and others are approaching these problems from many angles. But I shall try to show that the research effort has so far made insufficient use of the potential of ethology. Anthropologists, for instance, are beginning to look at animals, but they restrict their work almost entirely to our nearest relatives, the apes and monkeys. Psychologists do study a larger variety of animals, but even they select mainly higher species. They also ignore certain major problems that we biologists think have to be studied. Psychiatrists, at least many of them, show a disturbing tendency to apply the *results* rather than the *methods* of ethology to man.

None of these sciences, not even their combined efforts, are as yet parts of one coherent science of behavior. Since behavior is a life process, its study ought to be part of the mainstream of biological research. That is why we zoologists ought to "join the fray." As an ethologist, I am going to try to sketch how my science could assist its sister sciences in their attempts, already well on their way, to make a united, broad-fronted, truly biological attack on the problems of behavior.

I feel that I can cooperate best by discussing what it is in ethology that could be of use to the other behavioral sciences. What we ethologists do not want, what we consider definitely wrong, is uncritical application of our results to man. Instead, I myself at least feel that it is our method of approach, our rationale, that we can offer (6), and also a little simple common sense, and discipline.

The potential usefulness of ethology lies in the fact that, unlike other sciences of behavior, it applies the method or "approach" of biology to the phenomenon behavior. It has developed a set of concepts and terms that allow us to ask:

1) In what ways does this phenomenon (behavior) influence the survival, the success of the animal?

2) What makes behavior happen at any given moment? How does its "machinery" work?

3) How does the behavior machinery develop as the individual grows up?

4) How have the behavior systems of each species evolved until they became what they are now?

The first question, that of survival value, has to do with the effects of behavior; the other three are, each on a different time scale, concerned with its causes.

These four questions are, as many of my fellow biologists will recognize, the major questions that biology has been pursuing for a long time. What ethology is doing could be simply described by saying that, just as biology investigates the functioning of the organs responsible for digestion, respiration, circulation, and so forth, so ethology begins now to do the same with respect to behavior; it investigates the functioning of organs responsible for movement.

I have to make clear that in my opinion it is the comprehensive, integrated attack on all four problems that characterizes ethology. I shall try to show that to ignore the questions of survival value and evolution—as, for instance, most psychologists do—is not only shortsighted but makes it impossible to arrive at an understanding of behavioral problems. Here ethology can make, in fact is already making, positive contributions.

Having stated my case for animal ethology as an essential part of the science of behavior, I will now have to sketch how this could be done. For this I shall have to consider one concrete example, and I select aggression, the most directly lethal of our behaviors. And, for reasons that will become clear, I shall also make a short excursion into problems of education.

Let me first try to define what I mean by aggression. We all understand the term in a vague, general way, but it is, after all, no more than a catchword. In terms of actual behavior, aggression involves approaching an opponent, and, when within reach, pushing him away, inflicting damage of some kind, or at least forcing stimuli upon him that subdue him. In this description the effect is already implicit: such behavior tends to remove the opponent, or at least to make him change his behavior in such a way that he no longer interferes with the attacker. The methods of attack differ from one species to another, and so do the weapons that are used, the structures that contribute to the effect.

Since I am concentrating on men fighting men, I shall confine myself to intraspecific fighting, and ignore, for instance, fighting between predators and prey. Intraspecific fighting is very common among animals. Many of them fight in two different contexts, which we can call "offensive" and "defensive." Defensive fighting is often shown as a last resort by an animal that, instead of attacking, has been fleeing from an attacker. If it is cornered, it may suddenly turn round upon its enemy and "fight with the courage of despair."

Of the four questions I mentioned before, I shall consider that of the survival value first. Here comparison faces us right at the start with a striking paradox. On the one hand, man is akin to many species of animals in that he fights his own species. But on the other hand he is, among the thousands of species that fight, the only one in which fighting is disruptive.

In animals, intraspecific fighting is usually of distinctive advantage. In addition, all species manage as a rule to settle their disputes without killing one another; in fact, even bloodshed is rare. Man is the only species that is a mass murderer, the only misfit in his own society.

Why should this be so? For an answer, we shall have to turn to the question of causation: What makes animals and man fight their own species? And why is our species "the odd man out"?

Causation of Aggression

For a fruitful discussion of this question of causation I shall first have to discuss what exactly we mean when we ask it.

I have already indicated that when thinking of causation we have to distinguish between three subquestions, and that these three differ from one another in the stretch of time that is considered. We ask, first: Given an adult animal that fights now and then, what makes each outburst of fighting happen? The time scale in which we consider these recurrent events is usually one of seconds, or minutes. To use an analogy, this subquestion compares with asking what makes a car start or stop each time we use it.

But in asking this same general question of causation ("What makes an animal fight?") we may also be referring to a longer period of time; we may mean "How has the animal, as it grew up, developed this behavior?" This compares roughly with asking how a car has been constructed in the factory. The distinction between these two subquestions remains useful even though we know that many animals continue their development (much slowed down) even after they have attained adulthood. For instance, they may still continue to learn.

Finally, in biology, as in technology, we can extend this time scale even more, and ask: How have the animal species which we observe today—and which we know have evolved from ancestors that were different—how have they acquired their particular behavior systems during this evolution? Unfortunately, while we know the evolution of cars because they evolved so quickly and have been so fully recorded, the behavior of extinct animals cannot be observed, and has to be reconstructed by indirect methods.

I shall try to justify the claim I made earlier, and show how all these four questions—that of behavior's survival value and the three subquestions of causation—have to enter into the argument if we are to understand the biology of aggression.

Let us first consider the short-term causation; the mechanism of fighting. What makes us fight at any one moment? Lorenz argues in his book that, in animals and in man, there is an internal urge to attack. An individual does not simply wait to be provoked, but, if actual attack has not been possible for some time, this urge to fight builds up until the individual actively seeks the opportunity to indulge in fighting. Aggression, Lorenz claims, can be spontaneous.

But this view has not gone unchallenged. For instance, R. A. Hinde has written a thorough criticism (7), based on recent work on aggression in animals, in which he writes that Lorenz's "arguments for the spontaneity of aggression do not bear examination" and that "the contrary view, expressed in nearly every textbook of comparative psychology . . ." is that fighting "derives principally from the situation"; and even more explicitly: "There is no need to postulate causes that are purely internal to the aggressor" (7, p. 303). At first glance it would seem as if Lorenz and Hinde disagree profoundly. I have read and reread both authors, and it is to me perfectly clear that loose statements and misunderstandings on both sides have made it appear that there is disagreement where in fact there is something very near to a common opinion. It seems to me that the differences between the two authors lie mainly in the different ways they look at internal and external variables. This in turn seems due to differences of a semantic nature. Lorenz uses the unfortunate term "the spontaneity of aggression." Hinde takes this to mean that external stimuli are in Lorenz's view not necessary at all to make an animal fight. But here he misrepresents Lorenz, for nowhere does Lorenz claim that the internal urge ever makes an animal fight "in vacuo"; somebody or something is attacked. This misunderstanding makes Hinde feel that he has refuted Lorenz's views by saying that "fighting derives principally from the situation." But both authors are fully aware of the fact that fighting is started by a number of variables, of which some are internal and some external. What both authors know, and what cannot be doubted, is that fighting behavior is not like the simple slot machine that produces one platform ticket every time one threepenny bit is inserted. To mention one animal example: a male stickleback does not always show the full fighting behavior in response to an approaching standard opponent; its response varies from none at all to the optimal stimulus on some occasions, to full attack on even a crude dummy at other times. This means that its internal state varies, and in this particular case we know from the work of Hoar (8) that the level of the male sex hormone is an important variable.

Another source of misunderstanding seems to have to do with the stretch of time that the two authors are taking into account. Lorenz undoubtedly thinks of the causes of an outburst of fighting in terms of seconds, or hours—perhaps days. Hinde seems to think of events which may have happened further back in time; an event which is at any particular moment "internal" may well in its turn have been influenced previously by external agents. In our stickleback example, the level of male sex hormone is influenced by external agents such as the length of the daily exposure to light over a period of a month or so (9). Or, less far back in time, its readiness to attack may have been influenced by some experience gained, say, half an hour before the fight.

I admit that I have now been spending a great deal of time on what would seem to be a perfectly simple issue: the very first step in the analysis of the short-term causation, which is to distinguish at any given moment between variables within the animal and variables in the environment. It is of course important for our further understanding to unravel the complex interactions between these two worlds, and in particular the physiology of aggressive behavior. A great deal is being discovered about this, but for my present issue there is no use discussing it as long as even the first step in the analysis has not led to a clearly expressed and generally accepted conclusion. We must remember that we are at the moment concerned with the human problem: "What makes men attack each other?" And for this problem the answer to the first stage of our question is of prime importance: Is our readiness to start an attack constant or not? If it were—if our aggressive behavior were the outcome of an apparatus with the properties of the slot machine—all we would have to do would be to control the ex-ternal situation: to stop providing three-penny bits. But since our readiness to start an attack is variable, further studies of both the external and the internal variables are vital to such issues as: Can we reduce fighting by lowering the population density, or by withholding provocative stimuli? Can we do so by changing the hormone balance or other physiological variables? Can we perhaps in addition control our development in such a way as to change the dependence on internal and external factors in adult man? However, before discussing development, I must first return to the fact that I have mentioned before, namely, that man is, among the thousands of other species that fight, the only mass murderer. How do animals in their intraspecific disputes avoid bloodshed?

The Importance of "Fear"

The clue to this problem is to recognize the simple fact that aggression in animals rarely occurs in pure form; it is only one of two components of an adaptive system. This is most clearly seen in territorial behavior, although it is also true of most other types of hostile behavior. Members of territorial species divide, among themselves, the available living space and opportunities by each individual defending its home range against competitors. Now in this system of parceling our living space, avoidance plays as important a part as attack. Put very briefly, animals of territorial species, once they have settled on a territory, attack intruders, but an animal that is still searching for a suitable territory or finds itself outside its home range withdraws when it meets with an already established owner. In terms of function, once you have taken possession of a territory, it pays to drive off competitors; but when you are still looking for a territory (or meet your neighbor at your common boundary), your chances of success are improved by avoiding such established owners. The ruthless fighter who "knows no fear" does not get very far. For an understanding of what follows, this fact, that hostile clashes are controlled by what we could call the "attack-avoidance system," is essential.

When neighboring territory owners meet near their common boundary, both attack behavior and withdrawal behavior are elicited in both animals; each of the two is in a state of motivational conflict. We know a great deal about the variety of movements that appear when these two conflicting, incompatible behaviors are elicited. Many of these expressions of a motivational conflict have, in the course of evolution, acquired signal function; in colloquial language, they signal "Keep out!" We deduce this from the fact that opponents respond to them in an appropriate way: instead of proceeding to intrude, which would require the use of force, trespassers withdraw, and neighbors are contained by each other. This is how such animals have managed to have all the advantages of their hostile behavior without the disadvantages: they divide their living space in a bloodless way by using as distance-keeping devices these conflict movements ("threat") rather than actual fighting.

Group Territories

In order to see our wars in their correct biological perspective one more comparison with animals is useful. So far I have discussed animal species that defend individual or at best pair territories. But there are also animals which possess and defend territories belonging to a group, or a clan (10).

Now it is an essential aspect of group territorialism that the members of a group unite when in hostile confronta-tion with another group that approaches, or crosses into their feeding territory. The uniting and the aggression are equally important. It is essential to realize that group territorialism does not exclude hostile relations on lower levels when the group is on its own. For instance, within a group there is often a peck order. And within the group there may be individual or pair territories. But frictions due to these relationships fade away during a clash between groups. This temporary elimi-nation is done by means of so-called appeasement and reassurance signals. They indicate "I am a friend," and so diminish the risk that, in the general flare-up of anger, any animal "takes it out" on a fellow member of the same group (11). Clans meet clans as units, and each individual in an intergroup clash, while united with its fellow-members, is (as in interindividual clashes) torn between attack and with-drawal, and postures and shouts rather than attacks.

We must now examine the hypothesis (which I consider the most likely one) that man still carries with him the ani-mal heritage of group territoriality. This is a question concerning man's evolu-tionary origin, and here we are, by the very nature of the subject, forced to speculate. Because I am going to say something about the behavior of our ancestors of, say, 100,000 years ago, I have to discuss briefly a matter of methodology. It is known to all biolo-gists (but unfortunately unknown to most psychologists) that comparison of present-day species can give us a deep insight, with a probability closely ap-proaching certainty, into the evolution-ary history of animal species. Even where fossil evidence is lacking, this comparative method alone can do this. It has to be stressed that this compari-son is a highly sophisticated method, and not merely a matter of saying that species A is different from species B

(12). The basic procedure is this. We interpret differences between really allied species as the result of adaptive divergent evolution from common stock, and we interpret similarities between nonallied species as adaptive convergencies to similar ways of life. By studying the adaptive functions of species characteristics we understand how natural selection can have produced both these divergencies and convergencies. To mention one striking example: even if we had no fossil evidence, we could, by this method alone, recognize whales for what they are—mammals that have returned to the water, and, in doing so, have developed some similarities to fish. This special type of comparison, which has been applied so successfully by students of the structure of animals, has now also been used, and with equal success, in several studies of animal behavior. Two approaches have been applied. One is to see in what respects species of very different origin have convergently adapted to a similar way of life. Von Haartman (13) has applied this to a study of birds of many types that nest in holes—an anti-predator safety device. All such hole-nesters center their territorial fighting on a suitable nest hole. Their courtship consists of luring a female to this hole (often with the use of bright color patterns). Their young gape when a general darkening signals the arrival of the parent. All but the most recently adapted species lay uniformly colored, white or light blue eggs that can easily be seen by the parent.

An example of adaptive divergence has been studied by Cullen (14). Among all the gulls, the kittiwake is unique in that it nests on very narrow ledges on sheer cliffs. Over 20 peculiarities of this species have been recognized by Mrs. Cullen as vital adaptations to this particular habitat.

These and several similar studies (15) demonstrate how comparison reveals, in each species, systems of interrelated, and very intricate adaptive features. In this work, speculation is now being followed by careful experimental checking. It would be tempting to elaborate on this, but I must return to our own unfortunate species.

Now, when we include the "Naked Ape" in our comparative studies, it becomes likely (as has been recently worked out in great detail by Morris) that man is a "social Ape who has turned carnivore" (16). On the one hand he is a social primate; on the other, he has developed similarities to wolves, lions and hyenas. In our present context one thing seems to stand out clearly, a conclusion that seems to me of paramount importance to all of us, and yet has not yet been fully accepted as such. As a social, hunting primate, man must originally have been organized on the principle of group territories.

Ethologists tend to believe that we still carry with us a number of behavioral characteristics of our animal ancestors, which cannot be eliminated by different ways of upbringing, and that our group territorialism is one of those ancestral characters. I shall discuss the problem of the modifiability of our behavior later, but it is useful to point out here that even if our behavior were much more modifiable than Lorenz maintains, our cultural evolution, which resulted in the parceling-out of our living space on lines of tribal, national, and now even "bloc" areas, would, if anything, have tended to enhance group territorialism.

Group Territorialism in Man?

I put so much emphasis on this issue of group territorialism because most writers who have tried to apply ethology to man have done this in the wrong way. They have made the mistake, to which I objected before, of uncritically extrapolating the results of animal studies to man. They try to explain man's behavior by using facts that are valid only of some of the animals we studied. And, as ethologists keep stressing, no two species behave alike. Therefore, instead of taking this easy way out, we ought to study man in his own right. And I repeat that the message of the ethologists is that the methods, rather than the results, of ethology should be used for such a study.

Now, the notion of territory was developed by zoologists (to be precise, by ornithologists, 17), and because individual and pair territories are found in so many more species than group territories (which are particularly rare among birds), most animal studies were concerned with such individual and pair territories. Now such low-level territories do occur in man, as does another form of hostile behavior, the peck order. But the problems created by such low-level frictions are not serious; they can, within a community, be kept in check by the apparatus of law and order; peace within national boundaries can be enforced. In order to understand what makes us go to war, we have to recognize that man behaves very much like a group-territorial species. We too unite in the face of an outside danger to the group; we "forget our differences." We too have threat gestures, for instance, angry facial expressions. And all of us use reassurance and appeasement signals, such as a friendly smile. And (unlike speech) these are universally understood; they are cross-cultural; they are species-specific. And, incidentally, even within a group sharing a common language, they are often more reliable guides to a man's intentions than speech, for speech (as we know now) rarely reflects our true motives, but our facial expressions often "give us away."

If I may digress for a moment: it is humiliating to us ethologists that many nonscientists, particularly novelists and actors, intuitively understand our sign language much better than we scientists ourselves do. Worse, there is a category of human beings who understand intuitively more about the causation of our aggressive behavior: the great demagogues. They have applied this knowledge in order to control our behavior in the most clever ways, and often for the most evil purposes. For instance, Hitler (who had modern mass communication at his disposal, which allowed him to inflame a whole nation) played on both fighting tendencies. The "defensive" fighting was whipped up by his passionate statements about "living space," "encirclement," Jewry, and Freemasonry as threatening powers which made the Germans feel "cornered." The "attack fighting" was similarly set ablaze by playing the myth of the Herrenvolk. We must make sure that mankind has learned its lesson and will never forget how disastrous the joint effects have been—if only one of the major nations were led now by a man like Hitler, life on earth would be wiped out.

I have argued my case for concentrating on studies of group territoriality rather than on other types of aggression. I must now return, in this context, to the problem of man the mass murderer. Why don't we settle even our international disputes by the relatively harmless, animal method of threat? Why have we become unhinged so that so often our attack erupts without being kept in check by fear? It is not that we have no fear, nor that we have no other inhibitions against killing. This problem has to be considered first of all in the general context of the consequences of man having embarked on a new type of evolution.

460

Cultural Evolution

Man has the ability, unparalleled in scale in the animal kingdom, of passing on his experiences from one generation to the next. By this accumulative and exponentially growing process, which we call cultural evolution, he has been able to change his environment progressively out of all recognition. And this includes the social environment. This new type of evolution proceeds at an incomparably faster pace than genetic evolution. Genetically we have not evolved very strikingly since Cro-Magnon man, but culturally we have changed beyond recognition, and are changing at an ever-increasing rate. It is of course true that we are highly adjustable individually, and so could hope to keep pace with these changes. But I am not alone in believing that this behavioral adjustability, like all types of modifiability, has its limits. These limits are imposed upon us by our hereditary constitution, a constitution which can only change with the far slower speed of genetic evolution. There are good grounds for the conclusion that man's limited behavioral adjustability has been outpaced by the culturally determined changes in his social environment, and that this is why man is now a misfit in his own society.

We can now, at last, return to the problem of war, of uninhibited mass killing. It seems quite clear that our cultural evolution is at the root of the trouble. It is our cultural evolution that has caused the population explosion. In a nutshell, medical science, aiming at the reduction of suffering, has, in doing so, prolonged life for many individuals as well—prolonged it to well beyond the point at which they produce offspring. Unlike the situation in any wild species, recruitment to the human population consistently surpasses losses through mortality. Agricultural and technical know-how have enabled us to grow food and to exploit other natural resources to such an extent that we can still feed (though only just) the enormous numbers of human beings on our crowded planet. The result is that we now live at a far higher density than that in which genetic evolution has molded our species. This, together with long-distance communication, leads to far more frequent, in fact to continuous, intergroup contacts, and so to continuous external provocation of aggression. Yet this alone would not explain our increased tendency to kill each other; it would merely lead to continuous threat behavior.

The upsetting of the balance between aggression and fear (and this is what causes war) is due to at least three other consequences of cultural evolution. It is an old cultural phenomenon that warriors are both brainwashed and bullied into all-out fighting. They are brainwashed into believing that fleeing —originally, as we have seen, an adaptive type of behavior—is despicable, "cowardly." This seems to me due to the fact that man, accepting that in moral issues death might be preferable to fleeing, has falsely applied the moral concept of "cowardice" to matters of mere practical importance—to the dividing of living space. The fact that our soldiers are also bullied into all-out fighting (by penalizing fleeing in battle) is too well known to deserve elaboration.

Another cultural excess is our ability to make and use killing tools, especially long-range weapons. These make killing easy, not only because a spear or a club inflicts, with the same effort, so much more damage than a fist, but also, and mainly, because the use of long-range weapons prevents the victim from reaching his attacker with his appeasement, reassurance, and distress signals. Very few aircrews who are willing, indeed eager, to drop their bombs "on target" would be willing to strangle, stab, or burn children (or, for that matter, adults) with their own hands; they would stop short of killing, in response to the appeasement and distress signals of their opponents.

These three factors alone would be sufficient to explain how we have become such unhinged killers. But I have to stress once more that all this, however convincing it may seem, must still be studied more thoroughly.

There is a frightening, and ironical paradox in this conclusion: that the human brain, the finest life-preserving device created by evolution, has made our species so successful in mastering the outside world that it suddenly finds itself taken off guard. One could say that our cortex and our brainstem (our "reason" and our "instincts") are at loggerheads. Together they have created a new social environment in which, rather than ensuring our survival, they are about to do the opposite. The brain finds itself seriously threatened by an enemy of its own making. It is its own enemy. We simply have to understand this enemy.

The Development of Behavior

I must now leave the question of the moment-to-moment control of fighting, and, looking further back in time, turn to the development of aggressive behavior in the growing individual. Again we will start from the human problem. This, in the present context, is whether it is within our power to control development in such a way that we reduce or eliminate fighting among adults. Can or cannot education in the widest sense produce nonagressive men?

The first step in the consideration of this problem is again to distinguish between external and internal influences, but now we must apply this to the growth, the changing, of the behavioral machinery during the individual's de-

velopment. Here again the way in which we phrase our questions and our conclusions is of the utmost importance.

In order to discuss this issue fruitfully, I have to start once more by considering it in a wider context, which is now that of the "nature-nurture" problem with respect to behavior in general. This has been discussed more fully by Lorenz in his book *Evolution and Modification of Behaviour* (*18*); for a discussion of the environmentalist point of view I refer to the various works of Schneirla (see *19*).

Lorenz tends to classify behavior types into innate and acquired or learned behavior. Schneirla rejects this dichotomy into two classes of behavior. He stresses that the developmental process, of behavior as well as of other functions, should be considered, and also that this development forms a highly complicated series of interactions between the growing organism and its environment. I have gradually become convinced that the clue to this difference in approach is to be found in a difference in aims between the two authors. Lorenz claims that "we are justified in leaving, at least for the time being, to the care of the experimental embryologists all those questions which are concerned with the chains of physiological causation leading from the genome to the development of . . . neurosensory structures" (*18*, p. 43). In other words, he deliberately refrains from starting his analysis of development prior to the stage at which a fully coordinated behavior is performed for the first time. If one in this way restricts one's studies to the later stages of development, then a classification in "innate" and "learned" behavior, or behavior components, can be considered quite justified. And there was a time, some 30 years ago, when the almost grotesquely environmentalist bias of psychology made it imperative for ethologists to stress the extent to which behavior patterns could appear in perfect or near-perfect form without the aid of anything that could be properly called learning. But I now agree (however belatedly) with Schneirla that we must extend our interest to earlier stages of development and embark on a full program of experimental embryology of behavior. When we do this, we discover that interactions with the environment can indeed occur at early stages. These interactions may concern small components of the total machinery of a fully functional behavior pattern, and many of them cannot possibly be called learning. But they are interactions with the environment, and must be taken into account if we follow in the footsteps of the experimental embryologists, and extend our field of interest to the entire sequence of events which lead from the blueprints contained in the zygote to the fully functioning, behaving animal. We simply have to do this if we want an answer to the question to what extent the development of behavior can be influenced from the outside.

When we follow this procedure the rigid distinction between "innate" or unmodifiable and "acquired" or modifiable behavior patterns becomes far less sharp. This is owing to the discovery, on the one hand, that "innate" patterns may contain elements that at an early stage developed in interaction with the environment, and, on the other hand, that learning is, from step to step, limited by internally imposed restrictions.

To illustrate the first point, I take the development of the sensory cells in the retina of the eye. Knoll has shown (20) that the rods in the eyes of tadpoles cannot function properly unless they have first been exposed to light. This means that, although any visually guided response of a tadpole may well, in its integrated form, be "innate" in Lorenz's sense, it is so only in the sense of "nonlearned," not in that of "having

grown without interaction with the environment." Now it has been shown by Cullen (21) that male sticklebacks reared from the egg in complete isolation from other animals will, when adult, show full fighting behavior to other males and courtship behavior to females when faced with them for the first time in their lives. This is admittedly an important fact, demonstrating that the various recognized forms of learning do not enter into the programing of these integrated patterns. This is a demonstration of what Lorenz calls an "innate response." But it does not exclude the possibility that parts of the machinery so employed may, at an earlier stage, have been influenced by the environment, as in the case of the tadpoles.

Second, there are also behavior patterns which do appear in the inexperienced animal, but in an incomplete form, and which require additional development through learning. Thorpe has analyzed a clear example of this: when young male chaffinches reared alone begin to produce their song for the first time, they utter a very imperfect warble; this develops into the full song only if, at a certain sensitive stage, the young birds have heard the full song of an adult male (22).

By far the most interesting aspect of such intermediates between innate and acquired behavior is the fact that learning is not indiscriminate, but is guided by a certain selectiveness on the part of the animal. This fact has been dimly recognized long ago; the early ethologists have often pointed out that different, even closely related, species learn different things even when developing the same behavior patterns. This has been emphasized by Lorenz's use of the term "innate teaching mechanism." Other authors use the word "template" in the same context. The best example I know is once more taken from the development of song in certain birds.

As I have mentioned, the males of some birds acquire their full song by changing their basic repertoire to resemble the song of adults, which they have to hear during a special sensitive period some months before they sing themselves. It is in this sensitive period that they acquire, without as yet producing the song, the knowledge of "what the song ought to be like." In technical terms, the bird formed a *Sollwert* (23) (literally, "should-value," an ideal) for the feedback they receive when they hear their own first attempts. Experiments have shown (24) that such birds, when they start to sing, do three things: they listen to what they produce; they notice the difference between this feedback and the ideal song; and they correct their next performance.

This example, while demonstrating an internal teaching mechanism, shows, at the same time, that Lorenz made his concept too narrow when he coined the term "innate teaching mechanism." The birds have developed a teaching mechanism, but while it is true that it is internal, it is not innate; the birds have acquired it by listening to their father's song.

These examples show that if behavior studies are to catch up with experimental embryology our aims, our concepts, and our terms must be continually revised.

Before returning to aggression, I should like to elaborate a little further on general aspects of behavior development, because this will enable me to show the value of animal studies in another context, that of education.

Comparative studies, of different animal species, of different behavior patterns, and of different stages of development, begin to suggest that wherever learning takes a hand in development, it is guided by such *Sollwerte* or templates for the proper feedback, the feedback that reinforces. And it becomes clear that these various *Sollwerte* are of a bewildering variety. In human education one aspect of this has been emphasized in particular, and even applied in the use of teaching machines: the requirement that the reward, in order to have maximum effect, must be immediate. Skinner has stressed this so much because in our own teaching we have imposed an unnatural delay between, say, taking in homework, and giving the pupil his reward in the form of a mark. But we can learn more from animal studies than the need for immediacy of reward. The type of reward is also of great importance, and this may vary from task to task, from stage to stage, from occasion to occasion; the awards may be of almost infinite variety.

Here I have to discuss briefly a behavior of which I have so far been unable to find the equivalent in the development of structure. This is exploratory behavior. By this we mean a kind of behavior in which the animal sets out to acquire as much information about an object or a situation as it can possibly get. The behavior is intricately adapted to this end, and it terminates when the information has been stored, when the animal has incorporated it in its learned knowledge. This exploration (subjectively we speak of "curiosity") is not confined to the acquisition of information about the external world alone; at least mammals explore their own movements a great deal, and in this way "master new skills." Again, in this exploratory behavior, *Sollwerte* of expected, "hoped-for" feedbacks play their part.

Without going into more detail, we can characterize the picture we begin to get of the development of behavior as a series, or rather a web, of events, starting with innate programing instructions contained in the zygote, which straightaway begin to interact with the environment; this interaction may be discontinuous, in that periods of predominantly internal development alternate with periods of interaction, or sensitive periods. The interaction is enhanced by active exploration; it is steered by selective *Sollwerte* of great variety; and stage by stage this process ramifies; level upon level of ever-increasing complexity is being incorporated into the programing.

Apply what we have heard for a moment to playing children (I do not, of course, distinguish sharply between "play" and "learning"). At a certain age a child begins to use, say, building blocks. It will at first manipulate them in various ways, one at a time. Each way of manipulating acts as exploratory behavior: the child learns what a block looks, feels, tastes like, and so forth, and also how to put it down so that it stands stably.

Each of these stages "peters out" when the child knows what it wanted to find out. But as the development proceeds, a new level of exploration is added: the child discovers that it can put one block on top of the other; it constructs. The new discovery leads to repetition and variation, for each child develops, at some stage, a desire and a set of *Sollwerte* for such effects of construction, and acts out to the full this new level of exploratory behavior. In addition, already at this stage the *Sollwert* or ideal does not merely contain what the blocks do, but also what, for instance, the mother does; her approval, her shared enjoyment, is also of great importance. Just as an exploring animal, the child builds a kind of inverted pyramid of experience, built of layers, each set off by a new wave of exploration and each directed by new sets of *Sollwerte*, and so its development "snowballs." All these phases may well have more or less limited sensitive periods, which determine when the fullest effect can be obtained, and when the child is ready for the next step. More important still, if the opportunity for the next stage is offered either too early

or too late, development may be damaged, including the development of motivational and emotional attitudes.

Of course gifted teachers of many generations have known all these things (25) or some of them, but the glimpses of insight have not been fully and scientifically systematized. In human education, this would of course involve experimentation. This need not worry us too much, because in our search for better educational procedures we are in effect experimenting on our children all the time. Also, children are fortunately incredibly resilient, and most grow up into pretty viable adults in spite of our fumbling educational efforts. Yet there is, of course, a limit to what we will allow ourselves, and this, I should like to emphasize, is where animal studies may well become even more important than they are already.

Can Education End Aggression?

Returning now to the development of animal and human aggression, I hope to have made at least several things clear: that behavior development is a very complex phenomenon indeed; that we have only begun to analyze it in animals; that with respect to man we are, if anything, behind in comparison with animal studies; and that I cannot do otherwise than repeat what I said in the beginning: we must make a major research effort. In this effort animal studies can help, but we are still very far from drawing very definite conclusions with regard to our question: To what extent shall we be able to render man less aggressive through manipulation of the environment, that is, by educational measures?

In such a situation personal opinions naturally vary a great deal. I do not hesitate to give as my personal opinion that Lorenz's book *On Aggression,* in spite of its assertativeness, in spite of factual mistakes, and in spite of the many possibilities of misunderstandings that are due to the lack of a common language among students of behavior— that this work must be taken more seriously as a positive contribution to our problem than many critics have done. Lorenz is, in my opinion, right in claiming that elimination, through education, of the internal urge to fight will turn out to be very difficult, if not impossible.

Everything I have said so far seems to me to allow for only one conclusion. Apart from doing our utmost to return to a reasonable population density apart from stopping the progressive depletion and pollution of our habitat, we must pursue the biological study of animal behavior for clarifying problems of human behavior of such magnitude as that of our aggression, and of education.

But research takes a long time, and we must remember that there are experts who forecast worldwide famine 10 to 20 years from now; and that we have enough weapons to wipe out all human life on earth. Whatever the causation of our aggression, the simple fact is that for the time being we are saddled with it. This means that there is a crying need for a crash program, for finding ways and means for keeping our intergroup aggression in check. This is of course in practice infinitely more difficult than controlling our intranational frictions; we have as yet not got a truly international police force. But there is hope for avoiding all-out war because, for the first time in history, we are afraid of killing ourselves by the lethal radiation effects even of bombs that we could drop in the enemy's territory. Our politicians know this. And as long as there is this hope, there is every reason to try and learn what we can from animal studies. Here again they can be of help. We have already seen that animal opponents meet-

ing in a hostile clash avoid bloodshed by using the expressions of their motivational conflicts as intimidating signals. Ethologists have studied such conflict movements in some detail (26), and have found that they are of a variety of types. The most instructive of these is the redirected attack; instead of attacking the provoking, yet dreaded, opponent, animals often attack something else, often even an inanimate object. We ourselves bang the table with our fists. Redirection includes something like sublimation, a term attaching a value judgment to the redirection. As a species with group territories, humans, like hyenas, unite when meeting a common enemy. We do already sublimate our group aggression. The Dutch feel united in their fight against the sea. Scientists do attack their problems together. The space program —surely a mainly military effort—is an up-to-date example. I would not like to claim, as Lorenz does, that redirected attack exhausts the aggressive urge. We know from soccer matches and from animal work how aggressive behavior has two simultaneous, but opposite effects: a waning effect, and one of self-inflammation, of mass hysteria, such as recently seen in Cairo. Of these two the inflammatory effect often wins. But if aggression were used successfully as the motive force behind nonkilling and even useful activities, self-stimulation need not be a danger; in our short-term cure we are not aiming at the elimination of aggressiveness, but at "taking the sting out of it."

Of all sublimated activities, scientific research would seem to offer the best opportunities for deflecting and sublimating our aggression. And, once we recognize that it is the disrupted relation between our own behavior and our environment that forms our most deadly enemy, what could be better than uniting, at the front or behind the lines, in the scientific attack on our own behavioral problems?

I stress "behind the lines." The whole population should be made to feel that it participates in the struggle. This is why scientists will always have the duty to inform their fellowmen of what they are doing, of the relevance and the importance of their work. And this is not only a duty, it can give intense satisfaction.

I have come full circle. For both the long-term and the short-term remedies at least we scientists will have to sublimate our aggression into an all-out attack on the enemy within. For this the enemy must be recognized for what it is: our unknown selves, or, deeper down, our refusal to admit that man is, to himself, unknown.

I should like to conclude by saying a few words to my colleagues of the younger generation. Of course we all hope that, by muddling along until we have acquired better understanding, self-annihilation either by the "whimper of famine" or by the "bang of war" can be avoided. For this, we must on the one hand trust, on the other help (and urge) our politicians. But it is no use denying that the chances of designing the necessary preventive measures are small, let alone the chances of carrying them out. Even birth control still offers a major problem.

It is difficult for my generation to know how seriously you take the danger of mankind destroying his own species. But those who share the apprehension of my generation might perhaps, with us, derive strength from keeping alive the thought that has helped so many of us in the past when faced with the possibility of imminent death. Scientific research is one of the finest occupations of our mind. It is, with art and religion, one of the uniquely human ways of

meeting nature, in fact, the most active way. If we are to succumb, and even if this were to be ultimately due to our own stupidity, we could still, so to speak, redeem our species. We could at least go down with some dignity, by using our brain for one of its supreme tasks, by exploring to the end.

REFERENCES

1. A. Carrel, *L'Homme, cet Inconnu* (Librairie Plon, Paris, 1935).
2. AAAS Annual Meeting, 1967 [see *New Scientist* 37, 5 (1968)].
3. R. Carson, *Silent Spring* (Houghton Mifflin, Boston, 1962).
4. K. Lorenz, *On Aggression* (Methuen, London, 1966).
5. D. Morris, *The Naked Ape* (Jonathan Cape, London. 1967)
6. N. Tinbergen. *Z. Tierpsychol.* **20**, 410 (1964).
7. R. A. Hinde, *New Society* **9**, 302 (1967).
8. W. S. Hoar, *Animal Behaviour* **10**, 247 (1962).
9. B. Baggerman, in *Symp. Soc. Exp. Biol.* **20**, 427 (1965).
10. H. Kruuk, *New Scientist* **30**, 849 (1966).
11. N. Tinbergen, *Z. Tierpsychol.* **16**, 651 (1959); *Zool. Mededelingen* **39**, 209 (1964).
12. ———, *Behaviour* **15**, 1–70 (1959).
13. L. von Haartman, *Evolution* **11**, 339 (1957).
14. E. Cullen, *Ibis* **99**, 275 (1957).
15. J. H. Crook, *Symp. Zool. Soc. London* **14**, 181 (1965).
16. D. Freeman, *Inst. Biol. Symp.* **13**, 109 (1964); D. Morris, Ed., *Primate Ethology* (Weidenfeld and Nicolson. London, 1967).
17. H. E. Howard, *Territory in Bird Life* (Murray, London, 1920); R. A. Hinde *et al.*, *Ibis* **98**, 340–530 (1956).
18. K. Lorenz, *Evolution and Modification of Behaviour* (Methuen, London, 1966).
19. T. C. Schneirla, *Quart. Rev. Biol.* **41**, 283 (1966).
20. M. D. Knoll, *Z. Vergleich. Physiol.* **38**, 219 (1956).
21. E. Cullen, *Final Rept. Contr. AF 61 (052)-29.* USAFRDC. 1–23 (1961).
22. W. H. Thorpe, *Bird-Song* (Cambridge Univ. Press, New York, 1961).
23. E. von Holst and H. Mittelstaedt, *Naturwissenschaften* **37**, 464 (1950).
24. M. Konishi, *Z. Tierpsychol.* **22**, 770 (1965); F. Nottebohm, *Proc. 14th Intern. Ornithol. Congr.* 265–280 (1967).
25. E. M. Standing, *Maria Montessori* (New American Library, New York, 1962).
26. N. Tinbergen, in *The Pathology and Treatment of Sexual Deviation*, I. Rosen, Ed. (Oxford Univ. Press, London, 1964), pp. 3–23; N. B. Jones, *Wildfowl Trust 11th Ann. Rept.*, 46–52 (1960); P. Sevenster, *Behaviour, Suppl.* 9, 1–170 (1961); F. Rowell, *Animal Behaviour* **9**, 38 (1961).

Name Index

*Numbers in parentheses indicate that the author is referred to by that reference number, rather than by name, on the page indicated.

Subject Index